Learning

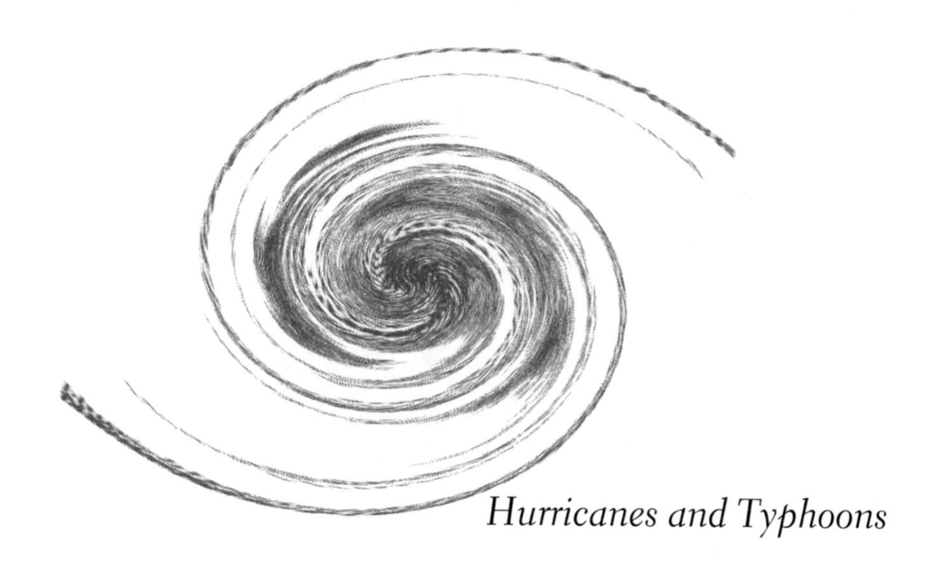

Hurricanes and Typhoons

Hurricanes and Typhoons

Past, Present, and Future

EDITED BY RICHARD J. MURNANE AND KAM-BIU LIU

COLUMBIA UNIVERSITY PRESS NEW YORK

COLUMBIA UNIVERSITY PRESS
Publishers Since 1893
NEW YORK CHICHESTER, WEST SUSSEX

Library of Congress Cataloging-in-Publication Data

Hurricanes and typhoons : past, present, and future / edited by Richard J. Murnane and Kam-biu Liu.
p. cm.
Includes bibliographical references and index.
ISBN 0-231-12388-4 (alk. paper)
1. Hurricanes. 2. Typhoons. 3. Cyclone tracks. 4. Paleoclimatology. I. Murnane, Richard J. II. Liu, Kam-biu.

QC944.H92 2004
551.55′2—dc22

2004048219

Columbia University Press books are printed on permanent and durable acid-free paper.
Printed in the United States of America
c 10 9 8 7 6 5 4 3 2 1

CONTENTS

PREFACE

This book presents a set of studies that examine the past, present, and future variability of tropical cyclones. We look at variability that occurs on time scales ranging from intraseasonal to millennial. The review is organized in a chronological sense; we start with the longest records of the past, move on to assess present-day variability and its relation to seasonal forecasts of tropical cyclone variability, and finish by considering potential future changes in tropical cyclone activity.

In addition to examining tropical cyclone variability on a variety of time scales, this book brings together for the first time a body of work that summarizes the results from the developing field of paleotempestology. The word "paleotempestology" suggests the study of past tempests, or storms; however, we use it to refer specifically to the study of past tropical cyclones. Landfalling tropical cyclones can produce a range of effects that may be preserved in the geologic or biologic record. In particular, storm surge from a tropical cyclone may leave overwash deposits in coastal lakes and marshes. There are other potential proxy techniques under development. For example, strong winds from a tropical cyclone can break off tree branches. The loss of the branches and foliage may reduce the photosynthetic capability of a tree to the extent that this reduction may be reflected in the tree ring record. Also, the oxygen isotopic composition of rain from tropical cyclones appears to have a distinct signature. Analyses of the oxygen isotopic content of corals or tree rings might provide a record of rainfall from tropical cyclones. The record of tropical cyclone landfall reconstructed from lake and marsh sediments currently provides the best information on prehistoric landfalls of tropical cyclones.

Our desire to present an overview of paleotempestology provided the original motivation for this book. The subject matter broadened to include the past, present, and potential future variability of hurricane and typhoon activity. The book includes a number of chapters on tropical cyclone variability over a range of time scales during the present and in the future to provide the proper context to appreciate results from the paleotempestological studies. This information is especially timely and important because it provides a context for assessing how natural variability in tropical cyclone activity compares to potential variability associated with anthropogenic climate change. The material in the book will be of interest to scientists, policy makers, and business people concerned with tropical cyclones and the social, environmental, and economic repercussions associated with their landfall.

Following the introduction, this book is divided into four parts that examine tropical cyclone variability over different time scales: centennial to millennial, annual to centennial, intraseasonal to interdecadal, and future changes on a CO_2-doubling time scale. The two paleotempestology studies in part I address tropical cyclone variability on the centennial to millennial time scales and summarize the state of the art in paleotempestology. In "Paleotempestology," Kam-biu Liu summarizes the existing methods used to analyze overwash deposits in coastal lakes and discusses the millennial scale variability in hurricane landfall frequency and location that is inferred from coastal lake sediments. Then, in "Backbarrier Sedimentary Records of Intense Hurricane Landfalls in the Northeastern United States," Jeffrey P. Donnelly and Thompson Webb III discuss work based on sand layers preserved in coastal marsh sediments and outline the issues that must be considered when developing a record of intense hurricane landfall from coastal marsh sediments. These chapters foreshadow the promise of paleotempestological studies for improving our understanding of large-scale changes in tropical cyclone landfall as a function of climate.

Part II deals with reconstructions of tropical cyclone landfall, track, and intensity using a variety of historical records. Four of the five examples of historical reconstructions are concerned with hurricanes in the North Atlantic Ocean and Caribbean Sea, and one examines landfalling typhoons in China. In "A Method for Reconstructing Historical Hurricanes," Emery R. Boose presents a summary of his work on hurricane landfalls in New England and Puerto Rico and demonstrates how historical information can be combined with model analyses to learn more about the track and intensity of hurricanes that have limited historical information. Cary J. Mock then gives an example of a careful analysis of a variety of records and provides information on newly identified storms that have struck the South Carolina coast in "Tropical Cyclone Reconstructions from Documentary Records." The following chapter, by Ricardo García Herrera and colleagues, "The Use of Spanish and British Documentary Sources in the Investigation of Atlantic Hurricane Incidence in Historical Times," describes the records available in Spanish and British archives and includes a preliminary reconstruction of some storms in the Caribbean region. Next, in "The Atlantic Hurricane Database Reanalysis Project," Christopher W. Landsea and colleagues discuss the reanalysis of the North Atlantic best-track data for the period 1851 to 1910. The chapter by Kin-sheun Louie and Kam-biu Liu, "Ancient Records of Typhoons in Chinese Historical Documents," summarizes the Chinese historical records as a source of information on typhoon activity. This work provides an example of using the county archives in Guangdong Province to reconstruct a 1,000-year record of typhoon landfalls in southern China. It also documents the earliest known description of typhoon as a distinct meteorological phenomenon. The historical reconstructions in the section hint at the information on changes in tropical cyclone activity on a variety of time scales that can be gained through archival work.

Part II closes with "The Importance of Best-Track Data for Understanding the Past, Present, and Future of Hurricanes and Typhoons" by Richard J. Murnane, describing the current status of best-track data, the official record of tropical cyclone tracks. As evidenced both by the efforts involved in the previous four chapters and by the analyses in the next four chapters, best-track data are essential for studying hurricane and typhoon variability. Users may, however, be unaware of the difficulties associated with developing modern best-track data sets.

Part III examines intraseasonal to interdecadal variability in tropical cyclone activity and its relevance for seasonal forecasts of tropical cyclone activity. It opens with "Variations of Tropical Cyclone Activity over the Western North Pacific" by Johnny C. L. Chan, who examines tropical cyclone variability on intraseasonal to interannual time scales and discusses an example of how this variability can be exploited for use in seasonal forecasts of tropical cyclone activity in the western North Pacific Ocean. Then, in "ENSO and Tropical Cyclone Activity," Pao-Shin Chu summarizes the varied response of tropical cyclone activity to changes in the El Niño–Southern Oscillation (ENSO) in different basins of the world ocean. The next chapter, by James B. Elsner and Brian H. Bossak, "Hurricane Landfall Probability and Climate," provides a detailed example of how climate variability can alter hurricane landfall probabilities along the U.S. coastline. It focuses on how landfall probability varies as a function of ENSO and the North Atlantic Oscillation (NAO). The section ends with "Dynamical Seasonal Forecasts of Tropical Storm Statistics," Frédéric Vitart's discussion of the potential for using dynamical models to forecast seasonal tropical cyclone activity. The most direct benefit of better understanding tropical cyclone variability on intraseasonal and longer time scales likely will come from improved seasonal forecasts of tropical cyclone activity and more precise estimates of conditional landfall probabilities.

Part IV examines potential future changes in tropical cyclones that could arise from anthropogenic climate change caused by increased levels of greenhouse gases (primarily carbon dioxide). In "Response of Tropical Cyclone Activity to Climate Change," Kerry Emanuel opens the section by first discussing the theoretical basis for alterations in tropical cyclone frequency and variability in response to climate change and then providing an analysis suggesting that a feedback relationship links tropical cyclone activity to climate. The feedback occurs through oceanic heat transport driven by tropical cyclone-induced upper-ocean mixing. Finally, in "Impact of Climate Change on Hurricane Intensities as Simulated Using Regional Nested High-Resolution Models," Thomas R. Knutson and colleagues discuss their work using a regional nested high-resolution model to examine changes in tropical cyclone statistics under a "high-CO_2" climate. Obviously, better knowledge of how tropical cyclone activity will change in response to future climate change will help meet a wide range of societal needs such as emergency planning and the development of building codes.

In the conclusion we summarize the material presented in previous chapters and discuss a number of topics of potential interest for future research.

We would like to express our gratitude to the numerous scientific reviewers that helped to improve the manuscripts included in this book. In addition, we are grateful to K. Emanuel for bringing our attention to the quote by Tor Bergeron. R. J. Murnane would like to acknowledge the financial support of the Risk Prediction Initiative.

ABBREVIATIONS

AMS	accelerator mass spectrometry
CAPE	Convective Available Potential Energy
COADS	Comprehensive Oceanic and Atmospheric Data Set
ECMWF	European Centre for Medium-range Weather Forecasts
ENSO	El Niño–Southern Oscillation
EXPOS	exposure model
FEMA	Federal Emergency Management Agency
GATE	Global Atmospheric Research Program Atlantic Tropical Experiment
GCM	General Circulation Model
GDEM	Generalized Digital Environmental Model
GFDL	Geophysical Fluid Dynamics Laboratory
GMT	Greenwich Mean Time
GPS	Global Positioning System
HKO	Hong Kong Observatory
HURDAT	North Atlantic hurricane database
IPCC	Intergovernmental Panel on Climate Change
JMA	Japanese Meteorological Agency
JTWC	Joint Typhoon Warning Center
LOI	loss-on-ignition
LTER	Long-Term Ecological Research
MHW	mean high water
MJO	Madden-Julian Oscillation
NAO	North Atlantic Oscillation
NAOI	North Atlantic Oscillation Index
NASA	National Aeronautics and Space Administration
NCDC	National Climatic Data Center
NCEP	National Centers for Environmental Prediction

NHC	National Hurricane Center
NMCC	National Meteorological Center of China
NOAA	National Oceanic and Atmospheric Administration
PAGASA	Philippine Atmospheric, Geophysical, and Astronomical Services Administration
ppm	part per million
ppt	part per thousand
QBO	Quasi-Biennial Oscillation
RAs	Regional Associations
RMW	radius of maximum winds
RPI	Risk Prediction Initiative
RSMC	Regional Specialized Meteorological Center
SOI	Southern Oscillation Index
SRI	Sahel regional rainfall index
UTC	coordinated universal time
WMO	World Meteorological Organization

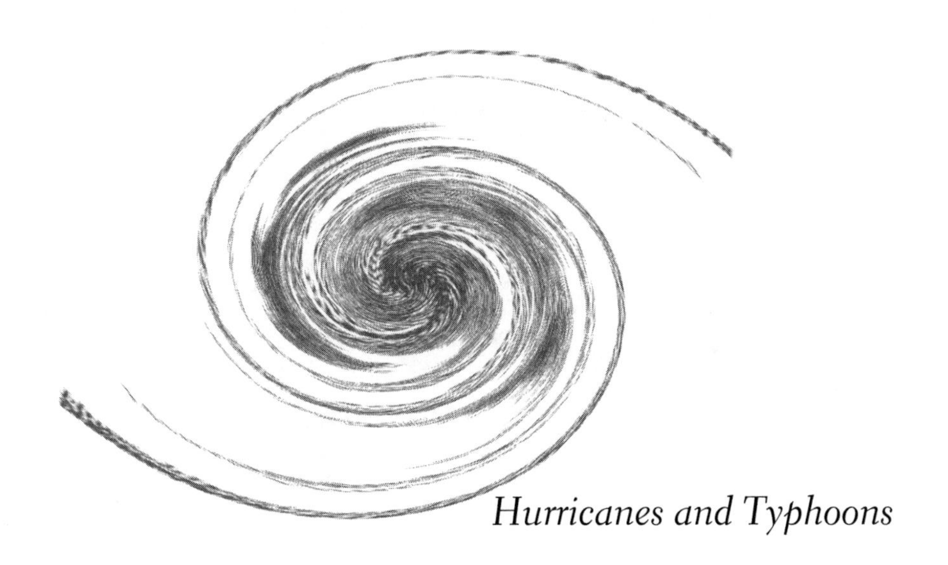

Hurricanes and Typhoons

1

Introduction

Richard J. Murnane

> Another problem, of much more far-reaching consequences, presents itself. What kind of secular changes may have existed in the frequency and intensity of the hurricane vortices of the Earth? And what changes may be expected in the future? We know nothing about these things, but I hope [to] have shown that even quite a small change in the different factors controlling the life history of a hurricane may produce, or may have produced, great changes in the paths of hurricanes and in their frequency and intensity. A minor alteration of the surface temperature of the sun, in the general composition of the earth's atmosphere, or in the rotation of the earth, might be able to change considerably the energy balance and the balance of forces within such a delicate mechanism as the tropical hurricane. During certain geological epochs, hurricanes may have been just as frequent as the cyclones of our latitudes, or they may have occurred all over the oceans and within all coastal regions, and they may have been even more violent than nowadays. During other periods they may have been lacking altogether. In studying paleo-climate and paleo-biological phenomena, especially along the coasts of previous geological epochs, it may be wise to consider such possibilities. Tor Bergeron (1954)

Tropical cyclones kill more people and cause more insured losses than any other natural disaster. The numbers can be staggering. For example, winds and flooding from a tropical cyclone striking Bangladesh in 1970 killed more than 300,000 people, and Hurricane Andrew sweeping through southern Florida and Louisiana in 1992 caused more than U.S.$20 billion (2002 dollars) in insured losses (Zanetti et al. 2003). These, of course, are extreme examples. Most tropical cyclones cause fewer deaths and produce less-extensive damage per event, but they are typically among the leading causes of loss of life and property in a given year. Better understanding the natural variability in tropical cyclone activity and how that activity has changed in the past and may change in the future will aid efforts to minimize the number of deaths and reduce the economic costs associated with tropical cyclone landfall. The purpose of this book is to provide an overview of studies aimed at understanding past, present, and future variability in tropical cyclone activity.

Over the past century the loss of life from hurricanes in the United States has fallen despite the dramatic increases in population along coastlines experiencing hurricane landfall (Simpson and Riehl 1981; Hebert, Jarrell, and Mayfield 1996). Continual improvements in forecast skill, and in the government's ability to effectively warn the public, demonstrably save lives (Pielke and Pielke 1997). Since the 1970s, the main cause of death from tropical storms and hurricanes in the United States has been inland freshwater flooding, not wind or storm surge produced by wind (Rappaport 2000).

While loss of life decreased over the last century, the financial losses due to hurricane landfall in the United States have increased dramatically; they currently average billions of dollars a year (Hebert et al. 1996). Yet, there has been no concomitant change in the number, size, or intensity of hurricanes making landfall. The increased losses are due to a combination of factors including inflation, increased personal wealth, and greater coastal populations (Pielke and Landsea 1998). To date, there has been no formal accounting for the total monetary loss from hurricanes striking the United States, although insured losses are commonly tracked by the insurance industry. A "rule of thumb" used to calculate the total loss from a hurricane striking the United States is that total losses are twice the insured losses.

It is sobering to note that for most countries the insured losses are only a small fraction of the total economic loss, and those losses ignore human suffering. Countries with less insurance and fewer societal resources are more susceptible to the effects of tropical cyclones and less able to recover from those effects. Consider the impact of Hurricane Mitch, the most powerful October hurricane ever recorded in the Atlantic Ocean, and possibly one of the most powerful Atlantic hurricanes ever (NCDC 1999). Mitch was "only" a tropical storm by the time it struck Central America in October 1998. Although Mitch did not have hurricane force winds while over Honduras and Nicaragua, it still produced tremendous flooding, killed more than 10,000 people, and caused more than U.S.$5 billion in damages ($4 billion in Honduras and $1 billion in Nicaragua). A $5 billion storm would be a big event in the United States, but it is catastrophic in

Central America. The loss nearly equaled the sum of the 1998 gross domestic products of Nicaragua ($2.1 billion) and Honduras ($5.3 billion).

The large number of deaths from a tropical storm such as Mitch and the small amount of insured property in developing countries stand in stark contrast to the impact of storms in the United States. The United States has an effective tropical cyclone forecast, warning, and emergency management effort. The last time more than 1,000 people died from a landfalling tropical cyclone in the United States was in 1928 when a hurricane struck southern Florida. In addition, the widespread use of insurance allows for a much quicker economic recovery. A better understanding of tropical cyclone variability would improve forecast, warning, and emergency management efforts and minimize the loss of life for all countries.

The societal need to better understand tropical cyclones and to reduce their lethality and economic cost is underscored by the amount of research supported by a number of governments and the activities of the private sector. For example, a better understanding of hurricane landfall (the intensity, wind structure, and precipitation associated with a storm) is one of the three initial major goals for the U.S. Weather Research Program.[1] The formation and activities of the Risk Prediction Initiative is an example of the need of the private sector to better understand tropical cyclone variability and the probability of tropical cyclone landfall.[2]

The financial and social impacts of tropical cyclones provide justification for studies aimed at understanding this natural hazard. Such studies cover a wide range of topics including the theoretical basis for tropical cyclones, the effect the environment surrounding a tropical cyclone has on its intensity and size, techniques for improving tropical cyclone forecasts, and analysis of tropical cyclone activity as a function of climate. For most of these studies, the observational record provides crucial information on how variables such as wind, pressure, and precipitation vary throughout the life of a tropical cyclone, as well as a baseline—a climatology—of tropical cyclone activity.

Best-track data provide the official observational records of tropical cyclone activity and are used to develop tropical cyclone climatologies. A shift in tropical cyclone activity away from the historical range in variability could have a dramatic impact on future losses and deaths. Such a shift could also have important implications for emergency preparedness, building codes, and disaster relief. Identifying a significant change in tropical cyclone activity will be difficult because the length and quality of the historical record is limited and the amount of data decrease with increasing age of the storm events. Additional factors not directly related to tropical cyclones such as rising sea levels, coastal development, and the carrying capacity of evacuation routes will also contribute to the extent of future losses.

Extending the historical and prehistoric record of tropical cyclone activity would enhance our understanding of the "climate-sensitivity" of tropical cyclones and support our ability to predict changes in activity that may occur in the future. Here we use tropical cyclone activity in a broad sense to account for a number of storm characteristics such as frequency, track, intensity, and size. It is, therefore, of both scientific and practical interest to study the variability of prehistoric, historic, current, and potential future tropical cyclones over a range of time scales.

Tropical cyclone activity is known to vary over time scales that range from days to millennia. In general, the historical record can be used to identify and study tropical cyclone variability on time scales that extend to decades. The highest frequency variability considered in this book is intraseasonal changes in tropical cyclone activity. The Madden-Julian Oscillation (MJO) is the best known, though incompletely understood, source of tropical atmospheric variability on intraseasonal time scales (Madden and Julian 1971).[3] Active phases of the MJO are associated with more frequent tropical cyclone formation (Maloney and Hartmann 2000a; Maloney and Hartmann 2000b; Hartmann and Maloney 2001; Maloney and Hartmann 2001). Once a tropical cyclone forms, other types of intraseasonal climate variability can have an impact on tropical cyclone track. For example the strength and phase of the North Atlantic Oscillation (NAO) appears to influence preferred hurricane tracks in the North Atlantic Ocean (Elsner, Liu, and Kocher 2000).

On interannual time scales the El Niño–Southern Oscillation (ENSO) appears to be the dominant factor controlling tropical cyclone activity (Nicholls 1979; Gray 1984; Chan 1985; Revelle and Goulter 1986; Chu and Wang 1997; Bove et al. 1998). ENSO alters tropical cyclone activity through its effects on atmospheric features (for example, vertical wind shear) and on ocean temperatures (Gray 1984; Shapiro 1987; Goldenberg et al. 2001). Additional factors such as the phase of the Quasi-Biennial Oscillation (Gray 1984; Shapiro 1989; Chan 1995; Elsner, Kara, and Owens 1999) and regional sea level pressure anomalies (Nicholls 1984; Gray et al. 1993; Gray et al. 1994; Knaff 1997) are correlated with tropical cyclone activity and appear to produce interannual variations in tropical cyclone activity.[4] Forecasts of these phenomena are used to produce seasonal forecasts of tropical cyclone frequency and landfall in different ocean basins.

Analyses of the historical record also reveal multi-decadal scale changes in tropical cyclone activity (Gray 1990; Landsea et al. 1992; Landsea 1993; Chan and Shi 1996; Landsea et al. 1996; Nicholls, Landsea, and Gill 1998; Elsner, Kara, and Owens 1999). Although not fully understood, the causes of the multi-decadal variations have been linked to factors such as changes in regional sea surface temperatures and pressure anomalies (Gray 1990; Landsea et al. 1994; Elsner, Kara, and Owens 1999) and large-scale patterns of atmospheric shear (Goldenberg et al. 2001). Multi-decadal variations in tropical cyclone activity are best defined in the North Atlantic Ocean because the record of hurricanes contained in the best-track data set for the North Atlantic Ocean is long and complete relative to the records for other ocean basins.

The initial motivation for developing a best-track data set (Jarvinen, Neumann, and Davis 1984) was to estimate the probability of tropical cyclone winds for rocket launches by the National Atmospheric and Space Administration (NASA).[5] Since then, the uses of the best-track data have expanded far beyond their original purpose. The data have been used for developing zoning regulations, creating emergency management plans, establishing insurance rates, designing hurricane risk models, and furthering scientific studies. An effort is now under way to enhance the best-track record for the North Atlantic and extend the record back to 1851 (Landsea et al., chapter 7 in this volume). This work is important because analyses of tropical cyclone best-track data form the foundation of most studies of tropical cyclone variability and the basis of decisions made by users in the private and public sectors.

Few studies of tropical cyclone variability consider storm events beyond the record in the best-track data sets, although a number of pioneering analyses of archival records and geological proxies are finding centennial and millennial-scale variations in tropical cyclone landfall. Several chapters in this book provide examples of the results of archival research: Mock (chapter 5); García Herrera et al. (chapter 6); and Louie and Liu (chapter 8). Earlier work provides examples of what are potentially among the longest possible records of tropical cyclone landfalls (Chan and Shi 2000; Liu, Shen, and Louie 2001). Archival studies exploit information in a wide range of sources including newspapers, personal diaries, official government records, and ships' logs. The quality and completeness of such records decrease as the records become older, but, fortunately, the historical record of tropical cyclone landfalls can be extended using geological proxies.

Most studies using geological proxies are based on sediment deposits produced by storm surge associated with a landfalling tropical cyclone (Liu and Fearn 1993; Liu and Fearn 2000a, 2000b; Donnelly et al. 2001a; Donnelly et al. 2001b). Other approaches use coral rubble deposited on land to reconstruct tropical cyclone landfall in Australia (Hayne and Chappell 2001; Nott and

Hayne 2001). Additional techniques for developing proxy records of tropical cyclone landfall are under development; they include oxygen isotope studies of corals and tree rings as well as conventional analysis of tree-ring variability.

Scientists have yet to determine convincingly the cause of centennial- and millennial-scale variations in hurricane and typhoon activity but they have shown that natural variability in the ocean and atmosphere system combine to produce a range of tropical cyclone activity over time. Fully understanding the range and character of the natural variability in tropical cyclones will help society answer an important question related to anthropogenically induced climate change. That is, how will human impacts on climate alter the range of natural variability in tropical cyclone activity and how can we identify when such changes occur?

The Intergovernmental Panel on Climate Change (IPCC), established by the United Nations, attempted to answer part of this question. The 2001 IPCC report examined, among other things, if and how climate change will alter the characteristics of future extreme events such as tropical cyclone activity (Houghton et al. 2001). The most relevant changes mentioned by the IPCC are that in some areas tropical cyclone peak wind intensity will "likely" increase by 5 to 10% and that mean and peak precipitation intensities will "likely" increase by 20 to 30%. Likely is quantified as equivalent to a 66 to 90% chance.

No matter how tropical cyclone activity changes in the future, it seems probable that the impact of a landfalling tropical cyclone will continue to increase as sea level rises and coastal populations grow (Changnon et al. 2000). Understanding the natural variability in tropical cyclone activity and the potential for future change in this variability is an important component of efforts to deal with the growing impacts of tropical cyclone landfall. The chapters that follow provide an overview of our understanding of the past, present, and future of hurricane and typhoon variability on a variety of time scales.

NOTES

1. The other two goals concerned heavy precipitation and flooding and societal and economic impacts.
2. The Risk Prediction Initiative (RPI) is based in Bermuda at the Bermuda Biological Station for Research, Inc. The RPI is supported by a number of insurance and reinsurance companies. For information, see http://www.bbsr.edu/rpi/
3. The Madden-Julian Oscillation (MJO) is the dominant source of tropical atmospheric variability at intraseasonal time scales (~20–90 days). The most visible man-

ifestation of the MJO is eastward propagation of large regions of convection and rainfall in the tropics over the Pacific and Indian Oceans.

4. The Quasi-Biennial Oscillation is an east-west oscillation in stratospheric winds above the equator with a period ranging between 22 to 32 months (Baldwin et al. 2001).
5. Best-track data contain estimates of the latitude, longitude, central pressure, and maximum wind speed of a storm at 6-hour intervals throughout the life of a storm.

REFERENCES

Baldwin, M. P., L. J. Gray, T. J. Dunkerton, K. Hamilton, P. H. Haynes, W. J. Randel, J. R. Holton, M. J. Alexander, I. Hirota, T. Horinouchi, D. B. A. Jones, J. S. Kinnersley, C. Marquardt, K. Sato, and M. Takahasi. 2001. The Quasi-Biennial Oscillation. *Reviews of Geophysics* 39:179–229.

Bergeron, T. 1954. The problem of tropical hurricanes. *Quarterly Journal of the Royal Geological Society* 80:131–64.

Bove, M. C., J. B. Elsner, C. W. Landsea, and J. J. O'Brien. 1998. Effect of El Niño on U.S. landfalling hurricanes, revisited. *Bulletin of the American Meteorological Society* 79:2477–82.

Chan, J., and J.-E. Shi. 1996. Long-term trends and interannual variability in tropical cyclone activity over the western North Pacific. *Geophysical Research Letters* 23:2765–68.

Chan, J. C. L. 1985. Tropical cyclone activity in the northwest Pacific in relation to the El Niño/Southern Oscillation phenomenon. *Monthly Weather Review* 113:599–606.

Chan, J. C. L. 1995. Tropical cyclone activity in the western North Pacific in relation to the stratospheric Quasi-Biennial Oscillation. *Monthly Weather Review* 123:2567–71.

Chan, J. C. L., and J. E. Shi. 2000. Frequency of typhoon landfall over Guangdong Province of China during the period 1470–1931. *International Journal of Climatology* 20:183–90.

Changnon, S. A., R. A. Pielke, Jr., D. Changnon, R. T. Sylves, and R. Pulwarty. 2000. Human factors explain the increased losses from weather and climate extremes. *Bulletin of the American Meteorological Society* 81:437–43.

Chu, P., and J. Wang. 1997. Tropical cyclone occurrences in the vicinity of Hawaii: Are the differences between El Niño and non-El Niño years significant? *Journal of Climate* 10:2683–89.

Donnelly, J. P., S. S. Bryant, J. Butler, J. Dowling, L. Fan, N. Hausmann, P. Newby, B. Shuman, J. Stern, K. Westover, and T. Webb III. 2001a. A 700 yr sedimentary record of intense hurricane landfalls in southern New England. *Bulletin of the Geological Society of America* 113:714–27.

Donnelly, J. P., S. Roll, M. Wengren, J. Butler, R. Lederer, and T. W. III. 2001b. Sedimentary evidence of intense hurricane strikes from New Jersey. *Geology* 29:615–18.

Elsner, J. B., A. B. Kara, and M. A. Owens. 1999. Fluctuations in North Atlantic hurricane frequency. *Journal of Climate* 12:427–37.

Elsner, J. B., K.-b. Liu, and B. Kocher. 2000. Spatial variations in major U.S. hurricane activity: Statistics and a physical mechanism. *Journal of Climate* 13:2293–2305.

Goldenberg, S. B., C. W. Landsea, A. M. Mestas-Nuñez, and W. M. Gray. 2001. The recent increase in Atlantic hurricane activity: Causes and implications. *Science* 293:474–79.

Gray, W. M. 1984. Atlantic seasonal hurricane frequency. Part I: El Niño and 30 mb Quasi-Biennial Oscillation influences. *Monthly Weather Review* 112:1649–68.

Gray, W. M. 1990. Strong association between West African rainfall and U.S. landfall of intense hurricanes. *Science* 249:1251–56.

Gray, W. M., C. W. Landsea, P. W. Mielke, Jr., and K. J. Berry. 1993. Predicting Atlantic basin seasonal tropical cyclone activity by 1 August. *Weather and Forecasting* 8:73–86.

Gray, W. M., C. W. Landsea, P. W. Mielke, Jr., and K. J. Berry. 1994. Predicting Atlantic basin seasonal tropical cyclone activity by 1 June. *Weather and Forecasting* 9:103–15.

Hartmann, D. L., and E. D. Maloney. 2001. The Madden-Julian Oscillation, barotropic dynamics, and North Pacific tropical cyclone formation. Part II: Stochastic barotropic modeling. *Journal of the Atmospheric Sciences* 58:2559–70.

Hayne, M., and J. Chappell. 2001. Cyclone frequency during the last 5000 years from Curacoa Island, Queensland. *Paleogeography, Paleoclimatology, Paleoecology* 168:201–19.

Hebert, P. J., J. D. Jarrell, and M. Mayfield. 1996. *The deadliest, costliest, and most intense U.S. hurricanes of this century.* NOAA Technical Memorandum, NWS TPC-1. Coral Gables, Fla: National Oceanic and Atmospheric Administration.

Houghton, J. T., Y. Ding, D. J. Griggs, M. Noguer, P. J. van der Linden, X. Dai, K. Maskell, and C. A. Johnson, eds. 2001. *Climate change 2001: The scientific basis.* Cambridge: Cambridge University Press.

Jarvinen, B. R., C. J. Neumann, and M. A. S. Davis. 1984. *A tropical cyclone data tape for the North Atlantic Basin, 1886–1983: Contents, limitations, and uses.* NOAA Technical Memorandum, NWS NHC Report No. 22. Coral Gables, Fla.: National Oceanic and Atmospheric Administration.

Knaff, J. A. 1997. Implications of summertime sea level pressure anomalies in the tropical Atlantic region. *Monthly Weather Review* 125:789–804.

Landsea, C. W. 1993. A climatology of intense (or major) Atlantic hurricanes. *Monthly Weather Review* 121:1703–13.

Landsea, C. W., W. M. Gray, P. W. Mielke, Jr., and K. J. Berry. 1992. Long-term variations of Western Sahelian monsoon rainfall and intense U.S. landfalling hurricanes. *Journal of Climate* 5:1528–34.

Landsea, C. W., W. M. Gray, P. W. Mielke, Jr., and K. J. Berry. 1994. Seasonal forecasting of Atlantic hurricane activity. *Weather* 49:273–84.

Landsea, C. W., N. Nicholls, W. M. Gray, and L. A. Avila. 1996. Downward trends in the frequency of intense Atlantic hurricanes during the past five decades. *Geophysical Research Letters* 23:1697–1700.

Liu, K.-b., and M. L. Fearn. 1993. Lake-sediment record of late Holocene hurricane activities from coastal Alabama. *Geology* 21:793–96.

Liu, K.-b., and M. L. Fearn. 2000a. Holocene history of catastrophic hurricane landfalls along the Gulf of Mexico coast reconstructed from coastal lake and marsh sediments. In *Current stresses and potential vulnerabilities: Implications of global change for the Gulf Coast region of the United States*, edited by Z. H. Ning and K. K. Abdollahi, 38–47. Baton Rouge: Franklin Press.

Liu, K.-b., and M. L. Fearn. 2000b. Reconstruction of prehistoric landfall frequencies of catastrophic hurricanes in northwestern Florida from lake sediment records. *Quaternary Research* 54:238–45.

Liu, K.-b., C. Shen, and K.-s. Louie. 2001. A 1000-year history of typhoon landfalls in Guangdong, southern China, reconstructed from Chinese historical documentary records. *Annals of the Association of American Geographers* 91:453–64.

Madden, R. A., and P. R. Julian. 1971. Detection of a 40–50 day oscillation in the zonal wind in the tropical Pacific. *Journal of the Atmospherica Sciences* 28:702–8.

Maloney, E. D., and D. L. Hartmann. 2000a. Modulation of eastern North Pacific hurricanes by the Madden-Julian Oscillation. *Journal of Climate* 13:1451–60.

Maloney, E. D., and D. L. Hartmann. 2000b. Modulation of hurricane activity in the Gulf of Mexico by the Madden-Julian Oscillation. *Science* 287:2002–4.

Maloney, E. D., and D. L. Hartmann. 2001. The Madden-Julian Oscillation, barotropic dynamics, and North Pacific tropical cyclone formation. Part I: Observations. *Journal of the Atmospheric Sciences* 58:2545–58.

National Climatic Data Center (NCDC). 1999. Special report: Hurricane Mitch—October 1998 (available at: http://www.ncdc.noaa.gov/oa/reports/mitch/mitch.html).

Nicholls, N. 1979. A possible method for predicting seasonal tropical cyclone activity in the Australian region. *Monthly Weather Review* 107:1221–24.

Nicholls, N. 1984. The Southern Oscillation, sea-surface temperature, and interannual fluctuations in Australian tropical cyclone activity. *Journal of Climatology* 4:661–70.

Nicholls, N., C. Landsea, and J. Gill. 1998. Recent trends in Australian region tropical cyclone activity. *Meteorological Atmospheric Physics* 65:197–205.

Nott, J., and M. Hayne. 2001. High frequency of "super-cyclones" along the Great Barrier Reef over the past 5,000 years. *Nature* 413:508–12.

Pielke, R. A., Jr., and C. W. Landsea. 1998. Normalized hurricane damages in the United States: 1925–1995. *Weather and Forecasting* 13:621–31.

Pielke, R. A., Jr., and R. A. Pielke, Sr. 1997. *Hurricanes: Their nature and impacts on society*. New York: Wiley.

Rappaport, E. N. 2000. Loss of life in the United States associated with recent Atlantic tropical cyclones. *Bulletin of the American Meteorological Society* 81:2065–74.

Revelle, C. G., and S. W. Goulter. 1986. South Pacific tropical cyclones and the Southern Oscillation. *Monthly Weather Review* 14:1138–45.

Shapiro, L. J. 1987. Month-to-month variability of Atlantic tropical circulation and its relationship to tropical storm formation. *Monthly Weather Review* 115:1598–614.

Shapiro, L. J. 1989. The relationship of the Quasi-Biennial Oscillation to Atlantic tropical storm activity. *Monthly Weather Review* 117:2598–614.

Simpson, R. H., and H. Riehl. 1981. *The hurricane and its impact.* Baton Rouge: Louisiana State University Press.

Zanetti, A., R. Enz, I. Menzinger, J. Mehlhorn, and S. Suter. 2003. Natural catastrophes and man-made disasters in 2002: High flood loss burdens. *Sigma,* no. 2/2003. Zurich: Swiss Reinsurance Company.

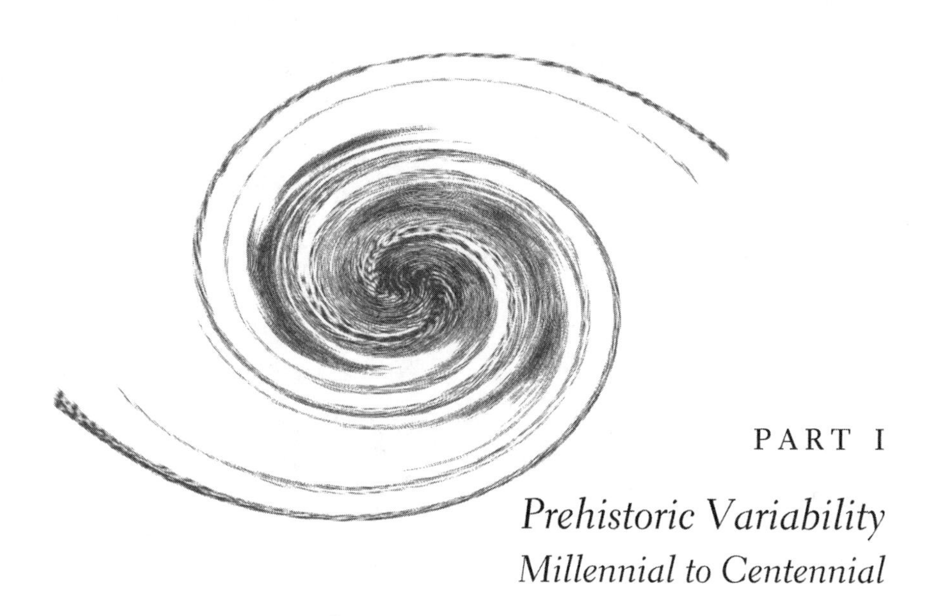

PART I

Prehistoric Variability

Millennial to Centennial

2

Paleotempestology:
Principles, Methods, and Examples from Gulf Coast Lake Sediments

Kam-biu Liu

Paleotempestology is an emerging field of science that studies past tropical cyclone activity mainly through the use of geological proxy techniques. Along the Gulf of Mexico coast, coastal lakes are often subjected to overwash processes during catastrophic hurricane strikes when their sandy barriers are overtopped by storm surges. Sand layers deposited by overwash processes can be identified in lake and marsh sediments along the U.S. Gulf Coast, thus providing proxy records of past catastrophic hurricane landfalls that may span the last 5,000 years. This chapter reviews the principles and methodology for using these overwash deposits as proxy records in paleotempestology. Case studies from Lake Shelby, Alabama, and Western Lake, Florida, are provided as examples.

The proxy records derived from a number of Gulf coastal lakes and marshes suggest a return period of about 300 years for catastrophic hurricanes of Category 4 or 5 intensity. For any specific coastal location, this is equivalent to a landfall probability of about 0.3% per year, although significant temporal variability occurs at time scales of centuries to millennia. The recent millennium (the last 1,000 years) has been a quiescent period of low hurricane activity, but it was preceded by a "hyperactive" period from about 3,400 to 1,000 years ago. The millennial-scale variability in hurricane activity may be related to long-term changes in the position of the Bermuda High and in the North Atlantic Oscillation (NAO).

Since Hurricane Andrew caused $30 billion worth of damage in Florida and Louisiana in 1992, there has been increasing societal concern over the vulnerability of the U.S. coasts to strikes by intense hurricanes, here defined as those of Categories 3, 4, and 5 according to the Saffir-Simpson intensity scale

(Pielke and Pielke 1997; Elsner and Kara 1999). Although intense hurricanes make up only 21% of landfalling tropical cyclones in the United States, they account for over 83% of the normalized damages after adjusting for inflation, wealth, and population increase (Pielke and Landsea 1998). Therefore, understanding the climatology and especially the landfall probabilities of these rare but highly destructive hurricanes is of great interest not only to climate researchers but also to policy-makers, socioeconomic planners, emergency response and risk managers, and the reinsurance industry (Liu 2004).

How vulnerable are the U.S. coastal areas to destructive strikes by intense hurricanes? The conventional approach to answering this question is to examine the historical record of intense hurricane landfalls, tally up the occurrences for any specific location (city, county, state, or region), and calculate the landfall probabilities by dividing the total number of strikes by the number of years of record. This approach, used extensively by climate scientists, risk managers, and actuaries working for insurance companies (e.g., Pielke and Pielke 1997; Elsner and Kara 1999), works well only if the frequency of occurrence is high enough to provide a statistically reliable sample. For the U.S. Gulf Coast, there are sufficiently large numbers of landfalls by minor hurricanes (defined as Categories 1 and 2 according to the Saffir-Simpson scale) during the past 100 years of instrumental record to permit a county-by-county estimation of landfall probabilities (Neumann et al. 1999). For the major or intense hurricanes, such estimates can still be made (e.g., Elsner and Kara 1999; Murnane et al. 2000), but the uncertainty is higher because there are fewer cases, and most of the cases used in the calculation come from the Category 3 hurricanes. The real problem is with the catastrophic hurricanes, here defined as those of Saffir-Simpson Categories 4 and 5 intensity. These "monster storms" are so rare that most localities on the U.S. coast have not been *directly* hit by them during the historical period. Obviously, the period of instrumental observation—essentially the last 150 years (Landsea et al., chapter 7 in this volume)—is not long enough to produce a realistic estimate of the landfall probabilities of these rare but most destructive storms.

For certain parts of the world, the historical record of tropical cyclone activity can be extended further into the past. In the United States and the Caribbean, historical records of hurricane activity have been compiled from newspapers, ships' logs, diaries, annals, and various British and Spanish government documents (Ludlam 1963; Millás 1968; Reading 1990; Caviedes 1991; Fernández-Partagás and Díaz 1996; Mock, chapter 5, and García Herrera et al., chapter 6 in this volume). The longest historical record, and the best potential for such data, exists in China, where written knowledge about the tropical cyclone (typhoon) as a distinct weather phenomenon began in the fifth century A.D. (Louie and Liu 2003, chapter 8 in this volume). A recent study using Chinese historical documentary records has produced a 1,000-year history of typhoon landfalls for the Guangdong Province of southern China (Liu, Shen, and Louie 2001). A common problem with these historical documentary records, however, is that data quality decreases and uncertainty increases as we go back in time.

Another approach for assessing hurricane landfall probabilities is to produce a geological record of past hurricane strikes, thereby extending the period of observation from a century to centuries or even millennia. This approach is now part of a new and exciting scientific field called *paleotempestology*. This term was coined by Kerry Emanuel of M.I.T. and has been defined as "a new field that studies past hurricane activities by means of geological proxy techniques" (Liu and Fearn 2000b:238). Several proxies have been explored for use in studying past hurricane activities. These include overwash deposits preserved in the sediments of coastal lakes (Liu and Fearn 1993, 2000a, 2000b; Liu, Lu, and Shen 2003) and coastal marshes (Liu and Fearn 2000a; Donnelly et al. 2001a, 2001b; Donnelly and Webb, chapter 3 in this volume); microfossils such as foraminifera, pollen, diatoms, dinoflagellates, and phytoliths contained in these coastal sediments (Gathen 1994; Li 1994; Fearn and Liu 1995; Zhou 1998; Collins, Scott, and Gaye 1999; Hippensteel and Martin 1999; Liu, Lu, and Shen 2003; Lu and Liu 2003), wave-generated or flood-generated sedimentary structures or deposits (tempestites) in marine or lagoonal sediments (Duke 1985; Davis, Knowles, and Bland 1989), storm-generated beach ridges (Hayne and Chappell 2001; Nott and Hayne 2001), oxygen isotopic ratios of hurricane precipitation recorded in shallow-water corals and speleothems (Scoffin, 1993; Lawrence and Gedzelman 1996; Malmquist 1997; Cohen 2001); and tree rings (Doyle and Gorham 1996; Reams and Van Deusen 1996). Although all of these proxies show some potential as research tools in paleotempestology, the first—overwash deposits preserved in coastal lake and marsh sediments—has so far proven to be the most useful in producing a millennial record of past hurricane landfalls.

This chapter introduces the principles and methods of paleotempestology based on the stratigraphic study of coastal lake sediments. Since 1989, we have applied these principles and methods in the paleotempestological investigations of more than 24 coastal lake and marsh sites along the Gulf of Mexico and Atlantic coasts (figure 2.1). Examples from two of these sites, Lake Shelby, Alabama, and Western Lake, Florida, on the Gulf Coast will be used to illustrate these principles and methods.

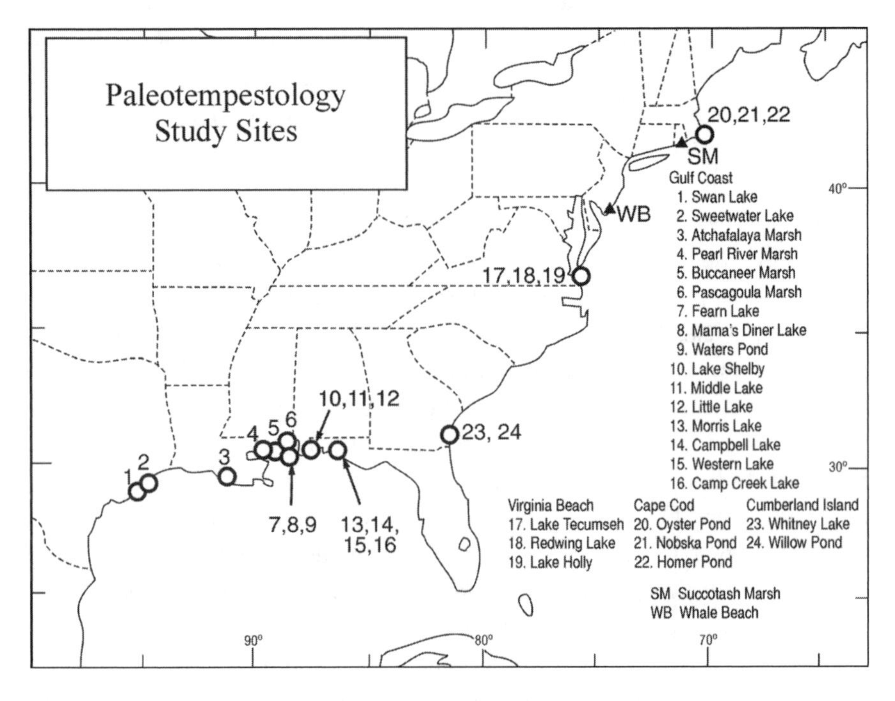

FIGURE 2.1 Coastal lake and marsh sites along the Gulf Coast and Atlantic coast investigated by the LSU paleotempestology research group (sites 1–24; Liu and Fearn 1993, 2000a, 2000b; Liu, Lu, and Shen 2003; Liu, unpublished data) and other researchers (SM, WB; Donnelly et al. 2001a, 2001b) for paleo-hurricane reconstruction studies. Lake Shelby and Western Lake are sites 10 and 15, respectively.

Tempestites in Coastal or Marine Sediments

Tempestites, or storm deposits, have been widely recognized from modern and ancient sediments in coastal and marine environments (e.g., Aigner 1985). For example, hummocky cross-stratification occurring in continental-shelf sediments has been attributed to wave dynamics associated with tropical cyclones (Duke 1985). Similarly, sedimentary structures and facies identified from cores of clastic sediments taken in tidally influenced lagoonal bays have also been interpreted as having been formed by hurricanes (Davis, Knowles, and Bland 1989). But there are at least three reasons why storm deposits and sedimentary structures formed in shallow-water marine environments are often poor candidates for providing high-resolution records of past hurricane strikes. First, in the more open, high-energy environments on the continental shelf or in lagoonal bays subjected to tidal flushing, tempestites can be reworked and obliterated by subsequent storms, leaving at best an incomplete proxy record or no record at all. Second, these features are typically formed in clastic, inorganic sediments that cannot be easily or reliably radiocarbon-dated, thus rendering a high-resolution chronology difficult to obtain. Third, sedimentary structures like hummocky cross-stratification cannot be easily detected from individual cores; they are typically identified from outcrops of ancient sediments.

By contrast, deposits formed by storm surges or overwash processes in the bottom of coastal lakes isolated from the sea are more likely to be preserved, because they will not be removed by tidal scouring or reworking. In addition, lake sediments often contain organic materials that can be used for radiocarbon dating, and storm deposits are relatively easy to identify in lake sediment cores. For coastal lakes situated behind sand barriers (barrier beaches, sand dunes, or beach ridges), overwash by hurricanes is an important mechanism that would result in the formation of a storm deposit in the lake bottom (Leatherman 1981). Such a storm deposit, usually occurring in the form of a sand layer sandwiched between organic lake sediments, can be identified in sediment cores taken in the part of the lake directly affected by the overwash. Thus overwash sand layers can be used as a proxy for past hurricane strikes.

Storm Surges, Overwash Processes, and Coastal Sand Barriers

Whether a storm surge will overtop a coastal sand barrier depends primarily on two factors—storm-surge height and the height of the barrier. As a rule of thumb, storm surge height is positively related to the intensity (as defined by central pressure), size, and forward speed of a landfalling hurricane. Generally, the stronger the hurricane, the higher the storm-surge height (Hoover 1957). Table 2.1 relates storm surge height to hurricane intensity according to the Saffir-Simpson scale. For example, the highest recorded storm surge, 24.2 feet

TABLE 2.1 The Saffir-Simpson Hurricane Scale

	Central pressure		Winds	Surge	
Category	(mb)	(inches)	(mph)	(feet)	Damage
1	>980	>28.94	74–95	4–5	Minimal
2	965–979	28.50–28.91	96–110	6–8	Moderate
3	945–964	27.91–28.74	111–130	9–12	Extensive
4	920–944	27.17–27.88	131–155	13–18	Extreme
5	<920	<27.17	>155	>18	Catastrophic

(7.2 m) above mean sea level, occurred when Hurricane Camille, a Category 5 storm, made landfall near Pass Christian on the Mississippi coast in 1969. Nevertheless, this unquestionably positive relationship between storm surge height and hurricane intensity may not be linear, as it may be complicated by local geomorphological factors such as the offshore bottom slope, configuration of the coastline, and supply of flood water from river discharge, as well as stochastic meteorological factors such as the timing and duration of the landfall relative to the astronomical tide or the angle of storm approach relative to coastline configuration (Jarvinen and Neumann 1985; Hubbert and McInnes 1999). It is also important to note that in the Northern Hemisphere the peak storm surge occurs along the forward-right quadrant of the landfalling hurricane, because here the seawater is pushed by the onshore wind of the cyclonic (anti-clockwise) circulation around the eye to pile up against the coastline. The storm surge is more subdued or even negative on the coastline to the left of the eye, mainly due to the prevalence of offshore winds during landfall (Simpson and Riehl 1980) (figure 2.2).

In many areas along the Gulf of Mexico coast, a sand barrier exists between the land and the sea. This sand barrier usually consists of a barrier beach of vary-

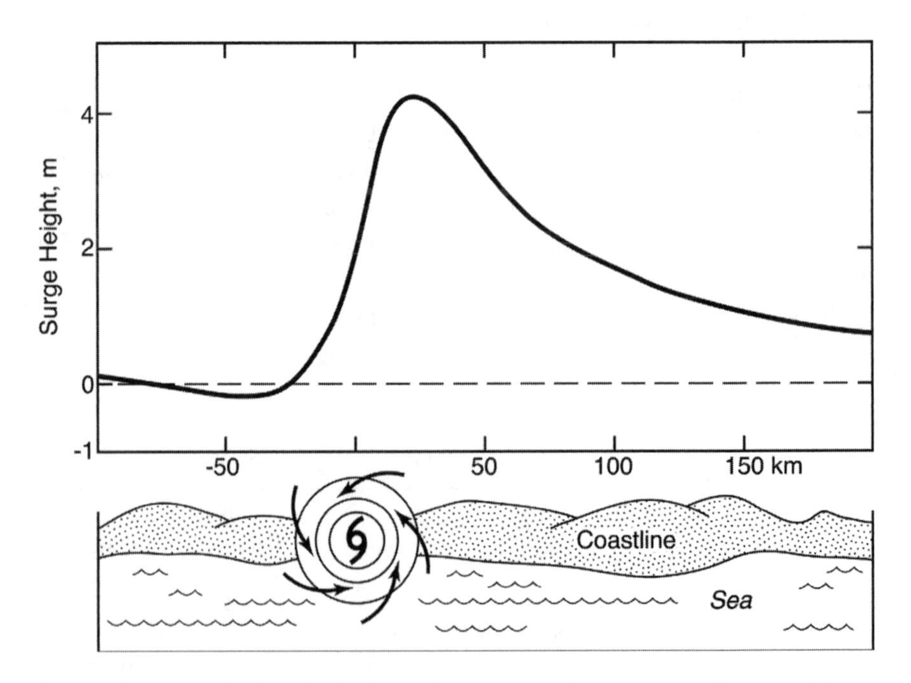

FIGURE 2.2 Hypothetical distribution of storm-surge heights along a coastline on an intense hurricane landfall (modified from Simpson and Riehl 1980). Storm-surge heights are greater on the right side of the hurricane because of the onshore winds.

ing length and width, on which beach ridges or sand dunes may develop. The height of the sand barrier may vary from 1 meter to more than 10 meters, the latter being exemplified by the high dune ridges near Western Lake in northwestern Florida. For a coastal lake completely enclosed behind a sandy barrier, the barrier height constitutes the *overwash threshold* of the coastline and thus determines the *paleotempestological sensitivity* of the lake, defined as the minimum intensity of a hurricane whose impacts will be recorded in the sediments of the lake. In the absence of barrier breaching, the higher the sand barrier, the more intense the hurricane will be required for the storm surge to exceed that threshold height and to wash sand into the lake. On time scales of decades to centuries, the sand barrier may be lowered or even breached by overwash processes during hurricane strikes (Leatherman 1981). Though few long-term measurements exist on post-hurricane recovery along the Gulf Coast, barrier heights probably recover fairly quickly after an overwash event, usually within a few years, or at the most a few decades for high dunes.

Ritchie and Penland (1987) envisioned a cycle of barrier development defined by the return period of hurricane impacts. They suggested that for low (~1.5 m above sea level) dunes typical of the Louisiana coastline, complete recovery to pre-storm heights occurs within five to seven years after an overwash event. Conceivably the recovery process may take longer for high dunes such as those of northwestern Florida, but it is reasonable to expect complete recovery within a few decades at the most. During the recovery phase, the coastline's overwash threshold and the lake's paleotempestological sensitivity will be temporarily lowered. This is a relatively short interruption for lakes with very high thresholds and hurricane return periods on the order of several centuries. During the relatively long interval between overwash events, the stability of the barrier will depend on a number of coastal morphodynamic and ecological factors such as sediment supply, eolian processes, and vegetation growth (Leatherman 1981; Ritchie and Penland 1987).

Proxy Record from Coastal Lake Sediments: General Principles and Methods

When a storm surge generated by a landfalling hurricane overtops the sand barrier, sand will be washed from the beach and dunes into the lake, forming an overwash fan (Leatherman 1981). The overwash fan, in the form of a sand layer, will be thicker near the barrier and thinner outward away from the shore. Starting with a simple model, we may expect that stronger hurricanes, which usually generate higher storm surges, will likely produce thicker, more extensive

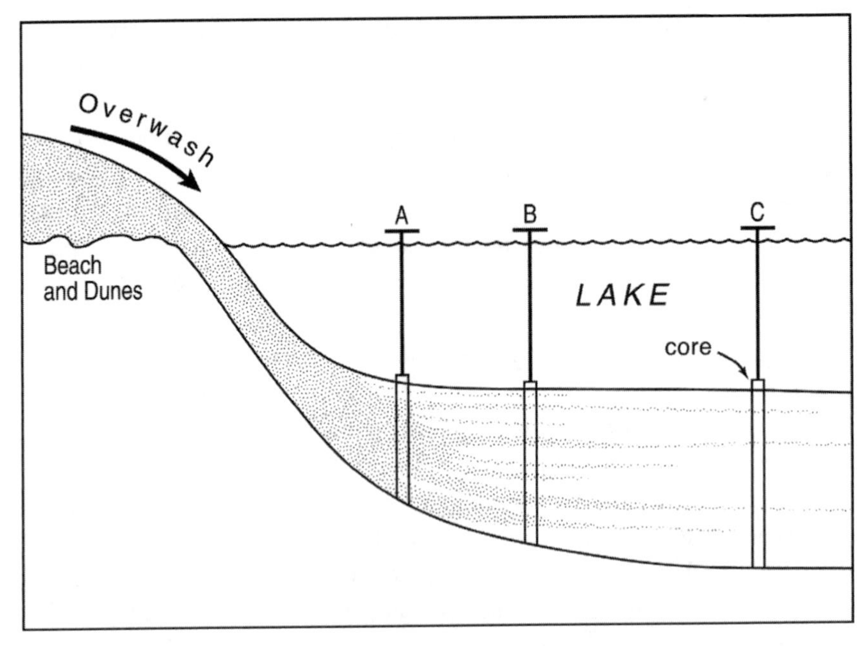

FIGURE 2.3 Hypothetical pattern of sand-layer deposition in a coastal lake subjected to repeated storm overwash events in the past. The overwash sand layers are thicker near the sand barrier and become thinner toward the lake center. A core taken from site B will contain more and thicker sand layers than one taken from site C. A core taken from site A, however, may consist of all sand without discrete layers.

overwash sand layers than those produced by weaker hurricanes. For a coastal lake site impacted repeatedly by landfalling hurricanes and their associated overwash processes, sediment cores taken from the lake are expected to contain multiple layers of overwash deposits sandwiched between normal organic lake sediments (figure 2.3). Each of these sand layers thus provides a proxy record of a past hurricane strike.

The Modern Analog Approach

As a first approximation, the horizontal extent and thickness of the overwash sand layer is expected to be positively correlated with the intensity of the landfalling storm, assuming that the strike is a "direct hit" and other factors are equal. This hypothesis can be tested by observing the impacts of recent or past hurricanes of known intensities on a given site, and comparing the sedimentary records of their overwash deposits. Once the relationship between storm intensity and sand layer size is known, it can provide a basis for inferring the intensity of prehistoric hurricanes from the size and thickness of their overwash sand layers. This is the same uniformitarian principle that underlies the modern analog approach commonly used in paleoenvironmental reconstruction (Birks and Birks 1980). In the case of Lake Shelby, the sand layer attributed to the strike by Hurricane Frederic, a High-Category 3 hurricane that devastated coastal Alabama in 1979, was found to be confined only to the nearshore sediments immediately impacted by the overwash. Using the assumption that the geomorphic setting of the lake has not changed drastically over the past 5,000 years, Liu and Fearn (1993, 2002) inferred that prehistoric hurricanes of Category 4 or 5 intensity produced the older sand layers that reached the center of the lake where the Frederic sand layer was absent. Similarly, the limited extent of the sand layer caused by Category 3 Hurricane Opal has provided the modern analog for inferring that the prehistoric sand layers found in the deep-water cores of Western Lake represent direct strikes by catastrophic hurricanes of Category 4 or 5 intensity (Liu and Fearn 2000b, 2002).

Fieldwork and Laboratory Methods

Typically, the lake sediments are cored by means of a piston corer operated on a wood-coring platform secured on two rubber boats. The piston corer, approximately 1.5 m long and constructed of clear-PVC tubes, is attached to a cable and pushed into the lake bottom by metal extension rods. After each core segment is retrieved, the coring tube containing the sediment is detached from the piston and sealed at both ends for transportation back to the laboratory. Care must be taken not to disturb the soft sediment, especially at the top of a core.

In the laboratory, the cores are examined visually for sand layers and other stratigraphic changes. The prominent or thick sand layers are usually clearly visible, but the thinner, more subtle layers may be overlooked. The subjectivity in detecting and identifying sand layers and other clastic laminations can be avoided by a high-resolution loss-on-ignition (LOI) analysis of the core at close sampling intervals. The LOI technique (Dean 1974) involves sampling the entire core at contiguous 1-cm intervals, and heating the sediment samples at 105°C, 550°C, and 1000°C to determine the water content, organic matter content, and carbonate content, respectively. On a LOI curve, sand layers can be recognized by having much lower water and organic matter contents than the organic lake sediments (Liu and Fearn 2000b) (figure 2.4). After the LOI analysis, the core is visually inspected again to confirm the occurrence of sand layers. Other sedimentary structures and sedimentological characteristics, such

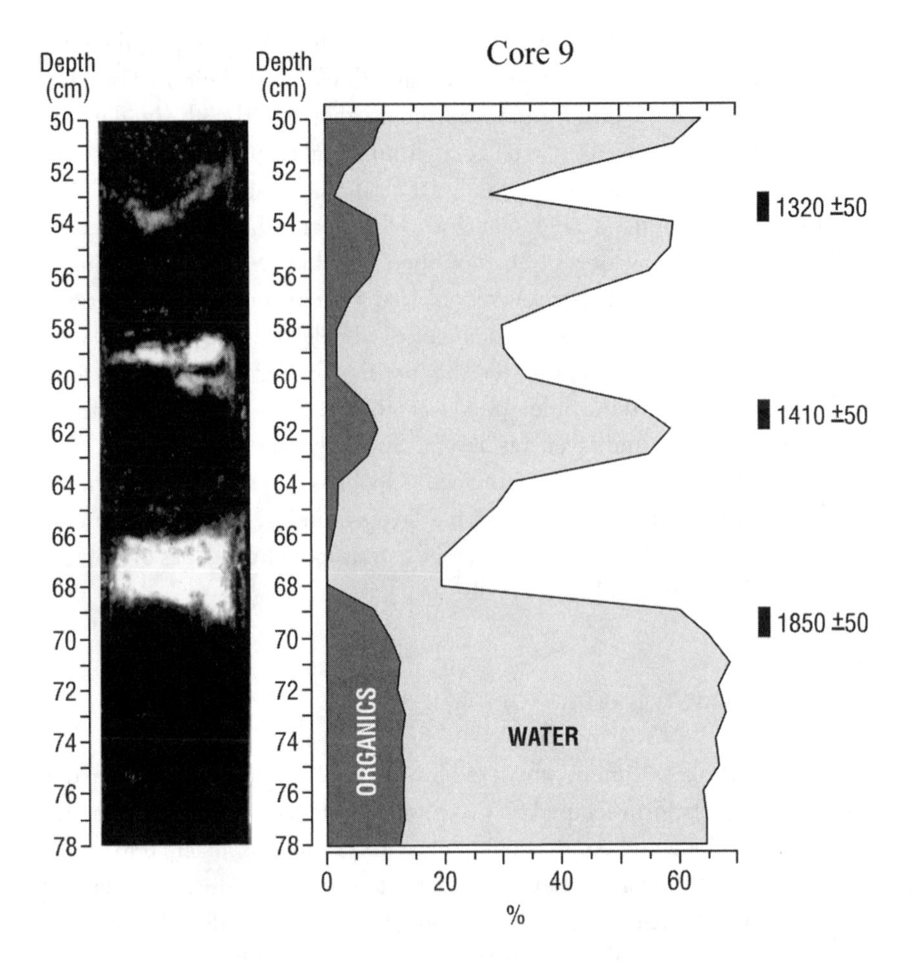

FIGURE 2.4 Photograph of three prominent sand layers in the 50- to 78-cm segment of core 9 from Western Lake, and the corresponding water and organic matter content curves determined by loss-on-ignition. Radiocarbon dates are on the right (after Liu and Fearn 2000b).

as texture, color, and lithology of the sand layers and their embedding sediment matrix, are also noted in the laboratory analysis.

Hurricanes are an important ecological agent that can cause significant ecological disturbance to coastal vegetation communities along the Gulf Coast (Conner et al. 1989; Walker et al. 1991; Batista and Platt 2003). Therefore, in addition to the overwash sand layers and other sedimentological parameters, paleolimnological and paleoecological evidence from the lake sediments can also provide useful supplementary information for detecting the impacts of past hurricane strikes or storm-surge events.

Theoretically, a hurricane-induced storm surge can cause saltwater intrusion into a freshwater lake or tidally connected oligohaline (low-salinity) lagoon without significant deposition of overwash sand at the coring site. Even in the absence of overwash sand layers in the core, storm-surge events can be detected by spikes of marine microfossils such as foraminifera, dinoflagellates, and marine diatoms in the sediment (Li 1994; Hemphill-Haley, 1996; Zhou 1998; Collins, Scott, and Gayes 1999; Hippensteel and Martin 1999; Liu, Lu, and Shen 2003).

It is well known that fossil pollen can be used as a sediment-dating technique by providing chronological marker horizons such as a rise in *Ambrosia* (ragweed) and a decline in *Castanea* (chestnut) (Fearn and Liu 1995; Donnelly et al. 2001a, 2001b). In addition, it can also be used to detect changes in aquatic vegetation communities caused by significant variations in the lake water salinity, as would be expected after a storm surge into a small freshwater lake. For example, in Momma's Diner Lake, a small freshwater pond on Horn Island off the Mississippi coast, pollen data reveal that the lake ecosystem changed episodically from freshwater to brackish during the past 600 years in a pattern consistent with the history of hurricane strikes reconstructed from the sand-layer evidence (Gathen 1994).

Opal phytoliths—microscopic silica particles derived from plant tissues—also have some promise in providing a proxy for overwash events because they can be found in certain oxidizing environments, such as sand dunes, where other microfossils are typically absent. Changes in phytolith assemblages associated with sand layers in coastal lake sediments may shed light on the provenance of the overwash sand (Lu and Liu 2003).

The age of the overwash sand layers can be estimated by means of radiocarbon dating of the organic sediments above or below the sand layer. For more recent sediments formed within the last 200 years or the past several decades, other isotopic dating techniques such as lead-210 (^{210}Pb) or cesium-137 (^{137}Cs) can be used (Donnelly et al. 2001a, 2001b; Donnelly and Webb, chapter 3 in this volume). The reliability of the sediment ages determined by these isotopic dating methods can be verified by comparison with the palynologically dated marker horizons such as the *Ambrosia* rise previously noted.

Site Selection: Geomorphic Setting

The sensitivity and completeness of the proxy record that can be obtained from coastal lake sediments depends on a number of factors. The first and foremost consideration is the physical or geomorphic setting of the lake. Freshwater or

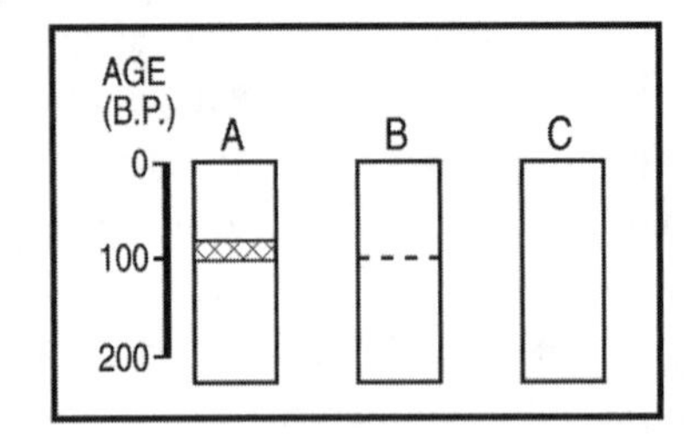

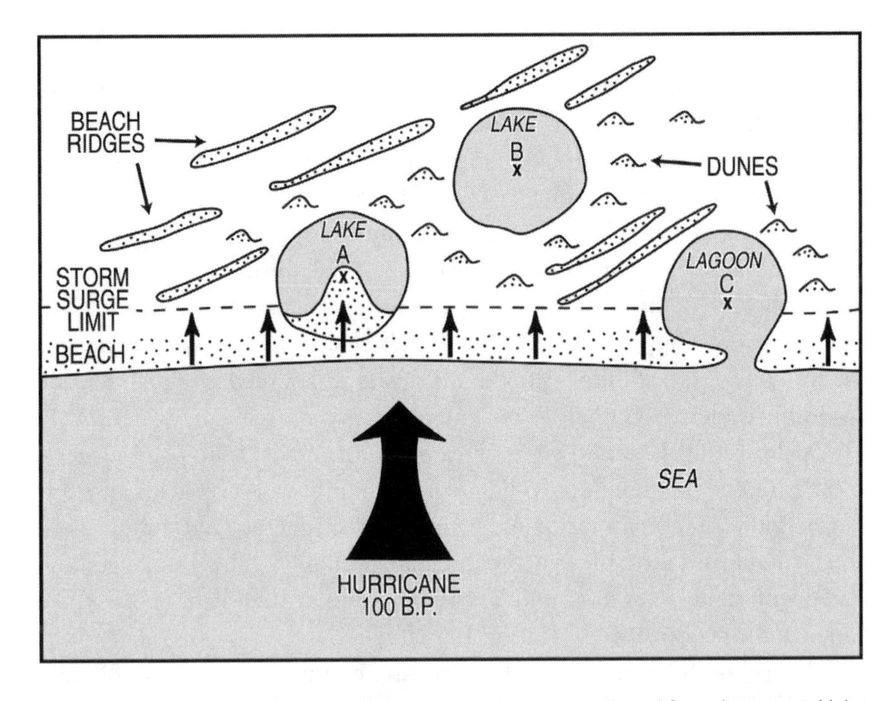

FIGURE 2.5 (*Top*) Hypothetical sedimentary records in cores collected from three coastal lakes (A, B, C) of different geomorphic settings that were impacted by a hurricane strike and the associated overwash 100 years ago. Thick and thin sand layers are represented by cross-shaded bands and dotted lines, respectively. (*Bottom*) Geomorphic settings of the three lakes in relation to coastal sand barriers (beach, dunes, beach ridges) and the spatial limit of the storm surge generated by the landfalling hurricane. Small arrows indicate waves overtopping the beach barrier, causing an overwash. An overwash fan is formed in Lake A.

oligohaline lakes that are completely isolated from the sea by a sand barrier are most likely to be sensitive to hurricane impacts and to preserve the evidence of overwash processes in the sediments. Figure 2.5 presents a hypothetical scenario in which a coastal location was struck by a landfalling hurricane 100 years ago. Suppose there are three coastal lakes with different geomorphic settings. Lake A, a closed-basin, back-barrier lake situated immediately behind a sandy beach, is directly affected by overwash. A core taken from Lake A is likely to contain a prominent sand layer reflecting that event. Lake B, situated farther away from the seashore and beyond the limit of the storm surge, will not yield an overwash sand layer in its core, although a thin or indistinct clastic layer may still be present due to eolian transport or slope wash of fine sand or silt from its shores or the adjacent sand dunes. Lake C is actually a lagoon that is tidally connected to the sea. Although this lagoon is also affected by the storm surge and sand is washed into its bottom from the adjacent beach and dunes, no sand layer may be preserved because the overwash sand is subsequently removed by tidal flushing. An alternative scenario for Lake C is that cores taken there may exhibit a complex stratigraphy consisting of multiple sand layers. This may happen if the tidal connection lowers the lagoon's paleotempestological sensitivity and allows frequent sand deposition into the lagoon by even minor storms passing through the region. In either case, cores from Lake C may not provide useful proxy records. Thus among the three sites, only Lake A would provide a clear, unambiguous record of that hurricane strike.

Many lakes along the Gulf and Atlantic coasts provide examples of the three types of coastal lakes described in the hypothetical case just described. Two of these, Lake Shelby in Alabama and Western Lake in northwestern Florida, exemplify the back-barrier lake in type A (figures 2.6 and 2.7). Both lakes have

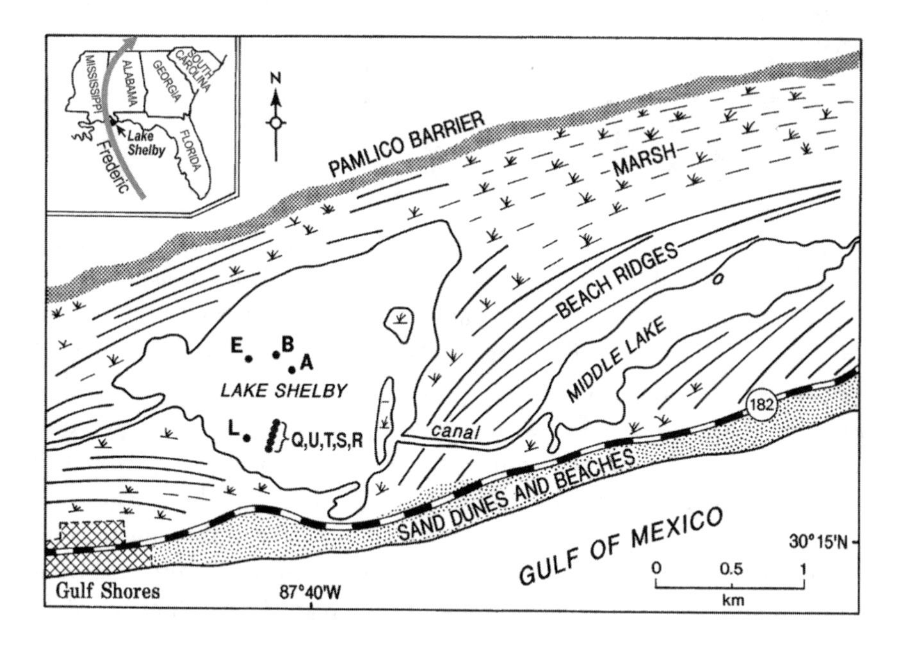

FIGURE 2.6 Geomorphic setting of Lake Shelby, Alabama, showing location of cores (black dots) mentioned in the text. The black and white line is Alabama State Highway 182. Inset shows location of Lake Shelby in relation to the path of Hurricane Frederic (modified from Liu and Fearn 1993).

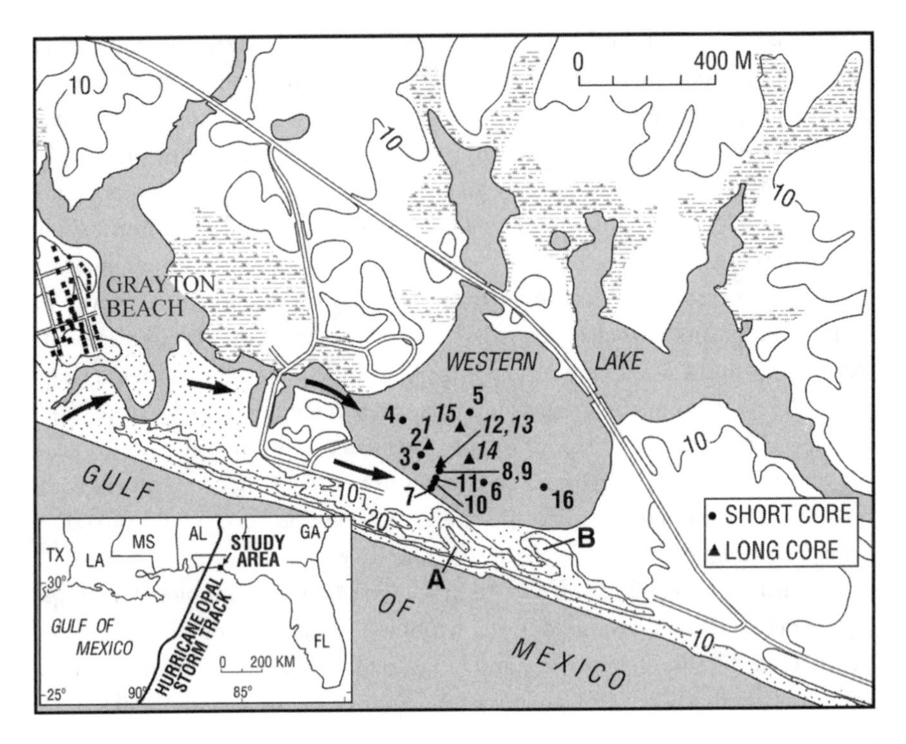

FIGURE 2.7 Geomorphic setting of Western Lake, northwestern Florida, showing location of long cores and short cores. Contour intervals are in meters. Arrows show direction of saltwater intrusion and sand transport caused by Hurricane Opal's storm surge. Areas marked A and B are two lobes of possible overwash deposits and associated sand dunes. Inset shows location of Western Lake in relation to the path of Hurricane Opal (after Liu and Fearn 2000b).

been subjected to intensive paleotempestological investigations. Lake Shelby is a freshwater lake (salinity 1.2 ppt) separated from the Gulf of Mexico by a complex system of beach ridges and sand dunes behind a barrier beach about 250 m wide (figure 2.6). Sediment cores taken from Lake Shelby contain at least 11 visibly distinct sand layers that were interpreted to have been caused by overwash events associated with direct strikes by catastrophic hurricanes over the last 3,200 years (Liu and Fearn 1993) (figure 2.8). Western Lake, situated behind a 150- to 200-m-wide barrier beach with a continuous dune ridge rising to over 6.2 m high above the sea (figure 2.7), has also yielded sediment cores that record at least 12 overwash events caused by catastrophic hurricane landfalls in the Florida Panhandle over the past 3400 years (Liu and Fearn 2000b) (figure 2.9).

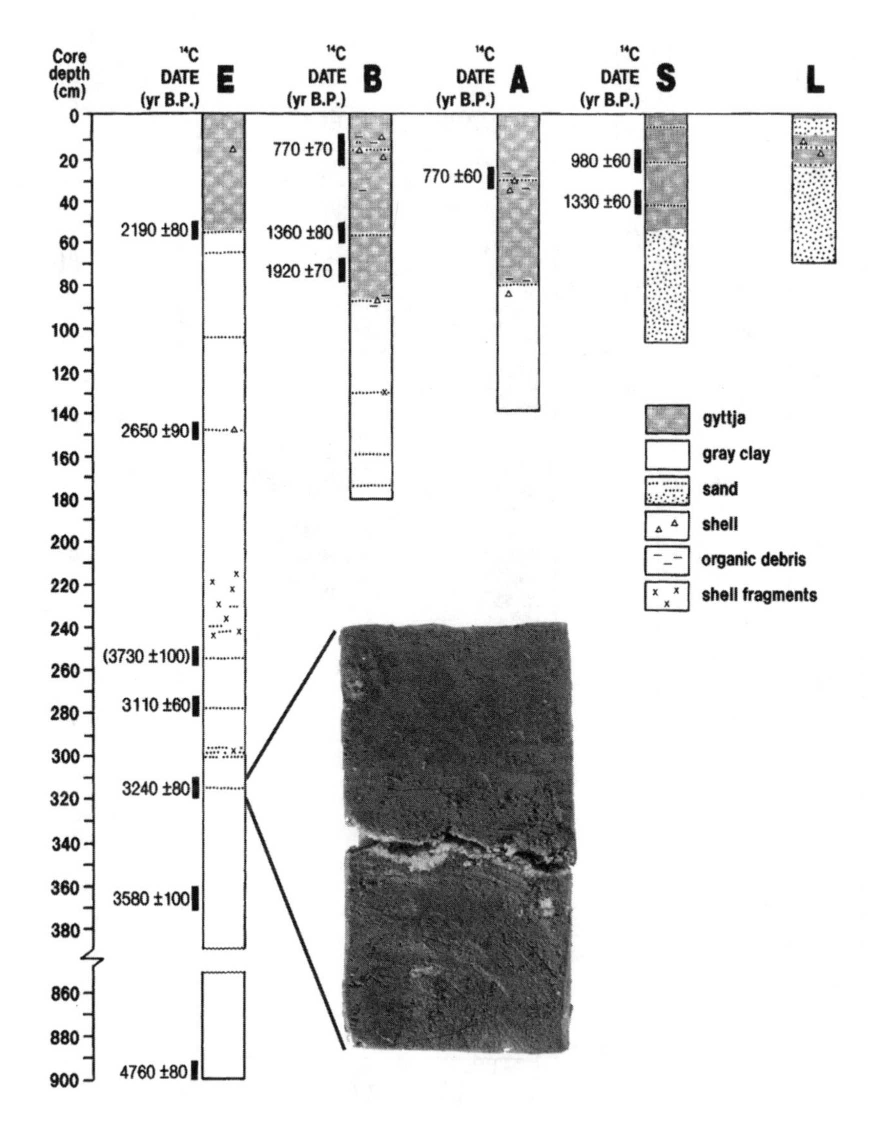

FIGURE 2.8 Radiocarbon-dated sediment stratigraphies of Lake Shelby cores E, B, A, S, L, showing positions of sand layers. Inset photograph shows sand layer in core E at 314 cm (after Liu and Fearn 1993).

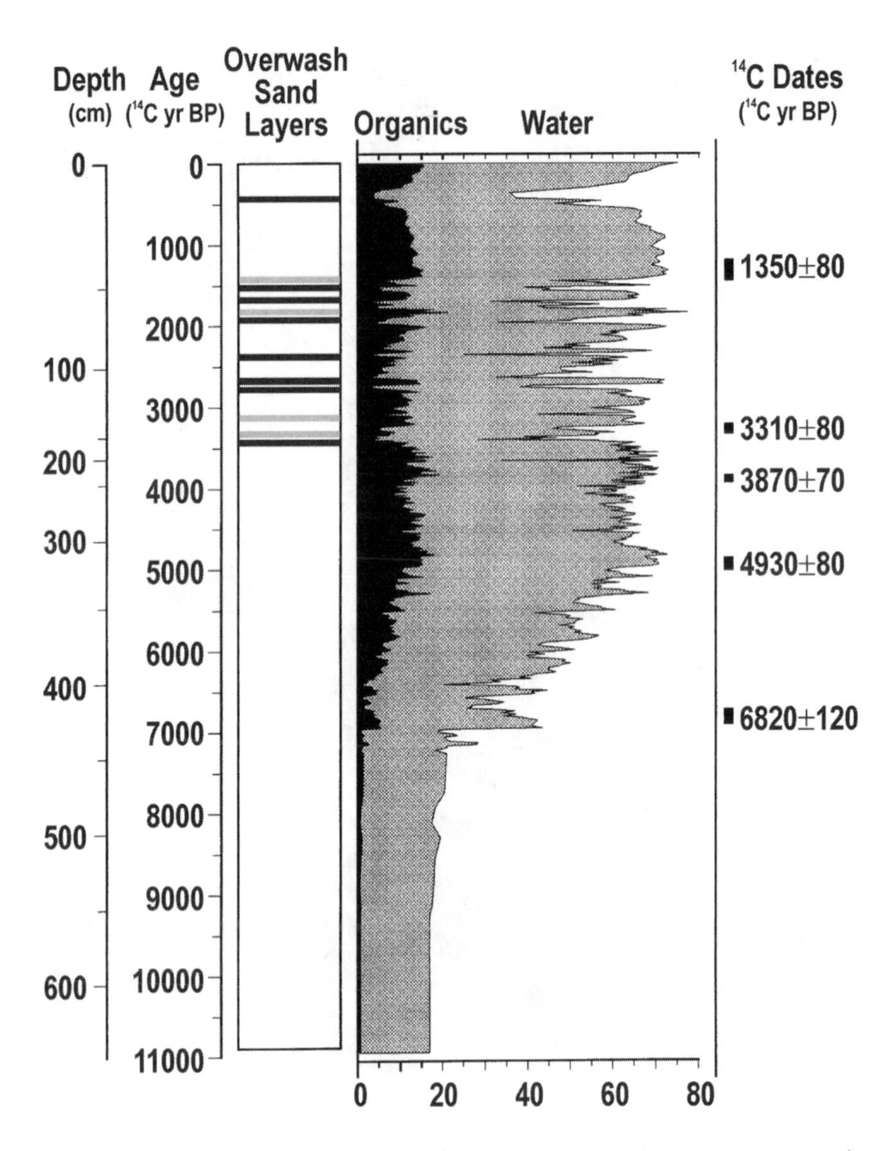

FIGURE 2.9 Loss-on-ignition curves for core 1 from Western Lake, showing water contents (% wet weight; gray curve) and organic matter contents (% dry weight; black curve). The stratigraphic column shows overwash sand layers occurring in the top 1.6 m (after Liu and Fearn 2000a).

Landfall Location and Meteorological Characteristics of Storm

The sensitivity of a lake site to hurricane strike is affected by a number of meteorological factors that collectively determine the wind speed, wind direction, and storm-surge height experienced at the site. These factors include, among others, storm intensity and forward speed, angle of the approaching storm relative to the configuration of the coastline, and direction and proximity of the landfalling storm relative to the lake site. Figure 2.10 shows a hypothetical case in which a stretch of a coastline was struck by landfalling hurricanes of various intensities six times during the last 3,000 years. Suppose a core is taken from a lake behind a line of barrier beach and sand dunes—the kind of site that, like Lake A in the scenario discussed earlier (figure 2.5), is expected to produce a well-preserved record of hurricane strikes.

Hurricane A, a Category 5 hurricane that struck 3,000 years ago, made landfall more than 200 km west of the site. Despite its high intensity, this hurricane did not leave a recognizable record in the sediments because the lake was outside of the zone of maximum winds and storm surge. Hurricane B, a Category 3 hurricane that struck 500 years later, was less intense than A, but it made landfall immediately to the west of the lake. Therefore, the storm surge experienced at the site was high enough to cause an overwash of the sandy barrier, leaving behind a layer of sand at the bottom of the lake. The most prominent sand layer in the core was deposited by Hurricane C, a Category 5 storm that directly struck the site 2,000 years ago. The storm surge experienced at the lake site from this storm would have been the highest and the overwash sand layer the most extensive. This is due not only to the high intensity of the landfalling storm, but also to the path of Hurricane C being situated immediately to the west (left) of the site, thereby exposing the lake to the brunt of the onshore winds and high surge driven by the cyclonic circulation around the eye.

By contrast, Hurricane D, a Category 1 hurricane, did not leave a sand layer in the sedimentary record because its storm surge was not high enough to overtop the sandy barrier, even though this storm scored a direct hit at the site. Hurricane E was a Category 5 hurricane that also struck the lake site directly. But unlike Hurricane C, this catastrophic hurricane made landfall immediately to the east (right) of the site, so that the lake was subjected to primarily offshore winds and somewhat lower storm-surge heights. An overwash sand layer would still have formed due to the severity of the storm impact, but it would have been somewhat thinner than the one formed by Hurricane C. Finally, no sand layer was formed by Hurricane F because this Category 3 storm struck the coast at some distance to the east (right) of the lake.

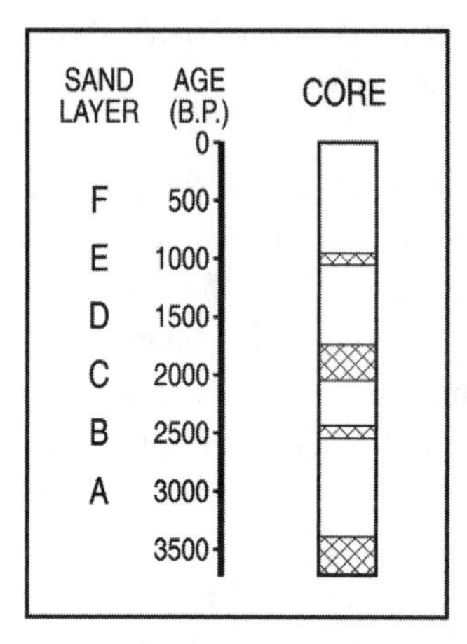

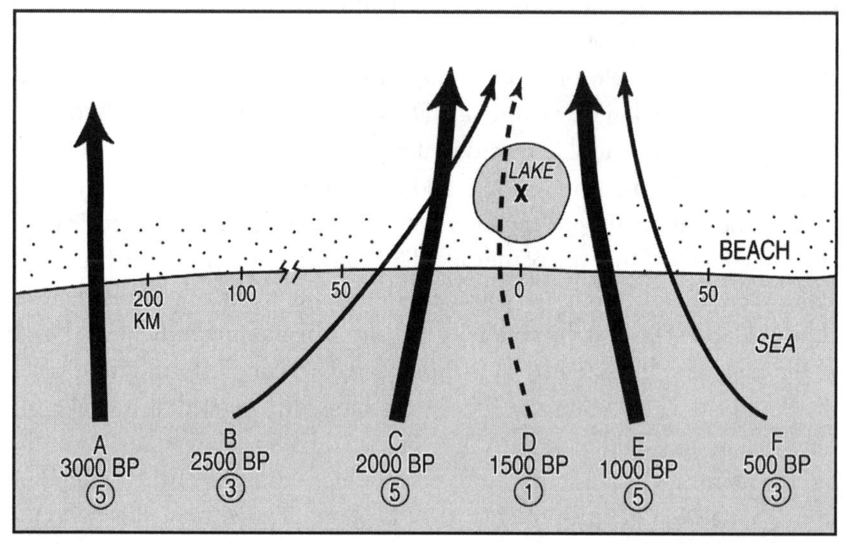

FIGURE 2.10 (*Top*) Hypothetical sedimentary record of a lake impacted by landfalling hurricanes and associated overwash events six times during the past 3,000 years. (*Bottom*) Location of the lake in relation to the paths, ages, and intensities of the six landfalling hurricanes (A–F). Thicknesses of the arrows are proportional to the intensity of the hurricanes according to the Saffir-Simpson scale, the latter also designated by the circled number below each arrow.

From this hypothetical scenario, it can be inferred that the proxy record from this lake is sensitive to *direct hits* (i.e., landfall within 50-km radius) by major hurricanes of Category 3 intensity or above. In particular, hurricanes that made landfall immediately to the west (left) of the lake were more likely to leave a distinct record. But based on the sedimentary record, it is not easy to distinguish a Category 3 hurricane that struck immediately to the west of the lake (e.g., Hurricane B) from a Category 5 hurricane that struck immediately to the east (e.g., Hurricane E) or one that made landfall to the west but somewhat farther away (not shown in figure 2.10), because these storms produced broadly similar impacts (e.g., storm-surge heights) at the site.

An example of decay in impact severity with distance from the location of landfall can be found in Lake Shelby. Camille, a Category 5 hurricane that caused a storm surge of more than 6.7 m and severely impacted coastal Mississippi in 1969, left a distinct sedimentary record in the lakes and marshes of coastal Mississippi (e.g., Horn Island lakes, Pearl River Marsh, Buccaneer Marsh, Pascagoula Marsh), but hardly any trace in the sediments of Lake Shelby. The lake was about 160 km east of Hurricane Camille's path of landfall (Liu and Fearn 1993, 2000a; Liu, unpublished data). By contrast, Hurricane Frederic, in 1979, was only a Category 3 storm with a 4.8-m-high storm surge, but it made landfall near Dauphin Island, Alabama, only 50 km west of Lake Shelby. Its impacts cannot be detected in the coastal sediments of the Mississippi sites west of the landfall location, but this storm left a prominent overwash sand layer in the nearshore sediments of Lake Shelby (Liu and Fearn 1993).

Coring Location and Overwash Sand Deposition

Within a lake, a suite of cores taken from different localities is necessary in order to reconstruct a complete record of past hurricane strikes. A sand layer, in the form of an overwash fan, does not spread out evenly over the bottom of a lake. Instead, it is thickest at the proximal end of the overwash fan near the coast, becoming thinner toward the distal end, until it disappears beyond the limit of the overwash deposition (figure 2.3). Therefore, within any core taken from a lake, the thickness of an overwash sand layer depends on the coring location relative to the horizontal extent of the overwash fan. The importance of obtaining multiple cores and cross correlating the sand layers among them is further illustrated by figure 2.11 (Liu and Fearn 2000b). Suppose a coastal lake site was directly struck by hurricanes of various intensities six times during the last 3,000 years. As a first approximation, it can be assumed that hurricanes of higher intensities (A, E) will produce larger overwash fans than weaker hurricanes (B, D, F). Thus, the Category 5 hurricanes will be registered in most cores, whereas the weaker hurricanes will only be recorded in cores taken close to shore (cores 2, 3, 8, 9) but absent in most other cores (cores 4, 5, 6, 7, 10, 11). In other words, a suite of cores taken from different parts of the lake may not

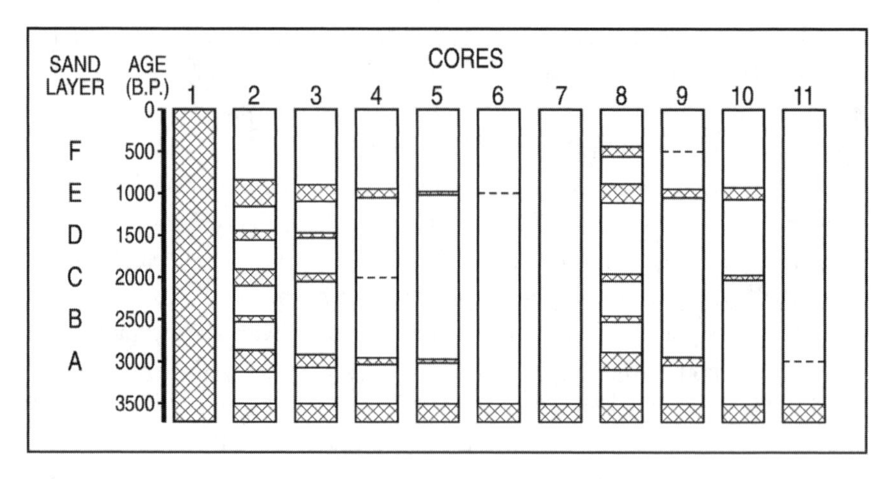

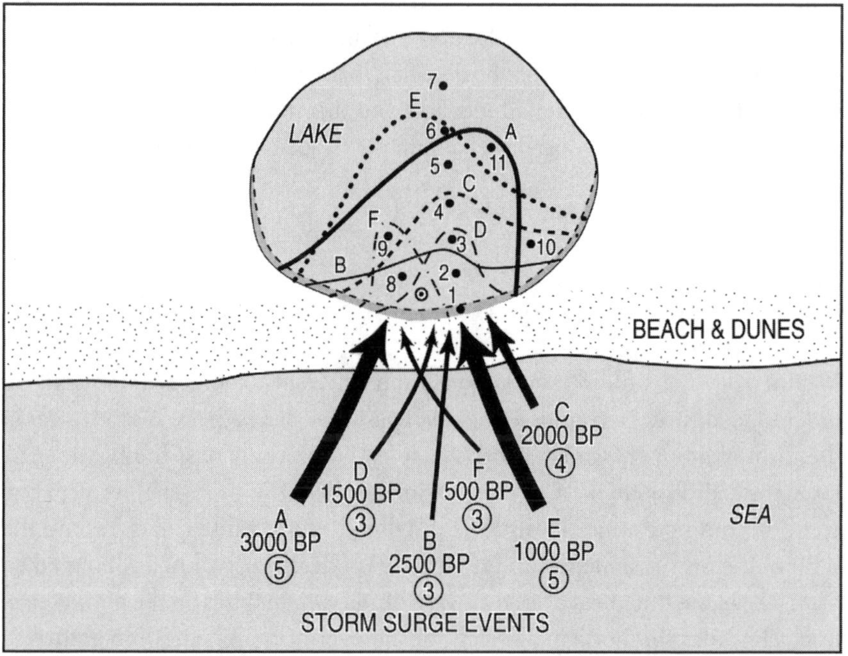

FIGURE 2.11 (*Top*) Hypothetical sedimentary stratigraphies of eleven cores (1–11) taken from different parts of a lake impacted by intense hurricanes and associated overwash events six times during the past 3,000 years. Thick and thin sand layers are represented by cross-shaded bands and dotted lines, respectively. (*Bottom*) Hypothetical pattern of overwash sand deposition in the lake where the 11 cores were taken. Solid, dotted, or dashed lines inside the lake denote the horizontal extents of the six overwash fans (A–F) corresponding to the six hurricane strikes and overwash events (arrows). Thicknesses of the arrows are proportional to the intensity of the hurricanes according to the Saffir-Simpson scale, the latter also designated by the circled number below each arrow. Numbered black dots (1–11) represent cores taken from the lake (after Liu and Fearn 2000b).

contain the same number or combination of sand layers. Cores taken close to shore will typically contain more, and thicker, sand layers than cores taken farther away from shore, because they contain records of all hurricane strikes that exceed a certain threshold (Category 3 in this case). By contrast, cores taken nearer the center of the lake will contain fewer and thinner sand layers that represent only the most intense storms. Cores taken too far away from shore (e.g., core 7) may lie outside the limits of all overwash fans entirely and may not contain any sand layer at all. Conversely, cores taken too close to shore (e.g., core 1) and in shallow waters may be composed of all sand, in which individual events may not be distinguishable.

Evidence for this pattern of overwash sand deposition can be found in Lake Shelby. When Category 3 Hurricane Frederic struck coastal Alabama in 1979, the 4.8-m storm surge overtopped the barrier beach and sand dunes and caused sand deposition in the southern part of the lake. A prominent and distinct sand layer attributed to this overwash fan was found in core L in the nearshore sediments of Lake Shelby (figure 2.12), but was absent in cores A, B, and E taken

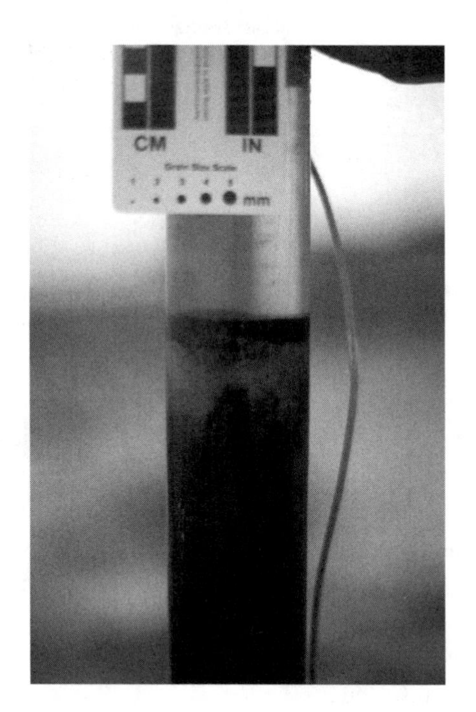

FIGURE 2.12 Photograph of sand layer near the top of core L from Lake Shelby that is attributed to Hurricane Frederic, as seen through the clear-PVC core tube. The core was taken in 1989. The thin layer of soft organic sediment overlying the Hurricane Frederic sand layer was accumulated over a period of 10 years since the Frederic strike in 1979.

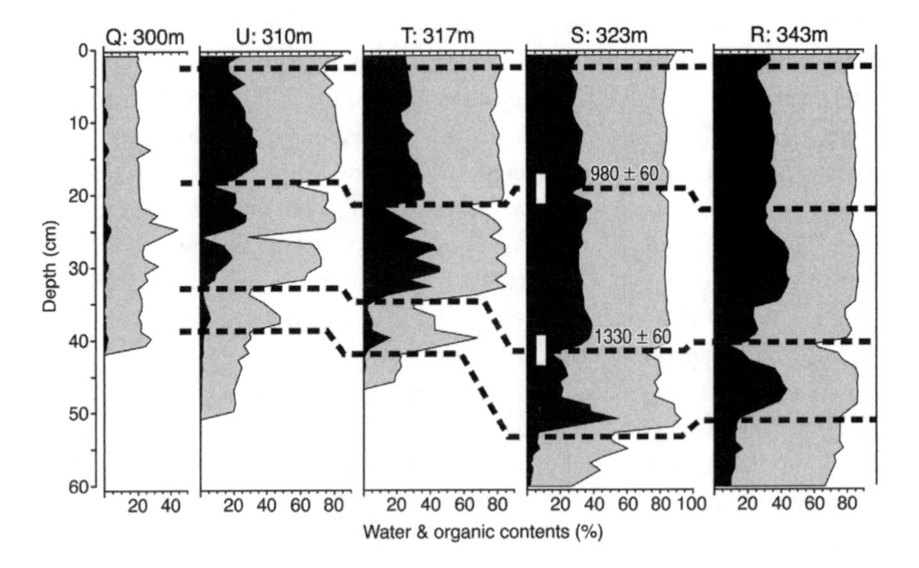

FIGURE 2.13 Loss-on-ignition curves showing water content (% wet weight; gray curve) and organic content (% dry weight; black curve) of five short cores (Q, U, T, S, R) collected along a transect near the south shore of Lake Shelby (for locations, see figure 2.6). Number at the top of each curve shows the distance of the core location from the south shore. Dotted lines between curves show stratigraphic correlation of sand layers. Short white bars on core S indicate stratigraphic levels of two radiocarbon dates.

near the center of the lake (Liu and Fearn 1993) (figure 2.6). A series of short cores taken along a transect reveal a steep depositional gradient along the edge of this overwash fan (figure 2.13). Core Q, collected at 300 m offshore, consists of all sand without any organic sedimentation. Cores U, T, S, and R, taken between 310 m and 343 m away from shore, contain several sand layers of various thicknesses. A very thin and indistinct band of sandy mud occurring near the top of core U may be the product of the Frederic event, but this layer is barely detectable in the other cores. A distinct sand layer occurs at a depth of around 20 cm in cores U and T, but becomes much thinner in core S and barely detectable in core R. A third sand layer occurs at 25 cm in core U only. A fourth sand layer can be detected between 30 cm and 40 cm in all cores, indicating a major event. The two distinct sand layers in core S have been C-14 dated to 980 ± 60 and 1330 ± 60 yr B.P., and can probably be correlated with the two sand layers dated to 770 ± 70 and 1360 ± 80 yr B.P. in core B, respectively (figure 2.8).

The sedimentary pattern of sand layers documented in this transect of cores is consistent with the hypothetical pattern described in our model. Specifically, cores taken from nearshore sediments contain more, and thicker, overwash sand layers than those taken farther away from shore, because these sand layers thin out toward the center of the lake (figure 2.13). The spatial pattern of sedimentation in Lake Shelby demonstrates clearly that the source of the sand is from the beach and dunes to the south of the lake. Therefore, the evidence supports the interpretation that the sand layers are produced by overwash processes, and refutes the notion that they are caused by some "intra-lacustrine depositional mechanism" or fluvial events as some have speculated (Otvos 1999, 2002; Liu and Fearn, 2002).

A similar pattern of overwash sand deposition was also observed in Western Lake. The Western Lake area was impacted by Hurricane Opal, a High-Category 3 storm that made landfall near Pensacola, Florida, about 75 km to the west. The nearly 6-m-high storm surge caused by Opal was not high enough to overwash the dune ridge fringing Western Lake, but seawater intruded into the lake through an intermittent tidal channel situated about 1 km to the west and inundated the south shore of the lake (Liu and Fearn 2000b, 2002). Consequently, sand was eroded from the base of the sand dunes on the south shore and washed into the lake. Sixteen cores have been taken from Western Lake for our paleotempestological study, including 6 (cores 1–6) taken before Opal and 10 (cores 7–16) after the event (figure 2.7). Among the latter cores, only core 11, a short core taken only 20 m away from the south shore, contains a sand layer at the top that can be attributed to Opal (Liu and Fearn 2000b, 2002) (figure 2.14). The very limited horizontal extent of the Opal sand layer attests to the fact that, in the absence of overwash processes, sand could not be transported very far from shore by the tidal floodwater alone. Except for cores 7 and 10, which are all sand (because they are too close to shore), all other cores contain multiple sand layers that were most likely deposited by overwash processes due to direct strikes by catastrophic hurricanes more intense than Opal. Figure 2.15 shows the water content curves for six of these cores. The sand layers vary somewhat in number and thickness among cores, but they can be broadly correlated. Generally, cores taken close to the south shore (e.g., core 11, 9) have more and thicker sand layers than those taken farther away from shore (e.g., core 14, 15), as our model predicts. The presence of prominent sand layers in core 16, a core taken near the far eastern end of the lake, confirms that the sand was deposited by overwash processes from the dunes and beach to the south, and not by tidal floodwater invading through the remote tidal channel to the west.

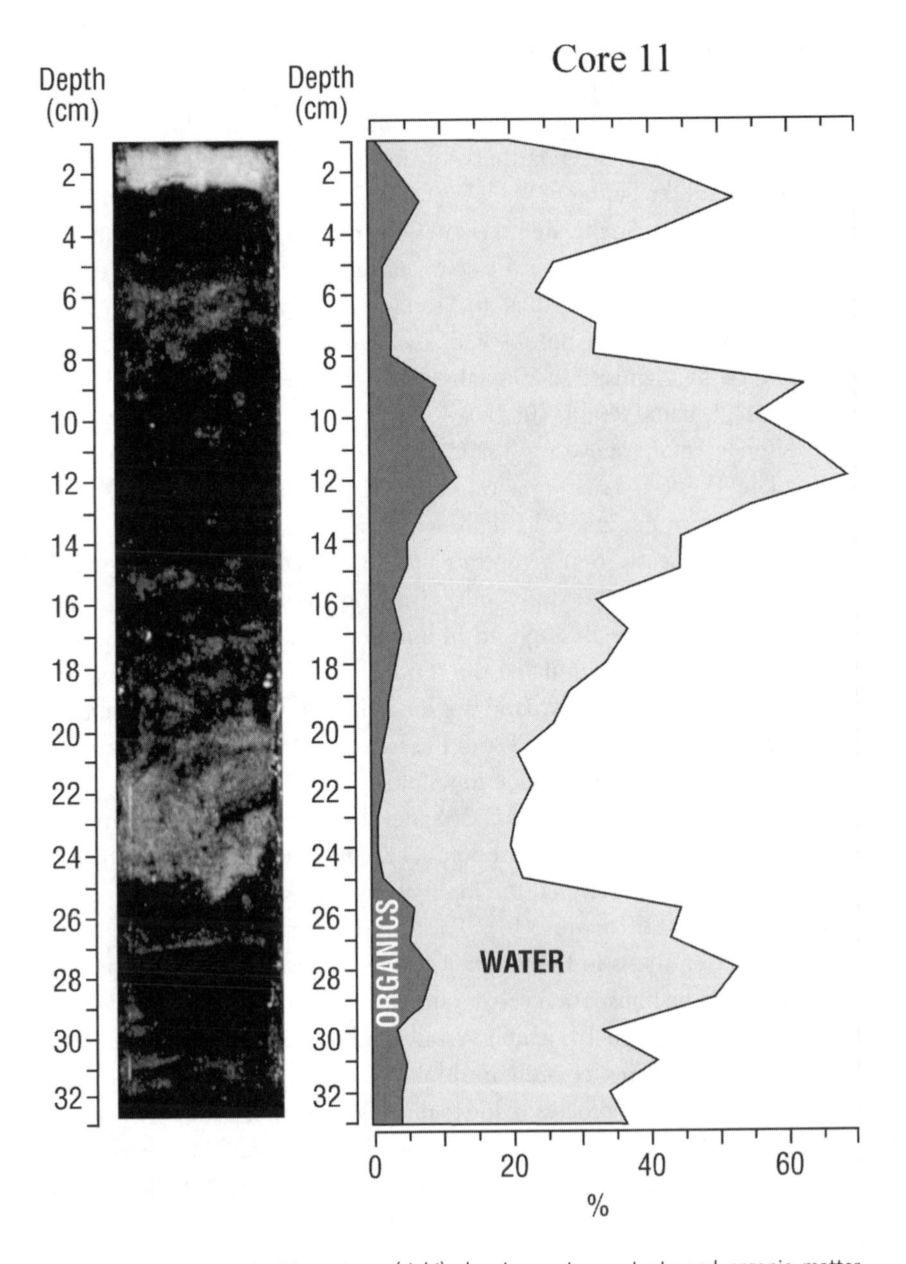

FIGURE 2.14 Loss-on-ignition curves (*right*) showing water contents and organic matter contents of the uppermost 33 cm of core 11 from Western Lake. Photograph of corresponding core segment (*left*) shows sand layer at the top attributed to saltwater intrusion and sand transport caused by Hurricane Opal of 1995. Two other sand layers are also present in this core segment (after Liu and Fearn 2000b).

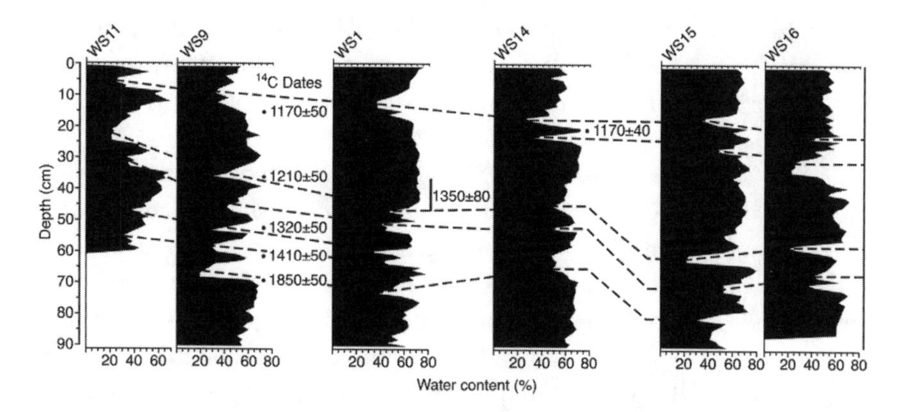

FIGURE 2.15 Loss-on-ignition curves showing water contents (% wet weight) of six cores (11, 9, 1, 14, 15, 16) taken from Western Lake (for core locations, see figure 2.7). Dotted lines between cores show probable stratigraphic correlation of sand layers. Dots and bar associated with the curves show stratigraphic locations of radiocarbon dates.

Long-term Changes in Lake Environment

In addition to the meteorological factors already discussed, the sensitivity of a lake in recording past hurricane strikes will also depend on a number of geomorphological factors related to the physical setting of the site. These factors include, for example, the height of the sandy barrier, which affects not only the supply of sand but also the threshold for overwash to occur. The higher the sand barrier, the less likely it is that weaker hurricanes will be recorded in the lake sediments. This is because only the stronger hurricanes can cause a storm surge high enough to overwash the high dune lines or beach ridges. Over time, the height of the sandy barrier may change, which would affect the sensitivity of the proxy record in relation to storm intensity. Another complicating factor is that the morphometry (e.g., size, shape, depth) and environmental setting of the lake may change over time. For example, a freshwater lake that occupies a closed basin today may have been evolved from a coastal lagoon with a tidal connection to the sea in the past. Caution should be used in interpreting the proxy record if past drastic changes in the lake's physical setting have altered the lake's sensitivity. Such changes, if any occurred, would almost certainly be reflected in the sedimentary or paleolimnological record of the lake. It is therefore advisable that other proxy data, such as those determined by loss-on-ignition, pollen, diatom, dinoflagellate, and foraminifera analyses, be obtained from the cores to supplement the reconstruction based on the sand-layer stratigraphies. Moreover, geomorphological changes such as breaching of a sandy barrier or variations in dune morphology are usually stochastic processes that occur at a local scale, and are unlikely to affect all the lakes on a stretch of coastline simultaneously. This consideration underscores the importance of obtaining parallel records from multiple sites to establish a comprehensive, regional record of past hurricane activity.

A 9-m core (core E) from Lake Shelby reveals a history of coastal environmental changes in Alabama during the last 4,800 years. The core consists of 55 cm of gyttja (organic lake mud) overlying gray lagoonal clay (figure 2.8). The boundary between these two sedimentary units was C-14 dated to 2190 ± 80 yr B.P. The sediment stratigraphy has been interpreted to suggest that Lake Shelby was a sheltered, quiet-water lagoon about 4,800 years ago, and that it changed to a freshwater lake isolated from the gulf around 2,200 years ago (Liu and Fearn 1993). The formation of the modern freshwater lacustrine environment might have been brought about by the lateral growth of the barrier beach, effectively cutting off the tidal connection between the sheltered lagoon and the sea. A sand layer occurring at the boundary between the two sedimentary units invites the speculation that a catastrophic hurricane strike 2,200 years ago might have triggered or facilitated the coastal morphological changes that sealed off Lake Shelby from the sea. In any case, the transition from a sheltered, low-energy lagoonal environment to a closed-basin lake probably occurred rather quickly, and required no major change in the lake's geomorphic setting relative to beach ridge height or sea level position. Under this scenario, this change in Lake Shelby's limnological environment might have only minor effects on the lake's sensitivity in recording prehistoric hurricane strikes.

During its lagoonal phase (4,800–2,200 yr B.P.) and lacustrine phase (2,200 yr B.P. to present), the depositional environment of Lake Shelby had been quite stable, as witnessed by the deposition of the nearly homogeneous clay and gyttja units, respectively, in core E. The only perturbation appears to be registered as a unit of sandy clay containing many small shell fragments and sand lenses between 210 and 240 cm in this core (figure 2.8). It is likely that this sedimentary unit reflects an interval during which the sand barrier in front of Lake Shelby was lowered or breached, allowing more tidal exchange between the lagoon and the sea. As a result, Lake Shelby became a less sheltered, more dynamic environment, and more susceptible to storm surges and overwash processes caused by less-intense hurricanes. The fact that this sandy unit overlies a sand layer suggests that the breaching of the sand barrier may have been caused by a catastrophic hurricane strike. In any case, this phase of breached barrier is inferred to have occurred around 2,900 to 3,000 yr B.P. and probably lasted for about a century before the barrier was sealed again and recovered its former height, as reflected by the resumption of pure organic clay deposition at the site. This example from Lake Shelby suggests that any local geomorphic changes that may conceivably affect a lake's sensitivity in recording past storm events may be detected in the stratigraphy and paleolimnology of its sediments.

Sea Level Changes

The sensitivity of a lake in recording past hurricane strikes may also be complicated by long-term changes in sea level. No consensus exists for the post-glacial sea level history of the Gulf of Mexico coast. Most studies seem to suggest that the sea level rose rapidly during the early Holocene to about 5 m below present datum about 5,000 to 6,000 years ago, and has risen only slowly since then (Pirazzoli 1991) (figure 2.16). This places an upper limit on the length of the proxy record that can be obtained from extant coastal lakes, because most of them were formed within the last 5,000 years. These changes in sea level in the late Holocene, though relatively minor, may still affect the sensitivity of the lake by altering the width of the sandy barrier and the proximity of the lake to the coast.

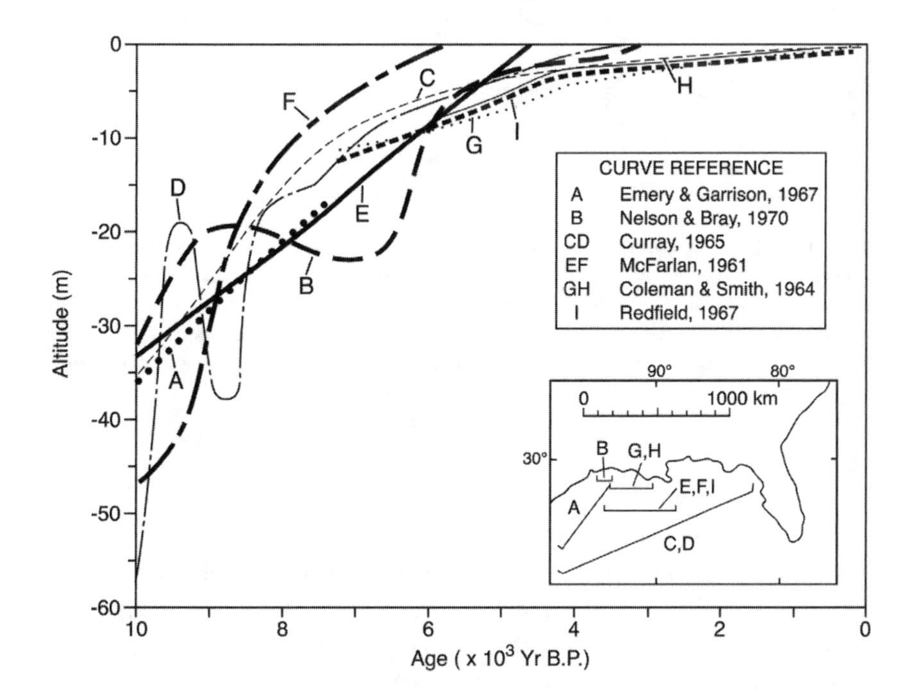

FIGURE 2.16 Curves of sea level changes for the Gulf of Mexico postulated by various authors. The inset map shows the coastal segment to which each curve applies (after Pirazzoli 1991).

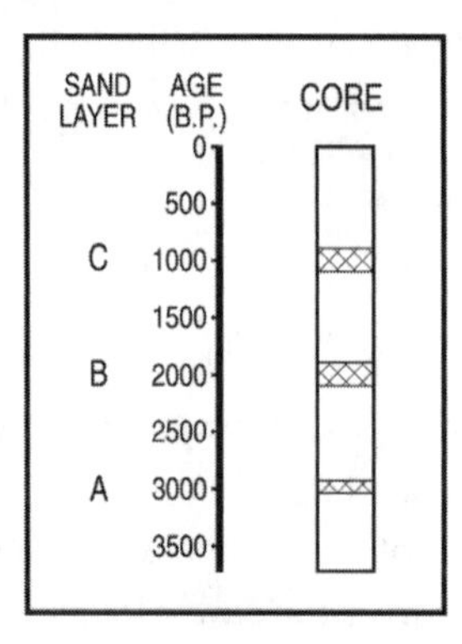

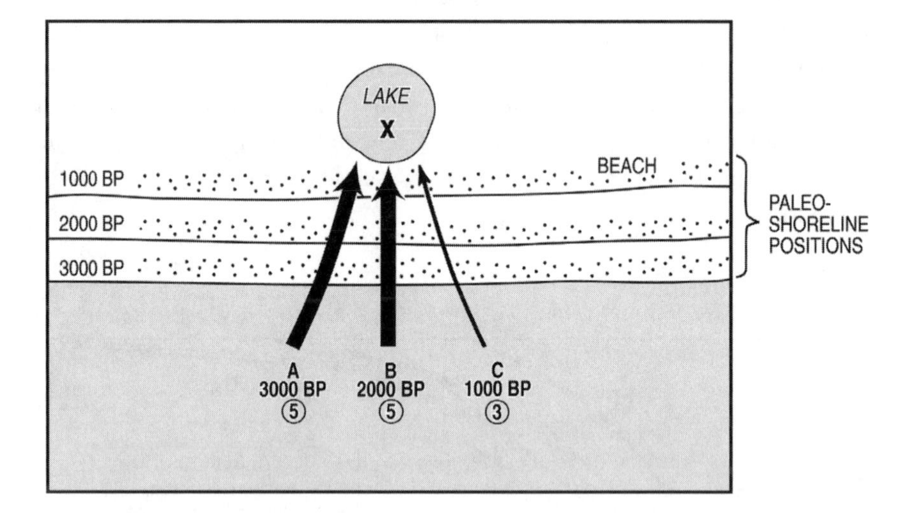

FIGURE 2.17 (*Top*) Hypothetical sedimentary record from a lake on a transgressive coast that was impacted by intense hurricanes three times during the past 3,000 years, illustrating the effect of sea level rise on the paleotempestological sensitivity of the lake. Sand layers are represented by cross-shaded lines. (*Bottom*) Location of the lake in relation to paleo-shoreline positions (horizontal lines) and associated barrier beaches (dotted patterns) during three time periods. Thicknesses of the arrows are proportional to the intensity of hurricanes. The circled number below each arrow denotes the intensity of the hurricane according to the Saffir-Simpson scale.

Figure 2.17 presents a hypothetical scenario in which a lake site on a transgressive coast was directly struck by intense hurricanes three times during the last 3,000 years. When Hurricane A (Category 5) struck 3,000 years ago, it left behind only a thin sand layer in the lake sediment because sea level was lower than at present and the lake was situated farther away behind the beach. By comparison, the overwash sand layer caused by Hurricane B, also Category 5, is much thicker, because by 2,000 yr B.P. the lake was closer to the coast, so that more sand was deposited at the coring site by the overwash. An equally prominent sand layer may have been deposited by Hurricane C, even though it is only a Category 3 hurricane, because by 1,000 yr B.P. the lake was situated very close to the retreating coast as a result of continued sea level rise. It should be noted, however, that if a lake's proximity to the coast or its environmental setting has changed drastically in the past, whether due to sea level change or other causes, such changes should be detectable in its sedimentological or paleolimnological records. In the preceding scenario, the lake might have been more fresh, or less saline, 3,000 years ago than at present due to its more inland location. That should be reflected by a different microfossil assemblage and a higher organic matter content in the sediments accumulated during that period.

In the case of Western Lake, there is no evidence to indicate that sea level change has been a significant factor affecting the lake's paleotempestological sensitivity over the past 5,000 years. The lake's environmental stability is reflected in the relatively uniform loss-on-ignition values of its organic lake sediments (figure 2.9). This uniformity is interrupted only by the presence of discrete sand layers, which usually have sharp contacts with the organic lake mud above and below (Liu and Fearn 2000b). Had the sea level been significantly lower in the past, the lake would have shifted to a different location relative to the coastline of that time, or at least would have had a different depositional environment at the coring site. There is no evidence in the sedimentological and paleolimnological record of Western Lake of environmental changes related to a significant change in sea level (Liu and Fearn 2000b, 2002).

Proxy Record from Coastal Marsh Sediments

In addition to coastal lakes, coastal marshes can also provide long and continuous proxy records of past hurricane strikes. Several estuarine marshes along the Gulf of Mexico coast have been studied for their paleotempestological records (Li 1994; Zhou 1998; Liu and Fearn 2000a) (figure 2.1). Notably, these estuarine marshes, which occur at the mouth of major rivers (e.g., Mississippi, Atchafalaya, Pearl, Pascagoula) on the Gulf Coast, differ somewhat from the back-barrier marshes described from the northeast Atlantic coast (Donnelly and Webb, chapter 3 in this volume). The principle of using these estuarine marshes for paleotempestological study has been explained in Liu and Fearn (2000a) and is discussed only briefly here.

Figure 2.18 presents a schematic diagram showing the stratigraphic pattern of storm deposits in an estuarine marsh. When an intense hurricane strikes a

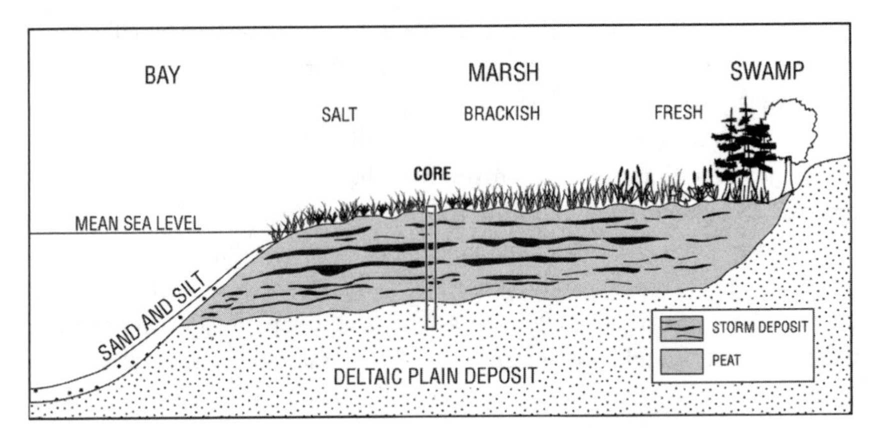

FIGURE 2.18 Hypothetical patterns of storm deposits in an estuarine marsh subjected to repeated hurricane strikes and storm surges in the past. A core taken from the marsh should contain multiple layers of storm deposits, which provide a proxy record of past hurricane strikes (after Liu and Fearn 2000a).

low-lying marshy coast, the storm surge is likely to inundate a large part of the coastal marsh with floodwater that contains clastic sediments derived from the marsh surfaces and the adjacent shallow bays and estuaries (Cahoon et al. 1995; Gunterspergen et al. 1995). Cores taken from estuarine marshes may contain multiple layers of storm deposits that provide a proxy record of past hurricane strikes. These storm deposits, usually in the form of clastic layers of sand, silt, or clay, may also contain distinct assemblages of pollen, diatoms, and marine microfossils that allow them to be readily identified from the embedding marsh peat or organic mud (Li 1994; Zhou 1998; Liu and Fearn 2000a).

Reconstruction of Late Holocene Hurricane Activity in Northwestern Florida

With these theoretical considerations and caveats in mind, the proxy record from Western Lake could be used to reconstruct a 5,000-year history of catastrophic hurricane activity for the Gulf Coast of northwestern Florida (Liu and Fearn 2000b). Core 1, a 6.3-m-long core from Western Lake, yielded a sedimentary record spanning the last 7,000 years (figure 2.9). From 3.3 m to 1.6 m, the sediment contains a few very thin (less than 1-mm thick) sand layers but no major ones, implying few catastrophic hurricane landfalls from 5,000 to 3,400 yr B.P. Twelve distinct sand layers occur in the core segment above 1.6 m. These 0.5- to 2.0-cm thick overwash sand layers are composed of medium to coarse sands that contain small shell fragments. This record has been interpreted to indicate a dramatic increase in the frequency of landfalls by catastrophic hurricanes after 3,400 yr B.P. Remarkably, 11 of the 12 sand layers present occur between 0.45 m and 1.6 m, suggesting a period of heightened hurricane activity on the northwestern Florida coast from 3,400 to 1,000 yr B.P. During this period, the Western Lake area was directly hit by Category 4 or 5 hurricanes about five times per millennium, indicating a return period of approximately 200 years, or a landfall probability of 0.5% per year. By contrast, only one landfall by catastrophic hurricanes occurred during the 1,000 to 0-yr B.P. period, implying a return period of 1,000 years for the recent millennium, or a landfall probability of only 0.1% per year. The recent millennium, therefore, has been a relatively quiescent period in terms of catastrophic hurricane landfalls on the Gulf Coast (Liu and Fearn 2000b).

Holocene Hurricane History of the Gulf Coast: A Regional Synthesis

The general chronological pattern of late Holocene hurricane activity found at Western Lake is corroborated by proxy records from other sites along the Gulf Coast. Besides Western Lake, three other sites have yielded sedimentary records spanning almost the last 5,000 years: Lake Shelby (Alabama), Pascagoula Marsh (Mississippi), and Pearl River Marsh (Louisiana) (Liu 1999). The latter two proxy records are derived from large estuarine marshes. The Pascagoula Marsh record is based on a 7.5-m core that contains 11 silt or clay layers deposited during the last 4,500 years. These clastic layers, sandwiched by marsh peat or organic clay, are interpreted to represent storm deposits laid down by storm surges during catastrophic hurricane strikes (Liu 1999). The Pearl River Marsh record comes from an 8.5-m core that contains nine clastic storm deposits laid down during the past 4500 years (Li 1994; Liu 1999; Liu and Fearn 2000a). At both sites a storm deposit occurring near the core top, inferred to represent the catastrophic strike by Hurricane Camille in 1969, was used as the modern analog for interpreting the intensity of the prehistoric hurricanes recorded in the older parts of the cores.

The chronostratigraphic patterns of sand layers (hence catastrophic hurricane strikes) found at these four sites along the Gulf Coast are collated in figure 2.19 (Liu 1999). A regionally coherent pattern can be observed, as summarized in two major findings.

First, over the last 3,400 years, each of these Gulf Coast sites was directly struck by catastrophic hurricanes about 9 to 12 times, thus implying a return

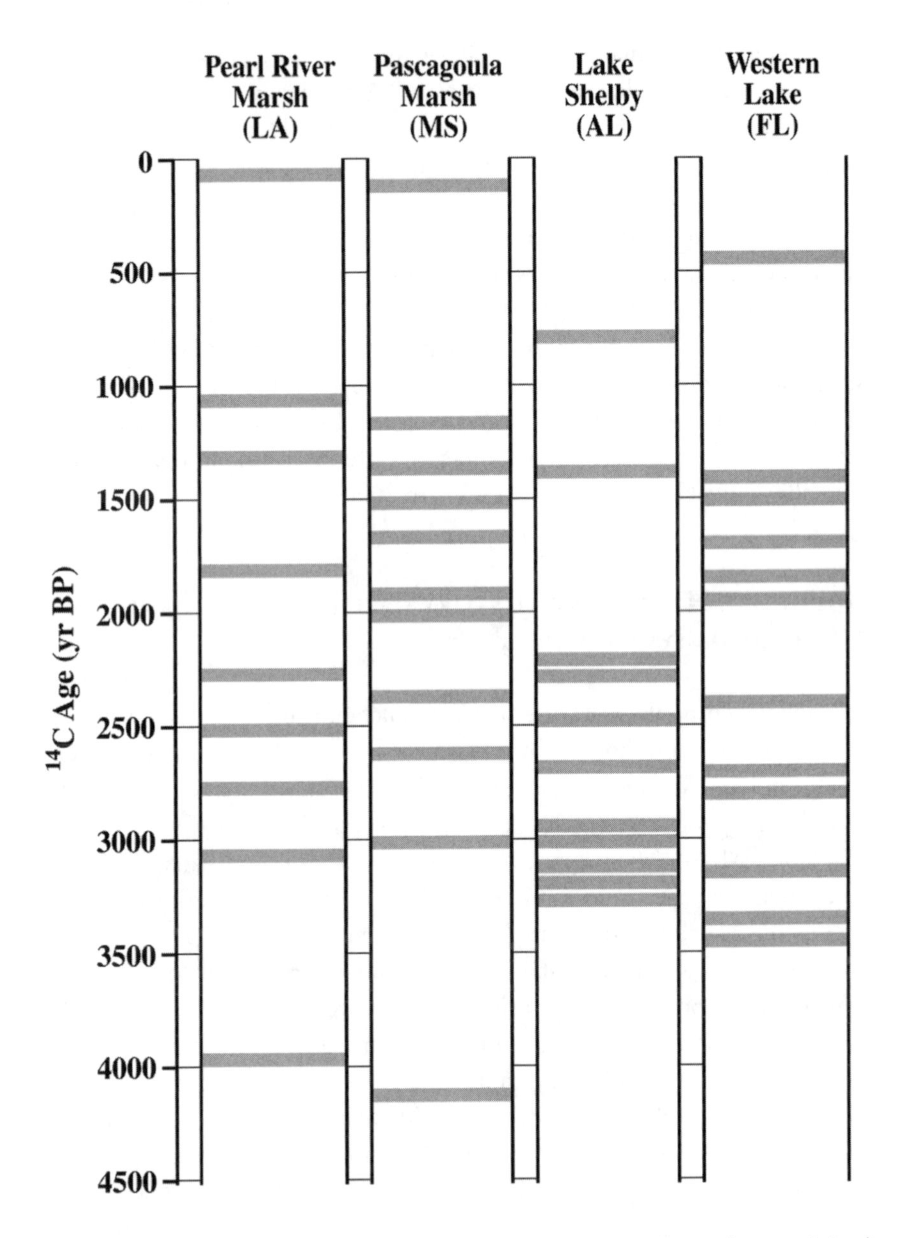

FIGURE 2.19 Chronostratigraphic patterns of overwash sand layers (horizontal bars) representing catastrophic hurricane strikes over the last 4,500 years documented at four sites along the Gulf Coast from west to east: Pearl River Marsh, Louisiana; Pascagoula Marsh, Mississippi; Lake Shelby, Alabama; and Western Lake, Florida (after Liu 1999).

period of approximately 300 years (or an annual landfall probability of 0.33%) for these rare but most destructive storms. These first empirical estimates have since been found to be consistent with predictions extrapolated from numerical models of wind speed probabilities for all coastal locations in the United States (Murnane et al. 2000).

Second, the chronological pattern across these sites reveals that there is significant *millennial-scale variability* in major hurricane activity over the past 5,000 years (figure 2.20). Few catastrophic hurricane strikes occurred during a quiescent period from 4,500 to 3,400 yr B.P., and again during the past 1,000 years, but the landfall frequency was about three to five times greater during the "hyperactive" period between 3,400 and 1,000 years ago. During the hyperactive period, the landfall probability increased to about 0.3 to 0.5% per year, compared with only 0.1% per year during the most recent millennium (Liu 1999; Liu and Fearn 2000b). Hurricane climatologists know well that multidecadal variability exists in the major hurricane activity of the past century (Landsea 1993; Landsea et al. 1996; Gray, Sheaffer, and Landsea 1997; Elsner and Kara 1999). But the discovery of millennial-scale variability based on the proxy records suggests that there are longer cycles superimposed on the shorter cycles that may be controlled by climate-forcing mechanisms operating at different time scales.

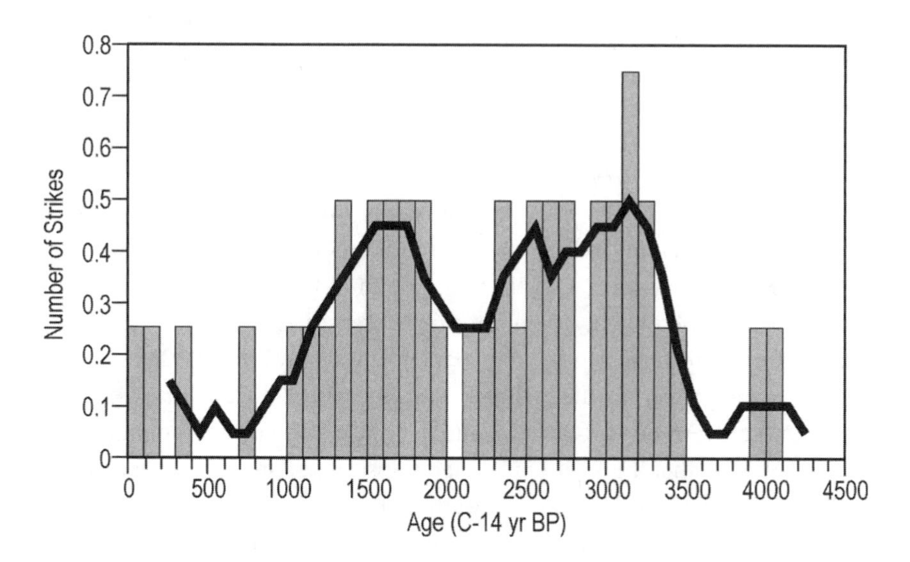

FIGURE 2.20 Summary diagram of catastrophic hurricane activity during the last 4,500 years along the Gulf Coast. Vertical bars show the mean number of strikes by catastrophic hurricanes per century obtained by averaging the proxy records of the four sites (figure 2.19) for every century. The continuous curve shows the 500-year moving average of the mean numbers of strikes per century.

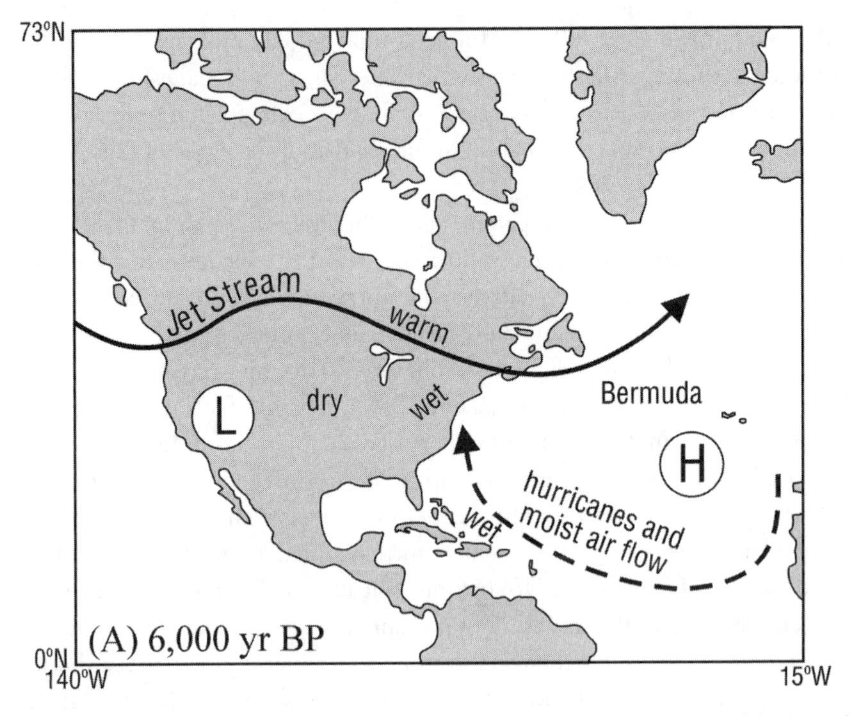

FIGURE 2.21 Postulated changes in paleoclimatic conditions and in the position of the Bermuda High—(*A*) around 6,000 yr B.P. (*B*) 3,000 yr B.P. (based on Forman et al. 1995)—and their impacts on the predominant tracks of catastrophic hurricanes (after Liu and Fearn 2000b). A more northeasterly position of the Bermuda High around 6,000 years ago would have resulted in more landfalls on the Atlantic coast. A shift of the Bermuda High to a more southeasterly position near the Caribbean around 3,000 years ago probably steered more catastrophic hurricanes to hit the Gulf Coast.

Liu and Fearn (2000b) have postulated that the observed millennial-scale variability of hurricane landfalls observed in the Gulf Coast proxy records can be explained by long-term shifts in the position of the Bermuda High (figure 2.21). The Bermuda High, a high pressure center situated over the subtropical Atlantic Ocean near Bermuda, has been known to play a pivotal role in steering the hurricanes from the tropical Atlantic toward the Atlantic and Gulf coasts of North America. Paleoclimatic data from eastern and central North America suggest that this subtropical high pressure system was displaced to the northeast during the mid-Holocene warm period around 5,000 to 6,000 years ago (Forman et al. 1995). According to the Bermuda High hypothesis (Liu and Fearn 2000b), due to the recurving of storm tracks around the southwestern

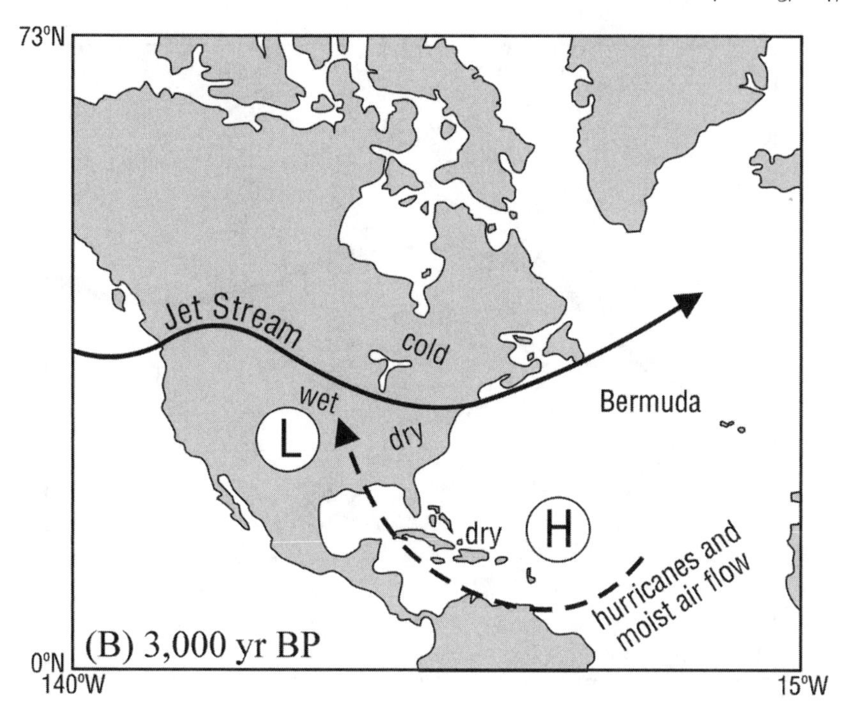

flank of the Bermuda High, more hurricanes were steered to the north during the mid-Holocene, resulting in more frequent landfalls on the Atlantic coast but fewer on the Gulf Coast (figure 2.21A). A southwestward displacement of the Bermuda High after 3,400 yr B.P. probably caused a major shift in predominant storm tracks from the Atlantic coast to the Gulf Coast, resulting in the dramatic increase in catastrophic hurricane landfalls along the Gulf Coast during the hyperactive period as observed in the proxy records (Liu 1999) (figure 2.21B). The decline in hurricane activity on the Gulf Coast during the past 1,000 years could be explained by a return of the Bermuda High to a more northeasterly position, thus steering more hurricanes to hit the Atlantic coast.

Recently, Elsner, Liu, and Kocher (2000) have expanded on the Bermuda High hypothesis by postulating that an anti-phase (seesaw) pattern in major hurricane activity between the Gulf and Atlantic coasts also exists on the interannual and interdecadal time scales, and that these north-south shifts are related to the intensity of the North Atlantic Oscillation (NAO) (figure 2.22)

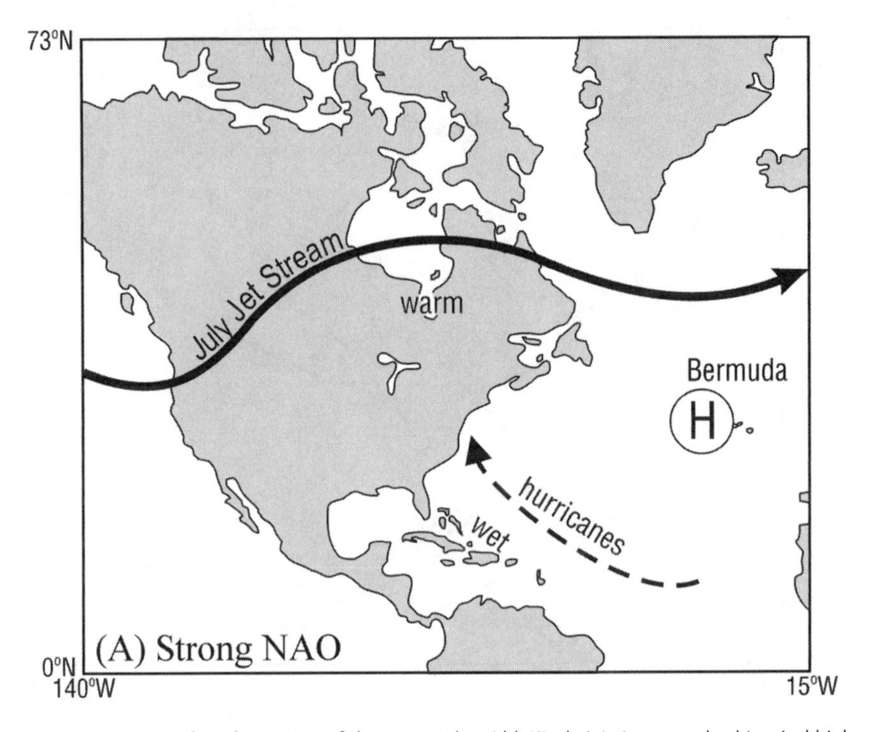

FIGURE 2.22 Inferred positions of the mean July mid-latitude jet stream and subtropical high for conditions of strong NAO (*A*) and neutral or weak NAO (*B*), and their impacts on the predominant tracks of Atlantic hurricanes (after Elsner, Liu, and Kocher 2000). Strong NAO tends to be associated with more frequent landfalls on the Atlantic coast, whereas weak or neutral NAO tends to be associated with more frequent Gulf Coast landfalls.

(see also Elsner and Bossak, chapter 12 in this volume). When the NAO is strong (highly positive values), as in the 1950s, major hurricanes are more likely to hit the Atlantic coast (figure 2.22A). Conversely, when the NAO is neutral or weak (slightly negative values), as in the 1960s, the Gulf Coast is more likely to get hit (Elsner, Liu, and Kocher 2000) (figure 2.22B). The Bermuda High is linked to the NAO because it is part of the same subtropical anticyclonic system over the North Atlantic that constitutes the Azores High, the southern wing of the North Atlantic sea level pressure seesaw comprising the NAO. Thus, the combined evidence from the historical and proxy records seems to suggest that spatial variations in major hurricane activity along the Gulf and Atlantic coasts can be explained by changes in the position and intensity of the Bermuda High and the NAO across millennial, decadal, and seasonal time scales.

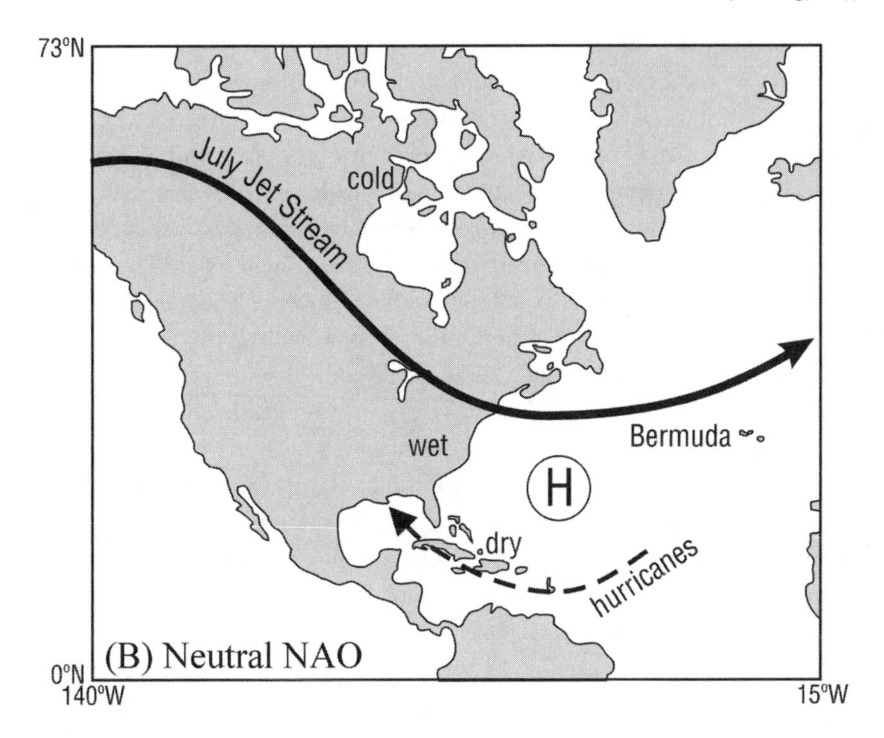

Discussions and Conclusions

Paleotempestology is a relatively young, but emerging, field of science. Its central mission is to reconstruct the spatial and temporal patterns of tropical cyclone (hurricane) activity of the past, especially at time scales of centuries and millennia. A long-term perspective is essential for understanding the low-frequency, high-amplitude variations in hurricane activity and to identify the physical mechanisms that control them, because the instrumental record for hurricane activity spans only the past 150 years—a period too short to encompass the full range of variability inherent in this climatic phenomenon. An extended record on century to millennial time scales is especially needed for estimating the return periods of catastrophic hurricanes of Category 4 and 5 intensities because these extreme events are very rare in the short period of instrumental observations.

Although written or documentary records can provide information on hurricane activity for the past several hundred years, geological proxy data are the only means by which a multi-millennial history of past hurricane activity can be reconstructed. Climatic proxies such as tree rings and oxygen isotopes in speleothems and corals offer some potential in recording the environmental impacts of past hurricane strikes, but overwash deposits preserved in the sediments of coastal lakes and marshes are the most promising and proven proxy that can provide a continuous record of past hurricane strikes spanning several millennia. The sedimentary evidence based on the storm deposits or overwash sand layers can be supplemented by other proxies such as pollen, diatoms, foraminifera, dinoflagellates, and phytoliths. These paleoecological or paleolimnological indicators can be useful in documenting the environmental impacts of past hurricane strikes, such as saltwater intrusion, coastal flooding, soil erosion, and vegetational change. They can also provide information about the coastal environmental background upon which the hurricane strike or overwash event occurred, such as the geomorphic setting of the lake in relation to past sea level positions. Thus, even though the main concerns of paleotempestology are climatological, the principles and methodologies of this field are interdisciplinary in nature, involving fields like sedimentology, coastal geomorphology, limnology, and paleoecology.

Although paleotempestology is part of the "paleo-sciences" that look into the past, it has significant practical applications and societal relevance (Liu 2004). The reinsurance industry, for example, has already incorporated the results of paleotempestology in their hazard models in order to calibrate their return period estimates for the extreme events (Michaels et al. 1997; Murnane et al. 2000). Figure 2.23 summarizes the basic principles, methods, and applications of the burgeoning field of paleotempestology (Liu 2004).

Some similarities exist between paleotempestology and its sister discipline—paleoseismology (Clague, Bobrowsky, and Hutchinson 2000). Both disciplines are grounded on geological principles, and both deal with natural hazards that have significant societal impacts. During the past decade, much progress has been made in paleoseismology, particularly in the use of coastal lake- and marsh-sediments in the reconstruction of prehistoric earthquake and tsunami events along the Pacific Northwest coasts of North America (Clague et al., 1999; Clague, Bobrowsky, and Hutchinson 2000; Hutchinson et al. 2000). There are significant similarities as well as differences between storm overwash and tsunami processes in terms of their geomorphic impacts and sedimentological mechanisms, which will offer fruitful grounds for comparative studies. At least in the continental United States and Canada, little overlap exists between coasts that are highly vulnerable to tsunami hazards (the Pacific coast) and those that are highly vulnerable to hurricane hazards (the Gulf and Atlantic coasts). It is, therefore, unlikely (though possible) to find overwash deposits and tsunami deposits in the same site. For other coastal zones such as

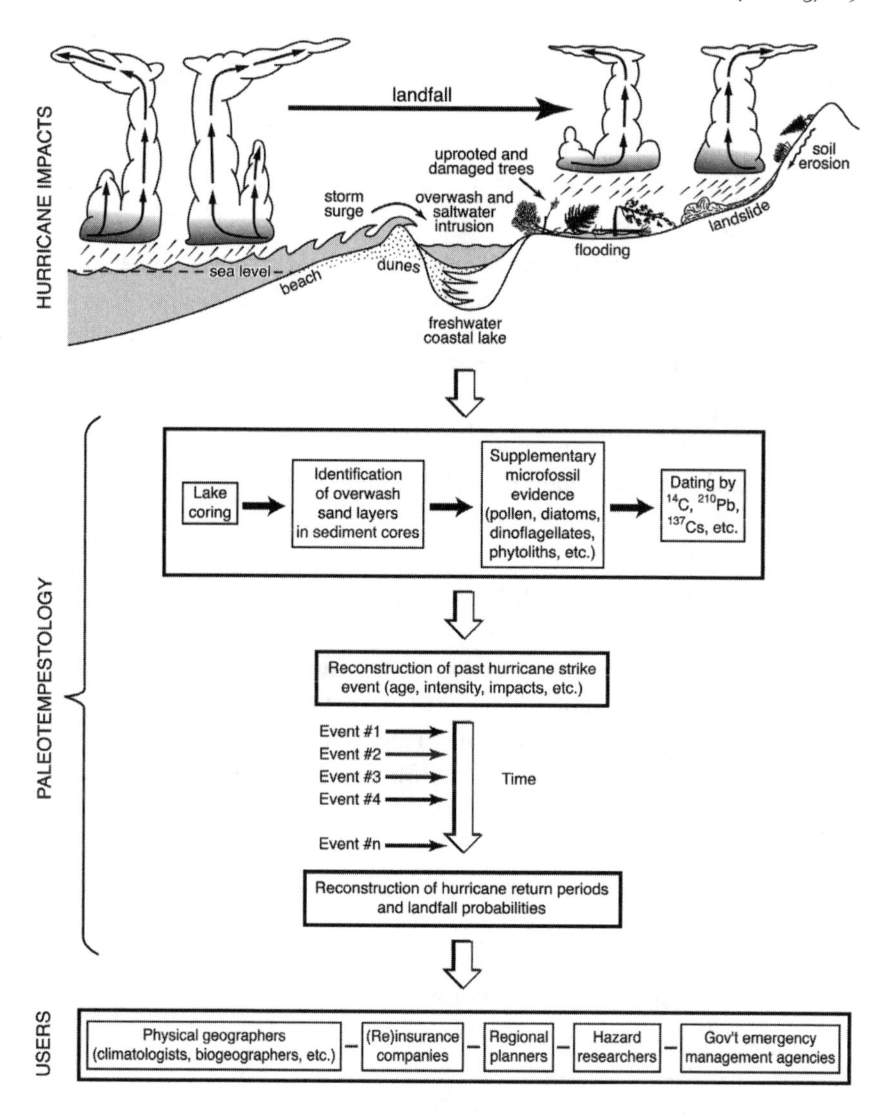

FIGURE 2.23 Schematic diagram showing the environmental impacts of catastrophic hurricanes (*top*), and the principles, methods (*middle*), and applications of paleotempestology (*lower half*) (after Liu 2004).

those in Hawai'i, Mexico, or the Caribbean, however, distinguishing between these two types of catastrophic events in the proxy record may become a critical research issue.

In this chapter, the author has discussed a number of theoretical and practical considerations related to site selection, record sensitivity, data interpretation, and research methodology. Based on some of these considerations, it can be concluded that the ideal coastal lakes for producing a paleotempestological record are those that satisfy the following conditions:

1. The lakes are situated behind a sandy barrier and close to the sea, so that they are susceptible to hurricane overwash and are adjacent to a source of sand.
2. They are freshwater or oligohaline lakes isolated from the sea, or at least occupying a sheltered basin immune from active tidal flushing.
3. The lakes are situated on morphologically and tectonically inactive coastlines that have a relatively stable sea level history during the late Holocene.
4. The lakes are small to medium in size and not fed by major rivers, so that their sedimentary record will not be complicated by fluvial or hydrological events unrelated to hurricane strikes.

Maximum length of the paleotempestological record may reach 5,000 years. Within a lake, multiple cores are necessary to establish the horizontal extents of the overwash deposits and to obtain a more complete record of past hurricane strikes. Data from multiple sites are also useful for validating the chronological pattern of hurricane activity in a region. In interpreting the past hurricane frequency and return period established for each site, it is important to keep in mind that the highest storm surge tends to occur within a short distance to the right of the eye of the landfalling hurricane, so that each overwash deposit is likely to be a record of a direct hit, or landfall fairly close to the site. A useful analogy is that each site is like a paleo-weather station that records past hurricane landfalls within a certain distance (e.g., a 50-km radius) from the site. Theoretically and ideally, a large number of closely spaced sites along the coast are necessary to reconstruct a more complete history of hurricane activity over a region.

More than two dozen coastal lakes and marshes along the Gulf and Atlantic coasts have been cored for paleotempestological investigations during the past decade. Two of the coastal lake sites, Lake Shelby and Western Lake, have yielded proxy records of catastrophic hurricane strikes spanning the last 5,000 years. The chronological pattern of overwash sand layers observed in these and other study sites reveals that distinct millennial-scale variability exists in catastrophic hurricane activity along the Gulf Coast, with a hyperactive period (3,400–1,000 yr B.P.) occurring between two relatively quiescent periods (5,000–3,400 yr B.P., 1,000–0 yr B.P.). These paleotempestological data have motivated the formulation of the hypothesis that the millennial-scale variations in catastrophic hurricane activity may have been controlled by long-term changes in the position and intensity of the Bermuda High and the NAO. Although efforts are being made to test this Bermuda High–NAO hypothesis, there is no doubt that such a long-term perspective on hurricane activity and climate variability would not have been possible without the contribution of paleotempestology.

Acknowledgments

Our research in paleotempestology has been supported by grants from the National Science Foundation (NSF) and the Risk Prediction Initiative (RPI) of the Bermuda Biological Station for Research. I thank Mimi Fearn for her collaboration and assistance in the developmental stages of the Gulf Coast work. Many students and research associates at LSU have assisted in fieldwork or laboratory analyses and have contributed to various phases of our paleotempestological research, especially Caiming Shen, Zuju Yao, Xinyu Zhou, Xu Li, Kari Gathen, Andy Reese, Jason Blackburn, and Houyuan Lu. Mary Lee Eggart, Clifford Duplechin, and Caiming Shen provided cartographic and technical assistance. We are also grateful to the staff of the Gulf State Park, Alabama, Grayton Beach State Park, Florida, and Gulf Coast National Seashore, Mississippi, for their cooperation and logistical support in the field.

References

Aigner, T. A. 1985. *Storm depositional systems: Dynamic stratigraphy in modern and ancient shallow-marine sequences*. Berlin: Springer-Verlag.

Batista, W. B., and W. J. Platt. 2003. Tree population responses to hurricane disturbance: Syndromes in a south-eastern USA old-growth forest. *Journal of Ecology* 91:197–212.

Birks, H. J. B., and H. H. Birks. 1980. *Quaternary palaeoecology*. London: Arnold.

Cahoon, D. R., D. J. Reed, J. W. Day, G. D. Steyer, R. M. Boumans, J. C. Lynch, D. McNally, and N. Latif. 1995. The influence of Hurricane Andrew on sediment distribution in Louisiana coastal marshes. *Journal of Coastal Research* [special issue] 21:280–94.

Caviedes, C. N. 1991. Five hundred years of hurricanes in the Caribbean: Their relationship with global climatic variabilities. *GeoJournal* 23:301–10.

Clague, J. J., P. T. Bobrowsky, and I. Hutchinson. 2000. A review of geological records of large tsunamis at Vancouver Island, British Columbia, and implications for hazard. *Quaternary Science Reviews* 19:849–63.

Clague, J. J., I. Hutchinson, R. W. Mathewes, and R. T. Patterson. 1999. Evidence for late Holocene tsunamis at Catala Lake, British Columbia. *Journal of Coastal Research* 15:45–60.

Cohen, A. 2001. Paleohurricanes day by day: Can we do it? In *Research update 2001, risk prediction initiative*. Proceedings of a workshop held on June 8, 2001. Hamilton, Bermuda: Risk Prediction Initiative.

Collins, E. S., D. B. Scott, and P. T. Gayes. 1999. Hurricane records on the South Carolina coast: Can they be detected in the sediment record? *Quaternary International* 56:15–26.

Conner, W. H., J. W. Day, R. H. Baumann, and J. M. Randall. 1989. Influence of hurricanes on coastal ecosystems along the northern Gulf of Mexico. *Wetland Ecology and Management* 1:45–56.

Davis, R. A., Jr., S. C. Knowles, and M. J. Bland. 1989. Role of hurricanes in the Holocene stratigraphy of estuaries: Examples from the Gulf Coast of Florida. *Journal of Sedimentary Petrology* 59:1052–61.

Dean, W. E. 1974. Determination of carbonate and organic matter in calcareous sediments and sedimentary rocks by loss on ignition: Comparison with other methods. *Journal of Sedimentary Petrology* 44:242–48.

Donnelly, J. P., S. S. Bryant, J. Butler, J. Dowling, L. Fan, N. Hausmann, P. Newby, B. Shuman, J. Stern, K. Westover, and T. Webb III. 2001a. A 700 yr sedimentary record of intense hurricane landfalls in southern New England. *Bulletin of the Geological Society of America* 113:714–27.

Donnelly, J. P., S. Roll, M. Wengren, J. Butler, R. Lederer, and T. Webb III. 2001b. Sedimentary evidence of intense hurricane strikes from New Jersey. *Geology* 29:615–18.

Doyle, T. W., and L. E. Gorham. 1996. Detecting hurricane impact and recovery from tree rings. In *Tree rings, environment, and humanity*, edited by J. S. Dean, D. M. Meko, and T W. Swetnam, 405–12. Radiocarbon 1996. Tucson: Department of Geosciences, University of Arizona.

Duke, W. L. 1985. Hummocky cross-stratification, tropical hurricanes, and intense winter storms. *Sedimentology* 32:167–94.

Elsner, J. B., and A. B. Kara. 1999. *Hurricanes of the North Atlantic: Climate and society*. New York: Oxford University Press.

Elsner, J. B., K-b. Liu, and B. Kocher. 2000. Spatial variations in major U.S. hurricane activity: Statistics and a physical mechanism. *Journal of Climate* 13:2293–305.

Fearn, M. L., and K-b. Liu. 1995. Maize pollen of 3500 B.P. from southern Alabama. *American Antiquity* 60:109–17.

Fernández-Partagás, J., and H. F. Díaz. 1996. Atlantic hurricanes in the second half of the nineteenth century. *Bulletin of the American Meteorological Society* 77:2899–906.

Forman, S. L., R. Oglesby, V. Markgraf, and T. Stafford. 1995. Paleoclimatic significance of Late Quaternary eolian deposition on the Piedmont and High Plains, central United States. *Global and Planetary Change* 11:35–55.

Gathen, K., 1994. Ecological impacts of historic and prehistoric hurricanes on Horn Island, Mississippi. M.Sc. thesis, Department of Geography and Anthropology, Louisiana State University, Baton Rouge.

Gray, W. M., J. D. Sheaffer, and C. W. Landsea, 1997. Climate trends associated with multidecadal variability of Atlantic hurricane activity. In *Hurricanes: Climate and socioeconomic impacts*, edited by H. F. Díaz and R. S. Pulwarty, 15–53. Berlin: Springer-Verlag.

Gunterspergen, G. R., D. R. Cahoon, J. Grace, G. D. Steyer, S. Fournet, M. A. Townson, and A. L. Foote. 1995. Disturbance and recovery of the Louisiana coastal marsh landscape from the impacts of Hurricane Andrew. *Journal of Coastal Research* [special issue] 21:324–39.

Hayne, M., and J. Chappell. 2001. Cyclone frequency during the last 5,000 years from Curacoa Island, Queensland. *Palaeogeography, Palaeoclimatology, Palaeoecology* 168:201–19.

Hemphill-Haley, E. 1996. Diatoms as an aid in identifying late-Holocene tsunami deposits. *Holocene* 6:439–48.

Hippensteel, S. P., and R. E. Martin. 1999. Foraminifera as an indicator of overwash deposits, Barrier Island sediment supply, and Barrier Island evolution: Folly Island, South Carolina. *Palaeogeography, Palaeoclimatology, Palaeoecology* 149:115–25.

Hoover, R. A. 1957. Empirical relationships of the central pressures in hurricanes to the maximum surge and storm tide. *Monthly Weather Review* 85:167–74.

Hubbert, G., and K. McInnes. 1999. A storm surge inundation model for coastal planning and impact studies. *Journal of Coastal Research* 15:168–85.

Hutchinson, I., J. -P. Guilbault, J. J. Clague, and P. T. Bobrowsky. 2000. Tsunamis and tectonic deformation at the northern Cascadia margin: A 3000-year record from Deserted Lake, Vancouver Island, British Columbia, Canada. *Holocene* 10:429–39.

Jarvinen, B. R., and C. J. Neumann. 1985. An evaluation of the SLOSH storm surge model. *Bulletin of the American Meteorological Society* 66:1408–11.

Landsea, C. W. 1993. A climatology of intense (or major) Atlantic hurricanes. *Monthly Weather Review* 121:1703–13.

Landsea, C. W., N. Nicholls, W. M. Gray, and L. A. Avila.1996. Downward trends in the frequency of intense Atlantic hurricanes during the past five decades. *Geophysical Research Letters* 23:1697–700.

Lawrence, J. R., and S. D. Gedzelman. 1996. Low stable isotope ratios of tropical cyclone rains. *Geophysical Research Letters* 23:527–30.

Leatherman, S. P. 1981. *Overwash processes*. Benchmark Papers in Geology. Stroudsburg, Pa.: Hutchinson and Ross.

Li, X. 1994. A 6200-year environmental history of the Pearl River Marsh, Louisiana. Ph.D. diss., Department of Geography and Anthropology, Louisiana State University, Baton Rouge.

Liu, K.-b. 1999. Millennial-scale variability in catastrophic hurricane landfalls along the Gulf of Mexico coast. In *Preprints of the 23rd Conference on Hurricanes and Tropical Meteorology*, 374–77. Boston: American Meteorological Society.

Liu, K.-b. 2004. Paleotempestology: Geographic solutions to hurricane hazard assessment and risk prediction. In *WorldMinds: Geographical perspectives on 100 problems*, edited by B. Warf, D. Janelle, and K. Hansen, 443–48. New York: Kluwer.

Liu, K.-b., and M. L. Fearn. 1993. Lake-sediment record of late Holocene hurricane activities from coastal Alabama. *Geology* 21:793–96.

Liu, K.-b., and M. L. Fearn. 2000a. Holocene history of catastrophic hurricane landfalls along the Gulf of Mexico coast reconstructed from coastal lake and marsh sediments. In *Current stresses and potential vulnerabilities: Implications of global change for the Gulf Coast Region of the United States*, edited by Z. H. Ning and K. Abdollahi, 38–47. Baton Rouge: Franklin Press.

Liu, K.-b., and M. L. Fearn. 2000b. Reconstruction of prehistoric landfall frequencies of catastrophic hurricanes in northwestern Florida from lake sediment records. *Quaternary Research* 54:238–45.

Liu, K.-b., and M. L. Fearn. 2002. Lake sediment evidence of coastal geologic evolution and hurricane history from Western Lake, Florida: Reply to Otvos. *Quaternary Research* 57:429–31.

Liu, K.-b., H. Lu, and C. Shen. 2003. Assessing the vulnerability of the Alabama Gulf coast to intense hurricane strikes and forest fires in the light of long-term climatic changes. In *Integrated assessment of the consequences of climate change for the Gulf Coast region*, edited by Z. H. Ning, R. E. Turner, T. Doyle, and K. Abdollahi, 223–30. Baton Rouge: Environmental Protection Agency.

Liu, K.-b., C. Shen, C., and K.-s. Louie. 2001. A 1000-year history of typhoon landfalls in Guangdong, southern China, reconstructed from Chinese historical documentary records. *Annals of the Association of American Geographers* 91:453–64.

Louie, K.-s., and K.-b. Liu. 2003. Earliest historical records of typhoons in China. *Journal of Historical Geography* 29:299–316.

Lu, H., and K.-b. Liu. 2003. Phytoliths of common grasses in the coastal environments of southeastern USA. *Estuarine, Coastal and Shelf Science* 58:587–600.

Ludlam, D. M. 1963. *Early American hurricanes, 1492–1870*. Boston: American Meteorological Society.

Malmquist, D. L., 1997. Oxygen isotopes in cave stalagmites as a proxy record of past tropical hurricane activity. In *Preprints of the 22nd Conference on Hurricanes and Tropical Meteorology*, 393–94. Boston: American Meteorological Society.

Michaels, A., D. Malmquist, A. Knap, and A. Close. 1997. Climate science and insurance risk. *Nature* 389:225–27.

Millás, J. C. 1968. *Hurricanes of the Caribbean and adjacent regions, 1492–1800*. Miami: Academy of the Arts and Sciences of the Americas.

Murnane, R. J., C. Barton, E. Collins, J. Donnelly, J. B. Elsner, K. Emanuel, I. Ginis, S. Howard, C. Landsea, K-b. Liu, D. Malmquist, M. McKay, A. Michaels, N. Nelson, J. O'Brien, D. Scott, and T. Webb III. 2000. Model estimates hurricane wind speed probabilities. *EOS, Transactions of the American Geophysical Union* 81:433–38.

Neumann, C. J., B. R. Jarvinen, C. J. McAdie, and G. R. Hammer. 1999. *Tropical cyclones of the North Atlantic Ocean, 1871–1998*. NOAA Historical Climatology Series 6-2. Ashville, N.C.: National Climatic Data Center.

Nott, J., and M. Hayne. 2001. High frequency of "super-cyclones" along the Great Barrier Reef over the past 5,000 years. *Nature* 413:508–12.

Otvos, E. G. 1999. Quaternary coastal history, basin geometry, and assumed evidence for hurricane activity, northeastern Gulf of Mexico coastal plain. *Journal of Coastal Research* 15: 438–43.

Otvos, E. G. 2002. Discussion of "Reconstruction of prehistoric landfall frequencies of catastrophic hurricanes in northwestern Florida from lake sediment records" (Liu and Fearn, 2000). *Quaternary Research* 57:425–28.

Pielke, R. A., Jr., and C. W. Landsea. 1998. Normalized hurricane damages in the United States, 1925–97. *Weather and Forecasting* 13:621–31.

Pielke, R. A., Jr., and R. A. Pielke, Sr. 1997. *Hurricanes: Their nature and impacts on society*. Chichester: Wiley.

Pirazzoli, P. A. 1991. *World atlas of Holocene sea-level changes*. Oceanography Series 58. Amsterdam: Elsevier.

Reading, A. J. 1990. Caribbean tropical storm activity over the past four centuries. *International Journal of Climatology* 10:365–76.

Reams, G. A., and P. C. Van Deusen. 1996. Detection of a hurricane signal in bald cypress tree-ring chronologies. In *Tree ring, environment, and humanity*, edited by J. S. Dean, D. M. Meko, and T. W. Swetnam, 265–71. Radiocarbon 1996. Tucson: Department of Geosciences, University of Arizona.

Ritchie, W., and S. Penland. 1987. When the hurricane blows. *Geographical Magazine* 59:177–82.

Scoffin, T. 1993. The geological effects of hurricanes on coral reefs and the interpretation of storm deposits. *Coral Reefs* 12:203–21.

Simpson, R. H., and H. Riehl. 1980. *The hurricane and its impact*. Baton Rouge: Louisiana State University Press.

Walker, L. R., N. V. L. Brokaw, D. J. Lodge, and R. B. Waide, eds. 1991. Ecosystem, plant, and animal response to hurricanes in the Caribbean. *Biotropica* [special issue] 23:313–521.

Zhou, X. 1998. A 4,000-year pollen record of vegetation changes, sea level rise, and hurricane disturbance in the Atchafalaya Marsh of Southern Louisiana. M.Sc. thesis, Department of Geography and Anthropology, Louisiana State University, Baton Rouge.

3
Back-barrier Sedimentary Records of Intense Hurricane Landfalls in the Northeastern United States

Jeffrey P. Donnelly and Thompson Webb III

Landfalling Category 3, 4, and 5 hurricanes often result in significant loss of life and property. Reliable historical records of these events go back to the late nineteenth century, and less complete records are available dating back to European settlement of the North American continent. The storm surges associated with these hurricanes remove sediments from beach and near-shore environments and deposit this material in normally quiescent back-barrier environments. These storm-induced deposits can provide a long-term record of past intense storms. In this chapter we address the sedimentary characteristics of storm-induced deposits and examine the geological record of intense historical hurricanes. In addition, we describe the various methods that have been used to reconstruct the history of overwash and storms at sites in Rhode Island and New Jersey and discuss the limitations and potential of using back-barrier sediments to develop long-term records of intense hurricane landfalls.

The sensitivity of a site to storm-induced deposition changes in response to variations in barrier position and height, sediment supply, changes in sea level, the stage of the astronomical tide at the time of landfall, the amount of time that has elapsed since the last storm surge event overtopped the barrier at the site, and the amount of wave energy accompanying landfall. A limitation of this method is the uncertainty associated with dating individual storm-induced deposits. And erosion of the stratigraphic record in back-barrier environments can result in an incomplete record of overwash deposition. Despite these complexities, the stratigraphic record preserved in back-barrier environments provides the best potential for developing long-term records of intense-hurricane activity. Comprehensive historical documentation of storms, local variations in geomorphology, and regional landscape changes can provide a framework for understanding and interpreting the sedimentary record. Multiple study sites can be used to minimize the influences of local changes in beach geomorphology on the interpretation of the sedimentary record of intense storms. Synchronous deposition of large-scale, storm-induced deposits at multiple sites within a region provide the most convincing evidence of past intense-hurricane strikes.

Intense hurricanes can modify coastal landforms and threaten lives and resources in heavily populated coastal regions. Although rare, intense hurricanes ≥ Category 3; sustained winds >50 m s^{-1}) that make landfall cause most of the damage and loss of life that result from tropical cyclones (Pielke and Landsea 1998). The deadliest hurricanes in the history of the United States (both Category 4 storms at landfall) were the 1900 Galveston, Texas, hurricane, which killed more than 8,000 people and the 1928 West Palm Beach, Florida, hurricane, which killed more than 1,800 people (Tannehill 1940). The most damaging hurricane in the United States was Hurricane Andrew (a Category 5 storm), which struck south Florida and Louisiana in 1992. This storm caused approximately $30 billion in damage. It has been estimated that if the Category 4 Miami, Florida, hurricane of 1926 were to strike today, it would cause damage in excess of $70 billion, as a result of inflation as well as increased population and wealth in the Miami area (Pielke and Landsea 1998).

Landfall of intense hurricanes is difficult to predict given the temporal and spatial variability of intense-hurricane occurrence. Records of past hurricane activity provide probabilistic information that can be used to estimate the likelihood of a future hurricane strike at any one location. Historical records of North American hurricanes date back several hundred years (Poey 1862; Tannehill 1940; Ludlum 1963); more detailed records maintained by the National Oceanic and Atmospheric Administration (NOAA) date back to the mid-nineteenth century (Neumann et al. 1993; Landsea et al., chapter 7 in this volume). The relative rarity of intense hurricanes and the short observational records result in significant uncertainties associated with using the historic record to estimate probabilities for intense hurricanes striking land (Murnane et al. 2000). Sedimentary records of past hurricane activity can extend our knowledge of hurricane occurrence into the prehistoric period and can be used to test the reliability of model-based strike probability estimates (Liu and Fearn 1993, 2000; Donnelly et al. 2001a, 2001b).

Given the potential for changes in the intensity and frequency of hurricanes due to carbon dioxide-induced warming of the global climate system (Emanuel, chapter 14, and Knutson et al., chapter 15 in this volume), decision makers, scientists, and the general public have become increasingly concerned about potential risks to coastal populations. Gaining an understanding of how changes in hurricane activity may link to changes in climate is important in order to forecast future changes and possibly mitigate socioeconomic impacts (Henderson-Sellers et al. 1998). Potential connections between the frequency and intensity of hurricanes and changes in the earth's climate system can be examined if long-term records of hurricane activity can be developed.

Overwash deposits (or washovers) and flood-tidal deltas produced by a combination of storm surge and waves are often preserved within back-barrier environments and provide a means of constructing a prehistoric record of intense storms. Overwash deposits in coastal sediments have provided the basis for constructing prehistoric records of intense-hurricane strikes in Alabama, Florida, Rhode Island, and New Jersey (Liu and Fearn 1993, 2000; Donnelly et al. 2001a, 2001b). In Alabama and Florida, sand deposited in coastal lakes provided evidence of apparent intense-hurricane strikes dating back several thousand years (Liu and Fearn 1993, 2000; Liu, chapter 2 in this volume). Higher-resolution records obtained from back-barrier marshes in Rhode Island (Donnelly et al. 2001a) and New Jersey (Donnelly et al. 2001b) contain evidence of intense-hurricane strikes dating back approximately 700 years.

In this chapter we address the evolution of barrier-beach complexes in the northeastern United States, the characteristics of storm-induced deposits, and the suitability of these environments for providing sedimentary records of intense storms. Specifically, we discuss the general sedimentary characteristics of overwash fans and flood-tidal deltas and examine the geological consequences of intense historical hurricanes on the northeastern coast of the United States in order to provide a framework for interpreting the prehistoric record. In addition, we describe the various methods that have been used to reconstruct the history of overwash and storms at sites in Rhode Island and New Jersey and compare and contrast intense hurricane landfall probabilities derived from both sediment and historical records. Finally, we discuss the limitations and potential of using back-barrier sediments to develop long-term records of intense hurricane landfalls.

Evolution of Barrier Beaches and Sedimentary Records of Storms

Barrier Beaches

Barrier beaches are shore-parallel accumulations of sediment sheltering back-barrier lagoons, bays, or ponds and are found on between 10 and 13% of earth's coastlines (Schwartz 1973). A barrier beach complex adjusts its position to changes in sea level and sediment supply, and sequences of facies representing the different depositional environments accumulate with time. Barrier complexes can migrate landward (transgressive barrier) or seaward (regressive barrier) over time, depending on the rate and direction of sea level changes and availability of sediment. If sea level drops or sediment supply is relatively high, the barrier complex will tend to prograde seaward. The barrier complex will tend to migrate landward if the sediment supply is low relative to the rate of sea level rise.

The barrier complexes of the northeast United States have been transgressing landward throughout the Holocene (0–10,000 years ago). Radiocarbon dates from drowned-back-barrier-marsh peat and lagoon deposits off the central New Jersey coast (Stuiver and Daddario 1963; Stahl, Koczan, and Swift 1974) and Long Island (Rampino and Saunders 1980) indicate that the barrier complexes of New Jersey and New York existed many kilometers seaward of their current position in the early Holocene. The age and depth of these back-barrier deposits preserved offshore mirror those of ancient corals used to estimate global sea level change in the Holocene (0–10,000 yrs. B.P.) (Fairbanks 1989, Lighty, MacIntyre, and Stuckenrath 1982) (figure 3.1), but are apparently rare or absent from the latest Pleistocene (10,000–20,000 yrs. B.P.). These barrier systems likely developed as global sea level rise slowed to less than 10 mm per year at the beginning of the Holocene (figure 3.1).

Like the more extensive barriers in New York and New Jersey, the barrier systems in New England formed during the Holocene (Redfield and Rubin 1962; McIntire and Morgan 1964; Redfield 1967). In contrast to the extensive and relatively continuous barriers of the New Jersey and New York coasts, the barrier systems of New England tend to be small spits and baymouth bars connected to glacially derived uplands (Dillon 1970; Boothroyd, Friedrich, and McGinn 1985; Donnelly et al. 2001a).

In order for barrier systems to persist and migrate landward during a period of sea level rise, barrier and nearshore sediment must be transported into the back-barrier environment. Storm-induced inlet and flood-tidal delta formation and overwash fan deposition are the two most significant mechanisms of transporting barrier sediments to the back-barrier environment (Pierce 1970; Dillon 1970).

Overwash Fans and Flood-Tidal Deltas

Overwash fans occur when wave energy combined with high water levels (storm surge) overtop or breach barrier beaches and transport nearshore and barrier sediments into the back-barrier environment (Schwartz 1975). If the

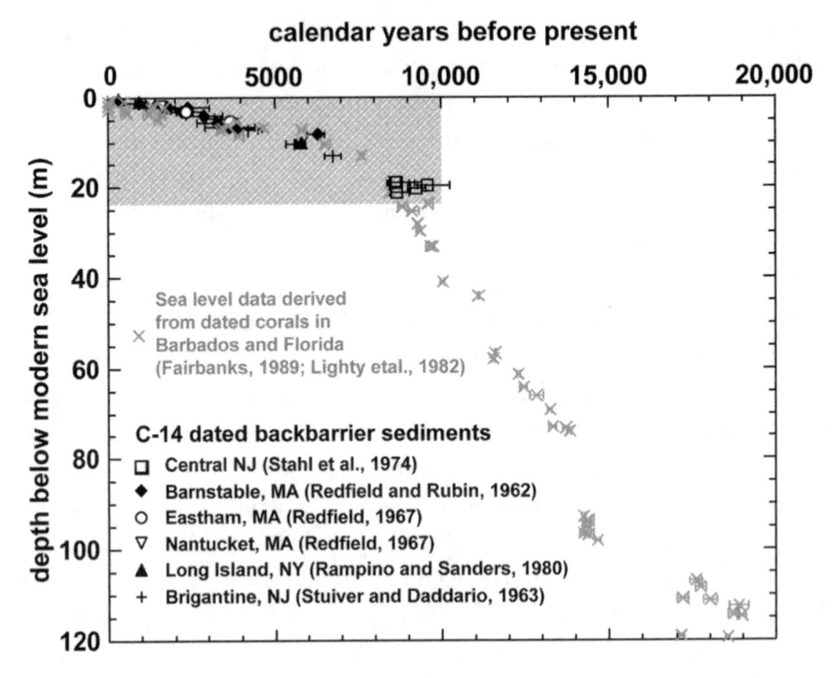

FIGURE 3.1 Age-depth plot of sea level change over the last 20,000 years based on dated corals from Barbados and Florida (gray) and radiocarbon-dated back-barrier sediments from the northeastern United States. Barrier island development began around 10,000 years ago (shaded) as the rate of sea level rise decreased to less than 10 mm per year.

barrier is breached, a lobate fan at the terminus of the breach will form. These lobes may coalesce if multiple breaches occur close together. When the entire barrier is overtopped by storm surge, sheet overwash fans that extend the length of the barrier are deposited. Between overwash events, low-energy estuarine deposits are typically deposited on top of overwash fans. If these deposits are not reworked during subsequent storms, the overwash fans will be preserved as laterally continuous horizons of sand within back-barrier sediments and provide a record of past storm surge (figure 3.2). If the barrier beach has not been breached to form a tidal inlet, back-barrier sediments will typically contain marsh or lagoonal deposits interbedded with overwash fans (figure 3.2, top).

An inlet through a barrier beach forms as a result of breaching by storm waves on the seaward side of the barrier and/or from the reversal of storm surge in the lagoon associated with a change from onshore to offshore winds as a storm passes (Pierce 1970). Inlets can be significant conduits for transporting sediment from the marine side of the barrier into back-barrier environments. Strong currents through an inlet can quickly erode previously deposited barrier

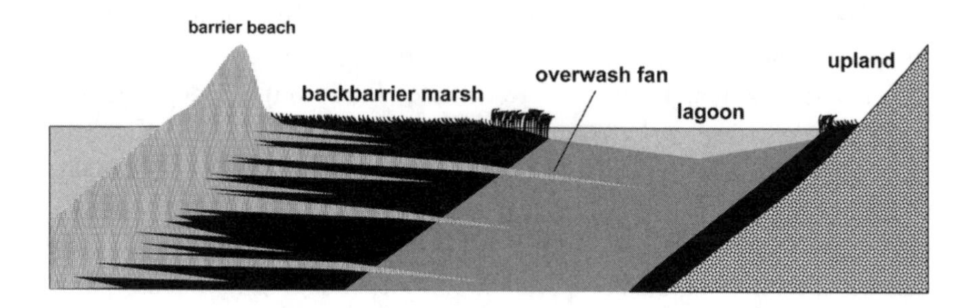

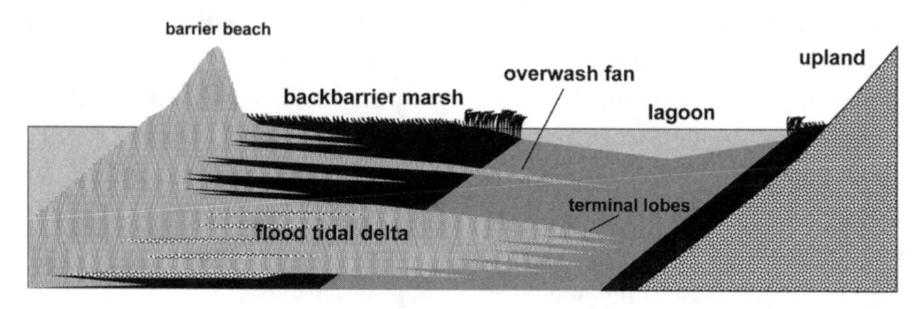

FIGURE 3.2 (*Top*) Cross-section of conceptual model of overwash deposition and the landward translocation of the barrier-marsh system in a regime of rising sea level. Overtopping of the barrier beach by storm surge results in overwash fan deposition across back-barrier marshes. Overwash fans are preserved as sea level increases and they are covered with marsh deposits. (*Bottom*) Cross-section of conceptual model when an inlet was present at the study site depositing a flood-tidal delta. Terminal lobes of the flood-tidal delta are interfingered with lagoonal deposits. Light stipple, sand; black, marsh deposits; gray, lagoon deposits; heavy stipple, upland sediments; vertical exaggeration, approximately 100 to 500 ×.

and back-barrier sediments. Once an inlet forms, it may close within a few days if there is (1) insufficient tidal flow to maintain an open inlet and (2) ample sediment supply. An inlet may remain open indefinitely if there is sufficient tidal flow. In this case the inlet may migrate laterally along the beach. Flood-tidal delta deposits are likely preserved within the back-barrier sediments if the barrier was breached in the past and an active inlet was maintained (figure 3.2, bottom). When the inlet is open, sand and gravel can be transported into back-barrier lagoons or bays by normal tides and especially during storms. While an inlet exists, storms with surge heights less than the height of the barrier are able to transport sediment into the back-barrier environment. During an intense storm the inlet may also provide an avenue for storm surge, resulting in the transport of sediments much greater distances into the back-barrier embayment than occurs during normal tides or less-intense storms. These terminal lobes of the flood-tidal delta may be capped with estuarine sediments and preserved (figure 3.2, bottom) as tidal flows return to normal after a storm.

Therefore, although much more complex than overwash deposits, relic flood-tidal delta deposits have some potential for preserving records of individual storm events. The initiation of deposits from a relic flood-tidal delta can provide a record of the intense storm that caused the initial breach in the barrier. One must, however, determine the lateral extent of the deposit to ascertain whether the inlet migrated to that position independent of an intense storm. For example, an inlet that migrated to where the relic delta deposit is would be laterally continuous and time transgressive. The interfingering of terminal lobes of flood-tidal deltas and estuarine mud can also provide a record of individual storm events (figure 3.2, bottom). But significant reworking of these deposits may occur, particularly at shallower depths within the back-barrier embayment. In addition, gauging the intensity of the storm that deposited the terminal lobes may be difficult because overtopping of the barrier is not required for their deposition. Therefore it is best to use overwash deposits, perhaps in combination with flood-tidal delta deposits, to construct a record of past storms from back-barrier sediments.

Overwash deposits are composed of locally derived material from the foredune, beach, and nearshore environments and are typically well- to poorly sorted, fine to coarse sand. The mean grain size within overwash deposits generally decreases as a function of distance from the barrier (Hennessy and Zarillo 1987). Overwash sediments often exhibit horizontal stratification, and if the deposit terminates in a coastal pond or lagoon they can exhibit medium- to small-scale delta foreset stratification (Schwartz 1975). Horizontal stratification is evident in the upper portion of an overwash deposit recovered in cores from the back-barrier marsh at Hick's Beach, New York (figure 3.3a).

Overwash deposits can also lack stratification. Cores from back-barrier environments in southern Rhode Island contain overwash deposits with little or no horizontal stratification (Boothroyd, Friedrich, and McGinn 1985; Donnelly et al. 2001a) (figure 3.3b). In addition, the lower portion of the overwash deposit from Hick's Beach lacks any significant horizontal stratification (figure 3.3a). Postdepositional processes like ice scour (Boothroyd, Friedrich, and McGinn 1985) and bioturbation (Hennessy and Zarillo 1987) may eradicate laminations. Given that both bioturbation and ice scour would work from the top down, these processes probably do not explain the lack of stratification in the lower portion of the Hick's Beach overwash fan (figure 3.3a).

Another characteristic of overwash deposits preserved in back-barrier environments is the abrupt nature of the contact with the underlying peat or estuarine mud (Donnelly et al. 2001a, 2001b) (figure 3.3). Evidence of soft-

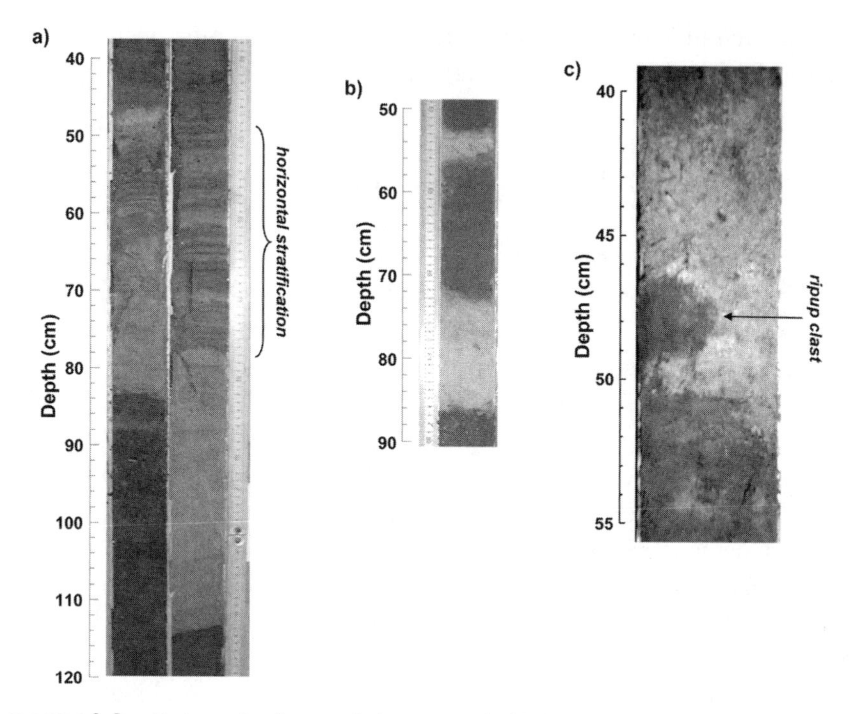

FIGURE 3.3 Photographs of overwash deposits interbedded with back-barrier marsh sediments: (*a*) Hick's Beach, New York; (*b*) Succotash Marsh, Rhode Island; (*c*) Whale Beach, New Jersey (note the rip-up clast at 46–50 cm).

sediment deformation is common where an overwash fan is deposited over saturated, fine-grained sediments (Klein 1986). In addition, rip-up clasts are frequently encountered at the basal contact of an overwash deposit (figure 3.3c) and are indicative of high-velocity currents associated with overwash deposition.

Like overwash deposits, flood-tidal delta deposits are composed of locally derived well- to poorly-sorted, fine to coarse sand, although coarse lag deposits consisting of gravel and shell hash often occur within channels (figure 3.2, bottom). The mean grain-size within flood-tidal delta deposits tends to be more spatially variable than in overwash sediments (Hennessy and Zarillo 1987). Flood-tidal delta deposits are dominated by ripple laminations, often with trough cross stratification from bi-directional paleocurrents (Hennessy and Zarillo 1987). In addition, deposits from flood-tidal deltas often contain abundant detrital organic laminae, shell hash layers, and disarticulated shells (Boothroyd, Friedrich, and McGinn 1985; Hennessy and Zarillo 1987; Donnelly et al. 2001a). Like overwash fans, the contact between the flood-tidal delta deposits and underlying estuarine deposits tends to be abrupt, often with evidence of soft-sediment deformation (Klein 1986; Donnelly et al. 2001a).

Changes in the Rate of Sea Level Rise

Sea level has risen approximately 23 m in New Jersey, Long Island, and southern New England over the last 10,000 years (Redfield and Rubin 1962; Stuiver and Daddario 1963; Donnelly 1998) (figure 3.1). On centennial to millennial time scales the rate of rise decelerated from about 3.5 mm per year in the early Holocene to about 1 mm per year in the last 1,000 years (figure 3.1). Smaller amplitude changes superimposed on the long-term trend may have occurred in the past but are not resolved in the available data set. Tide-gauge measurements in the northeastern United States indicate a return to sea level rise rates of close to 3 mm per year over the last 150 years (Emery and Aubrey 1991).

Several barriers in the northeastern United States are apparently narrowing and submerging in response to this recent increase in the rate of sea level rise (Leatherman 1983; Leatherman and Zaremba 1986). Submergence and narrowing of barrier beaches makes them more susceptible to breaching and overtopping. Consequently changes in the rate and direction of sea level change as well as sediment supply will likely cause changes in the frequency and magnitude of back-barrier sedimentation resulting from storms. As a result of the recent increase in the rate of sea level rise, barriers in the northeastern United States may be more susceptible to breaching and overtopping today than they were 500 years ago. Estimation of the frequency and intensity of past storms based on overwash and flood-tidal-delta deposits therefore requires knowledge of the rate and direction of sea level change.

Site Selection, Coring, and Dating Methods

Maps and Photos

Site selection is important in reconstructing the history of storms from overwash sediments. Modern charts and maps show the locations of depositional environments (marshes, lagoons, or ponds adjacent to barrier beaches) that are likely to preserve a record of storm surge. Historical charts, maps, and aerial photographs help to identify relatively stable barrier complexes without active inlets that may have eroded part or all of the sedimentary record. Historical written documents can provide information on human modifications to the barrier

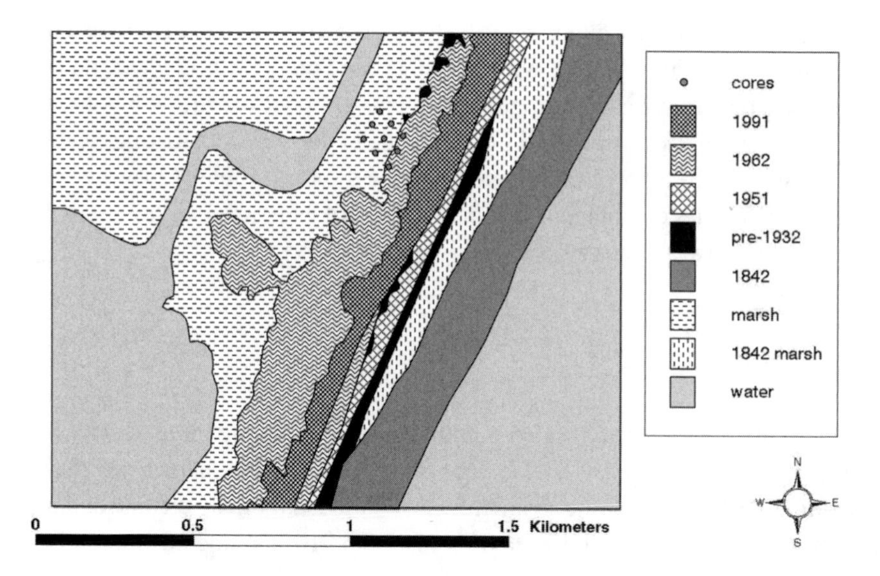

FIGURE 3.4 Shoreline change and overwash history map from Whale Beach for 1842 to 1991. Each pattern associated with a year denotes the location of exposed sand at that time. The beach has migrated landward approximately 400 meters since 1842.

(dredging, filling) that may not be apparent via inspection of modern maps, aerial photographs, and charts. Finally, historic charts, maps, and aerial photographs are extremely valuable in gaining an understanding of how the position of the shoreline may have changed. For example, an analysis of aerial photographs and historical charts for Whale Beach, New Jersey, revealed that the barrier beach there has migrated landward about 400 meters since 1842 (figure 3.4). If this rate of shoreline retreat (~2.5 m/year) is extrapolated over the last 1,000 years, an overwash deposit recovered from a back-barrier marsh that dates to A.D. 1000 may have been transported about 2500 meters farther landward than one deposited within the last few decades. In addition to site selection, historical research can be used to document evidence of overwash from historical storms of known intensity.

Probing and Coring

Once a potential site has been identified further reconnaissance work must be done. Probing the back-barrier sediments by driving a steel rod to refusal is useful for determining the depth of organic-rich sediments and therefore the length of any potential record. Overwash-sand layers interbedded within these organic sediments can often be easily detected as intervals of greater resistance encountered during probing. A probe's downward progress through a sand layer may also be audible as a grating, scratchy noise. In back-barrier environments containing sandy overwash and flood-tidal delta deposits, vibracoring methods (Lanesky et al. 1979) work best for acquiring sediment samples. Obtaining samples with a series of vibracores allows mapping of widespread sedimentary features like overwash deposits and flood-tidal deltas.

Radiometric Dating

A combination of several isotopic and stratigraphic dating methods can help establish age control for back-barrier stratigraphies. The activity of Pb-210 within marsh and estuarine sediments can be used to estimate the rate of marsh accretion during the last 100 to 150 years. Pb-210 is rendered immobile in anoxic organic sediments and decays with a half-life of 22.3 years (Faure 1986). Cs-137 produced by atmospheric testing of nuclear weapons provides two stratigraphic markers within the recent sediment record. The beginning of Cs-137 production occurred in 1954 and the maximum took place in 1963 (Delaune, Patrick, and Buresh 1978). Therefore, the first appearance of Cs-137 in the stratigraphic record is assumed to be the 1954 horizon and the peak in Cs-137 activity is defined as 1963.

Accelerator mass spectrometry (AMS) radiocarbon analysis of organic samples is useful in dating historical overwash deposits, particularly in conjunction with other stratigraphic dating tools. Radiocarbon results need to be calibrated to calendar years to account for secular changes in atmospheric C-14 concentrations (Stuiver et al. 1998). The nonlinear calibration curve often results in multiple calibrated calendar age ranges for each radiocarbon date. When this happens, stratigraphic dating methods can eliminate some of these calibrated age ranges and thus decrease some of the dating uncertainty (Donnelly et al. 2001a, 2001b). In addition, the mutual overlap of multiple calibrated radiocarbon dates from the same stratigraphic interval can be used to reduce uncertainty associated with radiocarbon dating (Donnelly et al. 2001a).

Stratigraphic Markers

Pollen preserved in sediments record the increase of native weeds ragweed (*Ambrosia*) and sorrel (*Rumex*) associated with European-style clearance in the northeastern United States around A.D. 1700 and provide a well-dated stratigraphic marker (Brugam 1978; Clark and Patterson 1985; Russell et al. 1993). Also evident in the pollen record is the decline of chestnut (*Castanea*), associated with a blight that was introduced in New York City in 1904 and quickly spread to southern New Jersey by 1920 (Anderson 1974; Brugam 1978; Clark and Patterson 1985).

Heavy metal concentrations in the sediments can provide a dated stratigraphic marker. For example, industrially produced lead, rendered immobile in anoxic marsh sediments, provides a stratigraphic time horizon. The increase in lead concentrations in coastal wetland sediments dates to the mid-1800s (McCaffrey and Thomson 1980; Bricker-Urso et al. 1989; Donnelly et al. 2001a).

Storms and Storm Surge

Hurricanes have repeatedly impacted the northeastern United States. New England and Long Island, New York, protrude into the western Atlantic close to the warm Gulf Stream current and are often in the path of fast-moving tropical storms and hurricanes as they track north. Intense hurricanes, however, are rare in New England. According to the National Oceanic and Atmospheric Administration (NOAA) "Best Track" data set, nine intense hurricanes threatened the northeastern United States and Maritime Canada in the last 149 years (Neumann et al. 1993, Landsea et al., chapter 7 in this volume) (figure 3.5). Only three of these made landfall as intense hurricanes: an 1869 hurricane that hit Long Island and New England, an 1893 event that came ashore in Nova Scotia, and the 1938 storm that devastated Long Island and New England. Two of the nine intense storms made landfall after weakening to Category 2 intensity—an 1896 storm that hit Maine and a 1969 storm that hit New Brunswick. The other four intense hurricanes recurved to the east and did not make landfall.

The cool sea-surface temperatures immediately south of New England typically result in significant weakening of these systems as they move north. An acceleration in the forward motion of hurricanes as they move up the eastern seaboard of the United States (as much as 18–27 m s^{-1} or 40–60 mph) can make up for this loss of intensity as it enhances the winds on the right side of the hurricane. In addition, the rapid forward motion shortens the length of time that a storm spends over cooler ocean waters, potentially preventing or slowing weakening.

Storm surge results from the strong winds driving ocean water onshore and, to a lesser degree, from the response of the sea surface to the extremely low

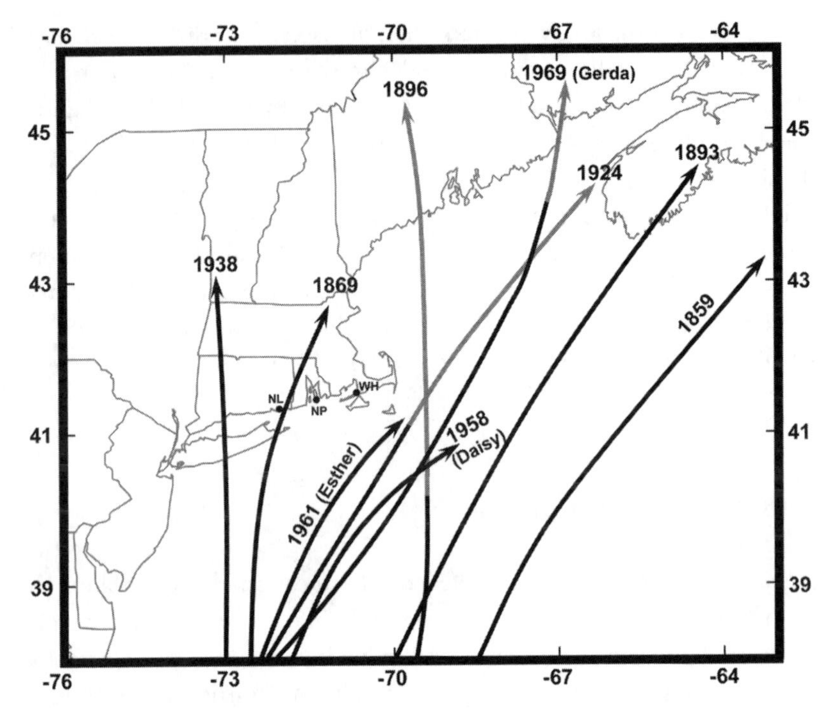

FIGURE 3.5 Tracks of intense hurricanes threatening the northeastern United States since 1851. The line type denotes storm intensity: black, ≥ Category 3; gray, Category 1 or 2. The location of tide gauges mentioned in the text are marked: NL, New London, Connecticut; NP, Newport, Rhode Island; WH, Woods Hole, Massachusetts.

atmospheric pressure of an intense storm. In studies examining the patterns of water level change that result from hurricanes, storm surge is typically defined as the rise in water level above the predicted astronomically driven level of the tide that results from strong onshore winds and low atmospheric pressure associated with a storm. As the highest water level attained is what governs the amount of coastal inundation and, in combination with wave energy, causes most of the damage and modifications to coastal landforms, determining the maximum water level associated with a historical storm is most useful for understanding the sedimentary record. For the purposes of this chapter, maximum storm surge represents the maximum water level recorded for any one storm. Tide-gauge measurements taken during hurricane strikes typically reveal a rapid but steady rise in water levels followed by an equally rapid fall in water levels (Redfield and Miller 1957). First-hand historical accounts, however, often describe storm surge accompanying hurricanes as coming in as a wave or series of waves (Tannehill 1927; Allen 1976). It is this wave energy in combination with coastal inundation that transports and deposits sediment in the coastal zone.

Hurricanes in the western Atlantic typically track northward along the eastern seaboard of the United States and, if they fail to recurve to the northeast, can strike New England or Long Island from the south. As a result, the strongest onshore southerly winds, and therefore maximum storm surge associated with hurricanes that make landfall in the northeastern United States, generally occur to the east of the storm track. Intense hurricane strikes in the northeastern United States have typically resulted in more than 3 meters of storm surge (Donnelly et al. 2001a, 2001b). Storm surge occurring in south-facing embayments can result in significantly heightened water levels (≥ 4 meters) at the heads of these bays (Redfield and Miller 1957).

Severe winter storms can also cause storm surge in the northeastern United States, although in general the water levels are considerably lower than storm surge that occurs with intense-hurricane strikes. Winter storms in this region often produce strong northeast winds. As a result the highest storm-surge levels associated with winter storms typically occur on north- or east-facing coastlines. Many severe winter storms have battered southern New England since European settlement, with some of the most infamous occurring in 1723, 1888, 1944, 1953, 1962, 1978, 1991, and 1993. Historical accounts of these storms confirm that damage from their storm surges was generally most severe on coastlines exposed to northeasterly winds (Snow 1943; Dickson 1978; Fitzgerald, van Heteren, and Montello 1994). Records from tide gauges in the region show that storm-surge associated with severe winter storms in the twentieth century typically reached heights of approximately 2 meters above mean sea level (Donnelly et al. 2001a, 2001b).

Historical Hurricanes and Their Geologic Record in the Northeast

Hurricanes of the Twentieth Century

Ten hurricanes made landfall in the northeastern United States in the twentieth century (figure 3.6). Hurricane Bob, which caused eight deaths and over $1.5 billion in damage, was the last hurricane to come ashore. This Category 2 storm passed over eastern Rhode Island and southeastern Massachusetts on August 19, 1991 (figure 3.6). The storm surge associated with Hurricane Bob was well documented by tide gauges in Woods Hole, Massachusetts; Newport,

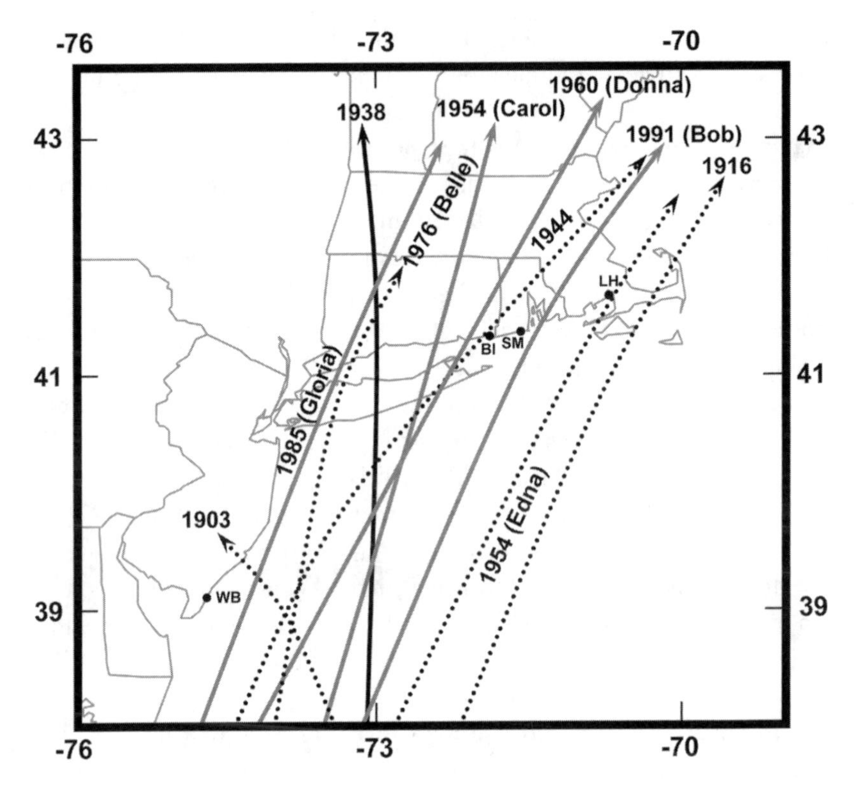

FIGURE 3.6 Hurricane landfalls in the northeastern United States in the twentieth century: black, Category 3; gray, Category 2; stippled, Category 1. Salt marshes with sediment analysis: WB, Whale Beach, New Jersey; BI, Barn Island, Connecticut; SM, Succotash Marsh, Rhode Island; LH, Little Harbor Marsh, Massachusetts.

Rhode Island; and New London, Connecticut (NOAA/NOS/ CO-OPS 2001; for tide gauge locations, see figure 3.5).

Water levels began to rise sharply at roughly the same time (~1600 GMT) and rate (0.55m/hr) at all three locations (figure 3.7). The Woods Hole tide gauge recorded the highest water level of the three stations at 1.55 m above mean high water (MHW) at 1851 GMT (figure 3.7). The maximum water level (1.34 m above MHW) at the Newport tide gauge (1818 GMT) occurred 33 minutes earlier than at the Woods Hole gauge. At New London, the water level reached a maximum of 1.03 m above MHW at 1742 GMT, 69 minutes before the maximum was reached at Woods Hole. Shortly after Hurricane Bob made landfall at Newport, the wind direction changed from south to northwest, and the offshore winds caused the water level to decrease. Following landfall, onshore winds continued for about another 30 minutes at Woods Hole, which

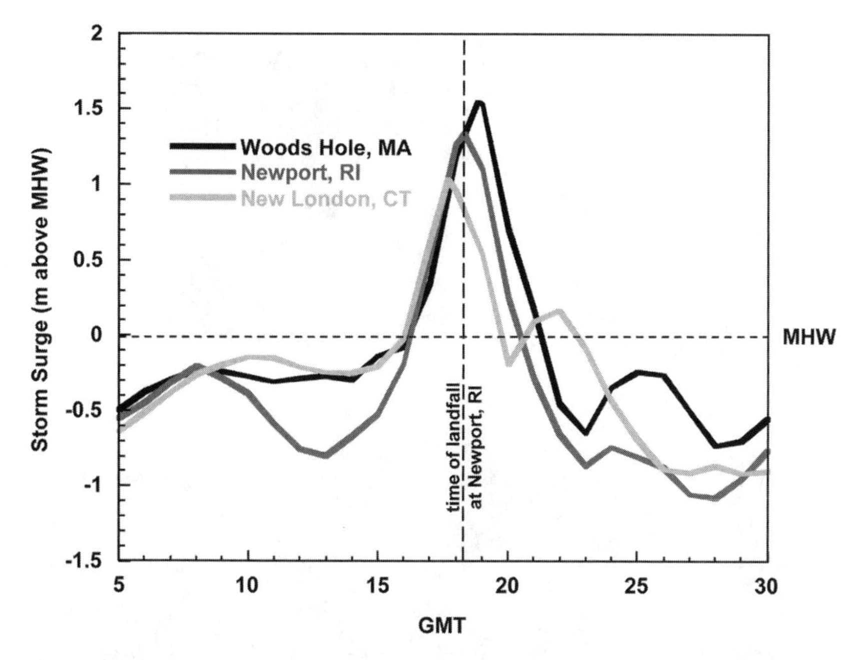

FIGURE 3.7 Water-level changes around mean high water accompanying Hurricane Bob, August 19, 1991, recorded by tide gauges in New London, Connecticut (light gray), Newport, Rhode Island (dark gray), and Woods Hole, Massachusetts (black). For tide-gauge locations, see figure 3.6. MHW, mean high water; GMT, Greenwich Mean Time.

is 55 km east of Newport, causing the water level to continue to rise until the wind shifted to offshore at that location. At New London, 70 km west of the storm's track, the wind changed from onshore to offshore about 35 minutes before this occurred at Newport, resulting in an earlier termination in the rise in water level and lower maximum storm surge.

Similar asymmetries in the pattern of storm surge favoring the eastern sector of the storm have been documented during other hurricane strikes in the region (Redfield and Miller 1957). The occurrence of the highest storm-surge levels to the right of the storm's track is likely the result of maximum onshore winds occurring to the right of the storm center, due to the addition of the forward speed of the storm to the winds on the right side and, perhaps more importantly as appears to be the case during Hurricane Bob, the duration of strong onshore winds.

The storm surge produced by a hurricane can overtop sandy barriers and deposit overwash fans across back-barrier environments. The upper 20 to 30 cm of the stratigraphy of several back-barrier marshes from southeastern New England is shown in figure 3.8. At Little Harbor Marsh located in Wareham,

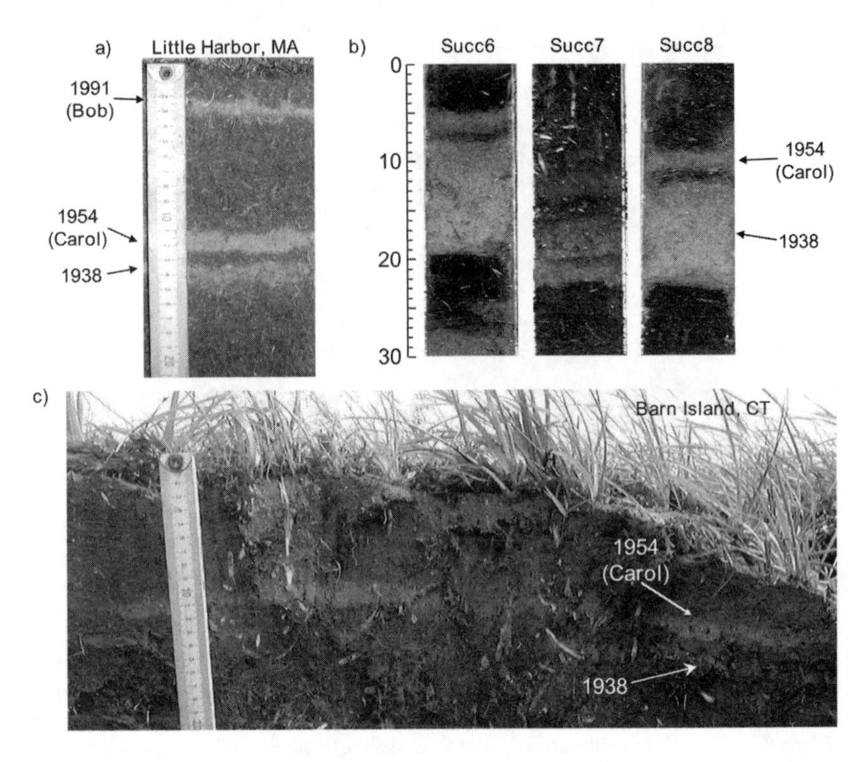

FIGURE 3.8 Photographs of the upper 20 to 30 cm of back-barrier marsh sediments containing overwash fans. For locations, see figure 3.6. (*a*) Three sand layers are evident in the upper 20 cm of Little Harbor Marsh, Wareham, Massachusetts. The upper sand layer (2–3 cm) was likely deposited by Hurricane Bob in 1991. Sand layers at 11–12 and 13–14 cm were likely deposited in 1954 by Hurricane Carol and the 1938 Hurricane, respectively. (*b*) Two sand layers deposited by Hurricane Carol (1954) and the 1938 Hurricane are evident in the upper 30 cm of cores 6, 7, and 8 from Succotash Marsh, Rhode Island (Donnelly et al. 2001a). (*c*) Two sand layers associated with Hurricane Carol (1954) and the 1938 Hurricane are also evident the upper 20 cm of Barn Island Marsh, Connecticut sediments.

Massachusetts, at the head of Buzzards Bay (figure 3.6), where storm-surge levels are typically very high due to focusing, a sand layer is apparent within the high-marsh peat, between 2 and 3 cm below the modern marsh surface. Given high-marsh environments in southern New England have been accreting at roughly 2.5 mm per year over the last 100 years (Bricker-Urso et al. 1989; Orson et al, Warren, and Niering 1998), this sand layer was most likely deposited about 10 years ago by Hurricane Bob. No overwash fans associated with Hurricane Bob are evident in the upper 3 cm of marsh sequences from Succotash Marsh in Rhode Island, or Barn Island Marsh in eastern Connecticut.

Three hurricanes made landfall in the northeastern United States between 1955 and 1990 (Neumann et al. 1993) (figure 3.6). Hurricane Gloria made landfall as a Category 2 storm in western Long Island, New York, and western

Connecticut on September 27, 1985, and produced storm surge heights of 1 meter or less above mean high water in southeastern New England (figure 3.9). On August 10, 1976, Hurricane Belle, a Category 1 storm, made landfall in western Long Island, New York, and produced storm surge of less than 0.5 m above mean high water. Hurricane Donna made landfall on eastern Long Island, eastern Connecticut, and Rhode Island on September 12, 1960, as a Category 2 storm, causing storm surge to rise roughly 1 m above mean high water in southeastern New England. None of these storms appear to have deposited overwash sediments at the three marshes shown in figure 3.8. Warren and Niering (1993) attribute a small sand layer of limited spatial extent at Barn Island, Connecticut to Hurricane Gloria, but little evidence of this sand layer was recovered in subsequent studies (Donnelly et al. 2000).

Hurricane Carol, which made landfall on August 31, 1954, caused significant storm surge in southern New England (figure 3.9). Although technically not an intense hurricane at landfall, Hurricane Carol was a strong Category 2 storm at landfall, with sustained winds of approximately 44 m s^{-1}. Carol made landfall on eastern Long Island and eastern Connecticut during a time of astronomical high tide. The timing of the landfall resulted in storm surges of between 2 and 3 meters above mean high water on the open coast from eastern Connecticut to southeastern Massachusetts, with lesser amounts to the west of the storm track (Redfield and Miller 1957) (figure 3.6). Focusing of storm surge in the south-facing Narragansett and Buzzards Bays resulted in storm-surge heights of over 4 m above predicted tide levels at the head of these bays (Red-

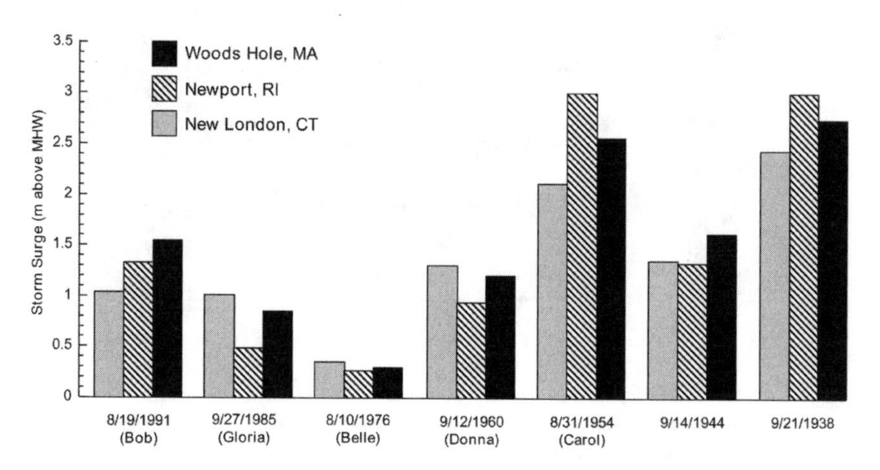

FIGURE 3.9 Maximum water levels associated with twentieth-century landfalling hurricanes in the northeastern United States recorded by tide gauges in New London, Connecticut; Newport, Rhode Island; and Woods Hole, Massachusetts.

field and Miller 1957). This storm caused over $400 million in damage and 60 deaths. Due to inflation and increased population and wealth in the areas impacted by Carol, if this hurricane occurred in 1995 it would have likely resulted in nearly $10 billion in damage (Pielke and Landsea 1998).

Aerial photographs taken immediately before and after Hurricane Carol at Succotash Marsh in East Matunuck, Rhode Island, document the deposition of an overwash fan (Donnelly et al. 2001a) (figure 3.6). The storm surge overtopped the entire barrier and destroyed a small cottage community. A stratigraphic study of 14 cores from Succotash Marsh (Donnelly et al. 2001a) reveals an overwash deposit at between 3 and 12 cm depth (figures 3.8 and 3.10) with an extent generally consistent with the overwash deposition from Hurricane Carol apparent in aerial photographs taken after August 1954. The initiation of Cs-137 associated with nuclear weapons testing within the sediments at the top of this layer in core 7 confirm the timing of the deposition of this overwash fan to about 1954 (figure 3.11). Sand layers deposited within salt-marsh sediments between 11 and 12 cm below the modern surface at Little Harbor, Massachu-

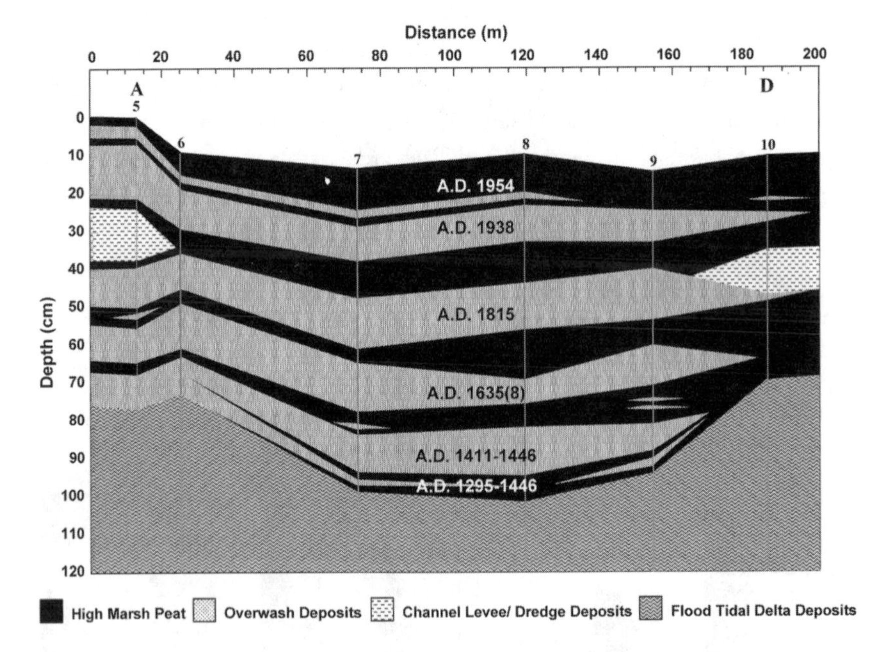

FIGURE 3.10 Stratigraphic cross-section of Succotash Marsh, Rhode Island, based on a transect of vibracores. The labels on the overwash deposits are the best estimate of the date of deposition. Vertical lines show core locations with core numbers previously noted. The 1635(8) date accounts for the fact that the resolution of the sediment record is too coarse to resolve overwash deposits from the 1635 and 1638 storms.

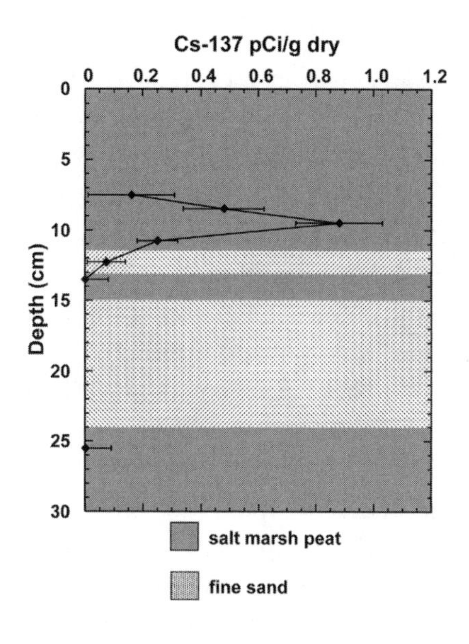

FIGURE 3.11 Cesium-137 activity profile and core log from the upper 30 cm of core 7 from Succotash Marsh. Note the initial rise in Cs-137 attributable to 1954 occurs in the sample between 12 and 13 cm. The peak in Cs-137 activity occurs in the sample between 9 and 10 cm and represents 1963.

setts, and 10 and 11 cm below the modern surface at Barn Island, Connecticut, are also consistent with deposition in the mid-1950s, assuming an accretion rate of about 2.5 mm per year, and were likely deposited by Hurricane Carol (Donnelly et al. 2000) (figure 3.8).

On September 14, 1944, a Category 1 hurricane made landfall on the eastern tip of Long Island and in southeastern New England (figure 3.6). Storm surge recorded by the tide gauges at Woods Hole, Newport, and New London ranged between 1.62 and 1.33 m above mean high water (MHW) (figure 3.9). This storm resulted in 26 deaths in New England and more than $100 million in damage. Pielke and Landsea (1998) estimate that this storm would have caused approximately $6.5 billion in damage if it had struck in 1995.

The most recent intense (Category 3 or greater) hurricane to strike Long Island and southern New England made landfall September 21, 1938, in central Long Island. It tracked north into Connecticut, Massachusetts, and Vermont (Brooks 1939; Minsinger 1988; Neumann et al. 1993) (figures 3.5 and 3.6). The 1938 Hurricane was moving north at 22 to 28 m s^{-1} (50–60 mph). Wind speeds to the right of the storm's track exceeded 53 m s^{-1} and a maximum wind gust of 83 m s^{-1} was recorded at the Blue Hills Observatory in Milton, Massachusetts. The lowest recorded barometric pressure was 946 mb at the Coast Guard Station, Bellport, New York. Storm surge and an astronomical high tide combined to cause the water level to rise more than 3 meters above normal spring tide levels along the open coast; Narragansett and Buzzards Bays experienced more than 4 meters of storm surge in some areas (Paulsen 1940; Redfield and Miller 1957) (figure 3.9). More than 600 lives were lost and property damage was estimated at approximately $400 million (Brooks 1939). A storm of similar intensity striking the region today would likely result in approximately $17 billion in property damage (Pielke and Landsea 1998).

Significant coastal modification and erosion occurred from Long Island to southeastern Massachusetts in 1938 as a result of the combined effect of storm surge and wave action (Nichols and Marston 1939; Wilby et al. 1939). Extensive sheet-overwash fans were deposited in the back-barrier environment from central Long Island to western Cape Cod as storm surge washed over nearly every barrier beach in the region.

A 1- to 20-cm thick overwash fan occurs between 7 and 23 cm depth in all cores at Succotash Marsh (figures 3.8b and 3.10) and is consistent with the extent of overwash deposition observed in aerial photographs taken following the 1938 Hurricane (Donnelly et al. 2001a) (figures 3.8b and 3.10). If the accretion rate of 2.8 mm per year derived from the 12.5 cm of accumulation following the onset of Cs-137 in 1954 (figure 3.11) is extended to the top of this sand layer in core 7 from Succotash Marsh, the initiation of peat accumulation above this overwash deposit dates to about 1945. As revegetation of the overwash fan and therefore peat formation likely did not occur for several years following the deposition of the fan, the age of this overwash deposit is probably slightly older than 1945 and is consistent with deposition associated with the 1938 Hurricane. Continuous sand layers are also evident at about 13 to 14 cm below the modern marsh surface at Barn Island, Connecticut, and Little Harbor, Massachusetts, and have been attributed to the 1938 Hurricane (Warren and Niering 1993; Donnelly et al. 2000) (figure 3.8c and a).

As expected, hurricanes in the twentieth century that produced significant storm surge (>3 m) across wide areas of the coast also left sand layers preserved in back-barrier marshes. The two hurricanes that resulted in the most storm surge in the twentieth century in southern New England, Hurricane Carol in 1954 and the 1938 Hurricane (figure 3.9), deposited overwash deposits from eastern Connecticut to Cape Cod, Massachusetts. Storms that result in lower storm surge tend not to deposit overwash fans, or they leave smaller, less extensive and localized deposits. For example, of the sites examined, only Little Harbor, Massachusetts, contains evidence of Hurricane Bob (figure 3.8a), and evidence of Hurricane Gloria has only been reported at Barn Island, Connecticut.

Hurricanes of the Nineteenth Century

The best-track HURDAT data set recently has been extended back to 1851. It provides a relatively complete list of tropical cyclones in the North Atlantic Basin (Landsea et al., chapter 7 in this volume). The record is incomplete before that, especially for less intense tropical storms and Category 1 hurricanes that received considerably less attention than the more intense storms. At least seven hurricanes made landfall in the Northeast in the nineteenth century (Ludlum 1963; Neumann et al. 1993; Landsea et al., chapter 7 in this volume) (figure 3.12). Four of these storms (1893, 1869b, 1858, and 1804) may have been of Category 2 intensity, and three (1869a, 1821, and 1815) may have been of Category 3 intensity (Boose, Chamberlin, and Foster 2001; Landsea et al.,

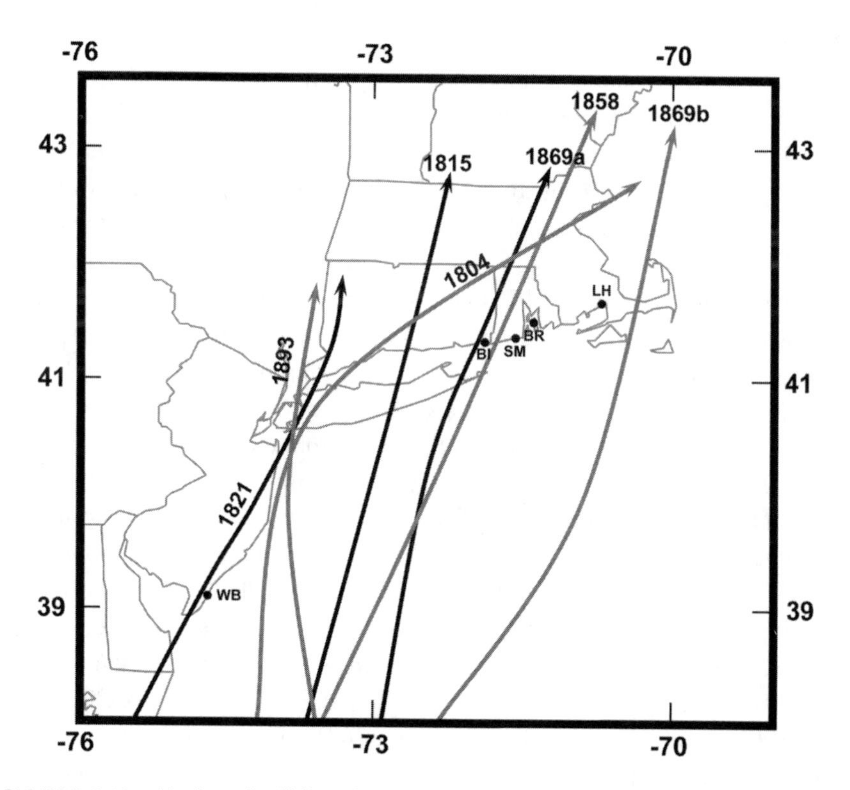

FIGURE 3.12 Hurricane landfalls in the northeastern United States in the nineteenth century: black, estimated Category 3; gray, estimated category 2; WB, Whale Beach, New Jersey; BI, Barn Island, Connecticut; SM, Succotash Marsh, Rhode Island; LH, Little Harbor Marsh, Massachusetts; BR, Bristol, Rhode Island.

chapter 7 in this volume). The latter three hurricanes caused significant loss of life and damage and are well documented in the historical record.

On September 8, 1869 (1869a in figure 3.12), a compact but intense hurricane struck southeastern New England (Ludlum 1963). The storm made landfall first at Montauk, Long Island, then again just to the west of Stonington, Connecticut. The short duration of hurricane force winds and the timing of landfall coincident with a relatively low tide combined to lessen the level of storm surge. A storm surge of approximately 2 meters above the normal high-tide level was noted at Bristol, Rhode Island, approximately 40 km northeast of Succotash Marsh (figure 3.12). Both Barn Island, Connecticut, and Little Harbor, Massachusetts, have a thin (~1 cm), but relatively extensive sand layer between 20 and 30 cm depth that was likely deposited during the 1869 Hurricane (Donnelly et al. 2000). No sedimentary record of the 1869 Hurricane was evident in Succotash Marsh in East Matunuck, Rhode Island (Donnelly et al. 2001a).

From 1815 to about 1900 an inlet was open adjacent to the Succotash Marsh study area. The open inlet allowed deposition of barrier and nearshore sand into the back-barrier marsh without the overtopping of the barrier beach. In sediment cores adjacent to the abandoned channels associated with the inlet, a complex unit consisting of fine-to-medium sand interbedded with silt and organic mud (core 1, 19 to 30 cm) dates to the time interval when the inlet was open (figure 3.13a and b). The diverse nature of the sediment suggests repeated deposition associated with numerous events of varying intensities and quiescent periods of fine-grained deposition during normal tidal inundation. The 1869 Hurricane likely is responsible for depositing some of these sediments, but it is impossible to discern this event from the many others recorded in the marsh sediments during this interval. In sediment cores taken from other areas of the Succotash Marsh away from the old inlet, no sand was deposited in the late nineteenth century, indicating that the storm surge from the 1869 Hurricane was not sufficient to overtop the barrier at this site.

On September 3, 1821, the most intense storm to strike New Jersey and western Long Island in the historical period made landfall. This hurricane tracked across the mid-Atlantic coastline and passed over New York City (Ludlum 1963; Boose, Chamberlin, and Foster 2001) (figure 3.12). Even though the hurricane struck at very low tide, reports of storm surge heights of approximately 3 meters came from southern New Jersey and the outer coast of the Delmarva Peninsula (Ludlum 1963). In New York City the storm raised water levels nearly 4 meters in one hour (Redfield 1831). Given that the elevation of the barrier islands on this coast are generally less than 3 meters above mean sea

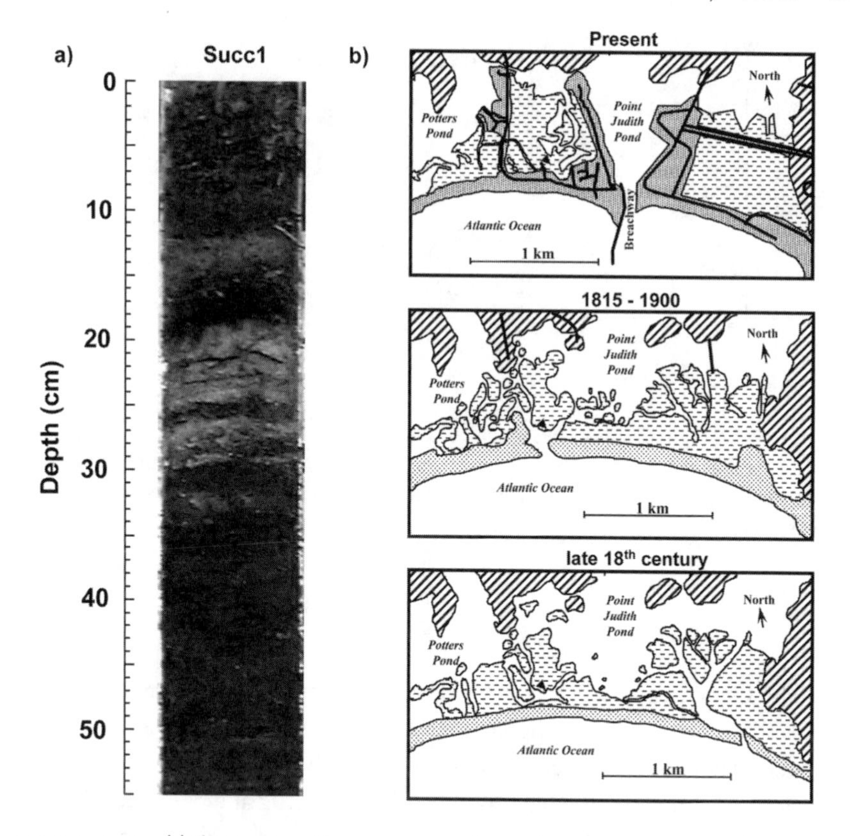

FIGURE 3.13 (*a*) Photographs of the upper 55 cm of core 1 from Succotash Marsh, Rhode Island. A complex unit of sand interbedded with mud and thin peat intervals between 19 and 30 cm are associated with a relic inlet. (*b*) Maps of historical inlet changes at Succotash Marsh. The inlet adjacent to the study site (denoted by a triangle) was opened during the 1815 Hurricane.

level, the 1821 Hurricane likely overtopped most of the barrier islands in New Jersey and western Long Island, New York, and deposited extensive overwash fans on many of the back-barrier marshes.

A series of sediment cores from the back-barrier marsh at Whale Beach, New Jersey, revealed two overwash fans deposited within the historical period (Donnelly et al. 2001b) (figure 3.14). The uppermost and smallest overwash fan was deposited during the 1962 Ash Wednesday Northeaster. Numerous calibrated radiocarbon-dated samples and pollen stratigraphic dates constrain the age of the overwash fan that was recovered at a depth of between 0.3 and 1 m to the late eighteenth or early nineteenth century. Donnelly et al. (2001b) attributed this fan to the 1821 Hurricane.

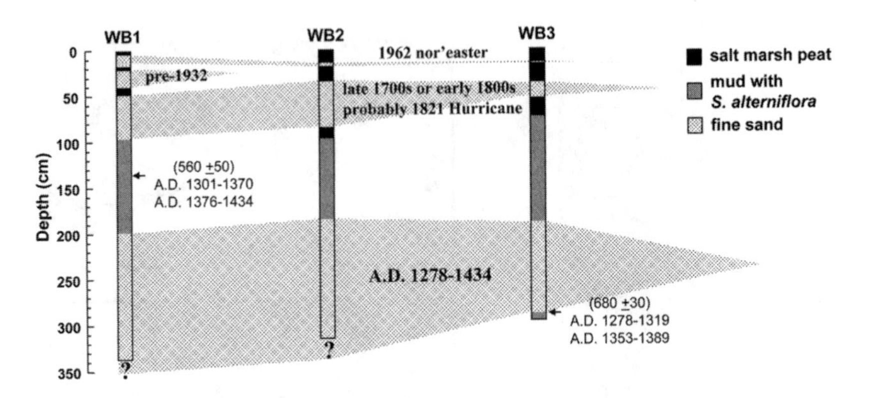

FIGURE 3.14 Stratigraphic cross-section of Whale Beach, New Jersey, based on a transect of three vibracores. Radiocarbon-dated samples are indicated with arrows. Samples have been calibrated using Calib 4.1 (Stuiver et al., 1998). Radiocarbon ages in years before present are noted in parentheses.

The Great September Gale of 1815 struck Long Island and southern New England on the morning of September 23, 1815 (figure 3.12). Historians have frequently equated the intensity of this storm with that of the 1938 Hurricane (Snow 1943; Ludlum, 1963; Minsinger 1988). Moving at close to 22 m s^{-1} (or 50 mph), it made landfall on Long Island near Center Moriches, less than 16 km to the east of the landfall location of the 1938 Hurricane and with a similar damage pattern (Boose, Chamberlin, and Foster 2001). The height of the storm surge at Providence, Rhode Island, was approximately 3.5 meters, nearly 70 cm below the level reached during the 1938 Hurricane. In Stonington, Connecticut, near the Rhode Island border, the storm surge reportedly exceeded 5 meters above the normal high-tide level (Snow 1943). As in 1938, the storm surge from the 1815 Hurricane overtopped many of the barrier beaches from Connecticut to Cape Cod (Ludlum 1963; Lee 1980; Perley 1891).

An extensive overwash fan is preserved at depths between 35 and 57 cm within the back-barrier-marsh sediments at Succotash Marsh, Rhode Island. Donnelly et al. (2001a) attributed the fan sediments to the 1815 Hurricane (figure 3.10). This sand layer is bounded above by organic sediments with high concentrations of lead and below by peat with background concentrations of lead (approximately 20 ppm) (figure 3.15). Pollen analysis of sediment samples from core 4 reveals an increase in the relative abundance of *Rumex* pollen that most likely dates to between 1700 and 1750, when Europeans had sufficiently cleared the southern New England landscape. *Rumex* pollen increases from less than 2 percent to over 4 percent within the peat interval directly below this sand layer in core 4 (figure 3.15). A radiocarbon date from the base of this sand layer in core 4 yielded calibrated ranges of 1645–1682, 1734–1806, and 1931–1947. The increase of *Rumex* pollen in this peat interval indicates deposition after about 1700. The 1645–1682 age range is, therefore, too old and can be eliminated. The 1931–1947 age range can be eliminated as this peat interval lacks evidence of lead pollution evident in the late nineteenth and twentieth centuries. The range of 1734 to 1806 is the only range compatible with the high *Rumex* values and the lack of lead pollution. This calibrated-age range combined with elevated lead levels within the sediments above the sand layer constrain the age of this fan to the late eighteenth or early nineteenth century. The age of this fan and our inference that complex deposits above this sand layer in some cores are associated with the opening of the inlet at the study site in the 1815 Hurricane, are consistent with this overwash fan also being associated with the hurricane of September 23, 1815.

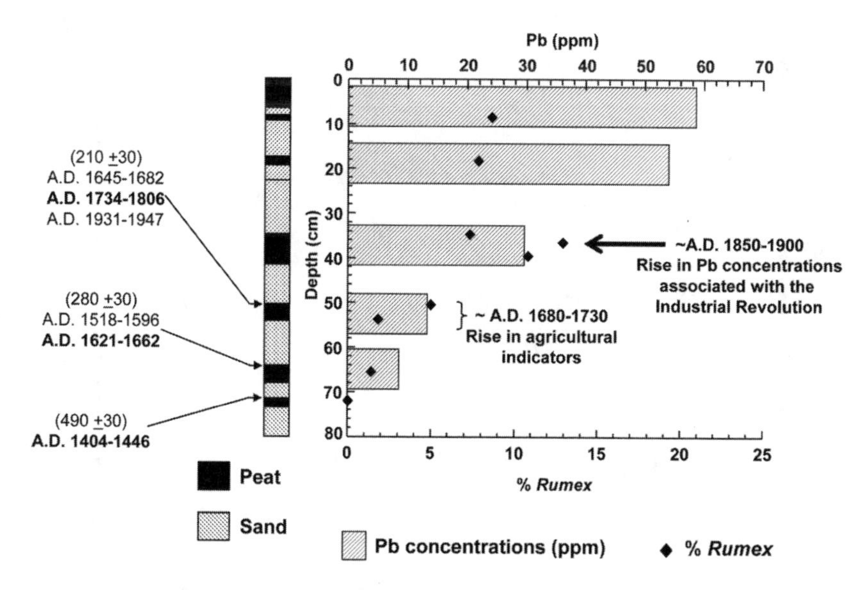

FIGURE 3.15 Percentage of *Rumex* pollen relative to total tree pollen, average lead (Pb) concentration per peat layer and radiocarbon samples from core 4 at Succotash Marsh, Rhode Island. The increase in *Rumex* to almost 5% at 52 cm indicates this interval dates to about A.D. 1700. The increase in lead concentration to 30 ppm between 33 and 42 cm suggests this sediment was deposited in the late eighteenth century. Calibrated radiocarbon age range in bold is the most likely range given the pollen and lead results. Radiocarbon ages in years before present are noted in parentheses.

Hurricanes of the Colonial Period: 1620 to Late Eighteenth Century

At least six hurricanes made landfall in the region between the time of earliest European settlement of the region (1620) and 1800 (figure 3.16). The most intense-hurricane strike during this interval is the Great Colonial Hurricane of August 25, 1635. Occurring 15 years after the settlement of Plymouth Plantation and 5 years after the establishment of the Massachusetts Bay Colony, this hurricane was well documented in the journals of Governors William Bradford of Plymouth Plantation and John Winthrop of Massachusetts Bay Colony (Ludlum 1963). John Winthrop wrote in his journal, "[t]he tide rose at Narragansett fourteen feet higher than ordinary, and drown [*sic*] eight Indians flying from their wigwams." Likewise, William Bradford recorded, "[i]t caused the sea

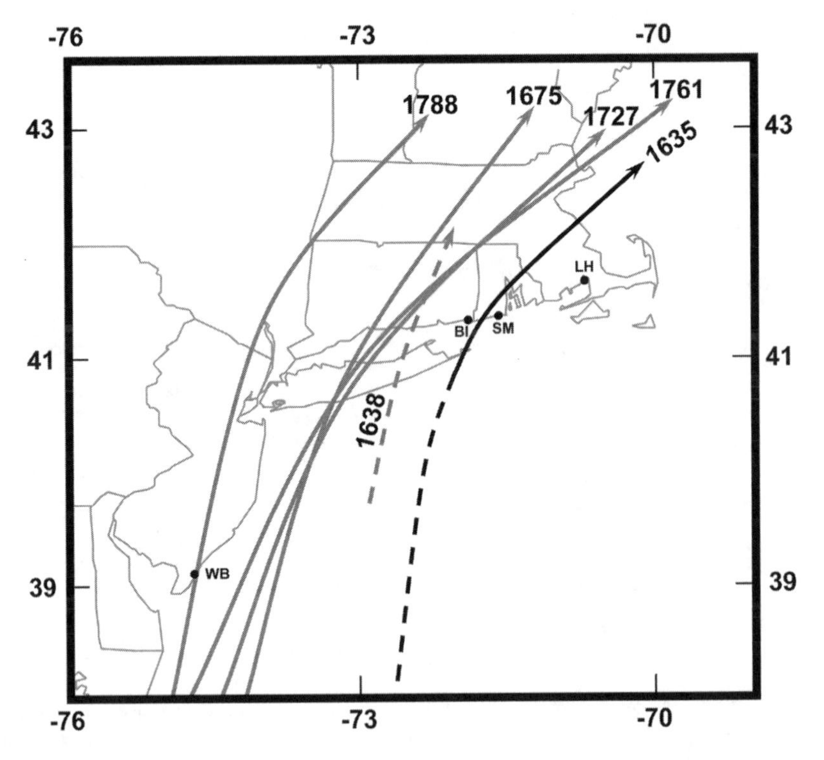

FIGURE 3.16 Hurricane landfalls in the northeastern United States in the colonial period. Based on historical accounts, only the 1635 Hurricane is estimated to have been Category 3 (black); the remainder were likely Category 1 or 2 storms. As a result of sparse information the track and intensity of the 1638 Hurricane are uncertain. WB, Whale Beach, New Jersey; BI, Barn Island, Connecticut; SM, Succotash Marsh, Rhode Island; LH, Little Harbor Marsh, Massachusetts.

to swell to the south wind of this place above 20 foot [*sic*] right up and down, and made many Indians to climb into trees for their safety." These accounts of significant storm surge and further accounts of extensive destruction of forests within the region indicate a storm of intensity similar to or greater than those of the hurricanes of 1815 and 1938 (Boose, Chamberlin, and Foster 2001).

John Winthrop also made note of a "very great tempest or hurricano" on August 13, 1638. It does not appear to have been nearly as intense as the storm three years earlier in Boston, but a storm surge of approximately 5 meters was noted at Narragansett, Rhode Island. This and a wind direction from the southwest at Boston, as noted by Winthrop (Ludlum 1963), suggests that this storm may have taken a track to the west of Rhode Island, similar to the 1938 Hurricane, resulting in extremely high tides in Narragansett Bay and less damage in eastern Massachusetts. Boose, Chamberlin, and Foster (2001) estimate the intensity of this storm at Category 2, but their estimate is based on only two reports. Given the sparseness of available historical data, areas of more extreme damage may have gone unreported. It is possible that this storm may have been more intense than a Category 2, as the report of tremendous storm surge in Narragansett may indicate.

An overwash fan preserved at Succotash Marsh, Rhode Island, was deposited in the seventeenth century (Donnelly et al. 2001a) (figure 3.10). The age of this fan is constrained by two radiocarbon dates and the increase in *Rumex* pollen abundance within the peat layer above the fan (figure 3.15). The first radiocarbon date produced the calibrated ranges A.D.1494–1500, 1515–1599, 1616–1671, 1779–1798, and 1945–1945 and the second in core 4 produced the calibrated ranges 1518–1596 and 1621–1662. The mutual overlap between these dates is 1518–1596 and 1621–1662. The elimination of the youngest two ranges from the first calibrated date are further supported by the lack of *Rumex* pollen within this peat layer, indicating this peat layer was deposited prior to 1700, the time that European-style agriculture became widespread. Although the 1518–1596 range cannot be eliminated, the facts that *Rumex* pollen increases within the peat interval above this fan and the intercepts with the calibration curve occur at 1642 and 1645 (intercepts have the highest probability) point toward the most likely age of this peat layer being between 1621 and 1662. The 1621–1662 age range is consistent with this fan being deposited by the Great Colonial Hurricane of 1635, or possibly the hurricane of August 13, 1638.

A radiocarbon date (300 ± 50) from below an overwash fan at Wells, Maine, calibrates to a similar age range (1470–1672) with an intercept of 1637 (Kelley, Gehrels, and Belknap 1995). Given the coastal track of the 1635 Hurricane and reports of hurricane force-northeast winds to the north of Boston (Snow 1943), the overwash fan preserved at Wells, Maine, may also be the result of the 1635 Hurricane.

The Prehistoric Period

In addition to overwash fans attributed to historic hurricanes, several back-barrier marshes in the region also contain prehistoric overwash fans. The prehistoric overwash fans are similar in character and extent to the overwash fans attributed to historic intense hurricanes. At Succotash Marsh in southern Rhode Island two prehistoric overwash fans were recovered (Donnelly et al. 2001a) (figure 3.10). The youngest prehistoric overwash fan at this site occurs between 52 and 100 cm in 13 of 14 cores. The mutual overlap of five calibrated radiocarbon dates from samples of marsh vegetation preserved immediately below this fan yields a best-estimate range of A.D. 1411–1446 for the age of the peat directly below this sand layer. This age combined with the seventeenth-century age of the peat deposit above the fan, constrains the age of this storm-induced deposit to the fifteenth or sixteenth century A.D. The reproducibility of the calibrated radiocarbon age of the peat deposit directly below this fan indicates that little erosion of the marsh surface took place as this fan was deposited. Therefore, the age of the radiocarbon samples likely represents the age of the marsh surface at the time of fan deposition and suggests that this overwash fan was most likely deposited by an intense hurricane between 1411 and 1446.

Clark and Patterson (1984) may have found additional evidence for this storm. A calibrated radiocarbon date from marsh sediments below a sand layer recovered from Long Cove, Fire Island, New York, yields a calibrated age range of 1403–1443 (at 1 σ) with an intercept at 1426. If the sand layer at Long Cove was deposited by the same hurricane that deposited the 1411–1446 overwash fan at Succotash Marsh, then this storm likely had a track through central Long Island similar to the hurricanes of September 22, 1815, and September 22, 1938.

A second prehistoric overwash fan recovered at Succotash Marsh occurs in 9 of 14 cores at a depth of between 69 and 107 cm. Two radiocarbon dates from samples of peat from the base of this fan yielded calibrated-age ranges of A.D. 1295–1436 and 1285–1407, respectively. The mutual overlap of the 2σ ranges from these dates produces a best-estimate age range of 1295–1407 for the age of peat deposited directly below this fan. Given that the age of this fan should be older than the peat layer above it, which dates to 1411–1446, the most likely age of this fan is between 1295 and 1446.

A prehistoric overwash fan was recovered at Whale Beach, New Jersey (Donnelly et al. 2001b) (figure 3.14). The age of this fan has been determined by two calibrated radiocarbon-dated samples, one below and one above the overwash fan. The 2σ calibrated radiocarbon age ranges of the sample from above the fan is A.D. 1301–1370 and 1376–1434 indicating that this fan was deposited prior to 1434. The calibrated radiocarbon age ranges of 1278–1319 and 1353–1389 for the sample immediately below the fan indicates that this overwash fan must have been deposited after 1278. In combination these two radiocarbon dates constrain the age of this deposit to between 1278 and 1434. Assuming a beachfront migration rate of about 2.5 meters per year derived from the last 150 years (figure 3.4) is applicable over the last 700 years, the beachfront when this oldest-overwash fan recovered at Whale Beach was deposited may have been more than a kilometer farther seaward than it was in 1821. As the sand layer deposited between 1278 and 1434 is thicker than the fan deposited in 1821 and was probably transported a greater distance, the 1278–1434 overwash fan was likely deposited during an intense-hurricane strike, potentially more powerful than the 1821 Hurricane.

Landfall Probabilities and Vulnerability

The combined historical and sedimentary evidence from Succotash Marsh indicates that at least six intense hurricanes made landfall in southern New England in the last 700 years, which yields an annual probability for intense hurricane landfall of about 0.9%. This estimate is somewhat higher than the estimate of 0.15%-0.81% for category 3 winds (>50 m s^{-1}) at Succotash Marsh derived from a wind-speed probability model (Murnane et al. 2000) constrained by the 1900–1997 HURDAT data set (Neumann et al. 1993). The annual probability of an intense-hurricane strike on the coast of New Jersey, based on the coupling of the 700-year sediment record from Whale Beach and historical records, is approximately 0.3%. This probability is generally consistent with the estimate (0.15–0.45%) derived from the Murnane et al. wind-speed probability model.

The lower probability of an intense-hurricane strike occurring in New Jersey relative to the probability of one occurring in Rhode Island reflects the tendency of hurricanes to track to the north and northeast under the influence of prevailing westerly winds in the western North Atlantic. Given this tendency for hurricane track recurvature and the geometry of the East Coast of the United States (figure 3.5), New Jersey is somewhat sheltered from the more common hurricane paths. Conversely, New England protrudes out into the western Atlantic and is much more often in the path of hurricanes that are slow to recurve to the northeast.

Sediment records from the Gulf Coast of the United States (Liu and Fearn 1993, 2000; Liu, chapter 2 in this volume) indicate that the annual probability for Category 4 and 5 hurricane strikes (winds >59 m s^{-1}) on the Gulf Coast has been $<0.1\%$ in the past 1400 years. Therefore the northeastern United States may be between three and nine times more likely to be struck by a Category 3 or stronger hurricane than the Gulf Coast of the United States is likely to be struck by a Category 4 or 5 hurricane. This difference may reflect, at least in part, the relative rarity of Category 4 and 5 hurricanes compared to Category 3 hurricanes. Approximately 30% of landfalling hurricanes in the United States during the twentieth century were of Category 3 intensity, while only about 11% were of Category 4 or 5 intensity (Neumann et al. 1993).

The probabilities of a Category 3 hurricane striking many parts of the gulf and the northeastern U.S. coast based on the landfalls documented in the HURDAT dataset are similar. In terms of risk to lives and property, the threat from intense hurricane strikes is significantly greater in the northeastern United States. According to the 1990 U.S. census, over 35 million people were living in coastal counties in the northeastern United States (in Maryland, Delaware, New Jersey, New York, Connecticut, Rhode Island, and Massachusetts). Conversely, the 1990 population of coastal counties along a similar length of coast in the Gulf of Mexico (Texas, Louisiana, Mississippi, and Alabama) was just over 10 million people. Similarly, insured coastal property in these northeastern states totaled about $1.6 trillion, while insured coastal property in the Gulf states totaled only $300 billion (IIPLR/IRC 1995).

For example, the coast-parallel track of the 1821 Hurricane resulted in the center of the storm passing over several northeastern states (Maryland, Delaware, New Jersey, and New York). The coasts in all of these states were developed extensively during the twentieth century. Coastal populations in the region today are more than 31 million and insured coastal property exceeds $1 trillion (IIPLR/IRC 1995). The financial loss associated with a hurricane similar to the 1821 storm striking this region today would likely far exceed that of Hurricane Andrew in 1992, the nation's most costly insured natural disaster to date. Many lives may also be lost because of difficulties evacuating this densely populated region in advance of a fast-moving intense hurricane.

Promise and Limitations

Recent studies demonstrate that reconstructing the overwash history of back-barrier environments (like salt marshes) can provide convincing evidence of past intense storms. Succotash Marsh in Rhode Island contains six distinct overwash fans (Donnelly et al. 2001a). Four of these deposits date to the historical period and the ages are consistent with known hurricane strikes in 1635, 1815, 1938, and 1954. Two additional overwash fans at Succotash Marsh date to 1411–1446 and 1295–1446, and were probably deposited during prehistoric hurricane strikes. At Whale Beach, New Jersey, a large-scale overwash deposit is attributed to an intense-hurricane strike in 1821 (Donnelly et al. 2001b). An earlier land-falling hurricane probably deposited an older, larger-scale overwash fan at Whale Beach dated to between A.D.1278 and 1434.

These results indicate that mapping and dating of overwash fans can provide a record of historic and prehistoric storms. The interpretation of the stratigraphic record requires care, however, given the many uncertainties inherent in back-barrier sediment records. The sensitivity of a site to overwash deposition changes in response to variations in barrier position and height. After a barrier has been overwashed it may take years or decades before the barrier regains its former height. Storms that produce relatively minor storm surge may be able to overtop a barrier recently flattened by a large storm surge. In addition to these changes in sensitivity driven by the storms themselves, changes in barrier geometry associated with changes in the rate and direction of sea level fluctuations and sediment supply can alter a particular site's sensitivity to overwash.

The stage of the astronomical tide when a storm strikes is one important uncertainty associated with attempts to infer the intensity of a prehistoric storm from the sediment record. A storm that strikes at low tide may not result in sufficient storm surge to overtop or breach a barrier beach. But if that same storm were to strike at high tide, the combination of the astronomical tide and the storm surge may be sufficient to overtop or breach a barrier. As a result the stratigraphic record of a Category 2 storm that strikes at a high tide may be very similar to that of a Category 3 storm that strikes at low tide.

The wave energy that accompanies the storm surge is another important factor that may influence, not only whether a storm results in overwash deposition, but the extent of overwash deposition. Although two storms may have similar levels of storm surge, they may result in appreciably different amounts of overwash deposition if the energy delivered to the coast via waves is significantly different. This kind of difference in wave energies may be the reason that the overwash fan at Succotash Marsh, Rhode Island, attributed to Hurricane Carol in 1954 is much less extensive than the one attributed to the 1938 Hurricane, although the maximum water level attained during both of these storms was similar. In southern Rhode Island the wind speeds associated with Hurricane Carol (a Category 2 storm) were significantly less than wind speeds associated with the 1938 Hurricane. As a result the wave energy at the coast was likely less during Hurricane Carol than in 1938 and resulted in less sediment being transported into the back-barrier environment.

The uncertainties associated with dating individual overwash fans constitute another limitation of using this method to develop records of prehistoric hurricane strikes. Under the best conditions the uncertainties in the dating of individual fans deposited prior to European settlement is on the order of decades. As a result it may be difficult to state conclusively that two prehistoric overwash deposits from two adjacent sites that may date to roughly the same 30- or 50-year interval were deposited by the same storm. It is possible that the overwash fans could have been deposited by storms that struck within a few decades of one another, for example, the 1938 and 1954 events. In addition, overwash fans from two storms that occur within a few years of one another (e.g., the 1635 and 1638 events) may not be distinguishable within the sediment record if the sediment sources were similar and not enough time has passed for marsh deposition to occur. In this case, the second overwash fan would likely be composed of reworked sediment from the first overwash fan and other additional sediment of similar character. Therefore, multiple intense storms within a brief period may be difficult to resolve using these techniques.

Erosion of the stratigraphic record in back-barrier environments can result in an incomplete record of overwash deposition. Most commonly, the migration of inlets results in this type of erosion and loss of the stratigraphic record. The presence of an inlet also changes the susceptibility of the back-barrier environment to storm-induced deposition. The activation of the inlet at Succotash Marsh, Rhode Island, in the nineteenth century increased the sensitivity of some sections of the study site to overwash deposition. Although overwash deposits that date to before and after the activation of the inlet can be attributed to historic intense hurricane landfalls, it is much more difficult to interpret the many sand layers deposited while the inlet was open.

Other types of storms and tsunamis may also deposit overwash deposits. A danger exists that overwash deposits associated with severe winter storms may also be contained in the sedimentary record. But historically severe winter storms in the northeastern United States have resulted in significantly lower storm surge and less overwash deposition than intense hurricane strikes. Careful selection of sites that are less susceptible to winter storms can minimize the likelihood of inclusion of winter storms in the sedimentary record. Tsunamis can also result in overwash deposition (Armes et al. 1995; Delaney and Devoy 1995) and the eastern seaboard of the United States may have experienced tsunamis associated with submarine slope failures (Driscoll et al. 2000). The relative rarity of tsunamis in the northeastern United States compared to intense hurricane strikes, however, makes it unlikely for a tsunami deposit to be misinterpreted as a hurricane deposit.

In spite of these limitations the stratigraphic record preserved in back-barrier environments provides the best potential for developing long-term records of intense-hurricane activity. A comprehensive historical record of storms, local geomorphology, and regional landscape changes is necessary to provide a framework for understanding and interpreting the sedimentary record. Multiple study sites can be used to minimize the influences of local changes in beach geomorphology on the interpretation of the sedimentary record of intense storms. Overwash deposits at multiple sites within a region, dated to the same time interval, provide the most compelling evidence of past intense-hurricane strikes.

Acknowledgments

Funding for this research was provided by the Risk Prediction Initiative, Bermuda Biological Station for Research. In addition the Postdoctoral Scholar Program at the Woods Hole Oceanographic Institution, with funding from the United States Geological Survey, provided support for J. Donnelly. We thank Bryan Shuman, Paige Newby, Phil Leduc, Jack Williams, Neal Haussman, Jen Dowling, Linda Fan, Sarah Bryant, Erin Bryant, Marj Bryant, Jeff Bryant, Karlyn Westover, Jen Stern, Jess Butler, Chris Hepner, Stuart Roll, Micah Wengren, Peter Cleary, Bob Ettinger, Rick Lederer, Nat Logar, Justine Owen, Phil Furtado, and Elyse Scileppi for their assistance with data collection and analysis. In addition we would like to thank Kam-biu Liu, Rick Murnane, and two anonymous reviewers for useful comments and suggestions that improved this manuscript. This article is Contribution No. 10881 of the Woods Hole Oceanographic Institution.

References

Allen, E. S. 1976. *A wind to shake the world: The story of the 1938 hurricane.* Boston: Little, Brown.

Anderson, T. W. 1974. The chestnut pollen decline as a time horizon in lake sediments in eastern North America. *Canadian Journal of Earth Sciences* 11:678–85.

Armes, C. J., M. L. Cates, C. W. Wobus, B. M. Allen, and P. D. Lea. 1995. Comparison of Holocene tsunami and modern storm-overwash deposits, northern Bristol Bay, southwestern Alaska. *Abstracts and Programs,* 27. Geological Society of America, Northeastern Section, 30th annual meeting, Hartford, Conn.

Boose, E. R., K. E. Chamberlin, and D. R. Foster. 2001. Landscape and regional impacts of historical hurricanes in New England. *Ecological Monographs* 71:27–48.

Boothroyd, J. C., N. E. Friedrich, and S. R. McGinn. 1985. Geology of microtidal coastal lagoons: Rhode Island. *Marine Geology* 63:35–76.

Bricker-Urso, S., S. W. Nixon, J. K. Cochran, D. J. Hirschberg, and C. Hunt. 1989. Accretion rates and sediment accumulation in Rhode Island salt marshes. *Estuaries* 12:300–317.

Brooks, C. F. 1939. Hurricane into New England. *Geographical Review* 29:119–27.

Brugam, R. B. 1978. Pollen indicators of land-use change in southern Connecticut. *Quaternary Research* 9:349–62.

Clark, J. S., and W. A. Patterson III. 1984. Pollen, Pb-210 and opaque spherules: An integrated approach to dating and sedimentation in the intertidal environment. *Journal of Sedimentary Petrology* 54:1249–63.

Clark, J. S., and W. A. Patterson III. 1985. The development of a tidal marsh: Upland and oceanic influences. *Ecological Monographs* 55:189–217.

Delaney, C., and R. J. N. Devoy. 1995. Evidence from sites in western Ireland of late Holocene changes in coastal environments. *Marine Geology* 124:273–87.

Delaune, R. D., W. H. Patrick, Jr., and R. J. Buresh. 1978. Sedimentation rates determined by ^{137}Cs dating in a rapidly accreting salt marsh. *Nature* 275:532–33.

Dickson, R. R. 1978. Weather and circulation of February 1978: Record or near-record cold east of the continental divide with a major blizzard in the Northeast. *Monthly Weather Review* 106:746–51.

Dillon, W. P. 1970. Submergence effects on a Rhode Island barrier and lagoon and inferences on migration of barriers. *Journal of Geology* 78:94–106.

Donnelly, J. P. 1998. Evidence of late Holocene post-glacial isostatic adjustment in coastal wetland deposits of eastern North America. *Georesearch Forum* 3–4:393–400.

Donnelly, J. P., R. Ettinger, and P. Cleary. 2000. Sedimentary evidence of prehistoric hurricane strikes in southern New England. *Abstracts and Programs.* Geological Society of America annual meeting, Reno, Nev.

Donnelly, J. P., S. S. Bryant, J. Butler, J. Dowling, L. Fan, N. Hausmann, P. N. Newby, B. Shuman, J. Stern, K. Westover, and T. Webb III. 2001a. A 700-year sedimentary record of intense hurricane landfalls in southern New England. *Bulletin of the Geological Society of America* 113:714–27.

Donnelly, J. P., S. Roll, M. Wengren, J. Butler, R. Lederer, and T. Webb III. 2001b. Sedimentary evidence of intense-hurricane strikes from New Jersey. *Geology* 29:615–18.

Driscoll, N. W., J. K. Weissel, and J. A. Goff. 2000. Potential for large-scale submarine slope failure and tsunami generation along the U.S. Mid-Atlantic coast. *Geology* 28:407–10.

Emery, K. O., and D. G. Aubrey. 1991. *Sea levels, land levels, and tide gauges*. New York: Springer-Verlag.

Fairbanks, R. G. 1989. A 17,000-year glacio-eustatic sea level record: Influence of glacial melting rates on the Younger Dryas event and deep-ocean circulation. *Nature* 342:637–42.

Faure, G. 1986. *Principles of isotope geology*. Second edition. New York: Wiley.

Fitzgerald, D. M., S. van Heteren, and T. M. Montello. 1994. Shoreline processes and damage resulting from the Halloween Eve Storm of 1991 along the north and south shores of Massachusetts Bay, U.S.A. *Journal of Coastal Research* 10:113–32.

Henderson-Sellers, A., H. Zhang, G. Berz, K. Emanuel, W. Gray, C. Landsea, G. J. Holland, J. Lighthill, S. Shieh, P. Webster, and K. McGuffle. 1998. Tropical cyclones and climate change: A post-IPCC assessment. *Bulletin of the American Meteorological Society* 79:19–38.

Hennessy, J. T., and G. A. Zarillo. 1987. The interrelation and distinction between flood-tidal delta and washover deposits in a transgressive barrier island. *Marine Geology* 78:35–56.

Insurance Institute for Property Loss Reduction/Insurance Resource Council (IIPLR/IRC). 1995. *Coastal exposure and community protection: Hurricane Andrew's legacy*. Boston: Insurance Institute for Property Loss Reduction/Insurance Resource Council.

Kelley, J. T., W. R. Gehrels, and D. F. Belknap. 1995. The geological development of tidal marshes at Wells, Maine, U.S.A. *Journal of Coastal Research* 11:136–53.

Klein, G. deV. 1986. Intertidal flats and intertidal sand bodies. In *Coastal sedimentary environments*, edited by R. A. Davis, 187–224. New York: Springer-Verlag.

Lanesky, D. E., B. W. Logan, R. G. Brown, and A. C. Hine. 1979. A new approach to portable vibracoring underwater and on land. *Journal of Sedimentary Petrology* 49:654–57.

Leatherman, S. P. 1983. Barrier dynamics and landward migration with Holocene sea-level rise. *Nature* 301:415–17.

Leatherman, S. P., and R. E. Zaremba. 1986. Dynamics of a northern barrier beach: Nauset Spit, Cape Cod, Massachusetts. *Bulletin of the Geological Society of America* 97:116–24.

Lee, V. 1980. *An elusive compromise: Rhode Island coastal ponds and their people*. Marine Technical Report 73. Narragansett: Coastal Resource Center, University of Rhode Island.

Lighty, R. G., I. G. MacIntyre, and R. Stuckenrath. 1982. *Acropora palmata* reef framework: A reliable indicator of sea level in the western Atlantic for the past 10,000 years. *Coral Reefs* 1:125–30.

Liu, K.-b., and M. L. Fearn. 1993. Lake-sediment record of late Holocene hurricane activities from coastal Alabama. *Geology* 21:793–96.

Liu, K.-b., and M. L. Fearn. 2000. Reconstruction of prehistoric landfall frequencies of catastrophic hurricanes in northwestern Florida from lake sediment records. *Quaternary Research* 54:238–45.

Ludlum, D. M. 1963. *Early American hurricanes, 1492–1870*. Boston: American Meteorological Society.

McCaffrey, R. J., and J. Thomson. 1980. A record of the accumulation of sediment and trace metals in a Connecticut salt marsh. In *Advances in geophysics*. Vol. 22, *Estuarine physics and chemistry: Studies in Long Island Sound*, edited by B. Saltzman, 165–235. New York: Academic Press.

McIntire, W. G., and Morgan, J. P., 1964, *Recent geomorphic history of Plum Island, Massachusetts and adjacent coasts*. Baton Rouge: Louisiana State University Press.

Minsinger, W. E. 1988. *The 1938 hurricane, an historical and pictorial summary*. Randolph Center, Vt.: Greenhills Books.

Murnane, R. J., C. Barten, E. Collins, J. Donnelly, J. Elsner, K. Emanuel, I. Ginis, S. Howard, C. Landsea, K.-b. Liu, D. Malmquist, M. McKay, A. Michaels, N. Nelson, J. O'Brien, D. Scott, and T. Webb III. 2000. Model estimates hurricane wind speed probabilities. *Eos, Transactions of the American Geophysical Union* 81:433, 438.

National Oceanic and Atmospheric Administration (NOAA)/National Ocean Service (NOS)/Center for Operational Oceanographic Products and Services (CO-OPS). 2001. Verified Historical Water Level Data (available at: http://www.opsd.nos.noaa.gov/data_res.html).

Neumann, C. J., B. R. Jarvinen, C. J. McAdie, and J. D. Elms. 1993. *Tropical cyclones of the North Atlantic Ocean, 1871–1992*. NOAA Historical Climatology Series 6-2. Asheville, N.C.: National Climatic Data Center.

Nichols, R. L., and A. F. Marston. 1939. Shoreline changes in Rhode Island produced by hurricane of September 21, 1938. *Bulletin of the Geological Society of America* 50:1357–70.

Orson, R. A., R. S. Warren, and W. A. Niering. 1998. Interpreting sea level rise and rates of vertical marsh accretion in a southern New England tidal salt marsh. *Estuarine, Coastal and Shelf Science* 47:419–29.

Paulsen, C. G. 1940. *Hurricane floods of September 1938*. U.S. Geological Survey Water-Supply Paper 867. Reston, Va.: U.S. Geological Survey.

Perley, S. 1891. *Historic storms of New England: Salem, Massachusetts*. Salem, Mass.: Salem Press.

Pielke, R. A., Jr., and C. W. Landsea. 1998. Normalized hurricane damages in the United States: 1925–95. *Bulletin of the American Meteorological Society* 13:621–31.

Pierce, J. W. 1970. Tidal inlets and washover fans. *Journal of Geology* 78:230–34.

Poey, A. 1862. *Table chronologique de quatre cents cyclones*. Paris: Dupont.

Rampino, M. R., and J. E. Sanders. 1980. Holocene transgression in south-central Long Island, New York. *Journal of Sedimentary Petrology*. 50:1063–79.

Redfield, A. C. 1967. Postglacial change in sea level in the western North Atlantic Ocean. *Science* 157:687–92.

Redfield, A. C., and A. R. Miller. 1957. Water levels accompanying Atlantic Coast hurricanes. *Meteorological Monographs* 2:1–22.

Redfield, A., C., and M. Rubin. 1962. The age of salt marsh peat and its relations to recent change in sea level at Barnstable, Massachusetts. *Proceedings of the National Academy of Science* 48:1728–35.

Redfield, W. C. 1831. Remarks on the prevailing storms of the Atlantic coast of the North American states. *American Journal of Science* 20:17–51.

Russell, E. W. B., R. B. Davis, R. S. Anderson, T. E. Rhodes and D. S. Anderson. 1993. Recent centuries of vegetational change in the glaciated northeastern United States. *Journal of Ecology* 81:647–64.

Schwartz, M. L. 1973. *Barrier islands.* Benchmark Papers in Geology. Stroudsburg, Pa.: Hutchinson and Ross.

Schwartz, R. K. 1975. *Nature and genesis of some washover deposits.* Technical Memorandum No. 61. Fort Belvoir, Va.: Army Coastal Engineering Research Center.

Snow, E. R. 1943. *Great storms and famous shipwrecks of the New England coast.* Boston: Yankee.

Stahl, L., J. Koczan, and D. Swift. 1974. Anatomy of a shoreface-connected sand ridge on the New Jersey shelf: Implications for the genesis of the surficial sand sheet. *Geology* 2:117–20.

Stuiver, M., and J. J. Daddario. 1963. Submergence of the New Jersey coast. *Science* 142:451.

Stuiver, M., P. J. Reimer, E. Bard, J. W. Beck, G. S. Burr, K. A. Hughen, B. Komar, F. G. McCormac, J. v. d. Plicht, M. and Spurk. 1998. INTCAL98 Radiocarbon age calibration 24,000–0 cal BP. *Radiocarbon* 40:1041–83.

Tannehill, I. R. 1927. Some inundations attending tropical cyclones. *Monthly Weather Review* 55:453–56.

Tannehill, I. R. 1940. *Hurricanes, their nature and history, particularly those of the West Indies and the southern coasts of the United States.* Princeton: Princeton University Press.

Warren, R. S., and W. A. Niering. 1993. Vegetation change on a northeast tidal marsh: Interaction of sea-level rise and marsh accretion. *Ecology* 74:96–103.

Wilby, F. B., G. R. Young, C. H. Cunningham, A. C. Lieber Jr., R. K. Hale, T. Saville, and M. P. O'Brien. 1939. Inspection of beaches in path of the hurricane of September 21, 1938. *Shore and Beach* 7:43–47.

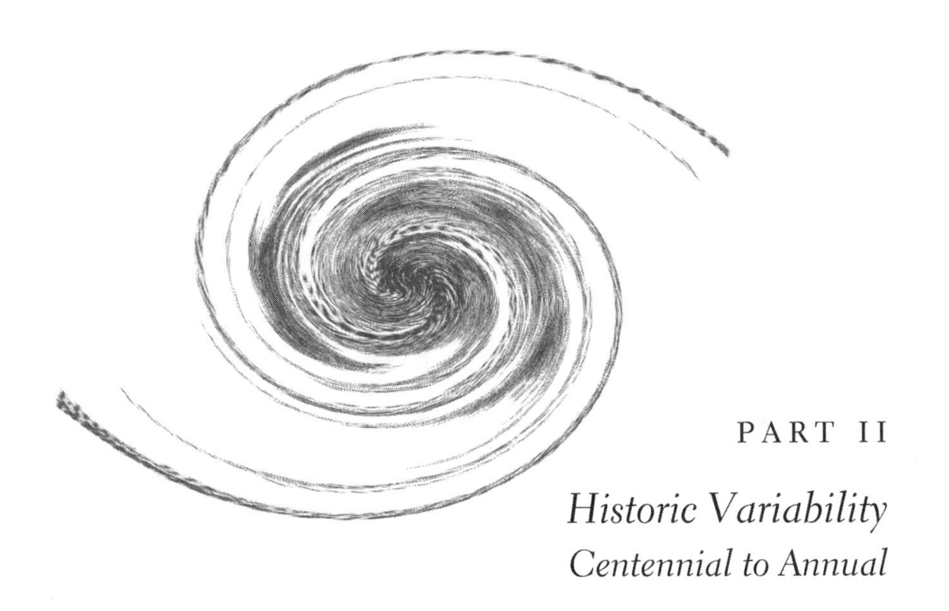

PART II

Historic Variability
Centennial to Annual

4
A Method for Reconstructing Historical Hurricanes

Emery R. Boose

In many coastal areas of the world the historical record provides evidence of landfalling hurricanes over a period of centuries. Using a combination of wind damage assessment and meteorological modeling, this record can be interpreted to reconstruct spatial and temporal patterns of hurricane impacts over the historical period. The resulting information can enhance understanding of natural ecosystems, improve assessment of hurricane risks, and help calibrate techniques for detecting prehistoric hurricanes. This historical-modeling method is discussed with illustrations from earlier studies in New England and Puerto Rico.

Understanding the impacts of hurricanes on natural and human systems requires a long-term perspective and an estimate of the spatial variation in individual storms. For the North Atlantic Basin, the HURDAT (Hurricane Data) database maintained by the U.S. National Hurricane Center currently provides track and maximum wind-speed data for all known hurricanes since 1851 (Jarvinen, Neumann, and Davis 1984; Landsea et al. 2002; Landsea et al., chapter 7 in this volume). These data must, however, be combined with modeling techniques to reveal the spatial gradient of impacts created by a landfalling hurricane. And the period covered (150 years) is not sufficient to study long-term trends or the frequency of the most intense hurricanes. For that purpose much can be learned from the historical record, which, for this basin, covers a period of 300 to 500 years since European settlement (e.g., Mock, chapter 5; García Herrera et al., chapter 6, and Landsea et al., chapter 7 in this volume).

In earlier studies of the ecological impacts of hurricanes on the forests of New England and Puerto Rico (Boose, Foster, and Fluet 1994; Boose, Chamberlin, and Foster 2001; Boose, Serrano, and Foster 2004), we developed an approach that combines historical research and computer modeling. The resulting historical-modeling method involves six steps:

1. Identify all historical hurricanes with damaging winds in the selected region.
2. Collect wind damage reports and meteorological observations for each storm.
3. Analyze and compile the damage reports into regional maps of actual damage.
4. Parameterize and test a simple meteorological model (HURRECON) with data from selected recent hurricanes.
5. Use the parameterized model to reconstruct each storm.
6. Compile the results of individual storm reconstructions.

The long-term effects of topography on a landscape scale can then be examined with a simple topographic exposure model (EXPOS).

Details of this method are discussed here, with illustrations from our earlier studies of hurricane impacts at two Long-Term Ecological Research (LTER) sites, Harvard Forest in central Massachusetts and the Luquillo Experimental Forest in northeastern Puerto Rico, and their respective regions, New England and Puerto Rico.[1]

Historical Data

Most of the time and effort required by this approach are devoted to historical research. The first tasks are to create a list of hurricanes for which there is evidence of wind damage in the study region and to locate historical documents for each storm. In our studies we relied on the works of other scholars to identify significant hurricanes in the early period. Our assessment of the impacts of each storm was based, wherever possible, on contemporary accounts. As expected, the number of historical reports was greater for recent or severe hurricanes. Efforts focused on obtaining good regional coverage for each storm.

For New England, newspapers proved to be the best source of information for hurricanes since 1700 (when newspapers first appeared), especially the *Boston Globe* and the *New York Times* for the period since 1871, and various local newspapers (depending on the area of impact) for the period 1700–1870. Personal diaries and town histories (especially those found at the American Antiquarian Society, Worcester, Massachusetts) provided contemporary evidence for seventeenth-century storms. Sixty-seven hurricanes during the period from 1620 to 1997 were selected for detailed study, including all hurricanes that approached within 200 km of New England since 1851 according to HURDAT, and all earlier hurricanes for which Ludlum (1963) presented evidence of F1+ damage on the Fujita scale in New England.

For Puerto Rico, contemporary Puerto Rican newspapers proved to be the best source of information for hurricanes since 1876. For earlier storms we relied mainly on secondary studies (especially Salivia 1950 and Millás 1968), supplemented wherever possible by contemporary documents (especially from the Archivo de Indias at the University of Puerto Rico in Rio Piedras and the General Archives of Puerto Rico in San Juan). Eighty-five hurricanes during the period from 1508 to 1997 were selected for detailed study, including all hurricanes for which we found historical evidence of F0+ damage on the island.

Wind Damage Assessment

Actual wind damage in each hurricane is then classified using Fujita's (1971, 1987) system for assessing wind damage in tornadoes and hurricanes. This system has been used by the U.S. National Weather Service for tornadoes since the early 1970s (Grazulis 1993). Fujita's damage classes extend from F0 (minor damage caused by gale or storm force winds) to F5 (extreme damage in the most severe tornadoes). Each F-scale (Fujita scale) class is defined by specified levels of damage to common cultural and biological features of the landscape. The system was designed for rapid application in the field and does not require detailed engineering analysis. Note that this measure of local wind damage is not equivalent to overall hurricane intensity (as measured, for example, by the Saffir-Simpson scale). For example, intense storms that remain offshore may cause lower levels of wind damage over land.

In our studies minor modifications of Fujita's system were required for application to historical materials from New England and Puerto Rico (table 4.1). These changes were based in part on the work of Grazulis (1993) and in part on historical evidence of comparable damage in the hurricanes studied (e.g., town halls in New England tended to suffer damage comparable to churches and barns, in the same location and the same storm). Some modifications were required by the nature of the historical materials. For example, when one or more trees were blown down the damage was classified as F1, even though Fujita regarded the pushing over of shallow-rooted trees as F0. That is because, in most cases, it was impossible to determine the condition of the tree from the historical reports. Similarly, when part or all of a roof was blown off a

TABLE 4.1 The Fujita Scale of Wind Damage Modified for Application to Historical Materials from New England and Puerto Rico

	F0 Damage	F1 Damage	F2 Damage	F3 Damage
Sustained wind speed (m/s)[1]	18–25	26–35	36–47	48–62
Trees	Leaves and fruit blown off, branches broken, trees damaged	Trees blown down	Extensive blowdowns	Most trees down
Crops	Damaged or blown down			
Masonry buildings	Minor damage	Roof peeled, windows broken, chimneys down	Unroofed	Blown down or destroyed
Wood houses[2]	Minor damage	Roof peeled, windows broken, chimneys down	Unroofed or destroyed	3+ blown down or destroyed in same town
Unspecified buildings, wood-zinc houses[3]	Minor damage	Unroofed or damaged	Blown down or destroyed	50% or more blown down or destroyed in same town[4]
Barns, churches, town halls, cottages[5]	Minor damage	Unroofed, steeple blown down, damaged	Blown down or destroyed	
Shacks, sheds, outbuildings, warehouses	Minor damage	Unroofed, blown down or destroyed		
Huts[6]	Damaged	Blown down or destroyed		
Furniture, bedding, clothes	Not moved	Blown out of building		
Masonry walls, radio towers, traffic lights	No damage	Blown down		
Utility poles	Wires down	Poles damaged or blown down, high-tension wires down		

	F0 Damage	F1 Damage	F2 Damage	F3 Damage
Signs, fences	Damaged	Blown down		
Autos	No damage	Moving autos pushed off road	Stationary autos moved or pushed over	Heavy autos lifted and thrown
Trains	No damage	Pushed along tracks	Boxcars pushed over	Trains overturned
Marinas, small airplanes	Minor damage	Destroyed		
Small boats	Blown off mooring	Sunk		
Missiles	None	None	Light objects, metal roofs	

[1] Corresponding sustained wind speed values are derived from Fujita's equations (1971), assuming a wind gust factor of 1.5 over land.
[2] Described as well-constructed or owned by a wealthy person (Puerto Rico [PR]). Also municipal buildings (PR).
[3] Constructed with light wood frame and metal roof (PR).
[4] F2 assigned if buildings described as rural or poor (PR).
[5] Includes schools, sugar mills, commercial buildings, and military buildings (PR).
[6] Constructed of palm leaves or similar materials (PR).
Sources: Adapted from Boose, Chamberlin, and Foster 2001; Boose, Serrano, and Foster 2004.

wood frame house, the damage was classified as F2, even though Fujita required that the entire roof be removed. Again, this is because it was usually impossible to determine exactly how much of the roof was gone. Damage to wood frame houses was assigned to F3 only if at least three houses in the same town were completely blown down, increasing the likelihood that at least one of the three houses was well-built and in good condition. For Puerto Rico, additional changes were required to account for different building practices and higher wind speeds. For example, unless described as well-constructed or owned by a wealthy person, houses were assumed to be constructed with a light wood frame and zinc-plated metal roof (or thatch before the late nineteenth century) and to sustain damage comparable to barns in Fujita's system. When at least half of the wood frame-metal roofed houses were completely blown down in a town, the damage was assigned an F3 category, unless the town was described as rural or poor.

Using the range of wind speeds proposed by Fujita (1971) for each damage class, the Fujita scale provides a means to quantify the level of wind damage caused by historical hurricanes and a link to the meteorological modeling described later in this chapter (table 4.1). Fujita's values were found to work well for the lower damage classes (F0 to F3) treated in our study, although some engineers have suggested that Fujita's wind speeds may be too high, especially for F3 to F5 damage (e.g., Twisdale 1978; Liu 1993). Much work is still needed to understand the forces generated on buildings in hurricane winds (Powell, Houston, and Reinhold 1994).

For each historical report that references a specific location, the following information is collected, summarized, and entered into a database: (1) hurricane, (2) location, (3) bibliographic source, (4) meteorological observations, (5) storm surge, (6) wind damage, (7) fresh-water flooding, (8) notes, and (9) Fujita scale. We found that nearly all reports referenced one or more individual towns. The town also proved to be a suitable geographic unit for mapping and modeling hurricane impacts. Each report that contains sufficient information is assigned an F-scale value based on the highest level of damage reported (table 4.2). Care must be taken to exclude coastal damage caused by the storm surge, valley damage caused by river flooding, or local damage caused by landslides. In our studies we collected a total of 2,710 reports for New England and 2,699 reports for Puerto Rico. In general, we found that the level of wind damage was consistent within reports and among reports for the same town; for example, if house roofs were blown off (F2), barns were also blown down (F2).

Maps of actual wind damage are then created for each hurricane, using the maximum F-scale value assigned for each town (plate 1). These maps provide a quantitative, spatial assessment of actual damage for each storm. In general, we found the regional patterns to be consistent with meteorological expectations; that is, damage was usually greater to the right of the storm track where wind velocities are normally higher (in the Northern Hemisphere), and the intensity of damage usually lessened along the storm track as the hurricane made landfall. In New England, the most intense hurricanes (Category 3 on the Saffir-Simpson scale) caused widespread F2 damage across southern and coastal areas, with scattered F3 damage. In Puerto Rico the most intense hurricanes (Category 5) caused widespread F3 damage across much of the island.

Temporal Variation

Information on actual wind damage can be used to study variation in hurricane impacts over the entire study region on a range of temporal scales, from seasonal to annual to decadal to centennial. For example, the multi-decadal vari-

TABLE 4.2 Historical Examples of F1, F2, and F3 Wind Damage in Puerto Rico

Hurricane Betsy, 1956	
Location	San Juan
Source	*El Mundo*, August 13, 1956, pp. 1, 25
Wind	"Along Ponce de Leon Ave. there were many downed trees and fallen power lines. Construction support structures, scaffolding, and zinc plates had fallen and blocked the streets. Showcases were shattered in the New York Department store. Chimneys fell at a bakery in Rio Piedras."
Fujita rating	F1
Hurricane Hugo, 1989	
Location	Luquillo
Source	*El Mundo*, September 20, 1989, p. 16
Wind	"In the barrios of Junquito, Pasto Viejo, Manbiche, and Anto Ruiz, where most of the houses were constructed of wood and zinc, everything blew away. The sports complex and a gymnasium lost their roofs and filled with water. In the barrio Canovanillas, several residences were knocked down."
Fujita rating	F2
Hurricane San Ciriaco, 1899	
Location	Yabucoa
Source	*La Corresponencia*, August 12, 1899, p. 3
Wind	"The town remains in ruins. Only the church resisted the cyclone, and its walls sheltered many people. One of the houses that fell down was of masonry, and it killed the family and other people who sought shelter within its walls. In the countryside, nothing remained standing. Several haciendas were completely destroyed."
Fujita rating	F3

Sources: Translated from the Spanish by M. Serrano.

ation that is well documented for North Atlantic hurricanes since 1871 was evident in both study regions over the entire historical period (e.g., Neumann, Jarvinen, and Pike 1987; Gray, Sheaffer, and Landsea 1997) (figure 4.1). On a centennial scale, the number of F3 hurricanes in both regions was fairly constant over the historical period, while the number of F1 and F2 hurricanes increased steadily until the nineteenth century. These trends were probably the

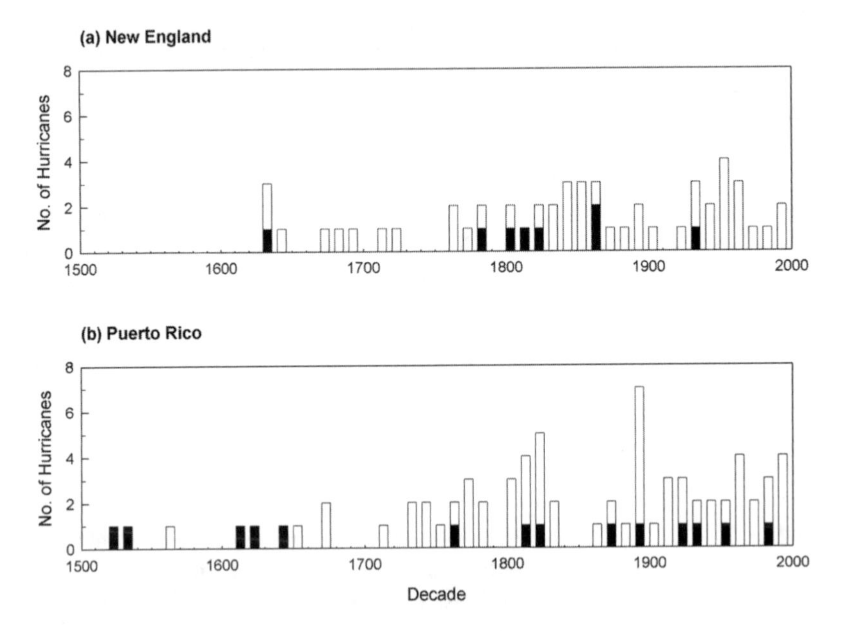

FIGURE 4.1 Number of hurricanes by decade with maximum reported damage equal to F1 to F2 (white) or F3 (black): (*a*) New England, 1620–1997; (*b*) Puerto Rico, 1508–1997 (adapted from Boose, Chamberlin, and Foster 2001; Boose, Serrano, and Foster 2004).

result of improvements in meteorological observations and records, and the natural tendency to retain records of the most damaging storms.

HURRECON Model

Individual hurricanes are then reconstructed with a simple meteorological model—HURRECON—based on published empirical studies of many hurricanes (Boose, Foster, and Fluet 1994; Boose, Chamberlin, and Foster 1997; Boose, Chamberlin, and Foster 2001). HURRECON uses information on the track, size, and intensity of a hurricane, as well as the cover type (land or water), to estimate surface wind speed and direction (figures 4.2 and 4.3). The model also estimates wind damage on the Fujita scale by using the correlation between maximum quarter-mile wind velocity (i.e., maximum wind velocity sustained over a quarter-mile distance) and wind damage proposed by Fujita (1971). The model complements actual wind damage assessment by providing informed estimates for sites that lack data as well as a complete regional picture of the impacts of each storm.

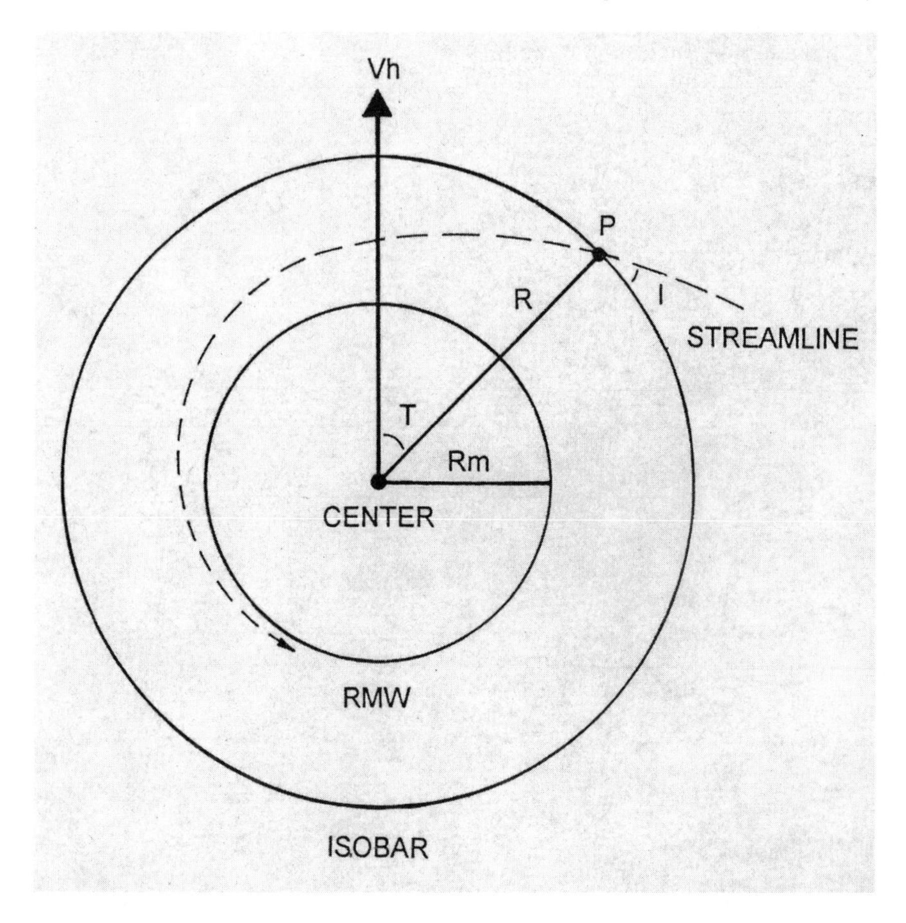

FIGURE 4.2 HURRECON model: wind direction. At the surface, air is drawn into the hurricane along spiral streamlines that cross the nearly circular isobars at inflow angle *I*. The estimated wind direction at point *P* is a function of *I* and the relative positions of *P* and the storm center (adapted from Boose, Chamberlin, and Foster 2001).

In the HURRECON equations provided here wind velocity and direction are measured relative to the Earth's surface, and angles are measured in degrees. Parameter values used for New England and Puerto Rico are given in parentheses. The sustained wind velocity (V_s) at any point P in the Northern Hemisphere is estimated as:

$$V_s = F\,[V_m - S(1 - \text{Sin}\,T)V_h\,/\,2]\,[(R_m\,/R)^B \exp(1 - (R_m\,/R)^B)]^{1/2} \qquad (1)$$

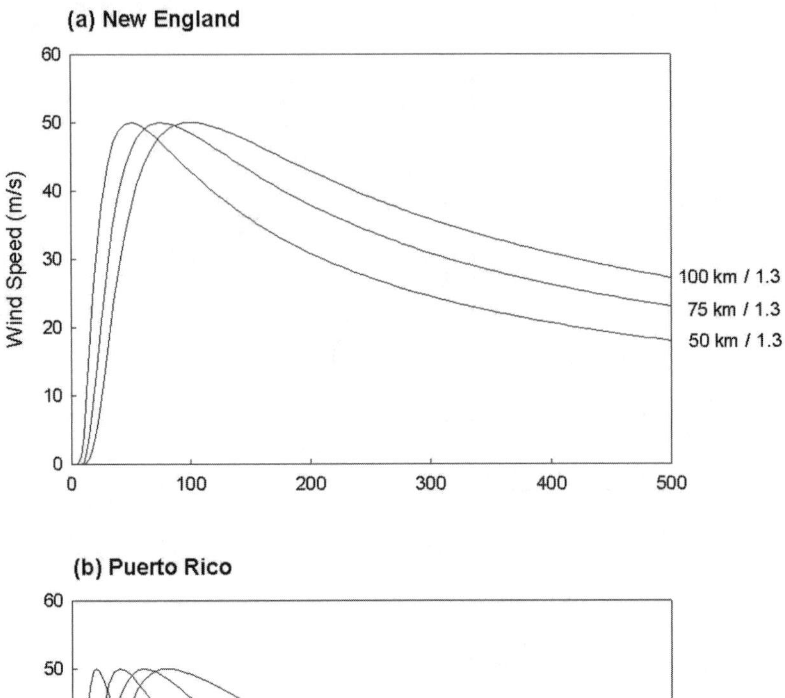

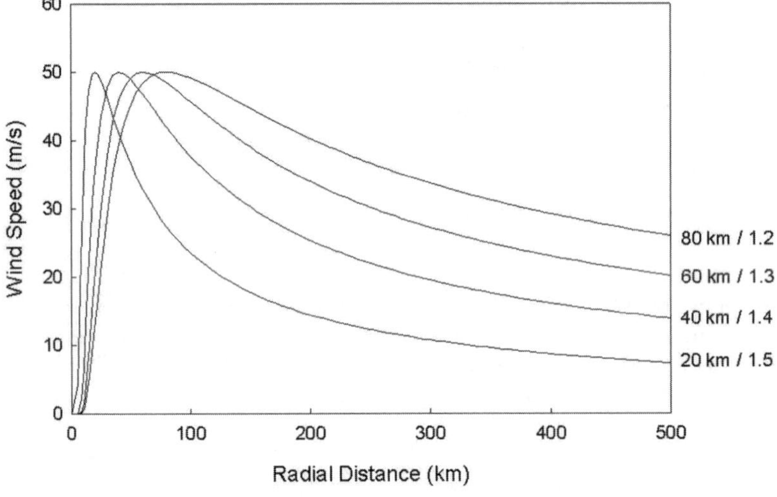

FIGURE 4.3 HURRECON model: wind speed. The estimated wind speed along a radial line outward from the hurricane center is a function of the radius of maximum winds (R_m), the wind speed at that radius (V_{rm}), and the scaling parameter B that controls the shape of the curve. Different combinations of R_m and B were selected to represent typical wind profiles for (*a*) New England and (*b*) Puerto Rico, shown here for an arbitrary value of V_{rm} = 50 m/s (adapted from Boose, Chamberlin, and Foster 2001; Boose, Serrano, and Foster 2004).

where F = scaling parameter for effects of friction (water = 1.0, land = 0.8), V_m = maximum sustained wind velocity over water anywhere in the hurricane, S = scaling parameter for asymmetry due to forward motion of storm (1.0), T = clockwise angle between forward path of hurricane and a radial line from hurricane center to point P, V_h = forward velocity of hurricane, R_m = radius of maximum winds, R = radial distance from hurricane center to point P, and B = scaling parameter controlling shape of wind profile curve. This equation was adapted from Holland's equation for the cyclostrophic wind (Holland 1980, equ. 5).

The peak wind gust velocity (V_g) at point P is estimated from V_s as follows:

$$V_g = G\,V_s \tag{2}$$

where G = gust factor (water = 1.2, land = 1.5). The maximum 1/4-mile wind velocity (V_f) is estimated from V_s and G using Fujita's method (Fujita 1971, equ. 12).

Wind direction (D) at point P is estimated as:

$$D = A_z - 90 - I \tag{3}$$

where A_z = azimuth from point P to hurricane center and I = cross isobar inflow angle (water = 20°, land = 40°). In the Southern Hemisphere, where the wind circulation is clockwise around the center, T = counterclockwise angle between forward path of hurricane and a radial line from hurricane center to point P, and $D = A_z + 90 + I$.

The HURRECON model was parameterized and tested for our two regions as follows:

1. Parameters were assigned from the literature and adjusted as necessary in detailed studies of six to seven major hurricanes over the last 100 years. For each storm, model estimates were compared to actual wind damage observations. The goal was to find parameters or a range of parameters that worked well for all storms.
2. The model thus parameterized was tested by comparing actual and reconstructed damage for the remaining hurricanes since 1851, where damage data were independent of the (input) meteorological data.
3. For New England, the model was then applied to hurricanes before 1851, where damage data were used to help determine the (input) storm track and/or maximum wind speed.

Parameter values for F, G, and I were adopted directly from published sources (Dunn and Miller 1964; Fujita 1971; Simpson and Riehl 1981; Powell 1982, 1987); F and G were chosen so that peak gust speeds are the same over water and land. The value $S = 2.0$ reported in the literature (i.e., peak wind speed on right side minus peak wind speed on left side = $2V_h$) was found to consistently underestimate wind speed and damage on the left side of the storm; better results were obtained with $S = 1.0$. The width of the modeled storm (for a given value of V_m) is controlled by the parameters R_m and B. Because direct measurements of the radius of maximum winds (R_m) were unavailable for all but the most recent hurricanes (H. Willoughby, personal communication), and to test model sensitivity to these critical parameters, each storm was modeled for a range of values of R_m and B selected to represent typical wind profiles for each region (figure 4.3). The combination of R_m and B that produced the best agreement between actual and reconstructed regional damage was selected for the final results.

HURRECON provides estimates for individual sites (as tables) and for entire regions (as GIS maps in Idrisi format; Eastman 1997). In our studies model runs for individual sites were made using a time step of 5 minutes; output variables included peak wind speed and direction and maximum F-scale damage for each storm. The cover type for individual sites was assumed to be land. Regional estimates were made at 10-km (New England) or 3-km (Puerto Rico) resolution using a time step equal to the minimum time required for each hurricane to traverse one grid cell in the regional study window. Output maps included peak wind speed and maximum F-scale damage across the region for each storm.

Model reconstructions are tested by comparing actual and reconstructed F-scale wind damage on a regional scale (plate 2). Such comparisons are quantified by creating and analyzing a difference map (reconstructed damage minus actual damage) for each storm (plate 2c). The difference maps provide a measure of the overall accuracy of each reconstruction as well as the spatial pattern of agreement (e.g., reconstructed values might be too high or too low on one side of the track, or along the fringes of the storm).

Reconstructed wind speeds can also be tested against observed wind speeds at surface stations where such data are available (Boose, Foster, and Fluet 1994). Though desirable, we found such tests to be impractical for all but the most recent hurricanes. Accurate comparisons require careful correction of the observed wind speed for various factors including height of the anemometer, surface roughness over the approaching wind trajectory, and duration of measurement (Powell, Houseton, and Reinhold 1994). Such information was difficult or impossible to obtain in many cases. In addition, peak wind speeds were often missed in all but the most recent storms because observations were only made at fixed, infrequent intervals.

Meteorological Reconstructions

The range and quantity of meteorological data in the New World have, of course, increased dramatically since European settlement as a result of a more widely distributed population, better historical records, and steady improvements in technology (Ludlum 1963; Neumann, Jarvinen, and Pike 1987). HURDAT provides an invaluable source of meteorological data for hurricanes since 1851, though the database has known problems (including both systematic and random errors) and is currently under revision by NOAA (Neumann and McAdie 1997; Landsea et al. 2002; Landsea et al., chapter 7 in this volume). At the present time, meteorological data for earlier hurricanes must be estimated from historical records. Though actual measurements of wind speed are not available, early observers often left careful records of wind speed (in qualitative terms) and direction (eight points of the compass), noting the times of peak wind, wind shift, lulls, and changes in cloud cover and precipitation intensity.

In our studies we relied on HURDAT for track and maximum wind speed data for hurricanes since 1851. In most cases there was good agreement between observed and reconstructed F-scale damage. But in a few cases there were significant discrepancies, which were interpreted as stemming from problems with the HURDAT data and resolved by making conservative adjustments to the maximum sustained wind speed (V_m) values in HURDAT (Boose, Chamberlin, and Foster; 2001; Boose, Serrano, and Foster 2004). In addition, maps of actual wind damage in Puerto Rico showed evidence of storm weakening (at least at surface levels) in nearly all cases where hurricanes passed directly over the interior mountains. Such weakening was simulated for landfalling hurricanes by reducing V_m by 1.5 m/s (3 knots) for each hour that the storm remained over land, a rate consistent with empirical observations (Anthes 1982) and recent empirical models (Kaplan and DeMaria 1995).

For hurricanes before 1851 in New England, we used the meteorological observations and wind damage data we collected along with analyses by Ludlum (1963) to reconstruct track and maximum wind speed data for each storm. For Puerto Rico, such reconstructions were not attempted because the spatial coverage of the data collected was too small to create reliable estimates of hurricane tracks. Such estimates, which would require extensive analysis of historical reports from surrounding islands and ships at sea, may become available in the future (C. Landsea, personal communication) but were beyond the scope of our project.

After the individual model runs are completed, maps of reconstructed F-scale damage for each hurricane are compiled to generate maps showing the number of storms at a given minimum intensity (F0, F1, F2, or F3). Each frequency map is divided into several regions by hand, and an average return time is calculated for each region based on the average number of storms and the observation period. For New England, we used three observation periods:

1. F2 maps: the entire historical period (1620–1997).
2. F1 maps: an intermediate period characterized by improvements in meteorological records and newspaper coverage (1800–1997).
3. F0 maps: the modern period beginning with the establishment of the U.S. Signal Corps storm warning system (1871–1997).

This approach was designed to maximize the observation period while minimizing the likelihood that storms of a given intensity escaped historical notice. For Puerto Rico all analyses were based on the entire period covered by HURDAT (1851–1997). In both regions the frequency of F0 events was no doubt underestimated, because F0 damage could have resulted from storms not included in our study; for example, hurricanes that passed farther out to sea or tropical storms that did not attain hurricane strength.

In nearly all cases we found good agreement between reconstructed and actual F-scale damage by town. For New England, reconstructed F-scale damage equaled actual damage in 62% of the cases and was within one damage class in 99% of the cases, with a slight tendency to underestimate actual damage. For Puerto Rico, reconstructed F-scale damage equaled actual damage in 52% of the cases and was within one damage class in 92% of the cases, with a slight tendency to overestimate actual damage.

Spatial Variation

One advantage to the modeling approach is that it provides complete spatial coverage for the study region. As a result it can be used to study spatial patterns of hurricane impacts, as well as temporal patterns at sites that lack complete historical records. For example, in our two study areas we found significant

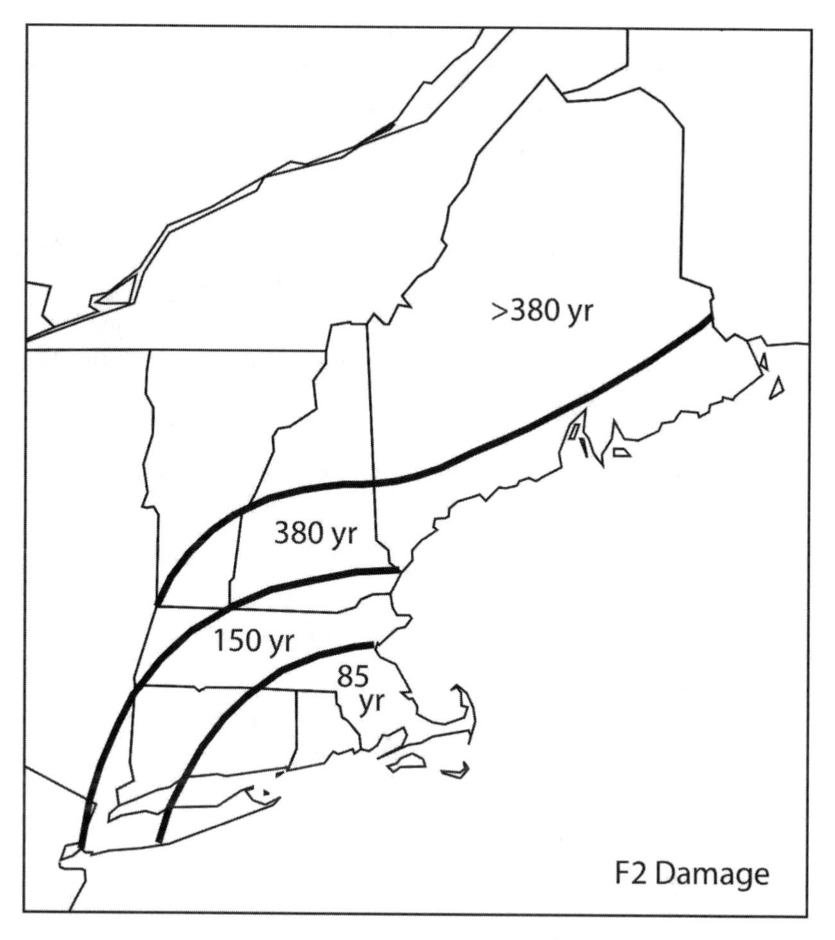

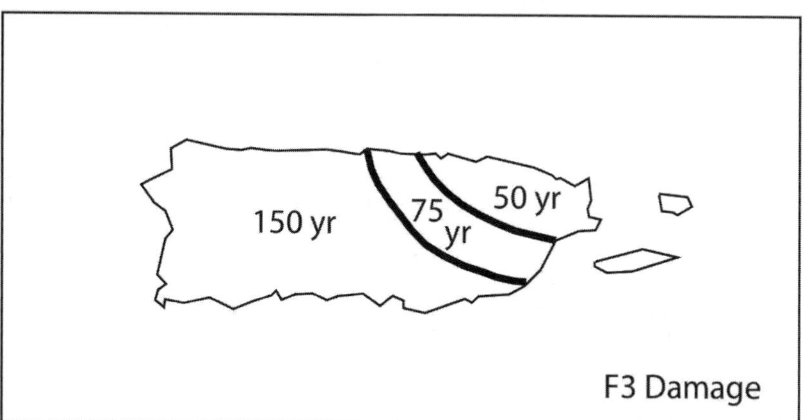

FIGURE 4.4 Regional gradients in reconstructed hurricane damage showing mean return intervals: (*top*) New England, F2 damage, 1620–1997; (*bottom*) Puerto Rico, F3 damage, 1851–1997 (adapted from Boose, Chamberlin, and Foster 2001; Boose, Serrano, and Foster 2004).

regional gradients in hurricane damage that resulted from the consistent direction of the storm tracks and the tendency for hurricanes to weaken as they moved over land or cold ocean water (New England) or over interior mountain ranges (Puerto Rico) (figure 4.4). At the site level there were also significant differences in hurricane frequency and intensity at the two LTER sites (figure 4.5). Topographic effects at a landscape level were studied using a simple exposure model and the reconstructed peak wind direction for each hurricane (Boose, Foster, and Fluet 1994) (figure 4.6). Results showed striking differences in reconstructed impacts on the north and south slopes of the Luquillo Mountains, creating a landscape-level gradient within the larger regional gradient (plate 3). Though none of these results were surprising, the historical-modeling method enabled us to quantify the regional gradients and evaluate site and landscape-level impacts at a level of detail and accuracy not previously possible.

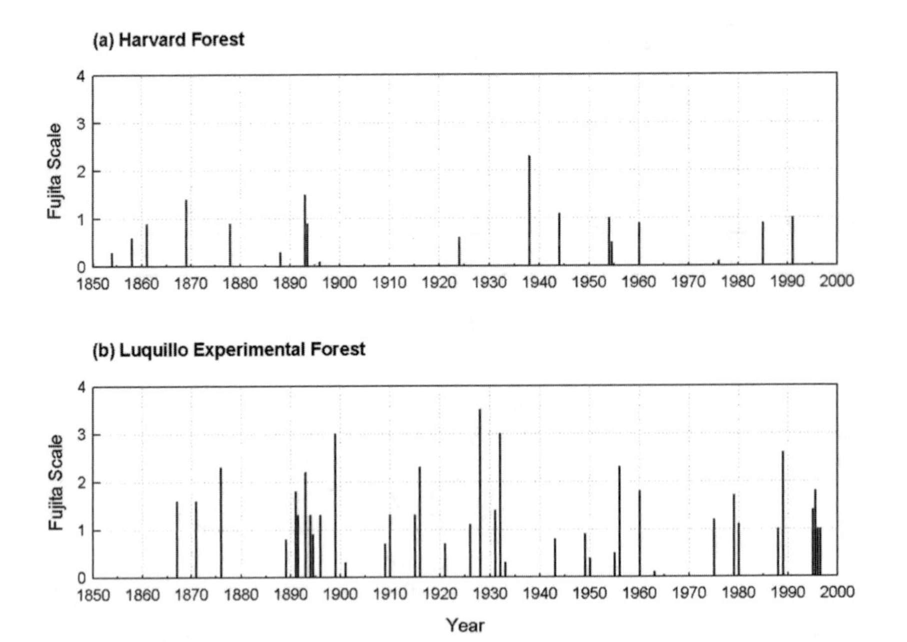

FIGURE 4.5 Reconstructed hurricane damage by year for 1851–1997: (*a*) Harvard Forest, central Massachusetts; (*b*) Luquillo Experimental Forest, northeastern Puerto Rico (adapted from Boose, Chamberlin, and Foster 2001; Boose, Serrano, and Foster 2004).

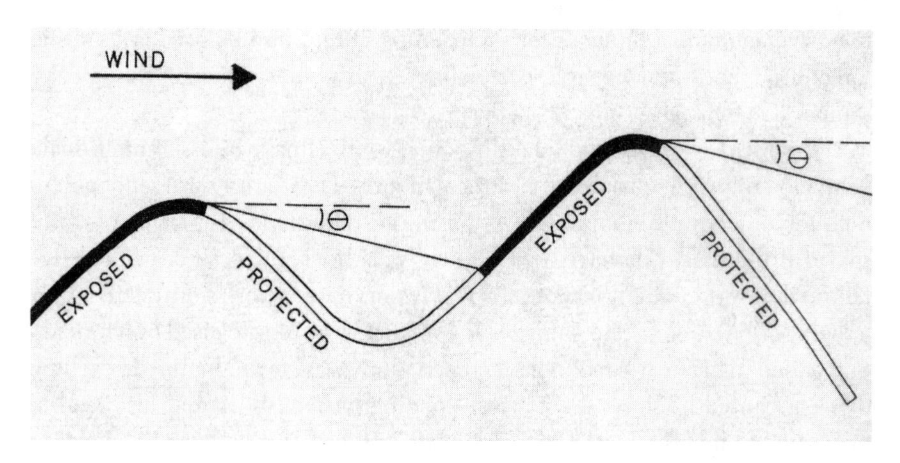

FIGURE 4.6 EXPOS model. For a given wind direction, each point in the study area is classified as protected or exposed, depending on whether or not the point falls within the wind shadow cast by points upwind (adapted from Boose et al. 1994).

Historical-Modeling Method

The approach described here relies on historical documents to assess and reconstruct the impacts of past hurricanes. The most difficult problem in interpreting results is estimating the completeness of the early records. It is quite possible, for example, that all records were lost for a New England storm in the first few decades after settlement, especially if the storm's impacts were not severe or were confined to sparsely populated areas. In both regions a number of early storms that reportedly did cause extensive damage were not included in our analyses because specific examples of damage were not given or the cause (wind, flooding, landslide) was not specified. Valuable information was no doubt overlooked in our studies, because historical searches are necessarily limited by time and resources. For Puerto Rico, especially, there are vast resources in the Archivo General de Indias in Seville that may someday shed more light on hurricanes during the colonial period (Marx 1983; García Herrera et al., chapter 6 in this volume).

One way to compensate for such incompleteness is to use more recent time periods for assessing weaker events, as we did in our analysis of New England hurricanes. Careful study may suggest that the historical record is complete for the most intense storms, at least in certain areas; for example, we concluded that few if any hurricanes that caused F2 damage in New England or F3 damage in San Juan escaped historical notice. This information was used to estimate average return intervals for such storms. It may also be used to examine long-term trends and possible correlations with climate change, though the sample size for such storms is small.

Fujita's system is used to quantify actual wind damage by assigning F-scale values to entire towns based on historical reports. There are a number of potential problems with this approach. For example, damage levels may be overestimated if the object damaged (e.g., a tree or a house) was weak or defective before the storm (which is more likely if two or more severe storms strike over a short period), or if severe damage on a smaller scale (e.g., caused by a tornado embedded in the hurricane) is generalized to an entire town. Failure to exclude damage caused by storm surge, river flooding, or landslides may also lead to overestimation of the level of wind damage in the affected areas. On the other hand, damage levels may be underestimated if suitable objects are not present in the area surveyed (e.g., if only barns and outbuildings are present, then the highest possible level of damage to buildings is F2), if examples of higher damage are not observed and reported (especially in sparsely populated areas), or if news reports are suppressed (as sometimes happened in Puerto Rico). F-scale values may also be higher for larger towns and cities than in the surrounding countryside because there are more observers, more property subject to potential damage, and better records. Systematic errors in damage assessment may occur because of differences in construction practices over time, or from place to place. The susceptibility of a particular building to wind damage is a complex function of building design and construction quality, as well as state of repair, wind direction, topographic position, surrounding wind breaks, and whether or not doors and windows are open, closed, or shuttered (Liu 1993). Such information is generally unavailable from historical sources. Finally, random errors may result from inaccuracies in the historical accounts. These problems arise mainly from the need to rely on written records and photographs for damage assessment.

The basic technique was found to work well for the purposes of our studies, probably because the Fujita damage classes are so broad. The overall tendency for the model to underestimate actual wind damage in New England may have resulted from the inclusion of all tree blowdowns as F1 and even partial roof removals as F2; while the overall tendency to overestimate actual wind damage in Puerto Rico may have resulted from the under-reporting of actual damage, especially from smaller towns.

Meteorological modeling is used to reconstruct a more complete picture of hurricane impacts than can be obtained from wind damage assessment alone. Here, too, there are a number of potential problems. For example, the HUR-

RECON model is based on an idealized wind profile that works best for intense hurricanes and less well for hurricanes that are weak or becoming extratropical storms. The model is not able to reconstruct multiple wind maxima or other mesoscale features (Willoughby 1995). Model estimates of wind damage are also based on peak quarter-mile wind speed following Fujita's method (1971), which assumes that the period of sustained wind required to produce specific damage is inversely proportional to wind speed. This approach yields wind durations appropriate for tree and building damage (e.g., 12 seconds for minimal hurricane force winds), but does not take into account fatigue and stress damage that may occur on a scale of minutes or hours. Nor does it account for damage associated with a shift in wind direction (Powell, Houston, and Ares 1995). Uncertainties in input data (hurricane track, size and intensity) are much greater for early hurricanes, especially in sparsely populated areas. Input data accuracy for New England, for example, was estimated to have increased by an order of magnitude over the historical period (Boose, Chamberlain, and Foster 2001).

Despite these problems, regional maps of actual and reconstructed wind damage were found to agree closely for the hurricanes modeled in both New England and Puerto Rico. This agreement was no doubt enhanced by the small number of predicted damage classes (no damage, F0, F1, F2, F3). A larger number of classes would provide a more robust test of the model but was not practical given the nature of the historical materials.

The historical-modeling method can be applied to any part of the world where good historical records survive. Ecologists can combine information obtained in this way with knowledge of other disturbance events to build a comprehensive disturbance history for a site or region. The same information can used to improve hurricane risk assessment for shoreline and inland areas. The historical-modeling method can also be used to help calibrate and interpret various techniques for studying prehistoric hurricanes, such as the stratigraphic analysis of salt marsh deposits (e.g., Donnelly et al. 2001). These techniques may provide the millennial-scale data needed to better assess long-term trends and correlations with climate change. The main disadvantage of the historical-modeling approach is the time and effort required to locate and interpret historical materials, especially for the early period.

Acknowledgments

The methods described here were developed in earlier studies of hurricanes in New England and Puerto Rico. The author thanks the many individuals who contributed to those studies, especially co-authors K. Chamberlin, M. Fluet, D. Foster, and M. Serrano. The research was supported by grants from the National Science Foundation (DEB-9318552, DEB-9411975, and DEB-9411973) and is a contribution from the Harvard Forest and Luquillo Long-Term Ecological Research Programs.

NOTE

1. The HURRECON and EXPOS models and the historical data used in our analyses are available on the Harvard Forest Web site (http://harvardforest.fas.harvard.edu).

REFERENCES

Anthes, R. A. 1982. *Tropical cyclones: Their evolution, structure and effects.* Boston: American Meteorological Society.

Boose, E. R., K. E. Chamberlin, and D. R. Foster. 1997. Reconstructing historical hurricanes in New England. In *Preprints of the 22nd Conference on Hurricanes and Tropical Meteorology,* 388–89. Boston: American Meteorological Society.

Boose, E. R., K. E. Chamberlin, and D. R. Foster. 2001. Landscape and regional impacts of hurricanes in New England. *Ecological Monographs* 71:27–48.

Boose, E. R., D. R. Foster, and M. Fluet. 1994. Hurricane impacts to tropical and temperate forest landscapes. *Ecological Monographs* 64:369–400.

Boose, E. R., M. I. Serrano, and D. R. Foster. 2004. Landscape and regional impacts of hurricanes in Puerto Rico. *Ecological Monographs* 74:335–52.

Donnelly, J. P., S. S. Bryant, J. Butler, J. Dowling, L. Fan, N. Hausmann, P. N. Newby, B. Shuman, J. Stern, K. Westover, and T. Webb III. 2001. A 700-year sedimentary record of intense hurricane landfalls in southern New England. *Bulletin of the Geological Society of America* 113:714–27.

Dunn, G. E., and B. I. Miller. 1964. *Atlantic hurricanes.* Baton Rouge: Louisiana State University Press.

Eastman, J. R. 1997. IDRISI for Windows, ver 2.0. Worcester, Mass: Clark University.

Fujita, T. T. 1971. *Proposed characterization of tornadoes and hurricanes by area and intensity.* SMRP Research Paper 91. Chicago: University of Chicago.

Fujita, T. T. 1987. *U.S. tornadoes: Part one, 70-year statistics.* SMRP Research Paper 218. Chicago: University of Chicago.

Gray, W. M., J. D. Sheaffer, and C. W. Landsea. 1997. Climate trends associated with multidecadal variability of Atlantic hurricane activity. In *Hurricanes: Climate and socioeconomic impacts,* edited by H. F. Díaz and R. S. Pulwarty, 15–53. Berlin: Springer-Verlag.

Grazulis, T. P. 1993. *Significant tornadoes, 1680–1991.* St. Johnsbury, Vt.: Environmental Films.

Holland, G. J. 1980. An analytical model of the wind and pressure profiles in hurricanes. *Monthly Weather Review* 108:1212–18.

Jarvinen, B. R, C. J. Neumann, and M. A. S. Davis. 1984. *A tropical cyclone data tape for the North Atlantic Basin, 1886–1983: Contents, limitations, and uses.* NOAA Technical Memorandum, NWS NHC Report No. 22. Coral Gables, Fla.: National Oceanic and Atmospheric Administration.

Kaplan, J., and M. DeMaria. 1995. A simple empirical model for predicting the decay of tropical cyclone winds after landfall. *Journal of Applied Meteorology* 34:2499–512.

Landsea, C., C. Anderson, N. Charles, G. Clark, J. Fernández-Partagás, P. Hungerford, C. Neumann, and M. Zimmer. 2002. Atlantic hurricane re-analysis project (available at: http://www.aoml.noaa.gov/hrd/hurdat).

Liu, H. 1993. Calculation of wind speeds required to damage or destroy buildings. In *The tornado: Its structure, dynamics, prediction, and hazards,* edited by C. Church, D. Burgess, C. Doswell, and R. Davies-Jones, 535–41. Geophysical Monograph 79. Washington, D.C.: American Geophysical Union.

Ludlum, D. M. 1963. *Early American hurricanes, 1492–1870.* Boston: American Meteorological Society.

Marx, R. F. 1983. *Shipwrecks in the Americas.* New York: Bonanza Books.

Millás, J. C. 1968. *Hurricanes of the Caribbean and adjacent regions, 1492–1800.* Miami: Academy of the Arts and Sciences of the Americas.

Neumann, C. J., and C. J. McAdie. 1997. The Atlantic tropical cyclone file: A critical need for a revision. In *Preprints of the 22nd Conference on Hurricanes and Tropical Meteorology,* 401–2. Boston: American Meteorological Society.

Neumann, C. J., B. R. Jarvinen, and A. C. Pike. 1987. *Tropical cyclones of the North Atlantic Ocean, 1871–1986.* NOAA Historical Climatology Series 6-2. Asheville, N.C.: National Climatic Data Center.

Powell, M. D. 1982. The transition of the Hurricane Frederic boundary-layer wind field from the open Gulf of Mexico to landfall. *Monthly Weather Review* 110:1912–32.

Powell, M. D. 1987. Changes in the low-level kinematic and thermodynamic structure of Hurricane Alicia (1983) at landfall. *Monthly Weather Review* 115:75–99.

Powell, M. D., S. H. Houston, and I. Ares. 1995. Real-time damage assessment in hurricanes. In *Preprints of the 21st Conference on Hurricanes and Tropical Meteorology,* 500–502. Boston: American Meteorological Society.

Powell, M. D., S. H. Houston, and T. A. Reinhold. 1994. Standardizing wind measurements for documentation of surface wind fields in Hurricane Andrew. In *Hurricanes of 1992: Lessons learned and implications for the future,* edited by R. A. Cook and M. Soltani, 52–69. New York: American Society of Civil Engineers.

Salivia, L. A. 1950. *Historia de los temporales de Puerto Rico (1508–1949).* San Juan, P.R.: Privately printed.

Simpson, R. H., and H. Riehl. 1981. *The hurricane and its impact.* Baton Rouge: Louisiana State University Press.

Twisdale, L. A. 1978. Tornado data characterization and windspeed risk. *Journal of the Structural Division, Proceedings of the American Society of Civil Engineers* 104:1611–30.

Willoughby, H. E. 1995. Mature structure and evolution. In *Global perspectives on tropical cyclones,* edited by R. L. Elsberry, chap. 2. Report No. TCP-38. Geneva: World Meteorological Organization.

5

Tropical Cyclone Reconstructions from Documentary Records: Examples for South Carolina, United States

Cary J. Mock

Tropical cyclones are commonly recurring natural hazards in South Carolina. Examination of the modern hurricane database (HURDAT) for events since 1851 indicate significant impacts on the region once every 1.8 to 2.0 years (based on tropical storm centers within 86.9 nautical miles [100 statute miles, or 161 km] of Charleston, South Carolina). This study extended this aspect back to the mid-eighteenth century by utilizing three different types of historical data to reconstruct tropical cyclone frequencies: newspapers, plantation diaries, and weather records. The record is continuous since 1778. Plantation diaries provide daily information that allows the detection of weak tropical storms. Newspapers generally provide more detailed information on the intensity and damage from tropical cyclones than plantation diaries, especially for stronger storms. Weather records from sources such as private records, the U.S. Army Surgeon General, and the Charleston Board of Health generally contain brief descriptive information, but sometimes also include valuable information regarding barometric pressure, temperature, wind direction, intensity, and precipitation data. Results from our reconstruction provide information on 68 storms from 1778 to 1870; 25 of which are newly documented. At least three new storms were found for the pre-1778 period. Temporal analysis of the 1778 to 2000 record, smoothed by a 5-year running mean, indicate substantial decadal variability that is unprecedented in the modern record for the region extending from South Carolina to the Caribbean, with a particularly active 1830s and 1890s. In particular, 1838 was the most active year for the entire period since 1778, with four tropical cyclones affecting South Carolina. The approach used in this reconstruction offers a useful framework for conducting historical tropical cyclone reconstructions at other locations of the Atlantic Basin, and for supplementing other proxy data sources.

During the past few decades, there has been increased concern over the temporal variability of Atlantic hurricane frequencies and landfall in the modern record (Gray, Schaeffer, and Landsea 1997; Malmquist 1997; Elsner and Kara 1999). Hurricane-return intervals at subregional scales are important because associated damage can dramatically affect insured property worth up to tens of billions of dollars (Clark 1997; Pielke and Landsea 1999). For example, insured losses for the Southeastern United States since 1950 due to climatic hazards exceed $25 billion, with hurricanes being responsible for approximately three-fourths of that amount (Changnon and Changnon 1992). Potential insured losses from a hurricane at the strength of Hurricane Andrew in 1992 (a Category 5 storm) are estimated at up to $100 billion (Díaz and Pulwarty 1997). For individual locations along the Atlantic and Gulf coasts, annual probabilities for a hurricane landfall vary from 4 to 16%, and probabilities for a major hurricane are generally less than 5% (Simpson and Lawrence 1971). Predictions notwithstanding, the gulf states experienced a greater frequency of landfalling major hurricanes during the first two decades of the twentieth century, as did the Eastern seaboard during the 1940s and 1950s (Landsea 1993; Landsea et al. 1999b; Smith 1999). The 1970s and 1980s were relatively inactive in terms of landfalling intense hurricanes compared to the 1950s and 1960s (Gray, Schaeffer, and Landsea 1997).

Detailed Atlantic Basin tropical cyclone records (particularly from aircraft reconnaissance) over the last five decades provide the most comprehensive of all tropical cyclone datasets. Hurricane climatology and variability of storm tracks and intensity since 1944 are very well known in the basin. Less detailed but still reliable Atlantic Basin tropical storm and hurricane track data (HURDAT) are available back to 1851 (Neumann et al. 1993; Landsea et al. 1999a). Future improvements in long-range hurricane prediction require that our current understanding of Atlantic hurricane variability based on the modern record be supplemented with a paleoclimatic perspective to investigate temporal and spatial hurricane variability (see, e.g., Liu, chapter 2 in this volume). During the late Little Ice Age of the mid-eighteenth to mid-nineteenth century, climatic forcing mechanisms differed from those of the modern record (Bradley and Jones 1993). These variations include increased explosive volcanism, reduced solar irradiance, changing amounts of trace gases, and periods of less—and more—frequent ENSO activity (Mann et al. 1998; Stahle et al. 1998). Some of these variations, for example ENSO frequency, would affect hurricane climatology in the Atlantic Basin. Only one major wet-dry cycle of Sahel rainfall and one strong mode of a vigorous Atlantic thermohaline circulation (strong oceanic circulation associated with warmer sea-surface temperatures) occurred within the twentieth century. Several are believed to have occurred during the nineteenth century (Gray and Sheaffer 1997).

Documentary records have been the most extensive data source for reconstruction of Atlantic hurricanes over the past 500 years. This high-resolution source consists of historical records such as ship logs, diaries, annals, and newspapers. Historical records provide a higher resolution than other available proxy data types such tree-rings and lake sediments, at times enabling detailed day-by-day reconstructions of hurricane activity (Rappaport and Ruffman 1999; Sandrik and Jarvinen 1999). Documentary reconstructions and compilations of localized histories of Atlantic hurricanes began more than a hundred years ago (Evans 1848; Poey 1855), with most work relating to storms in the Caribbean (e.g., Lapham 1872; Garriott 1900; Alexander 1902; Fassig 1913; Henry 1929; Millás 1968; Reading 1990; Walsh and Reading 1991). For other areas of the Atlantic Basin, including the United States, huge quantities of valuable and unexploited documentary data exist; but to date detailed studies generally go back at best to 1851 (Fernández-Partagás and Díaz 1995a, 1995b, 1996a, 1996b). Chenoweth (1998) demonstrated a high potential to utilize newspapers from the Bahamas for reconstructing tropical cyclones back to the early 1800s. Recent efforts have been conducted on documenting hurricanes from early newspapers for Texas, Louisiana, Virginia, Maryland, and Georgia (Sandrik and Jarvinen 1999; D. Roth and D. Blanton, personal communication).

This chapter describes a reconstruction of tropical cyclone history for South Carolina, extending our knowledge for the pre-1851 period in particular. It emphasizes the Charleston area of South Carolina and the nearby coastal environs because these are the regions where hurricanes have their most immediate impact (figure 5.1). The Charleston area is one of the first areas settled by European colonists in the southeastern United States (second only to St. Augustine, Florida), so written records are available back to the late seventeenth century. The study concentrates mostly on the post-1778 period, because after that point there is sufficient documentary information to provide a reliable continuous time series of tropical cyclones. This study focuses on tropical cyclone occurrence as opposed to magnitude, because the latter generally requires data coverage over a larger geographic area beyond the size of coastal South Carolina to provide reliable results. This time series of tropical cyclone frequency is the first of its kind to be reconstructed for a location in the United States. In addition to the reconstruction, this study investigates some synoptic-scale meteorological patterns that may explain the observed temporal variations of tropical cyclone frequency near South Carolina.

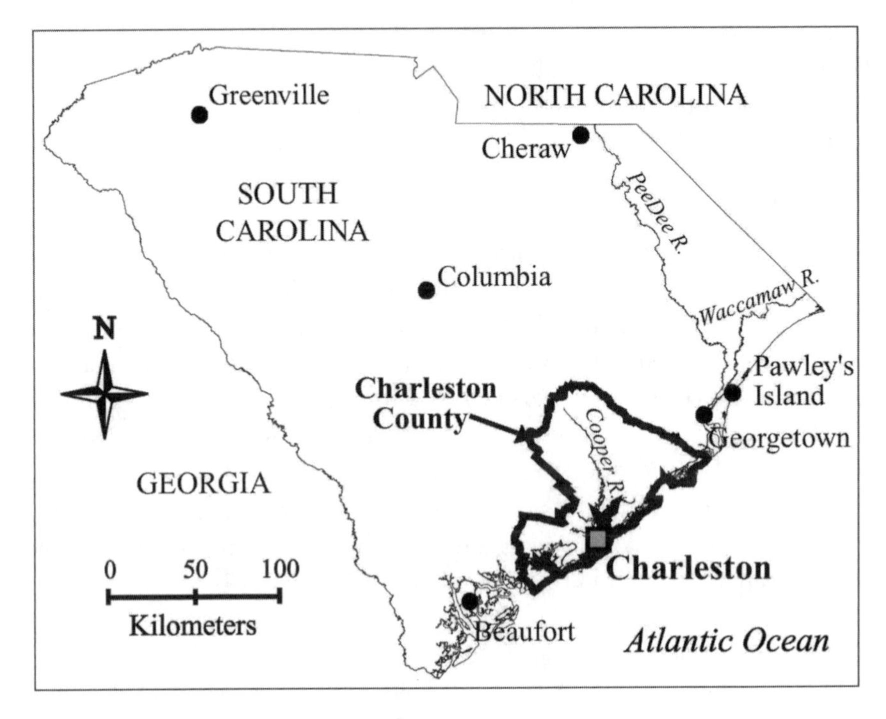

FIGURE 5.1 The state of South Carolina and specific locations mentioned in the text. Charleston County is emphasized on the map, as many plantation manuscripts came from this area.

Description of Data

Newspapers

Newspapers provide the most detailed documentary source of tropical cyclones, as previously demonstrated in Atlantic Basin reconstructions by Ludlum (1963) and Fernández-Partagás and Díaz (1996a). Information provided by newspapers includes descriptive aspects on the exact hourly timing of storm impact, wind direction, wind intensity, rainfall, height of tides, damage to buildings and trees, specifics on geographic extent of damage, and deaths. The amount of detail on tropical cyclones varied by newspaper and by storm intensity. Generally, the stronger storms received greater press coverage than weaker storms and newspapers usually had little or no information on many weaker hurricanes or tropical storms. Some newspapers, primarily those published before 1800, were produced weekly. Generally, they were biased toward news associated with the day that the newspaper went to press. Publication dates from four to six days after a landfalling hurricane may not have any mention of a storm because it is not "current news." Most newspapers had a section on the "shipping news" that sometimes included information on storms that potentially were tropical cyclones.

The newspapers perused in this study include the *South Carolina Gazette* (1732–1775), *Charleston Courier* (1803–present, name changed to the *Charleston Daily Courier* in 1852), *Charleston Mercury* (1825–1868), *South Carolina and American General Gazette* (1775–1781), *Gazette of the State of South Carolina* (1779–1780), *Royal South Carolina Gazette* (1781–1782), *Royal Gazette* (1782), *South Carolina State Gazette and General Advertiser* (1784), *State Gazette of South Carolina* (1785–1792), *Charleston Morning Post and Daily Advertiser* (1786–1787), *City Gazette and the Daily Advertiser* (1787–1803), *City Gazette* (1804–1805), *Times Charleston* (1800–1821), *Charleston Evening Post* (1815–1816), *Georgetown Gazette* (1798–1816), *Winyah Intelligencer* (1819–1835), *Winyah Observer* (1841–1852), *Pee Dee Times* (1853–1859), and *Southern Agriculturist* (1828–1839). Of these, the most useful were the *Charleston Courier* (beginning in 1803) and the *Charleston Mercury* (1825–1868), both of which published almost daily issues (excluding Sunday). The following excerpt from the *Charleston Courier* provides an example regarding a tropical cyclone on October 1, 1820:

> *Charleston Courier*, Tuesday, October 3: THE STORM.—On Sunday morning last, a Gale commenced at N.E. which blew with much violence during the day.—Towards evening, it veered round to East, and afterwards to S.E. at which point, about 8 o'clock, P.M. it blew with much violence, and excited the fears of all.—Fortunately little injury was done at our wharves—the schooner *Sarah & Hannah*, lying at Gibbs and Harper's wharf, the schooner *Live Oak*, at Chisolm's South wharf, and two or three wood droggers, at Dewees' wharf, were sunk—but no other material injury was sustained in the city.—We learn from Sullivan's Island, that no damage was done to any of the buildings there—a small wrecking vessel, called the *Sea Serpent*, was driven on shore and wrecked at that place.—As the Northern Mail arrived at an early hour, there is reason to hope that little injury was experienced in that direction.—The crops of Cotton and Rice, generally, must no doubt have suffered considerably.

Normally, the *Charleston Courier* and the *Charleston Mercury* contained four pages per daily issue, with the exception of periods during the War of 1812 and the American Civil War from 1861 to 1865. Tropical cyclone information usually appeared on page 2 under the "Charleston" heading, but occasionally it appeared on the front page and in the shipping news.

Plantation Diaries

Plantation diaries, journals, and letters provide the majority of the documentary manuscripts from the post-American Revolution up through the American Civil War. Because plantations were a commercial enterprise, careful record keeping was essential and often detailed. These manuscripts often contain large amounts of weather information as a farmer's lifestyle was very much related to the timing and location of weather events. They have been proven very useful for hurricane reconstructions for the Gulf Coast (Sullivan 1986). The manuscripts supplement newspaper information by providing daily weather activity during, just prior to, and after tropical cyclone events, although normally they do not provide quite as much detail as newspapers. The daily weather information in these manuscripts provided valuable information for reconstructing weaker tropical cyclones and hurricanes not documented in newspapers. The following is an example, also for the storm of October 1, 1820, by Daniel Cannon Webb. Webb resided near Charleston, and his diary from 1817 to 1850 provided an exceptionally detailed record of tropical cyclones:

> commenced blowing at N.E. in the morning—increased with rain about 9 o'clock by 11 blew a gale with Torrents of rain—continued a Severe Gale all Day & prodigious fall of rain—after night wind got to the East & then South East & at 10 to 11 blew very violently indeed, most so of any part of the Day past—The wind then shifted to South & S. West & began to abate . . . trees & fences blown down—this Gale was generally considered more Severe than that of 10 Sept.—[servants?] from Savannah say it was very Severe—more so probably than with us.

Seventy-four different diaries and letters were used in this study, with most coming from the coastal areas, because hurricanes have their greatest impact on these areas. Inland diaries were perused only for examining impacts from heavy rains. Temporal coverage of a typical plantation diary varied from several months to more than 50 years; the latter having more than 1,000 pages. Archives that were visited include the South Caroliniana Library, the South Carolina Historical Society, the Waring Historical Library in Charleston, the College of Charleston, Francis Marion University, and the University of Maryland. Some of the manuscripts were also taken from the microfilm series "Records of the Antebellum Southern Plantation" series, which include archives from the University of North Carolina at Chapel Hill (Southern Historical Collection) and Duke University.

Weather Records

Some early daily weather instrumental records in South Carolina were recorded during the late eighteenth and early nineteenth centuries (Aldredge 1940). Records included documentation of temperature, precipitation, wind speed, wind strength, cloudiness, written remarks, and, occasionally, pressure. These data, although limited in quantity, are invaluable for reconstructing hurricane patterns because they involve actual numerical weather data. Meteorological instrumental data were recorded by the U.S. Army Surgeon General beginning in the 1820s (Lawson 1855), the Smithsonian Institution in the late 1840s (Fleming 1990), and miscellaneous voluntary observers (Schott 1876, 1881; Havens 1958). In the U.S. Army Network, temperature records began in the 1820s due to interests in relating climate to health, but most precipitation records began after 1840. Several problems arise when comparing numerical data with those from the modern (twentieth-century) record. Temperature data problems include different routine observation times that affect the calculation of daily averages, and different thermometer exposure situations involving azimuth, shelters, and building material (Chenoweth 1992, 1993). Precipitation data problems include some high placements of rainfall gauges (8 feet high) that caused lower liquid accumulation due to increased wind speeds (Mock 2000).

As long as daily observation times were consistent, one can assess whether a potential storm may be tropical or extratropical by graphing trends of temperature. An extratropical system would generally exhibit sharper drops of temperature (sometimes to below 21°C [70° F]) and particularly sharp changes in wind directions (e.g., southwest to northwest for a cold front) associated with these drops. Tropical systems would not normally experience such dramatic falls of temperature. The differentiation between subtropical and tropical storms is not possible for most storms, given limited data at a point location. Thus all subtropical systems have been included with tropical systems in this study. One cannot simply use precipitation as an indicator of past tropical cyclone activity, but precipitation data provide some clues on the movement of rain bands and the speed of a tropical system. Wind-direction data were usually available and very informative because they provide excellent clues on the position of a tropical cyclone center and its movement. Wind speed normally was recorded on a relative numerical scale, which varied in the nineteenth century. The most common early format dealt with a scale from 1 to 6, with 1 indicating a very light breeze and 6 indicating a violent storm (Darter 1942). If barometric data were available, they provided valuable information on the strength and passage of a tropical cyclone from its V-shaped bar graph (Sandrik and Jarvinen 1999). Not all old barometric data were standardized, but plots still indicate relative information on storm existence and some idea of strength. Two examples for Charleston indicate the use of graphing barometric pressure data for the hurricanes of September 1854 and September 1861 (figure 5.2).

For this study on South Carolina, the primary early instrumental records found and used were taken at Beaufort by Belcher Noyes (1787–1789); at Charleston by Gabriel Manigault (1784–1792); at Charleston by Robert Wilson (1791–1808); at The Citadel (1830–1832), Castle Pinkney (1833–1835), Fort Moultrie (1822–1860), and Hilton Head (1862–1867) (all in the Charleston area) by the U.S. Army Surgeon General; at Charleston by numerous observers of the Charleston Board of Health (1824–1870); and near Pawley's Island by Plowden Weston (1829–1833) and the Reverend Alexander Glennie (1834–1880). Barometric pressure data were found primarily in the Charleston Board of Health and Alexander Glennie records beginning in the late 1840s. Additional detailed specifics concerning these records are available from the author.

HURDAT

The modern (post-1870) hurricane record was also examined in order to connect it with the South Carolina reconstruction to construct a continuous time series. The modern data, entitled HURDAT (North Atlantic hurricane database), consists of six-hour geographical coordinates (latitude and longitude) and intensities of all known tropical cyclones extending back to 1851. Maps of tropical cyclone tracks for each year were also constructed and available. HURDAT is the primary database used for hurricane climatic research (Landsea et al. 1999a, 1999b). Most of the 1851 to 1890 information, researched by Fernández-Partagás and Díaz (1996a), has been added to the database.[1] For this study, the author was only interested in utilizing the data as it pertains to South Carolina impact; this is expanded upon in the discussion of methodology later in this chapter.

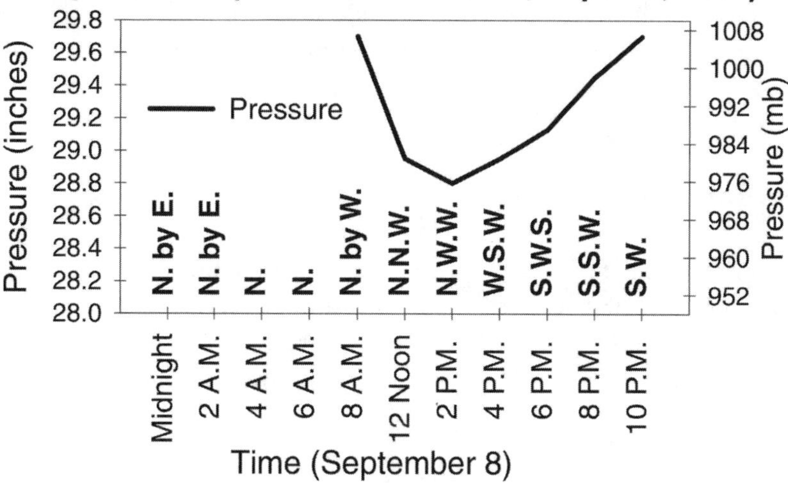

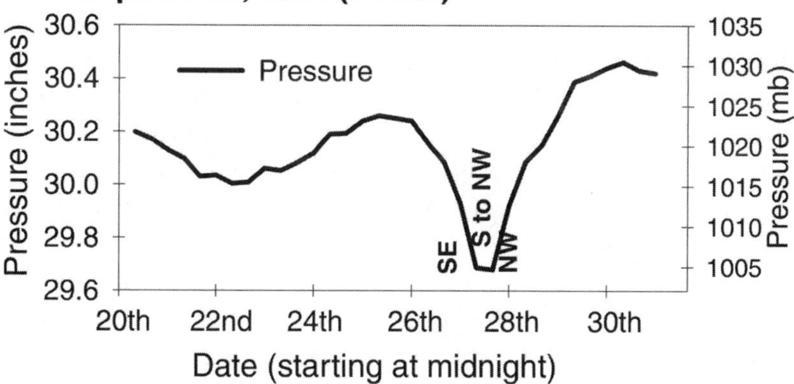

FIGURE 5.2 Examples of V-shaped bar graphs for Charleston.

Sea Level Pressure and 500-Millibars Data

Daily circulation data, covering the period from 1946 to 1994, were compiled for the days when tropical cyclones impacted Charleston in order to determine the synoptic climatic features responsible. The data were provided by the National Meteorological Center (now the National Centers for Environmental Prediction). Circulation data consist of gridded sea level pressure in millibars (mb) and 500-mb heights in geopotential meters (gpm) (Mass 1993). Sea level pressure data represent features such as the Bermuda High, and the 500-mb heights depict the main patterns of upper-level ridges and troughs. The data encompass a region from the western Pacific to western Europe to ensure that large-scale atmospheric circulation features can be described adequately.

Methodology

Because the bulk of the documentary data come from the Charleston region, the author concentrated on reconstructing the tropical cyclone frequency near Charleston. Focusing on Charleston alone enabled the construction of a continuous master time series. The methodology involved the following five steps.

Step 1: Analysis of HURDAT Data

Tropical cyclones can affect locations far away from the storm center; so this study also includes tropical cyclones that did not make landfall. No simple generalization exists on the relationship between storm size and intensity, but several scholars demonstrated that tropical cyclones generally affect a geographic location up to 86.9 nautical miles (100 statute miles, or 161 km) away from the center) and utilized this criteria for constructing maps of tropical cyclone climatology probabilities.[2] Purvis and McNab (1985) utilized a similar approach in constructing a tropical cyclone history for Charleston from 1886 to 1983. This study assessed all tropical cyclone tracks in the HURDAT database from 1851 to 2000. Tropical cyclone frequencies were compiled for each year, counting each different storm when it was centered within 86.9 nautical miles of Charleston. A storm had to be at least at tropical storm strength and subtropical storms were also included. Tropical depressions were not included in the frequency analysis.

Step 2: Analysis of Historical Tropical Cyclones

Analyses of newspaper, diary, and early instrumental data focused on Charleston in order to ensure that the documentary tropical cyclone frequencies are consistent with those from the HURDAT data. Under no circumstances were storms inferred as tropical if different records had conflicting implications, suggested extratropical characteristics, or if they did not affect

Charleston. To conclusively add a storm to the reconstruction, the data must show the classic signs of tropical systems. These include (1) sustained strong winds for longer than several hours, (2) specific wind directions that indicate the motion of tropical systems, (3) descriptions of tropical cyclone damage, and (4) information concerning storm surges. Documentary data for Charleston often described the limited extent of the storm surge to a particular street. Storm surges of greater than 3.10 m (10 feet) in downtown Charleston indicate the possibility of a major (Category 3) hurricane. Temperature and pressure data provide valuable information to verify the existence of a tropical cyclone. The following examples show how different records can be used together to verify the existence of a tropical cyclone on August 23, 1829:

Charleston Courier (Charleston), Monday, August 24
The Storm.—We had an old-fashioned autumnal storm yesterday, the wind blowing a gale from the North, with a heavy fall of rain throughout the day. The weather had put a threatening appearance for a day or two previous, and our fears of the consequences were, as usual at this critical period of the season, somewhat excited. . . .

We believe that little or no damage had been sustained, either in the city or harbor, when we put our paper to press last evening—but the Sea-Island Cotton crop, which was previously, from various circumstances, very unpromising, will unquestionably be more or less injured by it.

The Northern Mail did not come to hand; it not being thought safe, we suppose, to cross the harbor with it.

William John Connors (Clarendon County, Northeastern S.C.), August 24
It rained a great Deal yesterday evening & last night accompanied with very high winds, which injured the Cotton & Corn a little

Winyah Intelligencer (Georgetown), August 26
Sunday evening—stormy day from NE and continued rain and a dark strong gale till 12; wind shifted from NE at 1 to W at 3; tide came in unusually high in the evening and low in the morning; much damage done to rice; no mail from North in 2 days; swamps so full cannot travel through them; number of trees down is equal to that in 1822 but trifling on roads.

John Peyre Thomas (Mount Hope Plantation, northwest of Charleston), August 24
Cloudy with a little rain in the course of the forenoon. . . . The high wind continued during the night & all this morning. . . . I fear cotton is much injured, many forms are knocked off.

Davison McDowell (near Georgetown), August 24
we had a Severe Gale from N.N.E. which continued all day on the 23rd & until 2 1/2 P.M. on the 24 AM when it shifted to the W. We had a great deal of Rain when the Wind shifted it fell in Torrents: We are now suffering under the effects of it: a severe freshet in the River over the banks, very high water, & up to the areas of the Rice, in the old ground, & in the low spots over it: to gather with the Storm & freshet—our Crops will be Much injured—God only [knows] where.

Daniel Cannon Webb (Charleston), August 23
blowing fresh this morning at North—Light rain early & appearances make one rather anxious—8 o'clock A.M.—by 9 great increase Wind—1/2 past 9 commenced rainy and blowing heavily—increased to a gale & blew violently at North or a little West of North. Rained the whole day, with intervals of only a few minutes & very heavily—at 8 PM, wind was less—by 9 had moderated Somewhat—though still heavy wind—& blew heavily till towards morning when it got to N.W. & before day was quite moderate that ceased blowing—A good many trees & fences were blown down—and I fear [more?] from the Country will be unpleasant—Thermometer today 77—slept with a blanket. This was much such a day as 25 Aug. 1827—The Storm was not felt as far so as before. . . . Columbia . . . no injury done to the people.—At Georgetown, Santee South & East the Crops sustained considerable injury & the Gale was more Severe than here—at No. Island Severe Injury coincided the blow as great as 1822, but I trust they were mistaken. . . . The Cotton Crops in the Vicinity of Charleston suffered considerably. . . . Cotton is "injured much" by the blow.

Fort Moultrie Instrumental Record (Charleston), August 23
N. . . Rain . . . Heavy gale

James Pegues (Cheraw), August 24
Raing this Eving with the Appearance of a Storm

In general, it was possible to find at least eight different records for these nineteenth-century storms, making the results conclusive. Some storms prior to 1800, however, are based on no more than a few records due to the scarcity of data at times.

Step 3: Verification of the Historical Methodology

The HURDAT tropical cyclone frequencies compiled for Charleston from 1851 to 1870 are based primarily on items in the *Times* of London and the *New York Times*, and results are generally considered very reliable and complete concerning storms that affected land areas (Fernández-Partagás and Díaz 1996a). The *Times* and the *New York Times* contain news from ships traveling abroad from their home port. South Carolinians did not have quick access to these papers, and virtually none of the newspaper data in the *Times* and the *New York Times* originate from South Carolina. Thus, the larger-scale HURDAT data sources are mostly independent of the specific state-level South Carolina data sources used in this study. Therefore, the overlapping time period of the two different studies from 1851 to 1870 provide a unique opportunity to assess whether the South Carolina reconstruction method yields robust results similar to those by Fernández-Partagás and Díaz (1995a). Specific dates of all tropical cyclones for both data sets were simply compared with one another, and all of the historical descriptions for both data sets were analyzed on a case-study basis for each storm to determine the details that explain any discrepancies.

Step 4: Low-Pass Filtering of the Charleston Time Series of Tropical Cyclone Frequencies

A 5-year running mean of tropical cyclone frequencies was used to examine decadal-scale fluctuations. This procedure enabled direct comparison of the Charleston time series with a few long tropical cyclone time series from the Caribbean (Walsh and Reading 1991).

Step 5: Analysis of Sea Level Pressure and 500-mb Composite Anomaly Maps

Composite anomaly maps depict spatial climatic patterns that represent differences between averaged and anomalous values of circulation (Yarnal 1993). The author utilized the composite anomaly approach to examine the relationships between atmospheric circulation patterns and tropical cyclone strikes on Charleston, South Carolina. Composite anomaly maps of both sea level pressure and 500 mb were constructed based on the days of tropical cyclone impact on Charleston from the period from 1946 to 1994. A similar analysis was also done for one day prior to tropical cyclone landfall. The latter step was conducted in order to assess whether a characteristic atmospheric circulation exists prior to the day of tropical storm impact.

Tropical Cyclone Frequency

1871–2000

The entire continuous tropical time series is shown in figure 5.3. Discussion here first focuses on the modern—post-1870—HURDAT trends because of a greater frequency of data and the ability to cross reference with HURDAT data to understand how accurately the earlier data may represent tropical cyclone activity. The annual frequency of tropical cyclones ranges from 0 to 3. with a total of 96 tropical cyclones. The number of years with at least one tropical cyclone affecting Charleston since 1870 is 68, indicating that the region is usually affected by a tropical system once every 1.9 years, a return period close to the 1.8-year mean recurrence interval calculated by Purvis and McNab (1985).

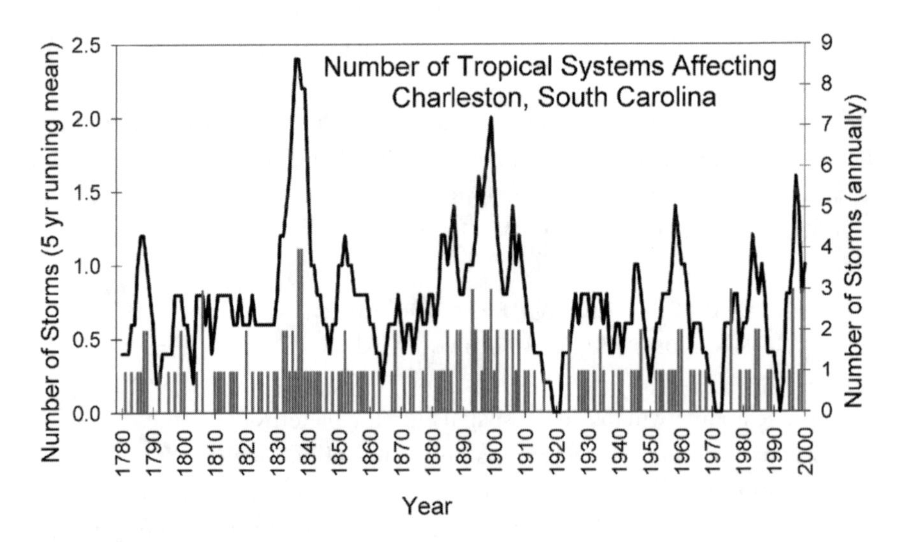

FIGURE 5.3 Time series of tropical cyclone frequencies for South Carolina, centered at Charleston, from 1778 to 2000. The vertical bars represent annual frequencies (right axis), and the line represents a 5-year running mean (left axis).

The 5-year moving average clearly indicates that the period from 1880 to 1910 was particularly active. The most active period of tropical cyclone activity occurred during the 1890s, which is unprecedented when compared to activity in the twentieth century (figure 5.3). Secondary peaks of tropical cyclone activity are evident for part of the 1950s and 1980s.

1851–1870

From 1851 to 1870, 10 storms were defined by analyzing documentary evidence for South Carolina (table 5.1). The exact dates of nine of these storms correspond exactly with the tropical cyclone reconstructions by Fernández-Partagás and Díaz (1996a), which are derived from a completely different database. This suggests that the historical reconstruction methodology for South Carolina tropical cyclones is reliable. The only storm that is not in the Fernández-Partagás and Díaz reconstruction occurred on June 22, 1867, and is believed to be a newly discovered storm. Newspapers from Charleston, sup-

TABLE 5.1 South Carolina Tropical Cyclones, 1851–1870

Date	Name (Ludlum 1963)	Fernández-Partagás and Díaz (1995a)
June 22, 1867		Not documented
September 17–18, 1863		Tracked offshore
September 27, 1861	Equinoctial Storm	Tracked offshore
September 15, 1858		Tracked offshore
September 12, 1857	S.S. *Central America* disaster	Tracked offshore
August 31, 1856	Southeastern States Hurricane	From Gulf of Mexico
September 7–9, 1854*	Great Carolina Hurricane	Hit near Savannah
October 9–10, 1852		From Gulf of Mexico
August 27, 1852*	Great Mobile Hurricane	Debatable tropical storm status in South Carolina
August 24, 1851*	Great Middle Florida Hurricane	Inland parallel to coast

Note: Specific comparisons between storms discussed in the text and in Fernández-Partagás and Díaz (1995a).

*Storm documented by Tannehill (1938), Garriott (1900), or Dunn and Miller (1960) that had an impact on South Carolina.

ported by several plantation diaries and several meteorological stations along the southeast coast down to south Florida, clearly indicate the existence of this storm, which may have been near hurricane strength at the time of landfall north of Charleston. Daily temperature trends at several South Carolina stations remained steady, not exhibiting signs of extratropical or frontal behavior. The following is an example of the description of this storm taken from the *Charleston Daily Courier* on Monday, June 24, 1867:

> THE WEATHER.—The heavy and continuous rains from Wednesday last, reported in Saturday's Courier, culminated Friday night and Saturday morning in the heaviest rains and one of the severest gales witnessed here for several years. On Friday night the shipping at the wharves had to be doubly secured, and between nine and ten o'clock, Saturday morning, the wind seemed to reach its highest, blowing almost a perfect hurricane from the Northeast, for nearly two hours. Its effects at sea, we fear, have been disastrous. The harbor presented a very black and terrible appearance. In the city a number of wharves sustained considerable damage, houses were unroofed, chimneys blown down, trees torn up by their roots, and large branches blown from others, blocking up streets and sidewalks.

Fernández-Partagás and Díaz (1995a) list the August 27, 1852, Great Mobile Hurricane as losing its tropical storm strength west of South Carolina. But a number of diaries and newspapers strongly suggest the possibility of some regeneration to minimal tropical storm strength as a result of its center going back over water east of Charleston. Documentary data from coastal Virginia also indicate perhaps up to near tropical storm force conditions (D. Blanton, personal communication). Another storm is recorded in the HURDAT database as occurring around mid-August 1860 at tropical storm status, with its center passing eastward just south of South Carolina. Although numerous plantation diaries indicate heavy rains, they indicate no signs of strong winds, and pressure data for the Charleston area are relatively high (29.92 inches or near 1013 mb). This storm may have been small and compact, and thus it was not included in the time series of this study. Overall, the documentary evidence in this study for both the August 1852 and August 1860 storms provide further details in refining the storm histories at a higher spatial resolution over South Carolina.

1778–1850

Sufficient documentary evidence dating back to 1778 allowed the construction of a reliable time series of annual tropical cyclone frequencies (figure 5.3). Fifty-eight storms were reconstructed for the 1778 to 1850 period, with 24 of them being newly documented. The number of storms per year varied from zero to four. The interannual variability in storm frequency during the 1778 to 1850 period was similar to interannual variability during the 1851 to 2000 period. A calculated mean recurrence interval for the 1778 to 1850 period suggests a significant tropical storm impact on Charleston every 1.24 years, a value that is shorter than for the 1851 to 2000 period. Closer inspection of the 5-year running mean, however, indicates that much of the greater activity is confined to the 1830s (figure 5.3). The activity of the 1830s exceeds the active 1890s, and is clearly the most active period for the entire time series. Excluding the 1830s, the mean recurrence interval for a tropical cyclone during the 1778 to 1850 period is 1.80 years, within the climatological range for the entire 1778 to 2000 record.

The decade of the 1830s includes the year with the greatest number of tropical cyclones: 1838 had four storms. The year 1838 is particularly interesting because two of the four storms occurred in June. The storms impacted Charleston on June 3–4 and June 16–17. Both June storms are well documented in plantation diaries and newspapers. Barometric data from the weather diary of Reverend Alexander Glennie, located at Pawley's Island, South Carolina, clearly exhibit V-shaped bar graphs for all four storms, and daily trends of temperatures do not show signs of extratropical or frontal activity. William Redfield (1839), one of America's foremost earliest hurricane researchers, also wrote briefly about the existence of the second June 1838 hurricane.

Also plotted in the time series are selected hurricanes that strongly affected Charleston, taken from Ludlum (1963) (figure 5.3). These hurricanes were likely of major status, based on records of storm surges that exceeded 3.1 m (10 feet), and of winds that uprooted trees and caused unmistakably widespread, severe building destruction. The primary point to emphasize is that the timings of major hurricanes correspond neither to peaks in annual tropical cyclone frequency nor peaks in the 5-year running mean. Landsea et al. (1999b) illustrated clearly that distinctive trends in major hurricanes are related to potential climatic forcings; however, those assertions were made for the entire Atlantic Basin. These assertions do not necessarily apply to individual events in a single location over a 220-year record.

Before 1778

The limited number of available documentary materials precludes the extension of the Charleston tropical cyclone frequency time series prior to 1778. The existing documents do, however, provide valuable information on the occurrences of storms. For this study, I analyzed some documentary evidence for all South Carolina storms listed by Ludlum (1963) and Tannehill (1938) and also perused diaries and the *South Carolina Gazette*. Further extension of the record will require more South Carolina archival work, but it must be augmented with the unexploited ships' logs archive from the Public Records Office in London (e.g., Chenoweth 1996; García Herrera et al., chapter 6 in this volume).

I did not analyze storms prior to 1740, but the existing records are listed for reference (table 5.2). It is worth noting that the storm of September 16, 1700, is at times incorrectly dated as 1699, the year of a well-known yellow fever epidemic in Charleston. I have found evidence for three new storms not previously documented and several more possible storms. The storms of September 1744 and June 1761, previously listed by Ludlum (1963) and Tannehill (1938), respectively, need further documentation. The mid-September hurricane of 1752 may be the strongest tropical cyclone ever known to impact Charleston, based on a potentially high storm surge that covered most of the downtown area and washed many ships onshore.

Comparison of the Record in South Carolina with Those of Other Regions

The much more active tropical cyclone frequencies of the 1830s and 1890s for South Carolina raise the question whether similar trends occur elsewhere in the Atlantic Basin. Published studies of continuous long-term tropical cyclone activity are only available for the Caribbean Sea region. Tropical cyclone time series for the Leewards/Virgin Islands, the French Islands/Dominica, and the Windward Islands were constructed by Walsh and Reading (1991). Inspection of tropical cyclone frequencies for the Leewards/Virgin Islands and French Islands/Dominica shows peaks of frequencies from 1806 to 1837 and 1876 to 1901 that are similar to the South Carolina record (figure 5.4). Interestingly, the 1765 to 1793 period exhibits even higher frequencies than the peaks in the nineteenth century. The tropical cyclone frequencies for the Windward Islands shows a peak for the 1876 to 1901 period, but activity for much of the remain-

TABLE 5.2 South Carolina Tropical Cyclones Before 1778

Cited in Ludlum (1963)	
Date	Name
September 15–16, 1686	Spanish Repulse Hurricane
September 16, 1700	Rising Sun Hurricane
September 16–17, 1713	
September 19–21, 1722?	
August 28, 1724	
August 13–14, 1728	
1730	

Analyzed in this study	
Date	Status
September 6, 1741?	Possibility
July 8, 1743?	Probably not
August 31, 1743	Yes, new addition
September 8, 1744	Probably not
Mid- to late October 1747	Yes, new addition
October 17–18, 1749?	Possibility
September 15, 1752	Yes
September 30–October 1, 1752	Yes
September 15, 1753	Yes
August 23, 1758	Yes
July 6, 1760?	Possibility
October 5–6, 1760	Yes, new addition
June 1, 1761?	Possibility
September 28–29, 1769	Yes
June 6–7, 1770	Yes
June 1, 1774?	Possibility

Note: Question marks indicate an approximate date, as suggested by documentary evidence; "Yes," additional support for a previously documented tropical cyclone; "Yes, new addition," a newly discovered undocumented storm; "Possibility," a possible date of a newly discovered undocumented storm; and "Probably not," no support for a tropical storm, even though it was suggested in earlier studies.

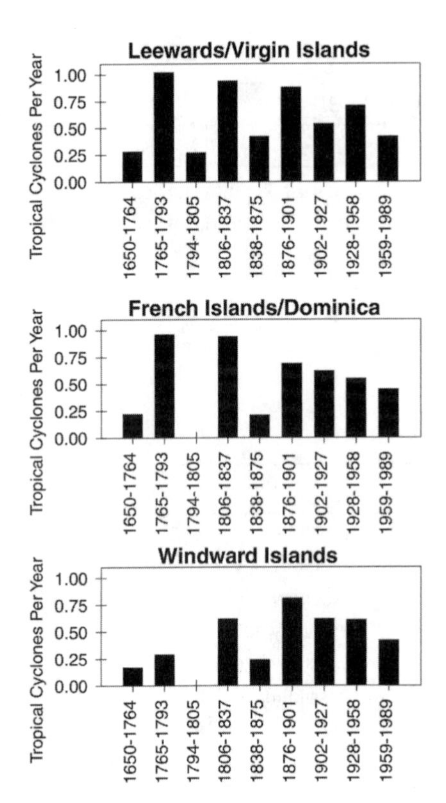

FIGURE 5.4 Tropical cyclone frequencies for selected sites in the Caribbean (data based on table in Walsh and Reading 1991).

der of the nineteenth century is not high. Perhaps tropical cyclones that affect Windward Islands are related to different synoptic controls than other areas of the Caribbean. Elsner and Kara (1999) compiled an analysis of tropical cyclone frequencies for Puerto Rico, Jamaica, and Bermuda for hurricanes only. Their plots indicate that Puerto Rico and Jamaica also experienced unprecedented hurricane activity during the first half of the nineteenth century as compared to the last 150 years, and that the 1750–1800 period was also very active. Conversely, they illustrate that Bermuda hurricane frequencies peaked around the early to mid-twentieth century; but hurricane landfalls there may be related to different synoptic controls than those in the Caribbean.

Synoptic Climatology of South Carolina Tropical Cyclones

Composite anomaly maps of sea level pressure and 500-mb heights for days when tropical cyclones impacted Charleston were based on the dates of 32 different storms from 1946 to 1994. The author also constructed composites based on the day before storm impact, but the maps were very similar to those of the day of actual impact, and thus only the latter are discussed here.

Sea level pressure patterns based on the 32 storm dates indicate a common summer-early fall appearance of the Bermuda High centered over the mid-North Atlantic, and the Pacific subtropical high centered over the North Pacific (figure 5.5). North of the Bermuda High is a somewhat enhanced Icelandic Low—both pressure centers represent a seesaw pressure pattern phenomena known as the North Atlantic Oscillation (Rogers 1984). The map of sea level pressure anomalies indicates generally that when Charleston is hit by a tropical cyclone, an area of positive anomalies is present off the northeast

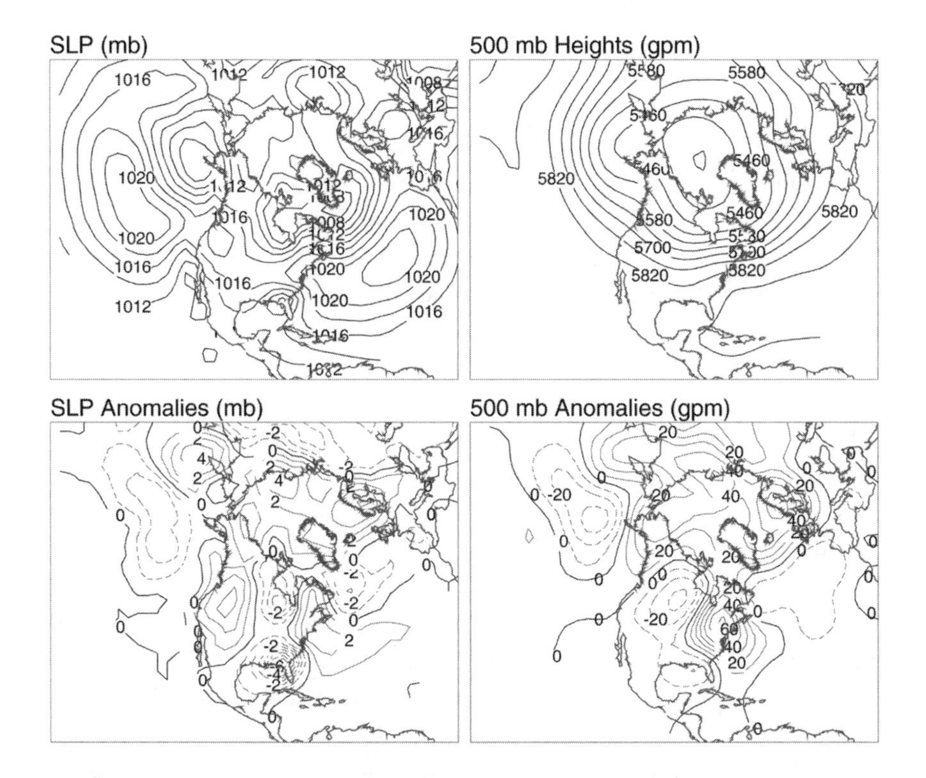

FIGURE 5.5 Daily sea level pressure (in millibars) and 500-mb height (geopotential meters) composite anomaly maps of South Carolina tropical cyclones. Solid contours indicate positive anomalies, and dashed lines indicate negative anomalies. SLP, sea-level pressure; gpm, geopotential meters.

United States, representing the westward expansion of the Bermuda High. Clockwise flow around the expanded Bermuda High creates anomalous easterly flow into South Carolina, conducive to steering any tropical storms toward the region. The composite of 500-mb heights indicates a trough extending into the western United States and some ridging along the East Coast. The 500-mb anomalies clearly indicate this enhanced trough and-ridge pattern (figure 5.5). The zone of 500-mb positive anomalies along the East Coast creates a situation where not much shearing of tropical cyclones would occur in the vicinity of South Carolina, and it also relates with the westward expansion of the Bermuda High to steer storms toward the southeast. Positive anomalies of 500-mb heights at high latitudes represent a somewhat weak circumpolar vortex, although anomalies are too weak to suggest anything conclusive for forecasting purposes.

The synoptic patterns described here may have been more prevalent during the early nineteenth century, especially the 1830s, when tropical cyclone frequencies were at their maximum. The idea of variations in the Bermuda High affecting the steering of tropical cyclones has been applied for sites in the Gulf Coast and New England (Elsner, Liu, and Kocher 2000; Liu and Fearn 2001; Liu, chapter 2, and Elsner and Bossak, chapter 12 in this volume). For this analysis, the synoptic situation of the Bermuda High would be important only during times when tropical cyclones were heading toward South Carolina. Other synoptic patterns may be more important for overall increased tropical cyclone activity at larger spatial scales. Lamb (1977), who examined July historical pressure data for the North Atlantic, noted that a trough tended to be more predominant around 70°W during the early nineteenth century. If this pattern was more predominant during summer and early fall, perhaps it could be more conducive for the formation of more baroclinic-enhanced tropical systems. The increased frequency of tropical cyclones in the Caribbean, however, suggests that tropical-only storms also had to increase in frequency.

Furthermore, the synoptic patterns discussed here are the composites of 32 storms. Some of these events may exhibit synoptic patterns quite different from the composite, and these deserve further study on a case-by-case basis. Some of the landfalling storms may relate to an anticyclone that is embedded in an upper-level longwave trough, moving east to southeast from Canada to a blocking position off the mid-Atlantic. This synoptic situation is common during fall; local meteorologists in northern Florida and southern Georgia refer to it as a "Wedge" (Sandrik 1999).

Conclusions

In this chapter, I have provided information on the longest continuous tropical cyclone frequency time series for the United States. The results are summarized as follows:

1. The methodology of constructing a tropical cyclone frequency history from South Carolina documentary records is robust, with sufficient data suggesting a significant new number of newly discovered undocumented storms.
2. A distinctive period of very active tropical cyclone activity is evident for much of the 1830s and the 1890s; this activity, as well as the extreme two-year period of 1837 to 1838, is unprecedented in the modern record for South Carolina. The increased activity in the early nineteenth century is also evident from several documentary records of the Caribbean, but it is not entirely representative for the entire Atlantic Basin.
3. Synoptic activity during the Little Ice Age, related to a westward expansion of the Bermuda High, may have led to an increased frequency of tropical cyclones affecting South Carolina, although baroclinically enhanced tropical systems may also have been responsible.

Opportunities still exist for extending the South Carolina reconstruction further back in time but it will be an arduous task to find additional useful South Carolina documents. The best records may be in the ships' logs archives in the Public Records Office in London (e.g., Chenoweth 1996; García Herrera et al., chapter 6 in this volume). Other sources may include Spanish and French materials, and perhaps integration with proxies such as tree-rings (Doyle and Gorham 1996). Numerous case study approaches to accurately reconstruct the tracks and magnitudes of particular storms, such as the major hurricanes of 1822 and 1854, are also possible from documentary evidence, but they require a larger spatial perspective beyond this study (e.g., Rappaport and Ruffman 1999; Sandrik and Jarvinen 1999). With caution, researchers can examine the specifics of hurricane damage and geographic extent of storm-surge flooding to derive detailed estimates of maximum wind speed. Most of all, a vast potential exists in the documentary archives to conduct similar reconstructions for other areas of the Atlantic Basin.

Acknowledgments

I gratefully acknowledge librarians at historical libraries and societies for helping me find materials, particularly Sam Fore and Henry Fulmer of the South Caroliniana Library. I also wish to thank Mike Chenoweth for his helpful suggestions, Jan Mojzisek and Michele McWaters for assistance with historical data collection, and helpful comments from two reviewers. This research was supported by NSF Grant ATM-9904383.

NOTES

1. The data for this study were taken from the Tropical Prediction Center's National Hurricane Center Web site (http://www.nhc.noaa.gov/pastall.html).
2. Information on the regional effect of hurricanes may be found in T. Kimberlain's work (http://www.aoml.noaa.gov/hrd/tcfaq/tcfaqG.html#G12).

REFERENCES

Aldredge, R. C. 1940. Weather observers and observations at Charleston, South Carolina from 1670–1871. In *Historical appendix of the year book of the city of Charleston for the year 1940*, 190–257. Self-published.

Alexander, W. H. 1902. *Hurricanes, especially those of Puerto Rico and St. Kitts*. Weather Bureau Bulletin 32. Washington, D.C.: Department of Agriculture.

Bradley, R. S., and P. D. Jones. 1993. "Little Ice Age" summer temperature variations: Their nature and relevance to recent global warming trends. *Holocene* 3:367–76.

Changnon, S. A., and J. M. Changnon. 1992. Temporal fluctuations in weather disasters: 1950–1989. *Climatic Change* 22:191–208.

Chenoweth, M. 1992. A possible discontinuity in the U.S. historical temperature record. *Journal of Climate* 5:1172–79.

Chenoweth, M. 1993. Nonstandard thermometer exposures at U.S. Cooperative Weather Stations during the late nineteenth century. *Journal of Climate* 6:1787–97.

Chenoweth, M. 1996. Ships' logbooks and "the year without a summer." *Bulletin of the American Meteorological Society* 77:2077–93.

Chenoweth, M. 1998. The early 19th century climate of the Bahamas and a comparison with 20th century averages. *Climatic Change* 40:577–603.

Clark, K. M. 1997. Current and potential impact of hurricane variability on the insurance industry. In *Hurricanes: Climate and socioeconomic impacts*, edited by H. F. Díaz and R. S. Pulwarty, 273–83. Berlin: Springer-Verlag.

Darter, L. J. 1942. *List of climatological records in the National Archives.* Washington, D.C.: National Archives.

Díaz, H. F., and R. S. Pulwarty. 1997. Decadal climate variability, Atlantic hurricanes, and societal impacts: An overview. In *Hurricanes: Climate and socioeconomic impacts,* edited by H. F. Díaz and R. S. Pulwarty, 3–14. Berlin: Springer-Verlag.

Doyle, T. W., and L. E. Gorham. 1996. Detecting hurricane impact and recovery from tree rings. In *Tree rings, environment, and humanity,* edited by J. S. Dean, D. M. Meko, and T. W. Swetnam, 405–2. Tucson: Department of Geosciences, University of Arizona.

Dunn, G. E., and B. I. Miller. 1960. *Atlantic hurricanes.* Baton Rouge: Louisiana State University Press.

Elsner, J. B., and A. B. Kara. 1999. *Hurricanes of the North Atlantic.* Oxford: Oxford University Press.

Elsner, J. B., K. B. Liu, and B. Kocher. 2000. Spatial variations in major U.S. hurricane activity: Statistics and a physical mechanism. *Journal of Climate* 13:2293–305.

Evans, S. J. 1848. A chronological list of hurricanes which have occurred in the West Indies since the year 1493; with interesting descriptions. *Nautical Magazine* 397, 453, 524.

Fassig, O. L. 1913. *Hurricanes of the West Indies.* Bulletin X. Washington, D.C.: Weather Bureau, Department of Agriculture.

Fernández-Partagás, J., and H. F. Díaz. 1995a. *A reconstruction of historical tropical cyclone frequency in the Atlantic from documentary and other historical sources.* Part I, *1851–1870.* Boulder, Colo: Climate Diagnostics Center, Environmental Research Laboratories, National Oceanic and Atmospheric Administration.

Fernández-Partagás, J., and H. F. Díaz. 1995b. *A reconstruction of historical tropical cyclone frequency in the Atlantic from documentary and other historical sources.* Part II, *1871–1880.* Boulder, Colo.: Climate Diagnostics Center, Environmental Research Laboratories, National Oceanic and Atmospheric Administration.

Fernández-Partagás, J., and H. F. Díaz. 1996a. Atlantic hurricanes in the second half of the nineteenth century. *Bulletin of the American Meteorological Society* 77: 2899–906.

Fernández-Partagás, J., and H. F. Díaz. 1996b. *A reconstruction of historical tropical cyclone frequency in the Atlantic from documentary and other historical sources.* Part III, *1881–1890.* Boulder, Colo.: Climate Diagnostics Center, Environmental Research Laboratories, National Oceanic and Atmospheric Administration.

Fleming, J. R. 1990. *Meteorology in America, 1800–1870.* Baltimore: Johns Hopkins University Press.

Garriott, E. B. 1900. *West Indian hurricanes.* Bulletin H. Washington, D.C.: Weather Bureau.

Gray, W. M., and J. D. Scheaffer. 1997. Role of the ocean conveyor belt as a cause of global multidecadal climate. In *Proceedings from a meeting on Atlantic climate variability,* edited by A. M. Wilburn, 42–50. Boulder, Colo.: National Oceanic and Atmospheric Administration and University Corporation for Atmospheric Research.

Gray, W. M., J. D. Sheaffer, and C. W. Landsea. 1997. Climate trends associated with multidecadal variability of Atlantic hurricane activity. In *Hurricanes: Climate and socioeconomic impacts,* edited by H. F. Díaz and R. S. Pulwarty, 15–53. Berlin: Springer-Verlag.

Havens, J. M. 1958. *An annotated bibliography of meteorological observations in the United States, 1715–1818.* Key to Meteorological Records Documentation No. 5-11. Washington, D.C.: Weather Bureau, Department of Commerce.

Henry, A. J. 1929. The frequency of tropical cyclones (West Indian Hurricanes) that closely approach or enter the continental United States. *Monthly Weather Review* 57:328–32.

Lamb, H. H. 1977. *Climate: Present, past and future.* Vol. 2, *Climatic history and the future.* London: Metheun.

Landsea, C. W. 1993. A climatology of intense (or major) Atlantic hurricanes. *Monthly Weather Review* 121:1703–13.

Landsea, C. W., C. Anderson, G. Clark, J. Fernández-Partagás, P. Hungerford, C. Neumann, and M. Zimmer. 1999a. The Atlantic Hurricane Database Re-analysis Project. In *Preprints of the 23rd Conference on Hurricanes and Tropical Meteorology,* 394–97. Boston: American Meteorological Society.

Landsea, C. W., R. A. Pielke, Jr., A.M. Mestas-Nuñez, and J. A. Knaff. 1999b. Atlantic Basin hurricanes: Indices of climatic changes. *Climatic Change* 42:89–129.

Lapham, I. A. 1872. List of the great storms, hurricanes, and tornadoes of the United States (1635–1870). *Journal of the Franklin Institute* 63:210–16.

Lawson, T. 1855. *Meteorological register for twelve years, from 1843 to 1854.* Washington, D.C.: War Department.

Liu, K.-b., and M. L. Fearn. 2001. Reconstruction of prehistoric landfall frequencies of catastrophic hurricanes in northwestern Florida from lake sediment records. *Quaternary Research* 54:238–45.

Ludlum, D. M. 1963. *Early American hurricanes, 1492–1870.* Boston: American Meteorological Society.

Malmquist, D., ed. 1997. *Tropical cyclones and climate variability.* Hamilton, Bermuda: Risk Prediction Initiative, Bermuda Biological Research Station.

Mann, M. E., R. S. Bradley, and M. K. Hughes. 1998. Global-scale temperature patterns and climate forcing over the past six centuries. *Nature* 392:779–87.

Mass, C. F. 1993. The application of compact discs (CD-ROM) in the atmospheric sciences and related fields: an update. *Bulletin of the American Meteorological Society* 74:1901–8.

Millás, J. C. 1968. *Hurricanes of the Caribbean and adjacent regions, 1492–1800.* Miami: Academy of the Arts and Sciences of the Americas.

Mock, C. J. 2000. Rainfall in the garden of the U.S. Great Plains during the late nineteenth century. *Climatic Change* 44:173–95.

Neumann, C. J., B. R. Jarvinen, C. J. McAdie, and J. D. Elms. 1993. *Tropical cyclones of the North Atlantic Ocean, 1871–1992.* NOAA Historical Climatology Series 6-2. Asheville, N.C.: National Climatic Data Center.

Pielke, Jr., R. A., and C. W. Landsea. 1999. La Niña, El Niño, and Atlantic hurricane damages in the United States. *Bulletin of the American Meteorological Society* 80:2027–33.

Poey, A. 1855. A chronological table comprising 400 cyclonic hurricanes which have occurred in the West Indies and in the North Atlantic within 362 years, from 1493–1855. *Journal of the Royal Geographical Society* 25:291–328.

Purvis, J. C., and A. McNab, Jr. 1985. *Hurricane vulnerability for Charleston County.* Columbia: South Carolina Water Resources Commission.

Rappaport, E. N., and A. Ruffman. 1999. The catastrophic 1775 hurricane(s): the search for data and understanding. In *Preprints of the 23rd Conference on Hurricanes and Tropical Meteorology*, 787–90. Boston: American Meteorological Society.

Reading, A. J. 1990. Caribbean tropical storm activity over the past four centuries. *International Journal of Climatology* 10:365–76.

Redfield, W. C. 1839. On the courses of hurricanes; with notices of typhoons of the China Sea, and other storms. *American Journal of Science* 35:201–23.

Rogers, J. C. 1984. The association between the North Atlantic Oscillation and the Southern Oscillation in the Northern Hemisphere. *Monthly Weather Review* 112:1999–2015.

Sandrik, A. 1999. A reevaluation of the Georgia and northeast Florida hurricane of 2 October 1898 using historical resources. National Weather Service, Jacksonville, Fla. (available at: http://www.srh.noaa.gov/jax/research/hurricanes/history/1898/index.html).

Sandrik, A., and B. Jarvinen. 1999. A reevaulation of the Georgia and northeast Florida tropical cyclone of 2 October 1898. In *Preprints of the 23rd Conference on Hurricanes and Tropical Meteorology*, 475–78. Boston: American Meteorological Society.

Schott, C. A. 1876. *Tables, distribution, and variation of atmospheric temperature in the U.S. and some adjacent parts of America.* Smithsonian Institution Contributions to Knowledge 21, Smithsonian Institution Publication No. 277. Washington, D.C.: Smithsonian Institution.

Schott, C. A. 1881. *Tables and results of the precipitation in rain and snow in the U.S. and at some stations in adjacent parts of North America and in Central and South America.* Smithsonian Institution Contributions to Knowledge 24, Smithsonian Institution Publication No. 353. Washington D.C.: Smithsonian Institution.

Simpson, R. H., and M. B. Lawrence. 1971. *Atlantic hurricane frequencies along the U.S. coastline.* NOAA Technical Memorandum, NWS SR-58. Springfield, Va:

National Technical Information Service Technology Administration, Department of Commerce.

Smith, E. 1999. Atlantic and East Coast hurricanes, 1900–98: A frequency and intensity study for the twenty-first century. *Bulletin of the American Meteorological Society* 80:2717–20.

Stahle, D. W., R. D. D'Arrigo, P. J. Krusic, M. K. Cleaveland, E. R. Cook, R. J. Allan, J. E. Cole, R. B. Dunbar, M. D. Therrell, D. A. Gay, M. D. Moore, M. A. Stokes, B. T. Burns, J. Villanueva-Diaz, and L. G. Thompson. 1998. Experimental dendroclimatic reconstruction of the Southern Oscillation. *Bulletin of the American Meteorological Society* 79:2137–52.

Sullivan, C. L. 1986. *Hurricanes of the Mississippi Gulf Coast.* Biloxi: Gulf.

Tannehill, I. R. 1938. *Hurricanes, their nature and history.* Princeton: Princeton University Press.

Walsh, R., and A. Reading. 1991. Historical changes in tropical cyclone frequency within the Caribbean since 1500. *Wurzburger Geographische Arbeiten* 80:199–240.

Yarnal, B. 1993. *Synoptic climatology in environmental analysis.* London: Belhaven Press.

6

The Use of Spanish and British Documentary Sources in the Investigation of Atlantic Hurricane Incidence in Historical Times

Ricardo García Herrera, Francisco Rubio Durán, Dennis Wheeler, Emiliano Hernández Martín, María Rosario Prieto, and Luis Gimeno

Si el viento te sopla furioso y valiente, enséñale al huracán los dientes.
[If the wind blows strongly and hard, it indicates a hurricane is imminent.]
Saying of Cuban fishermen (quoted in Haensch 2000)

Official documents from the Caribbean region are a rich source of information on hurricanes for the years following the establishment of the Spanish colonies. Some of these papers are from as early as the sixteenth century. One hundred years later, the growth of British political interest is reflected in the increasing frequency of logbooks from the ships of the Royal Navy deployed in the region. The Spanish and the British sources provide different, but detailed, information on hurricane events. Spanish reports are often storm-specific, describing in detail the onset, character, and consequences of individual events. British logbooks include references to hurricanes but within a wider body of information relating to day-to-day weather. Preliminary studies indicate that both sources permit the precise dating of hurricanes and, where several accounts of the same event exist, they provide information from which a reconstruction of the trajectories can be made. The results cited in this chapter are based on a limited sample of the many thousands of documents available in Spanish and British archives. An exhaustive study of these sources is demanding of time and resources but will yield detailed information from the past half millennium that adds to our knowledge of known hurricanes and identifies new events.

The use of documentary sources is a well-established high-resolution technique for reconstructing climate during past centuries when no instrumental data are available (Catchpole 1992; Martín Vide and Barriendos Vallvé

1995). Documentary material has been used to study a number of important climatic phenomena such as the El Niño–Southern Oscillation (ENSO) (Quinn, Neal, and Antúnez de Mayolo 1987; Ortlieb 2000) and the North Atlantic Oscillation (NAO) (Jones, Jonsson, and Wheeler 1997; Luterbacher et al. 1999). They have also been used to describe climate variability in a number of regions. Of particular relevance is the work of Lamb (1991), whose chronology of extra-tropical storms was based largely on such sources.

Hurricanes are particularly suitable for study through historical sources because extreme events with serious consequences for the communities that experience them are often recorded in written records. It is to be expected that severe hurricanes of the historical period would provide a legacy of documentary evidence describing their effects and consequences in different societies. Perhaps the oldest documentary records on hurricanes (taken here to include typhoons) are those kept in the Guandong province of south-eastern China in the form of local chronicles of natural disasters (Louie and Liu, chapter 8 in this volume). These documents provide information from as long ago as A.D. 975 (Liu, Shen, and Louie 2001). In the Atlantic Basin the first records come from the earliest years of the Spanish colonies. There is little doubt that Christopher Columbus experienced at least two hurricanes, one in 1495, the other in 1502 (Millás 1968).

The Spanish quickly became aware of the impact of hurricanes in the Caribbean area and promptly adopted the term *huracán* from the Carib language to describe the phenomenon. In the sixteenth century, for example, Fernandez de Oviedo wrote: "Huracán, in the language of this island, is precisely defined as a very excessive storm or tempest but being in reality nothing more than a very great wind with heavy and intense rainfall" (A.G.I.: Indif. Gral. 108-BIB. L.A. Siglo XVI-7). Since then countless documents have been produced containing information on hurricanes in the Atlantic Basin. The studies by Poey (1862), Tannehill (1940), Ludlum (1963), Dunn and Miller (1964), Millás (1968), Salivia (1950), Neumann et al. (1993), Rappaport and Fernández-Partagás (1997) and Fernández-Partagás and Díaz (1996) provide a comprehensive view of the information that can be obtained from such sources.

The topic continues to attract academic research and there are a number of current projects using old documents to investigate hurricanes.[1] Not surprisingly, most of the previously cited works have made use of documents kept in North America and the Caribbean. But a significant number of potentially relevant sources for the study of hurricanes in the past are still unexplored or have not been systematically examined. The majority of these documents are held in those European archives that contain information on the administration of former American colonies. So vast is the collection of such material that earlier attempts at their study were abandoned as too time-consuming. For example, Marx (1983) wrote in reference to the Spanish Archivo General de Indias: "If a team of one hundred researchers spent their whole lives searching through the more than 250,000 large legajos [bundles] in the Archive of the Indies (at Seville), I doubt that they could locate all the important documents concerning Spanish maritime history in the New World." Such warnings notwithstanding, some progress has been made in this direction and in this chapter we present an overview of the Spanish and British archives of potential interest for hurricane reconstructions. Some results are also provided from an exploratory but by no means exhaustive search of these archives.

Spanish Archives: Description and Previous Use for Climatic Purposes

The Spanish organized the settlement of their new American territories through the Consejo de Indias, which was established in 1523, soon after the Columbian discovery. By the mid-sixteenth century, most of present-day Mexico along the Gulf of Mexico and the Antilles were occupied by Spanish colonists. These colonies were followed by settlements in Florida and Louisiana, which were under Spanish control from the second half of the sixteenth century until the early 1800s. By virtue of the bureaucratic nature of the Spanish imperial system and its strictly formal decision-making procedures, this colonization resulted in the production of a large volume of information related to everyday life, especially in the coastal cities, and to naval trade. Fortunately, many of the millions of documents have survived and are today available for study. In Spain there are a number of archives that contain information with possible importance for hurricane studies. Of these, the most significant are the Archivo General de Indias in Seville; the Archivo Histórico Nacional, Archivo del Museo de la Marina, Biblioteca Nacional, and Real Académia de la Historia in Madrid; and the Archivo General de Simancas in Valladolid.

By far the richest source on the Spanish American colonies is the Archivo General de Indias (General Archive of the Indies; hereafter denoted by its initials, A.G.I.). The A.G.I. is the central repository for documents related to the colonial administration of Spanish America. It was founded in the late eighteenth century during the reign of Carlos III to house scattered collections of documents as well as material previously stored in the National Archive at Simancas, in Valladolid. The Lonja de Mercaderes (Merchant's Meeting House) of Seville was refurbished to house the new collection; a fitting choice as the Lonja served as headquarters for much of the commercial activity between Spain and the New World during the sixteenth, seventeenth, and eighteenth centuries. The first papers were received in 1785; since then the collection has grown to over 80 million pages of original writings encompassing all aspects of military, commercial, and cultural relations between Spain and its American colonies.

The A.G.I. collection is divided into "sections," each one covering a different area of colonial affairs. Table 6.1 shows the current organization of the A.G.I., together with the number of *legajos*, or manuscript bundles, contained in each section and the range of dates covered. Each *legajo* consists of a group of related documents and contains some 1,500 to 2,000 manuscript pages.

As noted previously, the A.G.I. has not been systematically searched for information on hurricanes, but it has been used to reconstruct climate in South America (Prieto 1998, 2000) and the circulation in the North Pacific (García et al. 2001). An overview of the potential uses of the A.G.I. for climatological studies can be found in García et al. (1999). This preliminary examination has identified those documents and sections most likely to provide information relevant to the study of hurricanes. These are briefly described:

- *Relaciones Geográficas* (Geographical Accounts), which usually describe the climate of an area or region

TABLE 6.1 Organization and Holdings of the Archivo General de Indias

Section	Years	Number of *legajos*
Patronato (Patronage)	1480–1790	306
Contaduría (Accounting)	1514–1778	2,126
Contratación (Contracts)	1492–1794	6,335
Justicia (Justice)	1515–1600	1,214
Gobierno (Administration)	1492–1858	18,760
Escribanía de Cámara (Chamber Clerks)	1525–1760	2,864
Secretaría Juzgado de Arribadas (Secretary of the Arrival Judge	1674–1822	648
Correos (Mail)	1620–1846	895
Estado (State)	1642–1830	110
Ultramar (Overseas)	1605–1870	1,013
Papeles de Cuba (Cuban Papers)	1712–1872	2,967
Consulados (Consulates)	1520–1870	1,903[1]
Títulos de Castilla (Titles from Castilla)	1700–1900	14
Tribunal de Cuentas (Accounting Court)	1851–1899	2,751
Diversos (Miscellaneous)	1492–1898	48
Mapas y Planos (Maps and plans)	1500–1900	6,457

[1] Plus 1,255 bound volumes.

- Logbooks, nautical courses, and accounts, which describe the events that occurred on board ship
- *Actas Capitulares* (Local Acts), periodic reports (usually weekly) from the local government officers
- Reports from scientific and military missions
- Accounting books, which are especially important for estimating damage
- Annals or diaries from priests, military or civil officers, and ordinary citizens
- Public and private correspondence sent by the colonial authorities to Spain
- Commercial and private correspondence.

As shown in table 6.1, the A.G.I. is organized according to administrative criteria; thus, when focusing on a certain geographical area, many different sections need to be investigated. In addition, the miscellaneous nature of the documents makes further demands and adds to an already time-consuming task that requires a high level of expertise.

British Archives: Description and Previous Use for Climatic Purposes

For nearly a century, the Spanish enjoyed unchallenged control of the western Atlantic and Caribbean areas that experienced hurricanes and tropical storms. Occasional, usually piratical, intrusions by English mariners left little useful information for climatologists. The picture changed in the latter half of the reign of Queen Elizabeth I (1558–1603), however, when the first faltering attempts at English colonization were made along the coast of what is today North Carolina and Virginia (Milton 2000). But it was another half century before those colonies became well established. Formal and significant English colonization in the Caribbean began in 1655 when the forces of Oliver Cromwell's Protectorate captured Jamaica. Even after colonization, the nature of the English government system did not demand the weighty administrative structures that the Spanish had erected. For these reasons English (later British) documentary sources can be thought of as starting only in the mid-seventeenth century and yielding a smaller volume of official papers. There is, however, one source that is both nearly continuous, voluminous, and of particular relevance to climatologists—the logbook collections from vessels engaged in trans-Atlantic voyages or in military activities in the Caribbean. With the notable exception of a few merchant ship logs (Wheeler 1995) this source is derived principally from the masters', captains' and lieutenants' logbooks kept on board Royal Navy ships. Admirals' journals also contain climatic information, but this is not set out in any standardized form and is usually presented in a narrative form woven into a more general account of the fleet's proceedings and actions. For this reason they are not to be regarded as logbooks in the strict sense, but as diaries. Many logbooks of the voyages of exploration have also survived. These are prepared in standard form but only rarely provide information for the Caribbean region.

All senior British officers were obliged to prepare a daily logbook when at sea. They also often maintained their logbooks when in harbor, albeit in briefer form. Upon return to England, the logbooks were deposited with the Admiralty and today many thousands of them are collected in two major archives. The captains' and masters' logbooks are to be found in the Public Records Office (PRO) in Kew, London, and the lieutenants' logbooks at the National Maritime Museum (NMM) in Greenwich. Admiral's journals and the logbooks of the voyages of exploration are also archived at the PRO. The logbooks at both archives form part of the Admiralty document collection under the catalogue heading ADM. At the NMM they are catalogued under ADM/L. At the PRO the principal collections of captains' logbooks are catalogued under ADM/51 and those of the masters under ADM/52. After the mid-nineteenth century the system of logbooks prepared by specific officers was gradually replaced by one requiring only one, more general, ship's logbook prepared by the succession of officers of each watch. Table 6.2 lists the quantity of logbooks and volumes under these various headings.

The scientific interest of these logbooks lies in their careful recording of the daily, sometimes hourly, passage of the weather. Interest in the weather was founded not on scientific curiosity but on the need to estimate the ship's "leeway," or drift, as a result of the effect of the wind on the course of the vessel (Taylor 1956). Such information was vital for safe navigation before methods existed for the accurate determination of longitude. Even when navigational methods were perfected in the nineteenth century, logbooks continued to include climatic information, as they do to the present day. An example of information from one of the rare merchant vessel logbooks is shown in figure 6.1. The logbook has written entries that are typical of the late eighteenth century (this logbook is, however, unusual in its inclusion of pen-and-ink sketches representing conditions on each day at sea). Information on the ship's speed is recorded every two hours and is often accompanied by information on wind direction and force. Other, more general, weather information relating to the state of sea or sky, precipitation and visibility was also recorded. The logbook is

TABLE 6.2 Summary of British Logbook Sources in Major Collections

Archive	Source	Catalogue reference	Date	Number of logbooks or volumes
NMM[1]	Lieutenants' logbooks	ADM/L/ . . .	1679–1809	5,300
PRO[2]	Admirals' journals	ADM/50/ . . .	1702–1916	413
PRO	Captains' logbooks	ADM/51/ . . .	1669–1852	4,563
PRO	Masters' logbooks	ADM/52/ . . .	1672–1840	4,460
PRO	Ships' logbooks	ADM/53/ . . .	1799–1971	172,412
PRO	Supplementary ships' logs	ADM/54/ . . .	1808–1871	337
PRO	Logs and journals of ships of exploration	ADM/55/ . . .	1757–1904	164

[1] The catalogue at NMM is available at http://www.nmm.ac.uk/.
[2] The catalogue at PRO, but not the documents themselves, is available at http://catalogue.pro.gov.uk/. The PRO collection of captains' and masters' logbooks are catalogued by "volumes." Each volume may contain up to 20 logbooks, so the stated number grossly underestimates the quantity of items that are available.

also typical in that it presents a summary of the navigational calculations for the day by which latitude and longitude were estimated.

Starting in the mid-nineteenth century, the crews of all major maritime nations also compiled instrumental records while at sea, recording them in their ships' logbooks. These sources have provided material for major databases such the COADS (Comprehensive Oceanic and Atmospheric Data Set) described in Woodruff et al. (1987), and the Kobe collection (Manabe 1999). The earlier English Officer's logbooks, particularly those from before 1800, lack such instrumental data but are nonetheless rich in weather accounts. These accounts focus on the three aspects previously noted: wind direction, wind force, and a general description of the weather for the day (table 6.3). Although such items are purely descriptive and not based on any form of instrumental observation, it is important to recall that they were recorded in response to a common requirement for safe navigation. They appear also to have been recorded to consistent standards using a conventional and well-understood vocabulary that, although it differed through time, was common to all contemporary records. Wind force was estimated by experience and by reference to the area and type of sail that a vessel could safely carry to secure maximum speed or, at slower speeds, by the effect of the wind on the water. Evidence to date indicates that wind direction was recorded with respect to true and not magnetic north. Methods for determining the variation of the latter were known

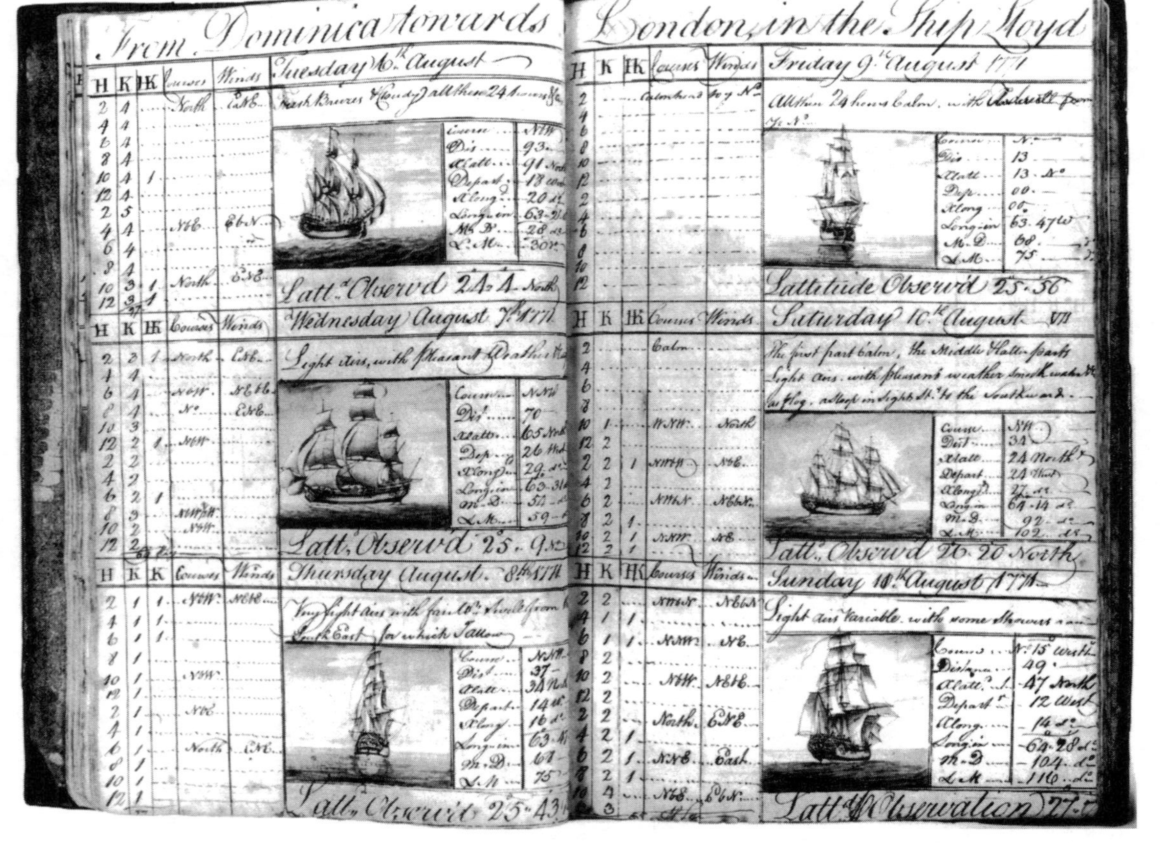

FIGURE 6.1 Pages from the logbook of the merchant vessel *Lloyd* for August 1771. This merchant vessel sailed regularly between England and the Caribbean region. The layout of the pages is typical of the time and differs only in the inclusion of pen-and-ink sketches by the ship's master, Nicholas Pocock, who later became a famous marine painter (reproduction courtesy of Bristol Records Office).

TABLE 6.3 Summary of Climatic Elements Recorded in Officers' Logbooks

Recorded element	Description
Wind force	Recorded by non-numerical description. By the early eighteenth century, the vocabulary was that later formalized by Francis Beaufort (Fry 1967). Over the preceding decades, the terminology only slowly had converged to this standard, and work remains to be completed for a reliable definition of the various terms that were used in earlier times.
Wind direction	Recorded from the sixteenth century on the standard 32-point compass, which still is in use.
Weather conditions	Rain, drizzle, fog, thunder, snow, and any other significant weather was recorded using, almost exclusively, terms that are still employed.

from the early seventeenth century (Hewson 1983). The direction itself was probably determined, as it is today when no instruments are available, by reference to streaming flags, wave and spray movement, and disturbances of the sea surface. Regrettably, no text from the periods in question make clear reference to the procedures, but the results of a practical, sea-going, experiment where wind force and direction were found to be reliably estimated from the open decks of a sailing vessel are reported in Wheeler (1988).

Terminology is a major problem in using these sources for climatic studies. In particular the search for the definition of terms such as "strong gales," "fresh gales," "easy gales," or even "indifferent gales" requires painstaking effort and has yet to be completed. It cannot be assumed that all terms were used in the past with the same meaning that they now have. The Oxford English Dictionary provides some guidance on the changing meaning of adjectival qualifiers, as well as of the words "gale," "breezes," and so on. More useful in this context are the various nautical dictionaries. Most recently the great authority is taken to be the *Sailor's Word Book* (Smyth 1867). Meanings change over time, however, so earlier word sources can be important. One of the most useful of the earlier items is Falconer's *An Universal Dictionary of the Marine*, published in 1780 and again in 1815. For yet earlier periods recourse can be made to

Boteler's Dialogues, edited by Perrin (1929) but covering the final decades of the seventeenth century. Mid-seventeenth-century terms are included in Sir Henry Mainwaring's *Nomenclator Navalis* (published in edited form by Manwaring and Perrin 1922); and Captain John Smith's *A Sea Grammar* first appeared as early as 1627 and has since been republished (Goell 1970). By these various means some interpretation can be made of the archaic terms employed particularly in the older logbooks.

Fortunately, when seeking information specifically relating to hurricanes this lexicographic problem is less acute because mariners quickly became aware of the character of these destructive weather elements; the specific word "hurricane" (precise spellings differ over time) was universally adopted to describe them. One of the earliest definitions in English is found in John Smith's *A Sea Grammer,* but a more interesting definition dates from the 1680s in *Boteler's Dialogues.* This book takes the form of an imaginary discussion between an uninformed "admiral" and a knowledgeable "captain." At one point the admiral inquires:

> Since we have here made mention of storms and tempests; tell me what your hurricanoes are, which are found so common in the West Indies.

The captain's response is informative:

> I shall, my Lord, and that in regard that this hurricanoe may be said to be the most enraged prince amongst them, and the lion of tempests. . . . [T]rue it is that these whirlwinds [*sic*] in the West Indies and those parts, are exceeding extraordinary, as well in regard of their violence as lasting. And it is very observable that in some places these devastating winds are found very frequent, and so extreme outrageous that, if reports misreport not, some ships that have been taken with them, near some of those coasts, have been rather thrown than driven, even far into the land. (Perrin 1929:164)

The captain goes on to describe how such hurricanes may endure at one place for two or three days and how they are to be found in the tropical latitudes rather than "in other parts where the sun looks not down so perpendicularly." The same items make it clear that the term "storm" was used and understood in very much the same way as it is today. In contrast, the arbitrary distinction currently made between "tropical storms" and "hurricanes" was unknown and unnecessary to the mariners of those distant times. In this sense, the present-day researcher must recognize that the precise definition employed today enjoyed a wider, but not necessarily imprecise, currency in the past.

Other, more practical, difficulties are encountered when using the British logbooks. The very large number of such items in the PRO and NMM are not all from vessels in the Caribbean and nearby waters. In order to render the task of data abstraction more efficient, some means is needed to identify those vessels in the Caribbean at any given time. Fortunately, documents exist to allow this. For dates from the mid-seventeenth century recourse can be made to the Admiralty List Books (PRO series ADM/8/. . .with a smaller collection of 'station of ship' records under ADM/7/. . .), in which the various stations are listed to which all Royal Navy vessels large and small were attached. These documents were prepared monthly, often give the sailing and arrival dates of individual vessels and although not always absolutely reliable, they provide great assistance in identifying those ships allocated to the hurricane-prone areas of which the Leeward Islands fleet and the Jamaica squadron are the most important. The quantity of logbooks is, however, variable from year to year and is determined largely by the political conditions of the day. The numbers of warships constructed and, to an even greater degree, the number in commission were much higher in wartime. Thus, for example, the Leeward Islands fleet of 1780 numbered 25 ships of the line with a commensurate number of frigates, sloops, brigs and support vessels perhaps doubling this number, but this was a time of conflict with France for possession of valuable territory in the Caribbean. During the Seven Years' War, the same fleet numbered just six with activity concentrated in Canadian waters (Lavery 1983). This lower figure would also be typical of times of peace but, most importantly, the overall number of logbooks is sufficient to ensure a continuity of record from the early eighteenth century onward.

It is therefore possible, using supplementary documents, to narrow the range of potentially useful logbooks by reference to the region or area and to a specific period of time. The selected logbooks will then provide the precise dates for each set of observations. But the accuracy with which the daily latitude and longitude were observed is another matter. The nature of navigation before approximately 1800 means that the day-to-day location of the ships is subject to error. This is marginal in relation to latitude, which, by the eighteenth century, could be determined with reasonable accuracy using octants or sextants. The reliable determination of longitude, on the other hand, excited much intellectual debate and might fairly be described as one of the great scientific challenges of the age. A review of the "longitude problem" can be found in Hewson 1983 while some of its more dramatic aspects were popularized by Sobel (1996). Error terms of tens of miles would not be uncommon in those days of "dead (deduced) reckoning" prior to the early nineteenth century, before the various solutions to the longitude problem were found and then widely adopted. Researchers must always keep in mind this possible source of error when using this data source in reconstruction exercises and when mapping out data. These characteristics, described for the British logbooks, apply equally to Spanish sources during the eighteenth and nineteenth centuries, a period during which most of the European logbooks exhibit a broadly common structure derived from identical observational procedures. Recent studies by Jackson, Jonkers, and Walker (2000) have, however, shown that reliable corrections using statistical methods can be made to latitude and longitude data from logbooks.

Even allowing for the vagaries of terminology and location, case studies using logbooks allied to other data have already demonstrated that such material can provide for the reconstruction of synoptic maps at a daily scale of temporal resolution. Although not focusing on hurricanes, at least one of these studies takes the case of a very severe storm as its theme (Wheeler 2000) and illustrates the potential of this source when a long series of data are available.

Some Examples of Hurricane Records

In this section, examples of the different forms of information obtained during some exploratory work, mostly at the Archivo General de Indias, is provided. A preliminary identification of hurricanes from Spanish historical documents can be found in García Herrera et al. (2000), where 12 hurricanes were identified from Spanish fleet reports. Unfortunately, García Herrera et al. (2000) used only secondary sources. The examples that follow include the more abundant and relevant categories of documents that can be used for hurricane reconstructions.

Reports from Spanish Ships

These reports can be found mostly, but not exclusively, in logbooks. A Royal Order issued in 1575 committed the masters and pilots of Spanish ships in the Carrera de Indias to produce accounts of each trans-Atlantic journey including, importantly, a detailed description of the voyage and of any geographical events, winds, currents, and hurricanes. The completed logbooks had then to be delivered to the Professor of Cosmology in the Casa de Contratación (A.G.I. Indif. Gral 1956, L.1, f266r–266v). A careful examination of even earlier documents, not specifically designed for the description of hurricanes, suggests that some years were particularly bad for navigation in the Caribbean (Navarrete 1901; Fernández Duro 1903; Marx 1983). For example, 16 ships were lost in a hurricane in 1553; in 1554 a number of shipwrecks were reported in Florida, Cuba and the Gulf of Mexico coast; a fleet was dispersed by a hurricane in the Bahamas in 1557; and 7 ships were lost in Nombre de Dios Bay, 5 in Campeche and 5 more in Bermuda in 1563. Eleven ships were lost in 1571 and in 1589 two fleets were struck by hurricanes in the Bahamas Strait. The next year 15 were lost at Veracruz. In 1622 the New Spain Fleet was dispersed by a hurricane in the Bahamas Strait and the Tierra Firme Fleet lost 1,000 men and dozens of ships. On July 31, 1715, a fleet was devastated by a hurricane and grounded on the reefs between Cape Canaveral and Fort Pierce in Florida; 11 ships were lost and 2,500 persons died in that storm.

The document A.G.I., Mexico 360, describes a shipwreck close to the Campeche Coast on October 20, 1620: "On October 20th a gale entered from the Southeast at nine in the night which seemed more than a hurricane. The night was so dark and full of thunder and lightning from the four quarters that we had to offer prayers to Our Lady Virgin of Carmen. . . . Later the hurricane veered to the north and a quarter, at dawn it blew away the foremast and jibsail." These words were sent to His Majesty by the Governor of Yucatan and formed part of a report that was originally produced in the dairy of Andrés de Aristizábal, Chief of the Tierra Firme, on board on *Nuestra Señora del Juncal* offshore of the Campeche Coast.

Information is also found in the claims for compensations for losses of merchandise in shipwrecks. An example is provided in A.G.I. Contratación, 730. Juan Ferrer, a ship's master, begged the Casa de Contratación that he be exonerated from any possible responsibility for the damage incurred by the *Santa Inés*, which sailed in the Nueva España Fleet in 1689: "being dismasted in the Bahamas channel between Cape Cañaveral and Saint Helena we survived a violent sea hurricane [*sic*] with north-east and east-north-east winds of such ferocity that they lasted six days during which time we thought ourselves lost together with the *naos* of the fleet."[2]

A.G.I. Contratación 5108 provides yet another description of hurricanes. In this case it occurred in 1589 close to the Florida coast and the account is contained in a letter from General Martín Pérez de Olázabal, issued at Sanlúcar de Barrameda. It informs His Majesty of the strong hurricane that affected the

Nueva España and Tierra Firme fleets on their way to Spain: "sailing out of the channel with the wind large on our quarter we encountered a great storm of wind from the east-north-east that, finding us in the narrows between Florida and the Bahamas, we were battered for five days during which we lost contact with the large naos."

Reports from English Ships

The logbooks of Royal Navy vessels were, in contrast, highly formalized documents, set out in a standard fashion with little latitude for personal embellishment or literary style. Given that seasoned officers who had experienced the worst the elements could hurl at them produced the logbooks, the rather matter-of-fact and laconic style in which they describe most storms, and even many hurricanes, should come as no surprise. Some of the following few examples make this point very clearly. They were abstracted from logbooks of vessels known to be in the Caribbean region during October 1780, a month of notable hurricane activity. The true value of such information lies not so much in the individual detail, which is routine, but in the collective picture and continuity of account that several such documents create. The mapping of the daily winds, their strengths and the direction of sea (swell) can also be used to help define the reconstructed path of the hurricane(s), to identify previously unknown events, and to confirm, or otherwise, independent reconstructions from other sources.

HMS *Hector* (74 gun battleship) was cruising off the east end of Hispaniola in early October 1780. On October 6, the logbook entry was:

> Winds: EbN (east-by-north), strong gales and squally, at 7 (pm) the gale increasing, at 12 (midnight) increasing to a hurricane. The sea running high.

The next day, the situation had moderated and the winds veered:

> Winds: W, hard gales. Heavy sea from the west.

The most severe of the month's hurricanes is thought to have approached Florida between October 11 and 18. On October 16, the entry was:

> Winds: EbN, strong gales. Heavy swell from the east

And on October 19:

> Winds: NbE, fresh breezes. Heavy swell from the north.

HMS *Ajax* (74 guns) was nearby off (British) Dominica. Its entry for October 12 reads:

> Winds: all around the compass. Strong gales and heavy squalls. At 6 (am) violent squalls of wind. At noon less wind. Very heavy sea. At noon the sea very high and confused. Lightning.

At the same time, HMS *Charon* (44 guns) was off Charleston. Here the worst of the weather, and by no means a hurricane, was recorded on October 17:

> Winds: NNE and NbE. Strong gales and heavy seas.

These examples show how fragments of evidence might be put together to provide a fuller picture. Such information from sea-going vessels provides also a valuable complement to land-based information. The problem of vocabulary is evident and the issue surrounding the precise meaning of terms such as "strong gales" has yet to be fully addressed.

Reports on Landfalling Hurricanes

There are abundant references to disasters on the coasts, mostly contained in Gobierno, Patronato, Ultramar, Secretaría de Guerra, Diversos, and Papeles de Cuba sections from the A.G.I. The Actas de Cabildo (Mayoralty Acts) are the written notes produced by the members of the mayoralties (councils) during their weekly meetings. They are an excellent source of information on any event interrupting the daily life of the city. Climate-related incidents, especially extreme events such as floods, droughts, and hurricanes affecting the city and its surroundings, were always noted in these documents. For Cuba and Puerto Rico the A.G.I. kept records for a period of more than four hundred years and they constitute a major source of environmental information for the study of hurricanes and their impacts. Of particular interest are the *Reports of the General Navy Command of the Antilles*, the *Gazetas* and other periodic publications such as the *Partes de tranquilidad*, the monthly papers produced by the authorities, or the reports from the Meteorological Services established in the area during the nineteenth century. These reports can be found in different A.G.I. sections. A.G.I. Patronato 181, R.25 for example provides an account of the damage caused by a hurricane on November 2, 1552, in San Juan de Ulúa (close to Veracruz) in which the mayor of the city described the storm damage and the actions taken to alleviate its consequences.

A copy of an informative document (A.G.I. Secretaría de Guerra 7241, exp. 26) shows the certificate of a lieutenant colonel supporting the claim of the repairs to a partially completed fort in Placaminas (Plaquemines) close to New Orleans (figure 6.2). The works were badly damaged by a hurricane on August 18, 1793. The contractor claimed against the Spanish government for additional funding to complete the fort. A copy of a report produced by the governor of Puerto Rico (AHN/ Ultramar 2007, exp. 2) on the damages caused by a hurricane that struck the island on July 26–27, 1825, is shown in figure 6.3. The report itemized for the different villages on the island the loss of lives, number of injured persons, and damage to houses, farming and livestock. AHN/ Ultramar 5066, exp. 14 and 15 contains a report on the simultaneous occurrence of a hurricane and earthquake affecting Puerto Rico and La Española on September 12, 1846.

The archival information is not limited to the gulf area, there is also evidence of hurricanes affecting Venezuela and Panamá. A.G.I./ Secretaría de Guerra 7171, exp. 22 contains a report of the General Captain based in Caracas on the "furious hurricane" in Margarita Island on August 11, 1790.

Some Detailed Descriptions

The combination of documents of a varied nature, such as those described previously, can provide detailed information on location, intensity and even trajectory of the hurricanes. Some examples are provided here.

The Hurricane of Mid-October 1768 in Cuba

This hurricane can be traced in two different documents. Firstly the logbook of *El Quirós* (A.G.I. Correos 271B R11), which describes how the mail ship encountered a frigate coming from Boston that had been affected by the hurricane on October 7. The logbook continues by referring to events on October 21 "at seven in the morning on passing through the Royal Passage I encountered much flotsam of wood from ships, boxes, windows and general stores."

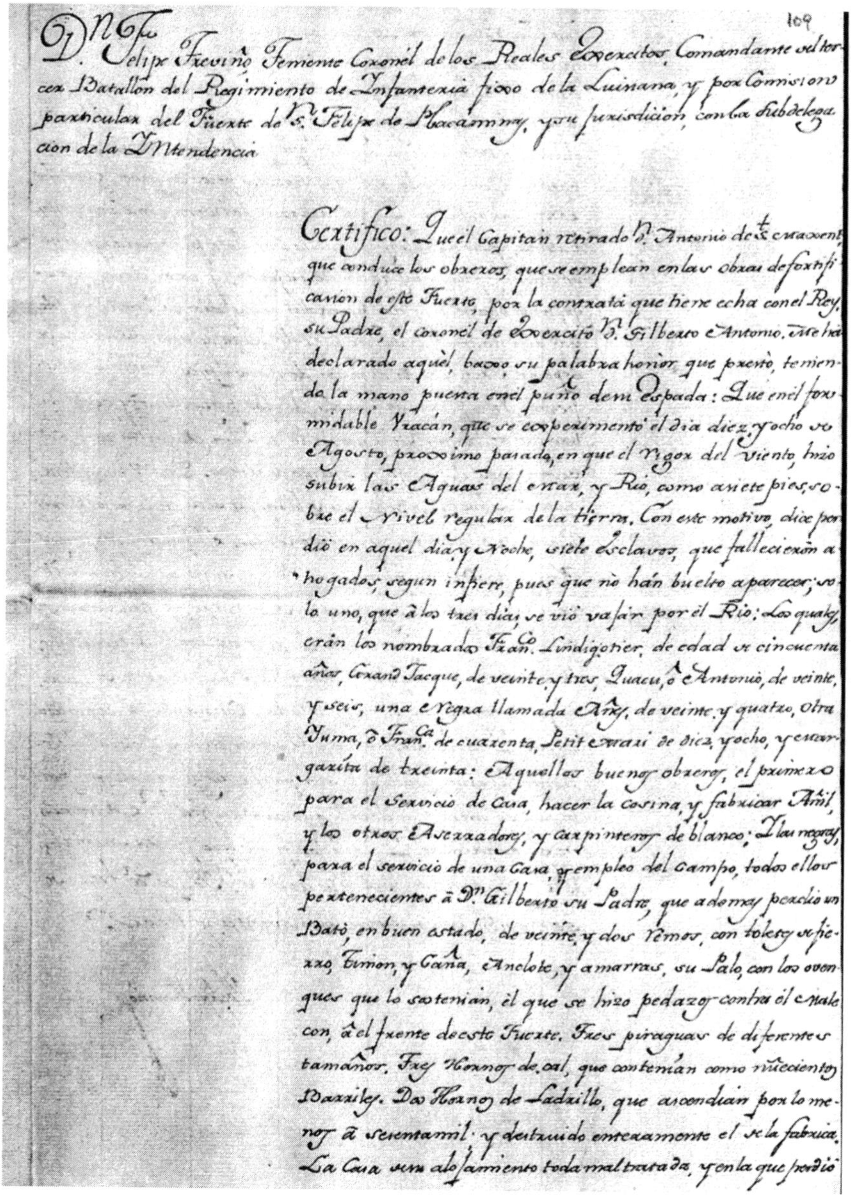

109

D.ⁿ Felipe Treviño Teniente Coronel de los Reales Exercitos, Comandante del tercer Batallon del Regimiento de Infanteria fixo de la Luisiana, y por Comision particular del Fuerte de S.ⁿ Felipe de Placamines, y su Jurisdiccion, con la Subdelegacion de la Intendencia

Certifico: Que el Capitan retirado D.ⁿ Antonio de S.ᵗ Maxent, que conduce los obreros, que se emplean en las obras de fortificacion de este Fuerte, por la contrata que tiene echa con el Rey su Padre, el Coronel de Exercito D.ⁿ Gilberto Antonio, me ha declarado aquel, baxo su palabra honor, que prestó, teniendo la mano puesta en el puño de su Espada: Que en el formidable Huracan, que se experimentó el dia diez y ocho de Agosto, proximo pasado, en que el rigor del viento, hizo subir las Aguas del Mar, y Rio, como a siete pies, sobre el nivel regular de la tierra, Con este motivo, dice perdió en aquel dia y Noche, siete Esclavos, que fallecieron ahogados, segun infiere, pues que no han buelto a parecer; solo uno, que a los tres dias se vió salir por el Rio: Los quales, eran los nombrados Fran.co Lindigotier, de edad de cincuenta años, Gerard Jacque, de veinte y tres, Luacu, ó Antonio, de veinte y seis, una Negra llamada Any, de veinte y quatro, otra Juma, ó Fran.ca de cuarenta, Petit Maxi de diez y ocho, y Margarita de treinta: Aquellos buenos obreros, el primero para el servicio de Casa, hacer la cosina, y fabricar Añil, y los otros Aserradores, y carpinteros de blanco; Las negras, para el servicio de una Casa, y empleo del campo, todos ellos pertenecientes a D.ⁿ Gilberto su Padre, que ademas perdio un Bató, en buen estado, de veinte y dos remos, con toletes de fierro, Timon, y Caña, Anclote, y amarras, su Palo, con los ovenques que lo sostenian, el que se hizo pedazos contra el Muelle con, a el frente de este Fuerte. Tres piraguas de diferentes tamaños. Tres Hornos de cal, que contenian como nuecientos Barriles. Dos Hornos de Ladrillo, que ascendian por lo menos a setentamil y destruido enteramente el de la fabrica. La Casa de su alojamiento toda maltratada, y en la que perdió

FIGURE 6.2 Report of an army officer supporting the claims of the builder of the fort in Plaquemines (close to New Orleans) for reimbursement for the damages of a hurricane (courtesy of the A.G.I.).

Ysla de S. Juan de Puerto Rico

Estado que manifiesta los estragos sufridos en dicha Ysla de resultas del temporal acaecido en ella la noche del 26 al 27 de Julio del año de 1825.

Pueblos	Desgracias: Personas muertas	Id. heridas	Id. contusas	Pérd. de casas: Casas caídas	Id. desarmadas	Pd. de plant. y siemb. de f.s men.: Cuerdas de plátanos	Cuerdas de cafetos	Id. de arroz	Id. de caña	Id. de maíz	Palmas de coco	Pérdidas de animales: Ganado mayor muerto	Id. menor muerto	Bestias muertas	Cerdos muertos	Aves muertas	Val. tot. Pesos
Patillas	24	177		368	47							34	37	23	38		
Maunabo	12			250	9							94		23			
Yabucoa	82	193	184	502	46	652	312½	487	54	385	739	225	201	73	182	5345	
Naguabo	2			500	175	384	187	342	73	165	320	57	4	13	78	622	
Humacao	141		200	200													
Piedras	5	139	126	430	31							172	96	60	51		
Juncos	11		72	162	143	408	172	242	40	107	73	138	181	60	109	7376	
Gurabo	11			147	27	452	377	162½	67½	98		113	90	22	42	8744	
Hato grande	17		81														
Caguas	20			464	292	13792	1013	327½	27	225	99	203	125	55	109	5075	
Río piedras	10																
Guaynabo	3		241	603	116							20	18	29	95	6000	
Bayamón	15	2		726	216	612	261	362	136	210	397	51	9	51	3		
Toa alta	7			468								11		5			
Toa baja	0																
Vega alta	2	3															
Vega baja	2	2		264													
Morovis	3			174								5		2			
Ciales				61	71	101	7	135½	7½	72	20	4					
Manatí	2	1										75					
Arecibo	3																
Hatillo	1		7	153													
Camuy				88								2					
Cayey	6			273													
Cidra				587		475	532	151	17	114		46		15	70	2501	
Trujillo bajo	1		1	58								255		27			
Aguada	1																
Barranquitas				250								80					
Corozal	3																
Fajardo				100													
Loíza	6			200								60		60			
Luquillo				75													
Utuado	2																
Total	279	477	884	6883	1179	25893	9963	9274	870½	1193½	1610	1796	960	606	858	24662	
Valor calculado				924160	29475	25825	103176	46110	27025	11335	3220	35920	1920	7580	4290	6165½	714090-6

Nota: En la Capital se sintieron la casa de gob.no, el lienzo de muralla que cae á la Marina, la casa blanca ó de Ponce de León, los Para-rayos de los Almacenes de Pólvora, el [illegible] y muchas casas sufrieron, lo mismo que los castillos en sus [illegible]
Otra: El resto de pueblos hasta 53 que componen el total no padecieron casi nada, los cuales están situados al sur y oeste de la Ysla.
Otra: [illegible]
Otra: [illegible]
Otra: [illegible]

FIGURE 6.3 Report of the governor of Puerto Rico describing the damages in the island due to the hurricane on July 26–27, 1825 (courtesy of the AHN).

TABLE 6.4 Summary of Damage in Cuba from the Hurricane of October 15, 1768

Location	Tile houses		Guano houses			
	Destroyed	Damaged	Damaged	Destroyed	Deaths	Injuries
Havana	34		488	125	11	31
Horcón, Jesús María y Guadalupe	15		289	70		
Guanabacoa y Arrabales	7	19	983	16	3	35
Partido de Buena Vista	1		411			
Bucaranao			193			
Cruz del Padre			421		1	10
Las Vegas			317			
Santa María del Rosario y Guaraco			456			3
Santuario de Regla			46	25	1	2
Santiago	28		254	69	4	4
Vejucal	6		225	68	1	
San Miguel	5		179	16	4	6
Guines			87	104	5	9
Managuana y Canoa			351	160	6	16

At the same time, a report was produced on the damage that occurred in Cuba on October 15 as the hurricane passed the island. An analysis of the damage to houses and persons in different cities and villages (table 6.4) can give additional evidence about the hurricane's intensity.

The Hurricane of September 13, 1876, in Puerto Rico

This hurricane is carefully described in A.G.I. AHN/Ultramar 374, exp. 2. The document, written by Leonardo de Tejada, Chief Engineer of Public Works, contains a theoretical introduction to the occurrence of hurricanes and a map with the trajectories of the strongest hurricanes that occurred in the nineteenth century (figure 6.4). A complete set of data for this storm is included, with hourly resolution for pressure, temperature, and wind, in figure 6.5. It can be seen that a maximum wind speed occurred at 8:30 A.M. (28 m/s), while the minimum pressure was 976.6 hPa at 6:00 A.M. The trajectory can be seen in figure 6.6; additional references to hurricane parameters are summarized in table 6.5.

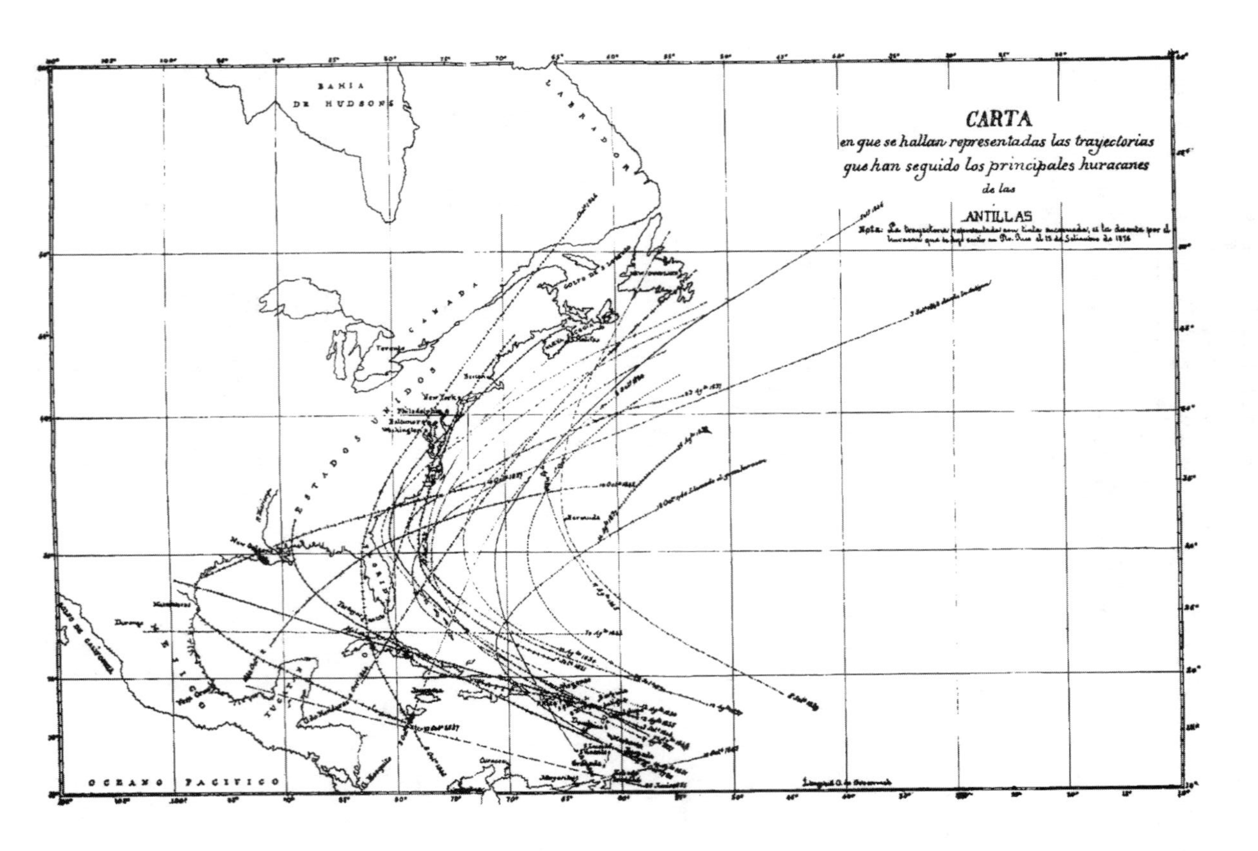

FIGURE 6.4 Trajectories of the strongest hurricanes affecting the Antilles during the nineteenth century, according to a report by Leonardo de Tejada, chief engineer of Public Works in Puerto Rico (courtesy of the AHN).

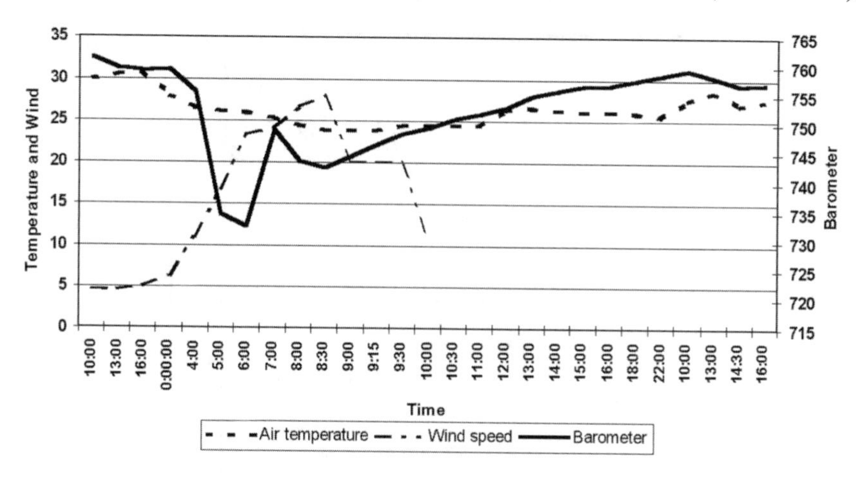

FIGURE 6.5 Hourly data for the hurricane of September 13, 1876, in Puerto Rico, according to the report by Leonardo de Tejada, chief engineer of Public Works in Puerto Rico (A.G.I. AHN/Ultramar 374, exp. 2) (courtesy of the AHN).

Perspectives and Future Development

The cases described in this chapter represent only a small fraction of those that have been identified in preliminary searches of the Spanish and British archives. They have been checked against the information contained in Evans (1848), Millás (1968), Salivia (1950), Tannehill (1940), Viñes (1877), Tuero (1860), and Ludlum (1963). It has been found that the preceding examples corresponding to the years 1552, 1589, 1620, and 1790 have not been identified in any of these previous works. Others, such as those in 1772 or 1876, are well known, but these A.G.I. sources provide complementary information. About 30% of the total storms identified in our preliminary searches can be considered as previously unreported. This confirms that there is significant potential for adding new hurricanes to present day chronologies.

The total sample obtained in the preliminary exploration contains 127 references from Spanish sources. The temporal and spatial distribution of these hurricanes concentrates in the second half of the eighteenth century (figure 6.7). It is still unclear if this concentration is the result of a higher frequency of hurricanes or greater availability of sources. Regarding the spatial coverage, Cuba, Puerto Rico, the open sea, and Florida and Louisiana are the areas providing more information. The information for the whole sample is currently being collated and will be included in a database to allow its free use for the scientific community.

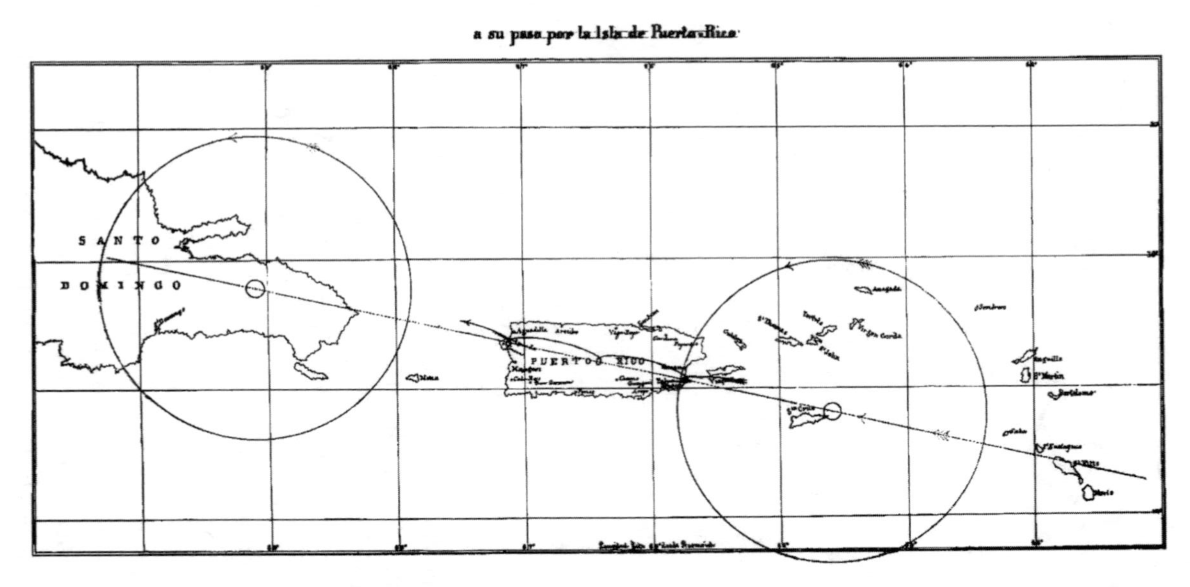

FIGURE 6.6 Trajectory of the hurricane of September 13, 1876, in Puerto Rico, according to the report by Leonardo de Tejada, chief engineer of Public Works in Puerto Rico (A.G.I. AHN/Ultramar 374, exp. 2) (courtesy of the AHN).

TABLE 6.5 Characteristics of the Hurricane of September 13, 1876, in Puerto Rico

Parameter	Value
Direction	ESE to WNW from St. Kitts to Puerto Rico
Speed	35 km/h
Diameter	280 km
Eye diameter	15 km
Maximum speed in San Juan	100–130 km/h
Temperature decrease	28 to 24°
Increase in relative humidity	71 to 94%
Minimum pressure	734.65 mm

Note: This information was described by Tejada in A.G.I. AHN/Ultramar 374, exp. 2.

This first analysis illustrates the potential of these sources, some of the problems associated with assembling these sources, and allows some conclusions to be drawn that have a bearing on future studies in this field.

- The Spanish and British archives contain significant volumes of information that are relevant for hurricane reconstructions and reanalysis. This information is especially valuable for the identification of "forgotten" hurricanes, particularly from the sixteenth to eighteenth centuries, an era when the information is more scarce in other local sources, and for the reduction of uncertainties related to the timing, trajectory, and/or intensity of already-known hurricanes.
- These sources are especially rich for the Caribbean area and in particular for Cuba, Puerto Rico and the Gulf Coast. These areas have been the most densely populated regions of the Caribbean since the beginning of the sixteenth century. Additional information can be obtained for other areas such as Panama, Colombia, Venezuela and the open sea. British records are abundant not only for the Caribbean region from the mid-seventeenth century, but also for the American seaboard (the present-day states of North and South Carolina, Georgia, Virginia, and lands to the north) at least until the War of Independence.
- The need to agree upon a strategy for future research presents a number of challenges. Logbooks provide the greatest geographical coverage, collectively cover a long time span, and pay particular attention to weather conditions. They do, however, have some limitations;

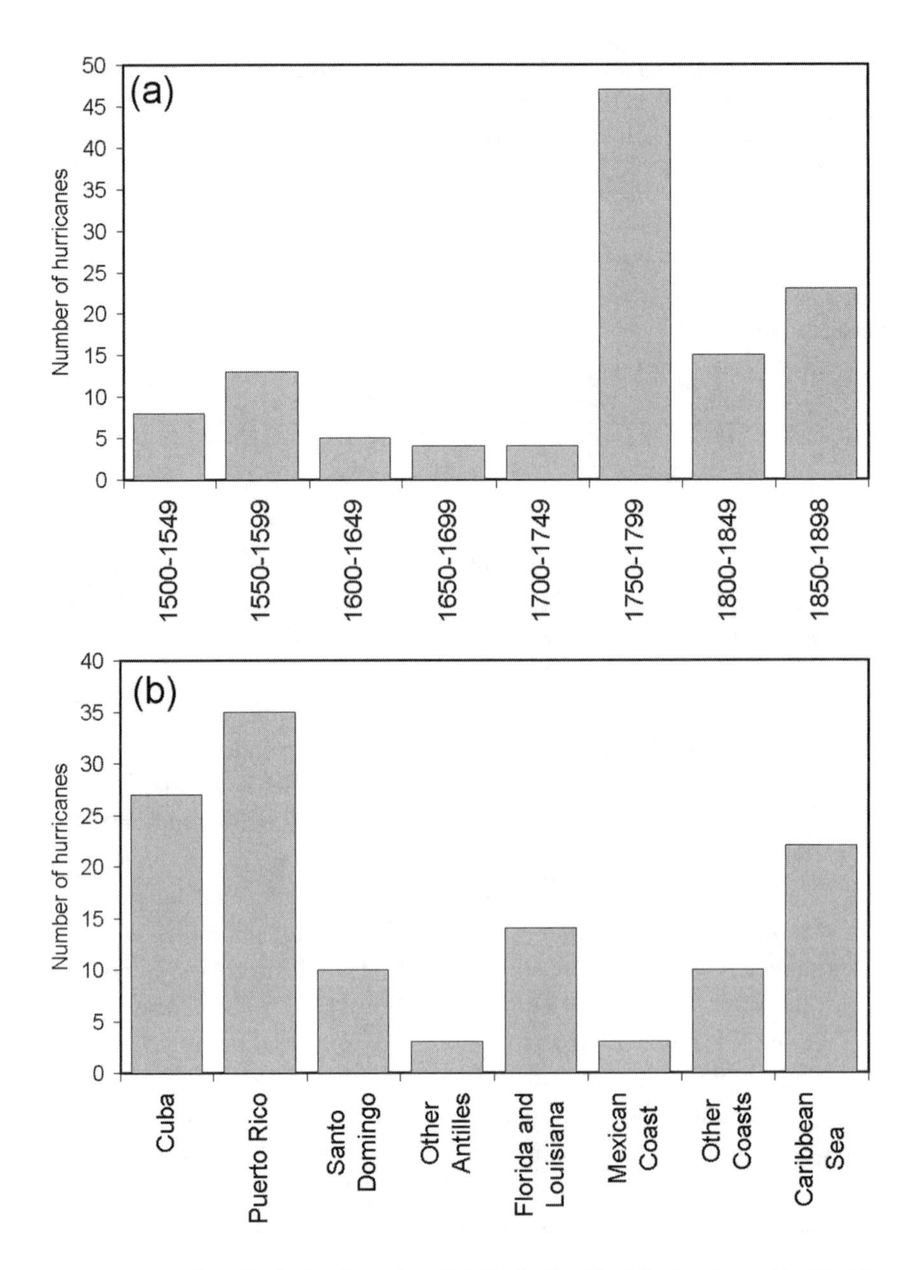

FIGURE 6.7 Time distribution (*a*) and spatial distribution (*b*) of the hurricanes identified by exploratory research in the Spanish Archives.

most importantly, unless close to shore a ship's logbook can provide little information on the landfall location of hurricanes. In addition many ships sought the refuge of harbors and anchorages during the hurricane season, and in these situations logbook entries are scarce or may even cease altogether. The logbooks may therefore yield only a partial view of individual events.

- Searches for hurricane landfall sites are circumscribed by the nature of the archive sources. In particular shifting political boundaries often caused documents to be cataloged under different headings. Even with greater experience and knowledge of these sources, future works will require significant amounts of time to undertake the unavoidably lengthy task of sifting through the vast volume of documents that have survived from earlier centuries.

NOTES

1. An overview of those pertaining to the United States can be found at http://www.ngdc.noaa.gov/paleo/hurricane/.
2. A *nao* was an ancient, high-sided sailing vessel. Saint Helena refers to St. Helena Sound, between Charleston and Savannah.

REFERENCES

A.G.I.	Archivo General de Indias, Seville
AHN	Archivo Histórico Nacional, Madrid
A.M.M.	Archivo del Museo de la Marina, Madrid
B.N.	Biblioteca Nacional, Madrid
R.A.H.	Real Académia de la Historia, Madrid
A.G.S.	Archivo General de Simancas, Valladolid.
PRO	Public Records Office, Kew, London
NMM	National Maritime Museum, Greenwich
ADM/L	Lieutenants' logbooks in the Admiralty document collection at the National Maritime Museum, Greenwich
ADM/51	Captains' logbooks at the Public Records Office, Kew
ADM/52	Masters' logbooks at the Public Records Office, Kew

Catchpole, A. J. W. 1992. Hudson's Bay Company ships' log-books as sources of sea ice data, 1751–1870. In *Climate Since A.D. 1500*, edited by P. D. Jones and R. S. Bradley, 17–39. London: Routledge.

Dunn, G. E., and B. I. Miller. 1964. *Atlantic hurricanes.* Baton Rouge: Louisiana State University Press.

Evans, S. J. 1848. A chronological list of hurricanes which have occurred in the West Indies since the year 1493; with interesting descriptions. *Nautical Magazine* 397, 453, 524.

Falconer, W. 1780. *An universal dictionary of the marine.* London: Cadell.

Fernández Duro, C. 1903. *La Armada Española desde la unión de los reinos de Castilla y Aragón.* Madrid: Estudios Tipográficos G. de Rivadeneyra.

Fernández-Partagás, J., and H. F. Díaz. 1996. *A historically significant revision of Atlantic tropical cyclone frequency, 1851 to 1890.* Final Report Supplement, ERL-NOAA Report 40 RANR 503516. Boulder, Colo.: National Oceanic and Atmospheric Administration.

Fry, H. T. 1967. The emergence of the Beaufort scale. *Mariner's Mirror* 53:311–13.

García, R. R., H. F. Díaz, R. García Herrera, J. Eischeid, M. R. Prieto, E. L. Hernández, E. L. Gimeno, F. Rubio, and A. M. Bascary. 2001. Atmospheric circulation changes in the tropical Pacific inferred from the voyages of the Manila Galleon in the 16th–18th centuries. *Bulletin of the American Meteorological Society* 82:2435–55.

García Herrera, R., L. Gimeno, E. Hernández, M. R. Prieto, and P. Ribera. 2000. Reconstructing the North Atlantic atmospheric circulation in the 16th, 17th, and 18th centuries from historical sources. *Climate Research* 14:147–51.

García Herrera, R., M. R. Prieto, L. Gimeno, E. Hernández, E. López, and R. Herrera. 1999. The Archivo General de Indias as a source of climatic marine surface information. In *Proceedings of the International Workshop on Digitization and Preparation of Historical Surface Marine Data and Metadata,* 61–68. WMO/TD No. 957. Geneva: World Meteorological Organization.

Goell, K., ed. 1970. *A sea grammar: With the plaine exposition of Smiths accidence for young seamen, enlarged.* London: Joseph.

Haensch, G. 2000. *Diccionario del español de Cuba; español de Cuba—español de España.* Madrid: Gredos.

Hewson, J. B. 1983. *A history of the practice of navigation.* Glasgow: Brown, Son and Ferguson.

Jackson, A., A. R. T. Jonkers, and M. R. Walker. 2000. Four centuries of geomagnetic secular variation from historical records. *Philosophical Transactions of the Royal Society of London* A 358:957–90.

Jones, P. D., T. Jonsson, and D. Wheeler. 1997. Extension of the North Atlantic Oscillation using early instrumental pressure observations from Gibraltar and south-west Iceland. *International Journal of Climatology* 17:1433–50.

Lamb, H. H. 1991. *Historic storms of the North Sea, British Isles and Northwest Europe.* Cambridge: Cambridge University Press.

Lavery, B. 1983. *The ship of the line.* Vol. 1, *The development of the battlefleet, 1650–1850.* London: Conway Maritime Press.

Liu, K.-b., C. Shen, and K.-s. Louie. 2001. A 1000-year history of typhoon landfalls in Guangdong, southern China, reconstructed from Chinese historical documentary records. *Annals of the Association of American Geographers* 91:453–64.

Ludlum, D. M. 1963. *Early American hurricanes, 1492–1870*. Boston: American Meteorological Society.

Luterbacher, J., C. Schmutz, D. Gyalistras, E. Xoplaki, and H. Wanner. 1999. Reconstruction of monthly NAO and EU indices back to AD 1675. *Geophysical Research Letters* 26:2745–48.

Manabe, T. 1999. The digitized Kobe collection, phase I: Historical surface marine meteorological observations in the archive of the Japan Meteorological Agency. *Bulletin of the American Meteorological Society* 80:2703–15.

Manwaring, G. E., and W. G. Perrin, eds. 1922. *The life and works of Sir Henry Mainwaring*. Vol. 2. London: Navy Records Society.

Martín Vide, J., and M. Barriendos Vallvé. 1995. The use of rogation ceremony records in climatic reconstruction: a case study from Catalonia (Spain). *Climatic Change* 30:201–21.

Marx, R. F. 1983. *Shipwrecks in the Americas*. New York: Bonanza Books.

Millás, J. C. 1968. *Hurricanes of the Caribbean and adjacent regions, 1492–1800*. Miami: Academy of the Arts and Sciences of the Americas.

Milton, G. 2000. *Big Chief Elizabeth: How England's adventurers gambled and won the New World*. London: Hodder and Stoughton.

Navarrete, A. 1901. Historia marítima militar de España. Armadas Españolas y marinas que le antecedieron en la Península Ibérica. *Sucesores de Rivadeneyra* 11.

Neumann, C. J., B. R. Jarvinen, C. J. McAdie, and J. D. Elms. 1993. *Tropical cyclones of the North Atlantic Ocean, 1871–1992*. NOAA Historical Climatology Series 6-2. Asheville, N.C.: National Climatic Data Center.

Ortlieb, L. 2000. The documented historical record of El Niño events in Perú: An update of the Quinn record (sixteen through nineteen centuries). In *El Niño and the Southern Oscillation: Multiscale variability and global and regional impacts*, edited by H. F. Díaz and V. Markgraf. Cambridge: Cambridge University Press.

Perrin, W. G., ed. 1929. *Boteler's dialogues*. London: Navy Records Society.

Poey, A. 1862. *Table chronologique de quatre cents cyclones*. Paris: Dupont.

Prieto, M. R. 1998. Austral climate and glaciers in the 16th century through the observations of the Spanish navigators. *Quaternary South American Antarctic Peninsula* 11:227–36.

Prieto, M. R. 2000. Archival evidence for some aspects of historical climate variability in Argentina and Bolivia during the 17th and 18th centuries. In *Southern Hemisphere paleo- and neoclimates*, edited by W. Volkheimer and P. Smolka. Berlin: Springer-Verlag.

Quinn, W. H., V. T. Neal, and S. E. Antúnez de Mayolo. 1987. El Niño occurrences over the past four and a half centuries. *Journal of Geophysical Research* 92:14449–61.

Rappaport, E. N., and J. Fernández-Partagás. 1997. History of the deadliest Atlantic tropical cyclones since the discovery of the New World in hurricanes. In *Hurricanes: Climate and socioeconomic impacts,* edited by H. F. Díaz and R. S. Pulwarty, 93–108. Berlin: Springer-Verlag.

Salivia, L. A. 1950. *Historia de los temporales de Puerto Rico (1508–1949).* San Juan, P.R.: Privately printed.

Sobel, D. 1996. *Longitude: The true story of a lone genius who solved the greatest scientific problem of his time.* London: Fourth Estate.

Smyth, W. H. 1867. *A sailor's word book.* London: Blackie.

Tannehill, I. R. 1940. *Hurricanes, their nature and history, particularly those of the West Indies and the sourthern coasts of the United States.* Princeton: Princeton University Press.

Taylor, E. G. R. 1956. *The haven-finding art: A history of navigation from Odysseus to Captain Cook.* London: Hollis & Carter.

Tuero, J. M. 1860. *Huracanes. Vda. de Calero.* Madrid.

Viñes, B. 1877. *Apuntes relativos a los huracanes de las Antillas en Septiembre y Octubre de 1875 y 76.* Havana: Iris.

Wheeler, D. A. 1988. Sailing ships logs as weather records: A test case. *Journal of Meteorology* 13:122–26.

Wheeler, D. A. 1995. The weather vocabulary of an eighteenth century mariner: The logbooks of Nicholas Pocock. *Weather* 50:298–304 [reprinted, 1997, *Marine Observer* 67:22–28].

Wheeler, D. A. 2000. The weather of the European Atlantic seaboard during October 1805: An exercise in historical climatology. *Climatic Change* 45:361–85.

Woodruff, S. D., R. J. Slutz, R. L. Jenne, and P. M. Steurer. 1987. A comprehensive ocean-atmosphere data set. *Bulletin of the American Meteorological Society* 68:1239–50.

7

The Atlantic Hurricane Database Reanalysis Project: Documentation for 1851–1910 Alterations and Additions to the HURDAT Database

Christopher W. Landsea, Craig Anderson, Noel Charles, Gilbert Clark, Jason Dunion, José Fernández-Partagás, Paul Hungerford, Charlie Neumann, and Mark Zimmer

A reanalysis of the Atlantic Basin tropical storm and hurricane database ("best track") for the period of 1851 to 1910 has been completed. The reworking and extension back in time of the main archive for tropical cyclones of the North Atlantic Ocean, Caribbean Sea, and Gulf of Mexico corrected systematic and random errors and biases in the original best-track data and incorporated into the reanalysis historical data by Fernández-Partagás and Díaz (1995a, 1995b, 1996a, 1996b, 1996c, 1997, 1999). Products of the reanalysis project include the revised tropical storm and hurricane database, a metadata file detailing individual changes for each tropical cyclone, a "center-fix" file of raw tropical cyclone observations, a collection of U.S. landfalling tropical storms and hurricanes, and comments from and replies to the National Hurricane Center's Best-Track Change Committee.

This chapter provides documentation of the first efforts to re-analyze the National Hurricane Center's (NHC) North Atlantic hurricane database (or HURDAT, also called best tracks because they are the "best" determination of track and intensity in a post-season analysis of the tropical cyclones). The original database of six-hourly tropical cyclone (i.e., tropical storm and hurricane) positions and intensities was assembled in the 1960s in support of the Apollo space program to help provide statistical tropical cyclone track forecasting guidance (Jarvinen, Neumann, and Davis 1984). Since its inception, this database has been used for many purposes: setting of appropriate building codes for coastal zones, risk assessment for emergency managers, analysis of potential losses for insurance and business interests, intensity forecasting techniques, verification of official and model predictions of track and intensity, seasonal forecasting, and climatic change studies.[1] Given its original purpose and limitations, the data in HURDAT may not be appropriate for all those uses.

There are many reasons why reanalysis of the HURDAT dataset was needed and timely. HURDAT contained systematic and random errors that needed correction (Neumann 1994). In addition, as our understanding of tropical cyclones developed, analysis techniques at NHC changed, leading to biases in the historical database that had not been addressed (Landsea 1993). Also, it was difficult to use the hurricane database in studies concerned with landfalling events because HURDAT had no data for a storm's location, time, and intensity at landfall. Finally, recent efforts led by José Fernández-Partagás have greatly increased our knowledge of previously undocumented historical tropical cyclones in the mid-nineteenth to the early twentieth century (Fernández-Partagás and Díaz 1996a). The reanalysis of the HURDAT dataset incorporates these advances, corrects known errors, and provides additional information on landfall, metadata, and raw data.

Currently, the HURDAT database is updated at the end of each year's hurricane season after the NHC hurricane specialists perform a post-season analysis of that year's storms. The most recent documentation generally available for the database is a NOAA Technical Memorandum by Jarvinen, Neumann, and Davis (1984). This reference is still valid for most descriptions of the tropical cyclone database, but it, too, is in need of revision. This chapter is designed to help provide a more up-to-date documentation for HURDAT.

A reanalysis of the Atlantic tropical cyclone database is justified by the need to address these deficiencies as well as to extend the historical record back in time. This chapter details the first efforts to improve the accuracy and consistency of HURDAT for the years 1886 to 1910 as well as to introduce an additional 35 years (1851–1885) into the archived database of Atlantic tropical storms and hurricanes.

Outline of Databases Provided in the Reanalysis

As part of the reanalysis effort, five files were made available:[2]

1. The revised Atlantic HURDAT: This contains six-hourly intensity and position estimates of all known tropical storms and hurricanes. The six-hourly intensity measurements are defined by the maximum-sustained 1-minute winds at the surface (10 m) and, when available, cen-

tral pressures. Position is evaluated to the nearest 0.1° latitude and longitude.

2. A HURDAT metafile: This documentation file has detailed information about each change in the revised HURDAT. Included are the original HURDAT values of position and/or intensity, the revised values in HURDAT, and the reasoning behind the changes.
3. A "center-fix" file: This file is composed of raw observations of tropical cyclone positions (thus "center fixes") and intensity measurements from either ships or coastal stations.
4. A U.S. landfalling tropical storm and hurricane database: This file contains information on the exact time, location, intensity, radius of maximum winds (RMW), environmental sea level pressure and storm surge for tropical storms and hurricanes that make landfall on the continental United States. This database also includes storms that affect land but whose centers do not make landfall.
5. NHC Best Track Change Committee comments: This file provides detailed comments from the NHC's Best Track Change Committee, a group tasked with approving alterations to the HURDAT database. Replies by the authors to the various comments and recommendations are also included.

The Work of José Fernández-Partagás

Efforts to digitize and quality control the work of Fernández-Partagás and Díaz (1995a, 1995b, 1996a, 1996b, 1996c, 1997, 1999) produced the largest additions and alterations to HURDAT. Fernández-Partagás and Díaz utilized a variety of sources for their research: ship reports in newspapers, individual and seasonal summaries published in the *Monthly Weather Review*, documents from government agencies, historical reviews and scientific publications (table 7.1). A distillation of this information by the reanalysis project led to the creation of completely new tropical cyclone tracks and intensities for the years 1851 to 1885 and the alteration of existing track and intensity data for the period of 1886 to 1910. The reanalysis effort also corrected many of the existing systematic and random errors that existed in the 1886 to 1910 portion of HURDAT. The improvements included: (1) corrected interpolations of winds near landfall, (2) more realistic speed changes at the beginning and/or end of the tropical cyclone track, (3) improved landfall locations, and (4) corrected the reduction of inland winds using Kaplan and DeMaria's (1995, 2001) methodology. A

TABLE 7.1 Sources Utilized by Fernández-Partagás and Díaz in Their Original Work

Sources
Ships' reports published in the *New York Times, the Times* of London, and *Gaceta de la Habana*
Individual storm and seasonal summaries in *Monthly Weather Review*
Historical Weather Maps series
Reports of the chief of the U.S. Weather Bureau
Other sources
Academia de Ciencias (1970)
Alexander (1902)
Cline (1926)
Dunn and Miller (1960)
Garcia-Bonnelly (1958)
Garriott (1900)
Gutierrez-Lanza (1904)
Ho et al. (1987)
Instituto Cubano de Geodesia y Cartografia (1978)
Ludlum (1963)
Martinez-Fortun (1942)
Mitchell (1924)
Neumann et al. (1993)
Ortiz-Hector (1975)
Rappaport and Fernández-Partagás (1995)
Rodriguez-Demorizi (1958)
Rodriguez-Ferrer (1876)
Salivia (1972)
Sarasola (1928)
Simpson and Riehl (1981)
Sullivan (1986)
Tannehill (1938)
Tucker (1982)
Viñes (1877)
Viñes (1895)

Sources: Sources utilized in the reanalysis effort, in addition to those listed in table 7.1, are Abraham, Parkes, and Bowyer (1998), Barnes (1998a, 1998b), Boose, Chamberlin, and Foster (2001), Boose, Serrano, and Foster (2004), Coch and Jarvinen (2000), Connor (1956), Doehring, Duedall, and Williams (1994), Ellis (1988), Hebert and McAdie (1997), Ho (1989), Hudgins (2000), Jarvinen (personal communication, 1990), Jarrell, Hebert, and Mayfield (1992), Neumann et al. (1999), Parkes et al. (1998), Perez, Vega, and Limia (2000), Roth (1997a, 1997b), Roth and Cobb (2000, 2001), Sandrik and Landsea (2003), and Sandrik and Jarvinen (1999).

number of sources beyond those utilized by Fernández-Partagás and Díaz were also used in the reanalysis work, which are detailed in table 7.1.

Jose Fernández-Partagás's painstaking and time-consuming research was detailed in full in the volumes from Fernández-Partagás and Díaz (1995a, 1995b, 1996b, 1996c, 1997 and 1999). An example of the documentation that he provided is shown here for the first storm of 1856. (The storm track mentioned as Fig. [1] in this quote is shown as Storm 1 in plate 4.)

> *Storm 1, 1856 (Aug. 10–11)*
> Tannehill (1938) has mentioned this storm as having occurred along the Louisiana coast. Dunn and Miller (1960) and Ludlum (1963) have also mentioned this storm. The author of this study has prepared the storm track which is displayed in Fig. [1] [plate 4, this volume].
>
> The *New-York Daily Times,* Aug. 16, 1856, p.1, col.1, published that there had been a storm in the New Orleans area on August 10 and that such a storm had been most disastrous at Last Island (Ile Derniere). A narrative of what had happened at Last Island included some meteorological remarks: Heavy N.E. winds prevailed during the night of August 9 and a perfect hurricane started blowing around 10 A.M. August 10. The water commenced to rise about 2 P.M. and by 4 P.M. currents from the Gulf and the Bay had met and the sea waved over the whole island (*The New-York Daily Times,* Aug. 21, p. 3, col. 4).
>
> The following information has been extracted from Ludlum (1963): The ship "C. D. Mervin" passed through the eye of the storm off the Southwest Pass. Captain Mervin checked the barometer at 8 A.M. Aug. 10 and noticed a reading of 28.20 inches, a 24-hr drop of 1.70 inches. At 9 A.M. the ship had a calm which lasted for 5 minutes. The sun shone and there was every appearance of clearing off but the wind suddenly struck the ship from the opposite direction. For two more hours, more a southerly hurricane struck the ship and then gradually abated. After the hurricane, the ship location was found to be only 60 miles to the W.S.W. of Southwest Pass.
>
> At Iberville, Parish of Vermillon, the Aug. 10–11 storm raged with terrific force but only gales were reported at New Orleans, where the maximum wind at observation time was force 8 on the Beaufort scale (39–46 miles per hour) from an easterly direction at 2 P.M. August 10 (Ludlum 1963).
>
> It can be inferred from the above information that Storm 1, 1856 was a hurricane which was moving on a northwesterly course as shown in Fig. [1].

Center-Fix Files

From the observations uncovered by Fernández-Partagás for this storm—Storm 1, 1856—the following data were archived, as shown in table 7.2. A center-fix position observation was unavailable for this storm, so a sample data point for Storm 5, 1852, is shown as an example in table 7.2.

The conversion from descriptive measures of winds to quantitative wind speeds, although quite subjective, is assisted by the usage of the Beaufort Scale. The Beaufort Scale was developed in 1805 by Admiral Francis Beaufort as a wind force scale for sailing ships. By 1838, the British Royal Navy had made it mandatory to use the scale in log entries (Kinsman 1969). Subsequently, the

TABLE 7.2 Wind, Pressure, and Location Data for Two Storms

Storm 1, 1856[1]					
Date	Time	Wind/ direction	Pressure[2]	Location	Source[3]
August 10, 1856	? UTC	40 kt/?	? mb	29.3N 89.9W	Fort Livingston
August 10, 1856	? UTC	60 kt/?	? mb	30.3N 91.4W	Iberville Parish
August 10, 1856	0900 UTC	70 kt/N-S	955 mb	28.6N 90.2W	*C. D. Mervin*
August 10, 1856	1400 UTC	40 kt/E	? mb	30.0N 90.1W	New Orleans
August 10, 1856	2100 UTC	70 kt/?	? mb	29.0N 90.9W	Last Island
August 10, 1856	2200 UTC	70 kt/?	? mb	29.7N 91.2W	Bayou Boeuf
August 11, 1856	? UTC	40 kt/?	? mb	30.4N 91.2W	Baton Rouge
August 11, 1856	? UTC	40 kt/?	? mb	32.2N 91.1W	New Carthage
August 11, 1856	? UTC	60 kt/?	? mb	31.6N 91.4W	Natchez
Storm 5, 1852 (center positions)[4]					
Date	Time	Wind/ direction	Pressure	Location	Source
October 9, 1852	? UTC	90 kt/ENE-SSW		25.6N 86.5W	*Hebe*

[1] No center-fix locations are available for this storm. The latitudes and longitudes refer to observation locations.

[2] In center-fix files, if the sea level pressure measurement was determined to be a "central pressure," C is indicated after the value. Otherwise, the pressure value was considered to be a peripheral (either eyewall or rainband environment of the storm) observation.

[3] Sources are either from coastal or inland station data or from ship data. Ship data are indicated by the name of the ship in italics.

[4] Latitude and longitude indicate center-fix location for this storm. This entry shows format for center-fix files in HURDAT reanalysis.

scale evolved into one associated with specific wind speed ranges as specified by interpretations of the state of the sea rather than the wind's impact on sails (table 7.3). Due to limitations at the top end of the Beaufort Scale, the center-fix and best-track data in the reanalysis generally list ship reports of "hurricane" force winds as 70 kt (36 m s^{-1}) winds. The listed wind speeds were boosted to 90 kt (46 m s^{-1}) when ship reports included terms such as "severe," "violent," "terrific," or "great hurricane." Hurricanes at sea were not assigned a best-track intensity value of major hurricane (Saffir-Simpson Scale Category 3, 4 or 5; 96 kt [50 m s^{-1}] or greater maximum sustained surface wind speeds) unless the intensity could be confirmed using corresponding central pressure data. Caution is warranted in the direct use of these Beaufort Scale wind estimates for tropical storm and hurricane intensity assignments due to lack of consistency and standardization in the scale during the late nineteenth and early twentieth centuries (Cardone, Greenwod, and Cane 1990). In many cases, however, these Beaufort Scale measurements are the only clues available for estimating the intensity of tropical cyclones of this era.

Occasionally, there were ship observations with no specific dates available. When other ship or land observations could help pinpoint the storm's timing, the undated ship observations were used to provide information about the track of the storm (e.g., a southwest gale noted by a ship's captain would indicate a tropical cyclone located to the northwest of the ship's position). Less frequently, "dateless" ship observations were also used in determining a storm's intensity.

There were generally two types of land-based observations of wind speed available during the second half of the nineteenth and the early twentieth century: visual estimates and the four cup Robinson anemometer (Ludlum 1963; Ho 1989). Visual estimates, though crude, were somewhat standardized by use of a 10-point scale used by volunteers of the Smithsonian Institution and U.S. Army observers at various forts (M. Chenoweth, personal communication, 2001) (table 7.4).

Of modestly more reliability was the four-cup anemometer, first developed by Robinson in the 1840s (Kinsman 1969). Primary difficulties with this instrument were calibration and its mechanical failure in high wind conditions. Even as late as 1890, the highest wind that could be reliably calibrated with this instrument was only about 30 kt (from a whirling machine), due to lack of a strict comparison with a known quantity of stronger winds (Fergusson and Covert 1924). By the early 1920s, wind tunnels allowed for calibration against much stronger winds. These showed that the winds from these early cup anemometers had a strong overestimation bias, which was most pronounced at

TABLE 7.3 The Beaufort Wind Scale

Beaufort number	Knots	Description	Specifications at sea
0	< 1	Calm	Sea like a mirror
1	1–3	Light air	Ripples with the appearance of scales, but without foam crest
2	4–6	Light breeze	Small wavelets, still short but more pronounced; crests have a glassy appearance and do not break
3	7–10	Gentle breeze	Large wavelets; crests begin to break; foam of glassy appearance; perhaps scattered white horses
4	11–16	Moderate breeze	Small waves, becoming longer; fairly frequent white horses
5	17–21	Fresh breeze	Moderate waves, taking a more pronounced long form; many white horses (chance of some spray)
6	22–27	Strong breeze	Large waves begin to form; white foam crests are more extensive everywhere (probably some spray)
7	28–33	Near gale	Sea heaps up; white foam from breaking waves begins to be blown in streaks in the direction of the wind
8	34–40	Gale	Moderately high waves of greater length; edges of crests begin to break into spindrift; foam is blown in well-marked streaks along the direction of the wind
9	41–47	Strong gale	High waves; dense streaks of foam along the direction of the wind; crests of waves begin to topple, tumble, and roll over; spray may affect visibility
10	48–55	Storm	Very high waves with long overhanging crests; the resulting foam, in great patches, is blown in dense white streaks along the direction of the wind; on the whole, the surface of the sea takes on a white appearance; the tumbling of the sea becomes heavy and shock-like; visibility affected
11	56–63	Violent storm	Exceptionally high waves (small and medium-size ships might be for a time lost to view behind the waves); sea completely covered with long white patches of foam lying along the direction of the wind; everywhere the edges of wave crests are blown into froth; visibility affected
12	> 63	Hurricane	Air filled with foam and spray; sea completely white with driving spray; visibility very seriously affected

Source: Fitzpatrick (1999).

TABLE 7.4 The Smithsonian Institution and Military Fort Wind Force Scale

Category	Description	Wind speed
1	Very light breeze	2 mph (2 kt)
2	Gentle breeze	4 mph (4 kt)
3	Fresh breeze	12 mph (10 kt)
4	Strong breeze	25 mph (22 kt)
5	High breeze	35 mph (30 kt)
6	Gale	45 mph (39 kt)
7	Strong gale	60 mph (51 kt)
8	Violent gale	75 mph (65 kt)
9	Hurricane	90 mph (78 kt)
10	Most violent	100 mph (87 kt)

Note: Values are estimates of the highest gusts (Ludlum 1963; Ho 1989; M. Chenoweth, personal communication, 2001).

very strong wind speeds (Fergusson and Covert 1924). For example, an indicated wind of minimal hurricane force (64 kt) in actuality was only about 50 kt. Moreover, most of these early four cup anemometers were disabled or destroyed by wind before the highest winds of hurricanes occurred. The strongest observed winds in an Atlantic hurricane by this type of anemometer was a 5-minute sustained wind measurement of 120 kt in Storm 2, 1879, just before the instrument was destroyed by this North Carolina-landfalling hurricane (Kadel 1926). A standard of 5 minutes was typically used for "maximum winds" in reports by the Weather Bureau of the Signal Corps of the U.S. Army due to instrumental uncertainties in values for shorter time period winds. With reliable calibrations available in the 1920s, this extreme wind's true velocity was only about 91 kt. Current understanding of gustiness in hurricane conditions suggests a boost of 1.05 to convert from a 5-min to a 1-min maximum sustained wind (Dunion et al. 2003), giving a best estimate of the maximum 1-min sustained wind of about 96 kt.

Coastal station wind data listed in the center-fix files are the original measurements provided. It is in the interpretation of these data for inclusion into the best track that these various biases and limitations (i.e., strong overestimation in high wind regime, conversion of 5-min to 1-min wind, and instrumental failure) are taken into account. More on the difficulties of the intensity estimations may be found in the Limitations and Errors section.

WIND-PRESSURE RELATIONSHIPS

Sea level atmospheric pressure measurements (either peripheral pressures or central pressures) can be used to estimate the maximum sustained wind speeds in a tropical cyclone, in the absence of in situ observations of the peak wind strength. In the case of Storm 1, 1856, the ship *C. D. Mervin* observed a peripheral pressure of 955 mb (table 7.2), likely while in the western eyewall. Central pressures of tropical cyclones can be estimated from such peripheral pressure measurements if relatively reliable values of the RMW and environmental (or surrounding) sea level pressure can also be obtained. Radius of maximum wind information was occasionally obtained from ships or coastal stations that were unfortunate enough to have the eye of the hurricane pass directly overhead. Careful notation of the times of the peak winds and the calm of the eye, along with the best estimate of the translational speed of the hurricane, allowed for direct calculation of the RMW. Another method for estimating RMW was to measure the mean distance from the hurricane's track to the location of the peak storm surge and/or the peak of wind-caused damages. Such RMW measurements or estimates were relatively rare over the open ocean and only somewhat more common as hurricanes made landfall over populated coastlines. Central pressure can then be estimated from the following equation (Schloemer, 1954; Ho 1989):

$$\frac{P_R - P_o}{P_n - P_o} = e^{\left(-\frac{RMW}{R}\right)}$$

where P_R is the sea level pressure at radius R, P_o is the central pressure at sea level, and P_n is the environmental (or surrounding) sea level pressure at the outer limit of a tropical cyclone where the cyclonic circulation ends.

Once a central pressure has been estimated, maximum sustained wind speeds can be obtained from a wind-pressure relationship. The current standard wind-pressure relationship for use in the Atlantic Basin by the NHC (OFCM 2001) is that developed by Dvorak (1984) as modified from earlier work by Kraft (1961).

The reanalysis developed new wind-pressure relationships (described later) to help derive winds from an observed (or estimated) central pressure only in the absence of reliable wind data. These relationships are not intended to give best-track wind estimates for hurricanes in the last few decades of the twentieth century. During this time, accurate flight-level wind measurements were commonly available from reconnaissance aircraft. The new wind-pressure relationship estimates should not supercede the use of any reliable, direct wind observations (rare in the nineteenth and early twentieth centuries), which may be available for a tropical cyclone. It is important to avoid situations where accurate in situ data are modified by estimates from a wind-pressure relationship.

The reanalysis used new wind-pressure relationships for four regions in the Atlantic Basin: the Gulf of Mexico (GMEX), southern latitudes (south of 25°N), subtropical latitudes (25°N–35°N), and northern latitudes (35°N–45°N). Regional wind-pressure relationships were developed because of a tendency for the association to differ depending upon latitude. The equations relating maximum sustained surface wind speeds to a corresponding central pressure as well as those for the Kraft and Dvorak formulations are shown here; representative values are displayed in table 7.5. The tabular wind values are based on the following regression equations:

For GMEX, Wind (kt) = $10.627*(1013-P_o)^{0.5640}$.
Sample size = 664; r = 0.991.

For < 25°N, Wind (kt) = $12.016*(1013-P_o)^{0.5337}$.
Sample size = 1033; r = 0.994.

For 25°N–35°N, Wind (kt) = $14.172*(1013-P_o)^{0.4778}$.
Sample size = 922; r = 0.996.

For 35°N–45°N, Wind (kt) = $16.086*(1013-P_o)^{0.4333}$.
Sample size = 492; r = 0.974.

For Kraft, Wind (kt) = $14.000*(1013-P_o)^{0.5000}$.
Sample size = 13.

The central pressure for these equations is given in units of millibars and *r* refers to the linear correlation coefficient. Dashes in table 7.5 indicate that the pressure is lower than any storm used to develop the wind speed-pressure relationship. Wind and pressure data used for the regression were obtained from the HURDAT file, 1970–1997. The developmental dataset excludes all overland tropical cyclone positions. Data for the < 25°N zone were obtained from longitudes of 62°W and westward. Data for the 25°N to 35°N zone are from 57.5°W and westward. Data for 35°N to 45°N include the longitudes of 51°W and westward. GMEX includes all over-water data west of a line from northeastern Yucatan to 25°N, 80°W. These locations were chosen based on their accessibility by aircraft reconnaissance that can provide both actual wind speed and pressure measurements.

When developing the wind-pressure relationships, attempts were first made to develop the equations with all of the available data for each region. But the overwhelming numbers of observations at the low wind speed ranges over-

TABLE 7.5 Newly Developed Regionally Based Wind-Pressure Relationships for the Atlantic Basin

	Wind speed[2]					
Pressure (mb)[1]	GMEX	< 25N	25–35N	35–45N	Kraft	Dvorak
1000	45	47	48	49	50	45
990	62	64	63	63	67	61
980	76	78	75	73	80	76
970	89	89	85	82	92	90
960	100	100	94	90	102	102
950	110	110	103	97	111	113
940	119	119	110	103	120	122
930	128	127	117		128	132
920	137	135	124		135	141
910	145	143			142	151
900	153	150			149	161
890		157				170

[1] Pressures are central pressures in millibars at sea level.
[2] Winds are maximum 1-minute sustained surface winds (10 m) in knots.

weighted the observations of the tropical storms and Category 1 hurricanes at the expense of the major hurricanes. When the derived equations were compared against the observations of wind and pressure at very high wind values (> 100 kt [51 m s^{-1}]), the fit was quite poor. This bias was overcome by binning the observations into 5-mb groups and then performing the regression. Using this methodology, the observations at the 981- to 985-mb range, for example, were weighted equally to those of the 931- to 935-mb range. After performing the regression this way, a more accurate set of regression equations for the wind and pressure estimates for the Category 3, 4, and 5 hurricane ranges was obtained. Because this method reduces the standard deviation of the sample as well as the sample size, the correlation coefficients are inflated.

In general, the Dvorak formulation is most similar to the Gulf of Mexico and southern latitude relationships. For example, a 960-mb hurricane is suggested to have 102 kt (52 m s^{-1}) sustained surface winds from Dvorak's relationship, which is quite close to the 100 kt (51 m s^{-1}) estimate provided by both the Gulf of Mexico and southern latitude relationships. There is, however, a tendency for the Dvorak wind values to be higher than winds provided by the Gulf of Mexico and southern latitude wind-pressure relationships for the extremely intense (< 920 mb) hurricanes, although the number of data points available for calibration of this end of the wind-pressure curves is quite low. In addition, the Dvorak wind-pressure relationship systematically overestimates the wind speeds actually utilized by NHC for the subtropical and northern latitude hurricanes with central pressures less than 975 mb. For the case of a hurricane with a 960-mb central pressure, the subtropical and northern latitude equations suggest 94 kt (48 m s^{-1}) and 90 kt (46 m s^{-1}), respectively. The weaker winds in higher latitudes can be explained physically with the following reasoning: As hurricanes move poleward, they encounter cooler sea surface temperatures and begin to evolve into an extratropical cyclone. In the process, the tight pressure gradients and resulting wind fields typically weaken and expand outward. This is due in part to structural evolution, but also to less efficient vertical momentum transport by convection in a more stable environment. In addition, increases in the Coriolis force causes a corresponding, but small, decrease in tangential wind speed (Holland 1987). Because these changes become more pronounced as the tropical cyclones move into higher latitudes, an even larger reduction in wind speed occurs poleward of 45°N. It is thus consistent that the Dvorak wind-pressure relationship overestimates winds in higher latitudes because the original formulation of Kraft is based primarily upon observations from the Caribbean Sea and the Gulf of Mexico.

The use of wind-pressure relationships to estimate winds in tropical cyclones has a few associated caveats. First, for a given central pressure, a smaller tropical cyclone (measured either by RMW or radius of hurricane or gale force winds) will produce stronger winds than a large tropical cyclone. Vickery, Skerlj, and Twisdale (2000) express the mean RMW (in km) of Atlantic tropical cyclones as a function of central pressure (P_o), environmental pressure (P_n), and latitude (L):

$$\ln(\text{RMW}) = 2.636 - 0.00005086*(P_o - P_n)^2 + 0.0394899*(L)$$

Tropical storms and hurricanes with observed or estimated RMWs that deviated by at least 25% from the average RMW values had wind speeds that were adjusted accordingly.

A second caveat concerns the translational speed of the tropical cyclone. In general, the translational speed is an additive factor on the right side of the storm and a subtractive factor on the left (Callaghan and Smith 1998). For example, a tropical cyclone moving westward in the Northern Hemisphere at 10 kt (5 m s^{-1}) with maximum sustained winds of 90 kt (46 m s^{-1}) on the west and east sides would produce approximately 100 kt (51 m s^{-1}) of wind on the north side and only 80 kt (41 m s^{-1}) on the south side. At low to medium translational speeds (less than around 20 kt [10 m s^{-1}]), the variation in storm winds on opposite sides of the storm track is approximately twice the translational velocity, although there is substantial uncertainty and non-uniformity regarding this impact on tropical cyclone winds. At faster translational speeds, this factor is somewhat less than two (Boose, Chamberlin, and Foster 2001). Storms that move significantly faster than the regionally-dependent climatological translational speeds (Neumann 1993; Vickery, Skerlj, and Twisdale 2000) have been chosen in the reanalysis to have higher maximum sustained wind speeds than slower storms with the same central pressure. Similarly, storms with slower than usual rates of translational velocity may have slightly lower winds for a given central pressure. Such alterations to the standard wind-pressure relationship were previously accounted for to some degree in the original version of HURDAT (Jarvinen, Neumann, and Davis 1984). The period from 1886 to 1910 was checked for consistency in the implementation of translational velocity impacts upon maximum sustained surface winds and changes were made where needed.

A third caveat of the wind-pressure relationships is that these algorithms were derived assuming over-water conditions. The use of the relationship for tropical cyclones overland must consider the increased roughness length of typical land surfaces and the dampening of the maximum sustained wind speeds that result. In general, maximum sustained wind speeds over open terrain exposures (with roughness lengths of 0.03 m) are about 5 to 10% slower than over-water wind speeds (Powell and Houston 1996), though for rougher terrain the wind speed decrease is substantially greater.

Finally, the derivation of the new regional wind-pressure relationships here is quite different from those originally analyzed by Kraft (1961) and Dvorak (1984). In these earlier efforts, observed central pressures were directly matched with observed maximum sustained surface winds. One substantial limitation in such efforts was in obtaining a sizable sample upon which to derive the wind-pressure equations. Here this limitation is avoided by using the actual HURDAT wind and central pressure values in recent years, which does provide a large dataset to work with. This approach lacks a degree of independence, however, as the NHC used the Kraft and Dvorak wind-pressure curves to provide estimates of maximum sustained surface winds from observed central pressures. This was especially the case during the 1970s, when aircraft flight-level winds were often discarded in favor of using the measured central pressure, because there was considerable uncertainty as to how to extrapolate flight-level winds to the surface (P. Hebert, personal communication, 2002). Such interdependence between recent HURDAT winds and central pressures may somewhat account for the close match between the Dvorak formulation to the Gulf of Mexico and southern latitude relationships. Despite these concerns, the development of regionalized wind-pressure relationships represents a step toward more realistic wind-pressure associations, though improvements beyond what has been presented here could certainly be achieved.

For many late-nineteenth- and early-twentieth-century storms, the central pressure could not be estimated from peripheral pressure measurements with the Schloemer equation because of unknown values for the RMW. Such peripheral pressure data were noted accordingly in the metadata file and used as a minimum estimate of what the best-track winds were at the time. In most of these cases, the best-track winds that were chosen were substantially higher than that suggested by the wind-pressure relationship itself. For Storm 1, 1856, maximum sustained winds consistent with the ship report of a 955-mb peripheral pressure measurement should be at least 105 kt (54 m s^{-1}) based on the Gulf of Mexico wind-pressure relationship (table 7.5). In this case, 130 kt (67 m s^{-1}) was chosen for the best track at the time of this ship report (for more details, see Metadata Files).

Best-Track Files

Tropical-cyclone positions and intensities in HURDAT have been added to and changed for the re-analyzed period of 1851 to 1910. Tracks added for the years of 1851 to 1870 were digitized from the work of Fernández-Partagás and Díaz (1995a). For the years 1871 to 1885, tracks for tropical cyclones that were unaltered by Fernández-Partagás and Díaz (1995b, 1996b) were digitized directly from Neumann et al. (1993). The intensity estimates for 1851 to 1885 were determined with consideration of available raw data found in Fernández-Partagás and Díaz (1995a, 1995b, 1996b), Ludlum (1963), Ho (1989), and other references, all of which have been recorded in the center-fix files. A large majority of the tropical cyclones for the years 1886 to 1910 were altered in their track and/or intensity based upon the work of Fernández-Partagás and Díaz (1996b, 1996c, 1997, 1999) and others listed in table 7.1. Additions and changes made to individual tropical cyclones and the references that were the basis for the alterations are listed in detail in the metafiles for the separate tropical cyclones.

Tropical cyclone positions were determined primarily by wind direction observations from ships and coastal stations and secondarily by sea level pressure measurements and reports of damages from winds, storm tides or freshwater flooding. Plate 5 illustrates, for an idealized case, how to estimate a tropical cyclone center from two ship observations. With these observations and the knowledge that the flow in a tropical cyclone is relatively symmetric (i.e., circular flow with an inflow angle of 20° [Jelesnianski 1993]), a relatively reliable estimate of the center of the storm can be obtained from a few peripheral wind direction measurements. Analysis of tropical cyclone intensity is much less straightforward. Intensity, described as the maximum sustained 1-minute surface (10 m) winds, of tropical cyclones for the period of 1851 to 1910, was based upon (in decreasing order of weighting) central pressure observations, wind observations from anemometers, Beaufort wind estimates, peripheral pressure measurements, wind-caused damages along the coast, and storm tide. The next section in the chapter goes into detail about limitations and possible errors in the HURDAT position and intensity estimates for this era.

Table 7.6 provides the best track for Storm 1, 1856, based upon the Fernández-Partagás and Díaz (1995a) track after conducting a critical independent assessment of their proposed positions and wind speeds (10 kt [5 m s^{-1}] increments) from known ship and land observations. This storm is a typical (though intense) example of one of the many newly archived tropical cyclones in the database. It is fully acknowledged that the best tracks drawn for tropical cyclones during the period 1851 to 1910 represent just a fragmentary record of what truly occurred over the open Atlantic Ocean. For this particular hurricane, the first six-hourly intensity given on August 9 at 00 UTC is 70 kt (36 m s^{-1}). It should not be inferred that this hurricane began its life cycle at 70 kt, but instead that data were lacking to make an estimate of its position and intensity before this date.

Occasionally, there are tropical cyclones in the best track for which only one six-hourly position and intensity estimate was available (single-point storms, such as Storm 1, 1851). This was typically due to one encounter of a tropical cyclone by a ship or the landfall of the system along the coast with no prior recorded contact with other ships or coastal communities. The position and intensity estimated for such tropical cyclones have more uncertainty than usual, because it was not possible to check for consistency between consecutive position or intensity estimates. Users are to be cautioned that these single-point storms will cause programming difficulties for versions of programs that are expecting at least two position or intensity estimates.

For the period of 1886 to 1930, the existing HURDAT was originally created from a once-daily (UTC) estimate of position and intensity (Jarvinen, Neumann, and Davis 1984). This caused some difficulty in situations of rapid intensification and rapid decay, such as the landfall of a tropical cyclone. For the latter case, the Kaplan and DeMaria (1995, 2001) models provided guidance

TABLE 7.6 Best-track Information for Storm 1, 1856

Standard HURDAT format

00820	08/09/1856 M = 4			1 SNBR =	29 NOT NAMED		XING = 1	SSS = 4				
00825	08/09*250	839	70	0*257	851	80	0*263	865	90	0*270	878	100	0
00830	08/10*277	891	110	0*282	898	120	0*287	905	130	0*292	911	130	934
00835	08/11*297	916	110	0*300	918	80	0*303	919	60	0*306	918	50	0
00840	08/12*309	916	40	0*313	910	40	0* 0	0	0	0* 0	0	0	0
00845	HR LA4												

"Easy-to-read" version

					Movement			Wind speed			
Month	Day	Hour	Latitude	Longitude	Direction	mph	km/hr	mph	km/hr	Pressure	Type
8	9	0 UTC	25.0N	83.9W	? deg	?	?	80	130	mb	H-Category 1
8	9	6 UTC	25.7N	85.1W	305 deg	13	22	90	150	mb	H-Category 1
8	9	12 UTC	26.3N	86.5W	295 deg	14	24	100	170	mb	H-Category 2
8	9	18 UTC	27.0N	87.8W	300 deg	14	24	120	190	mb	MH-Category 3
8	10	0 UTC	27.7N	89.1W	300 deg	14	24	130	200	mb	MH-Category 3
8	10	6 UTC	28.2N	89.8W	310 deg	8	12	140	220	mb	MH-Category 4
8	10	12 UTC	28.7N	90.5W	310 deg	8	12	150	240	mb	MH-Category 4
8	10	18 UTC	29.2N	91.1W	315 deg	8	12	150	240	934 mb	MH-Category 4 , landfall
8	11	0 UTC	29.7N	91.6W	320 deg	6	11	130	200	mb	MH-Category 3
8	11	6 UTC	30.0N	91.8W	330 deg	3	5	90	150	mb	H-Category 1
8	11	12 UTC	30.3N	91.9W	345 deg	3	5	70	110	mb	TS
8	11	18 UTC	30.6N	91.8W	15 deg	3	5	60	90	mb	TS
8	12	0 UTC	30.9N	91.6W	30 deg	3	5	50	70	mb	TS
8	12	6 UTC	31.3N	91.0W	50 deg	6	11	50	70	mb	TS

for determining wind speeds for the best track after landfall of a tropical cyclone, but only in the absence of observed inland winds. The models used by Kaplan and DeMaria begin with a maximum sustained wind at landfall and provides decayed wind speed values out to about two days after landfall. Kaplan and DeMaria (1995) was designed for landfalling tropical cyclones over the southeastern United States where nearly all of the region within 150 nmi (275 km) of the coast has elevations less than 650 ft (200 m). Therefore, the decay of winds over higher terrain areas such as Hispanola and much of Mexico predicted with the Kaplan and DeMaria (1985) model is inadequate (e.g., Bender, Tuleya, and Kurihara 1985). For these cases, a faster rate of decay than that given from this model (on the order of 30%-accelerated rate of decay) was utilized in the reanalysis.

Ho et al. (1987) also developed several relationships for the decay of tropical cyclone central pressure after landfall, which were stratified by geographic location and value of the pressure deficit (environmental pressure minus central pressure) at landfall. In general, for tropical cyclones striking the U.S. Gulf Coast, at 10 hours after landfall, the pressure deficit decreased by half. For Florida (south of 29°N) hurricanes at 10 hours after landfall, the pressure deficit decreased by only one-quarter. For U.S. hurricanes making landfall north of Georgia, the pressure deficit is 0.55 times that of the landfalling value at 10 hours after landfall. For extremely intense hurricanes, the rate of decay is somewhat faster. The relationships that Ho et al. (1987) developed are utilized here on occasion to derive an estimated central pressure at landfall from an inland central pressure measurement. The only deviation is for hurricanes traversing the marshes of southern Louisiana. In the Ho et al. (1987) study, Hurricane Betsy behaved anomalously, because it decayed much more slowly than most of the hurricanes striking the southeastern United States. It is hypothesized that this is due to enhanced sensible and latent heat fluxes available over the Louisiana marshes, relative to the dry land found throughout the rest of the region. Ho (1989) suggests using the Florida decay rate for these hurricanes (e.g., Storm 10, 1893), because this rate better matches decay rates for hurricanes similar to Betsy.

The best-track files for 1851 to 1870 do not include the tropical depression stages of tropical cyclones. Obtaining adequate information to document a storm's beginning and ending tropical depression stages would be extremely difficult, as most of the available observations focus upon gale force and stronger wind speeds. In addition, motivation for this work was to better document the tropical storm and hurricane stages, as these account for the large majority of impacts on society, through winds, storm surge, and inland flooding. For the years 1871 to 1898, however, the authors were able to add into

HURDAT the dissipating tropical depression stage for those tropical cyclones that decayed over land. The Kaplan and DeMaria (1995, 2001) inland decay models were utilized to calculate wind speed estimates after landfall, in the absence of in situ wind or pressure data. This was done to ensure that existing tracks indicated by Neumann et al. (1993, 1999) and the original HURDAT were not truncated because the tropical cyclones decayed from tropical storm to tropical depression status. Starting in 1899, both the formative and decaying tropical depression stages over water are included. This is consistent with the previous HURDAT methodology. Where possible, the transition to the extra-tropical storm stage was documented and included in the best track.

The period of 1886 to 1898 in the existing HURDAT contained rather generic peak intensities: most systems that were determined to have been tropical storms were assigned peak winds of 50 kt (26 m s^{-1}) and most hurricanes were assigned peak winds of 85 kt (44 m s^{-1}) (Hebert and McAdie 1997). In fact, of the 70 hurricanes from 1886 to 1898 in the original HURDAT, only one was Category 1, 59 were Category 2, 10 were Category 3, and none were Category 4 and 5. This compares to recent historical averages in which only about a fourth of all hurricanes are Category 2 (Pielke and Landsea 1998). In many of the tropical storms and hurricanes for this period, the available ship- and land-based observations were used to provide a more realistic peak intensity value, if possible.

For the years 1899 to 1910, Fernández-Partagás and Díaz (1996c, 1997, 1999) made extensive use of the Historical Weather Maps series, a reconstruction of daily surface northern hemispheric synoptic maps accomplished by the U.S. Navy and U.S. Weather Bureau in the late 1920s. This reconstruction of daily synoptic maps incorporated ship and coastal station data not available in the original tropical storm and hurricane track determinations. Thus, over 90% of the tracks for this 12-year period have been modified.

Limitations and Errors

The tropical storms and hurricanes that stayed out at sea for their duration and did not encounter ships (or tropical cyclones that sunk all ships that they overran) are, of course, undocumented for the time period of 1851 to 1910. It was estimated that the numbers of "missed" tropical storms and hurricanes for the 1851 to 1885 period are between zero and six per year. The estimate is a bit lower for the 1886 to 1910 period, when it is thought that between zero and four storms were missed. The higher detection for the latter period is due to increased ship traffic, larger populations along the coastlines, and more meteorological measurements being taken. By no means should the tropical cyclone record over the

Atlantic Ocean be considered complete for either the frequency or intensity of tropical storms and hurricanes for the years 1851 to 1910. More accurate and complete information is available for landfalling tropical cyclones along much of the U.S. coastline.

Tropical storms and hurricanes that remained out over the Atlantic Ocean waters in the 1851 to 1910 period had relatively few chances to be observed and thus included into this database. This is because, unlike today, the wide array of observing systems such as geostationary and polar orbiting satellites, aircraft reconnaissance, and radars were not available. Detection of tropical storms and hurricanes in the second half of the nineteenth century was limited to those tropical storms and hurricanes that affected ships and those that hit land. In general, the data should be slightly more complete for the years 1886 to 1910 than the preceding decades because of some improvements in the monitoring network during this period. Improvements in the monitoring of Atlantic tropical storms and hurricanes for the nineteenth and early twentieth centuries can be summarized in the following timeline (Fitzpatrick 1999; Neumann et al. 1999):

1800s	Ships' logs provided tropical cyclone observations (after returning to port).
1845	The first telegraph line completed from Washington, D.C., to Boston.
1846	The cup anemometer was invented by Robinson.
1848	The Smithsonian Institution volunteer weather observer network was started in United States.
1870	The U.S. national meteorological service began through the Army Signal Corps.
1875	The first hurricane forecasting system started by Benito Viñes in Cuba.
1890	The U.S. weather service was transferred to a civilian agency, the U.S. Weather Bureau.
1898	The U.S. Weather Bureau established observation stations throughout Caribbean.
1905	Ship observations of tropical storms and hurricanes began to be transmitted via radio.

Note that until the invention of radio in 1902, the only way to obtain ship reports of hurricanes at sea was after the ships made their way back to port. Observations from ship reports were not of use to the fledgling weather services in the United States and Cuba operationally, though some of them were available for post-season analyses of the tropical cyclone activity. These ship reports—many not collected previously—proved to be invaluable to Ludlum

(1963), Ho (1989) and Fernández-Partagás and Díaz (1995a, 1995b, 1996b, 1996c, 1997) and others in their historical reconstruction of past hurricanes.

Geographical positions of tropical cyclones in HURDAT were estimated to the nearest 0.1° latitude and longitude (~6 nmi or ~11 km), but the average errors in these measurements were typically much larger in the late nineteenth and early twentieth centuries than this precision might imply (table 7.7). Holland (1981) demonstrated that even with the presence of numerous ships and buoys in the vicinity of a strong tropical cyclone that was also monitored by aircraft reconnaissance, there were substantial errors in estimating its exact center position from the ship and buoy data alone. Based upon this, storms documented over the open ocean during the period of 1851 to 1885 were estimated to have position errors that averaged 120 nmi (220 km), with ranges of 180 to 240 nmi (330 to 440 km) errors being quite possible. In the later years of 1886 to 1910, this is improved somewhat to average position errors of around 100 nmi (185 km). At landfall, knowledge of the location of the tropical cyclone was generally more accurate, as long as the storm came ashore in a relatively populated region (table 7.7). Users should consult the corresponding center-fix files to see if there are actual location center-fixes available from ships or coastal observations. If so, the location error for the nearest six-hourly best-track position would be smaller—on the order of 30 nmi (55 km).

Storm intensity values for 1851 to 1885 were estimated to the nearest 10 kt (5 m s^{-1}), but were likely to have large uncertainty as well (table 7.7). Starting in 1886, winds were given in intervals of 5 kt (2.5 m s^{-1}), consistent with the previous version of HURDAT. Best-track intensity estimates for 1851 to 1910 were

TABLE 7.7 Estimated Average Position and Intensity Errors in Best Track, 1851–1910

Situation	Years	Position error	Intensity error (absolute)	Intensity error (bias)
Open ocean	1851–1885	120 nmi/220 km	25 kt/13 m s^{-1}	−15 kt/−8 m s^{-1}
	1886–1910	100 nmi/185 km	20 kt/10 m s^{-1}	−10 kt/−5 m s^{-1}
Landfall at sparsely populated area	1851–1885	120 nmi/220 km	25 kt/13 m s^{-1}	−15 kt/−8 m s^{-1}
	1886–1910	100 nmi/185 km	20 kt/10 m s^{-1}	−10 kt/−5 m s^{-1}
Landfall at settled area	1851–1885	60 nmi/110 km	15 kt/8 m s^{-1}	0 kt/0 m s^{-1}
	1886–1910	60 nmi/110 km	12 kt/6 m s^{-1}	0 kt/0 m s^{-1}

Note: Negative bias errors indicate an underestimation of the true intensity.

based mainly upon observations by ships at sea, which more often than not would not sample the very worst part of the storm (typically only 30 to 60 nmi [55–110 km] in diameter). Holland (1981) demonstrated that even in a relatively data-rich region of ship and buoy observations within the circulation of a tropical cyclone, the actual intensity was likely to be substantially underestimated. Figures 7.1 and 7.2 provide a graphic demonstration of this for Major

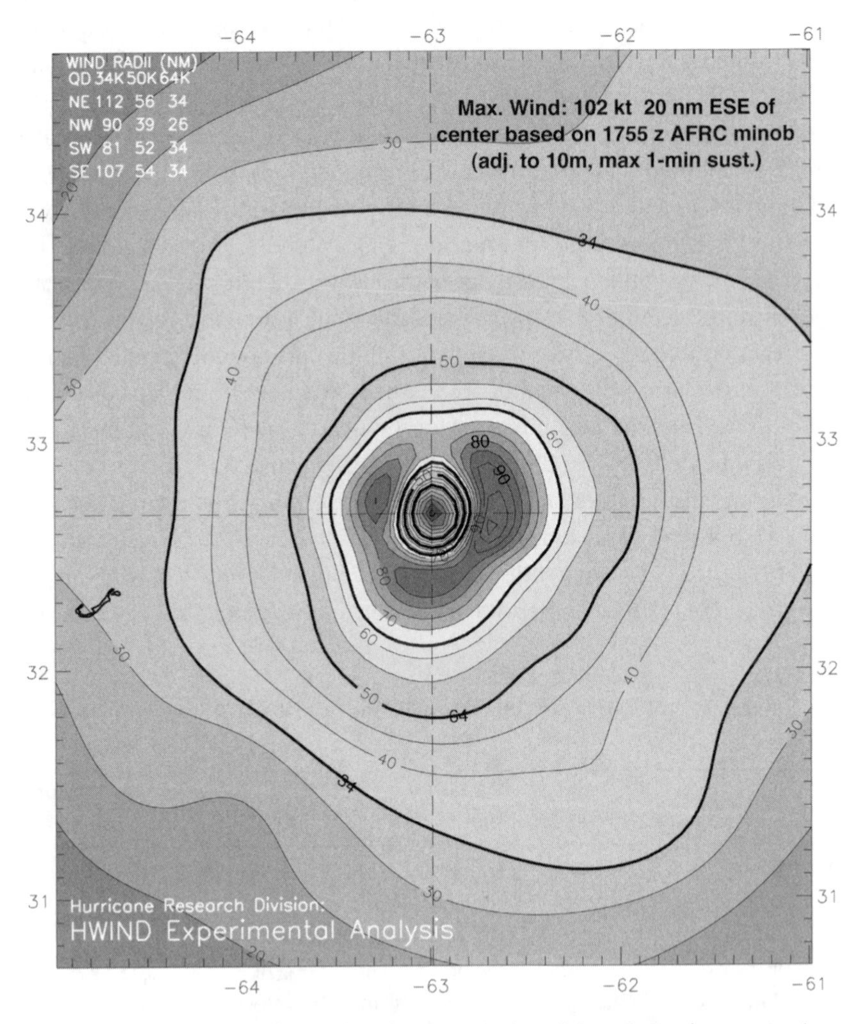

FIGURE 7.1 Surface wind-field analysis for Major Hurricane Erin on September 9, 2001, at 1930 UTC. This analysis utilizes all available surface and near surface wind data including surface-reduced aircraft reconnaissance winds, surface-reduced cloud-drift winds, and ship and buoy observations. These data are all storm-relative composited for the period of 1500 to 1900 UTC, September 9, 2001, and are adjusted to a standard maximum sustained surface (1 min, 10 m) measurement. Peak sustained winds are analyzed to be 102 kt (52 m s^{-1}) to the east-southeast of Erin's center at a radius of 20 nmi (37 km).

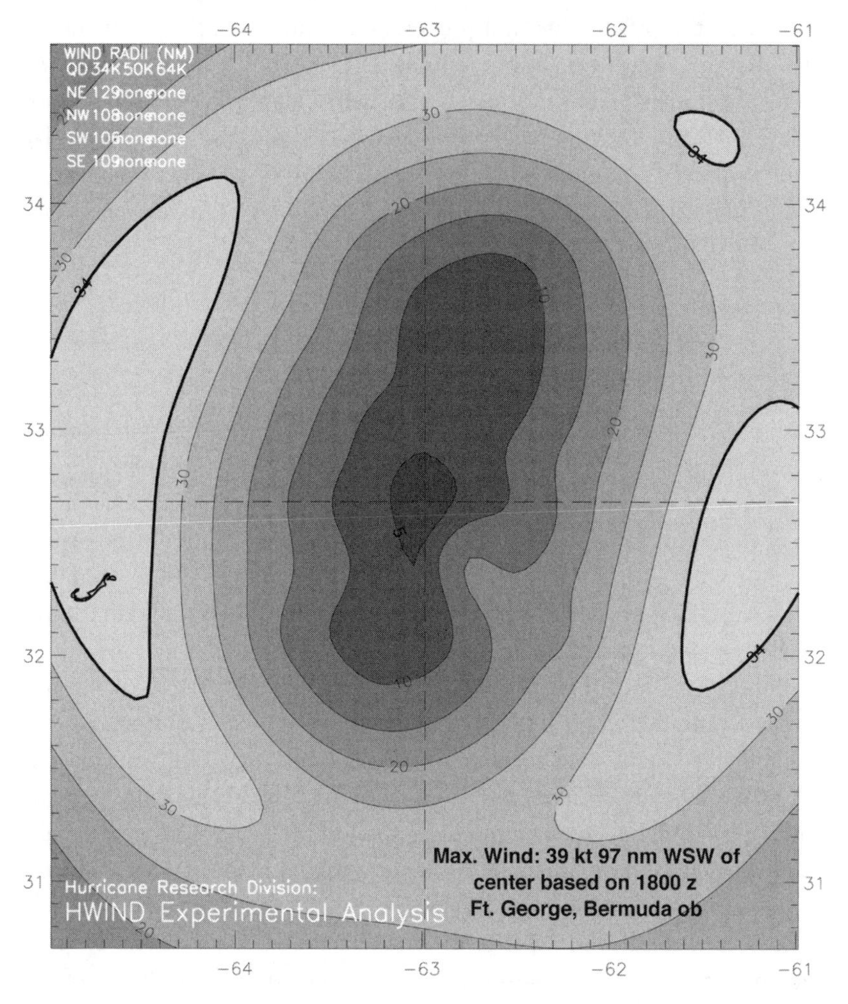

FIGURE 7.2 Same as figure 7.1, but without the benefit of surface-reduced aircraft reconnaissance flight-level winds. In this case, the highest analyzed surface winds were only 39 kt (20 m s^{-1}) based upon observations from Bermuda about 100 nmi (160 km) from Erin's center. Such an analysis is typical of data available before the advent of aircraft reconnaissance data in the mid-1940s and is illustrative of the underestimation bias that occurred for many tropical cyclones during the era of the late nineteenth and early twentieth centuries being reanalyzed.

Hurricane Erin of 2001 that made a close by-pass of Bermuda. Aircraft winds extrapolated to the ocean surface indicated maximum sustained surface winds of just above 100 kt (51 m s^{-1}) in Major Hurricane Erin (figure 7.1). Despite transiting within 85 nmi (160 km) of Bermuda, however, the highest observed surface winds from ships and coastal stations were only around 40 kt (20 m s^{-1}) (figure 7.2). Such an underestimation of tropical cyclone intensities was likely common in the pre-satellite and pre-aircraft reconnaissance era. It was estimated that the intensity measurements for 1851 to 1885 were in error an average of 25 kt (13 m s^{-1}) over the open ocean, with a bias toward underestimating the true intensity (table 7.7). For the later period of 1886 to 1910, this was slightly improved—to an average error of 20 kt (10 m s^{-1}) over the ocean. At landfall, intensity estimates were improved and show a negligible bias as long as the landfall occurs over a populated coastline (table 7.7).

Metadata Files

All Atlantic Basin tropical storms and hurricanes in the new best-track database are accompanied by a metadata file. This file consists of a descriptive paragraph about the particular storm of interest that provides information about the sources that went into creating the best track, whether or not a wind-pressure relationship was utilized, if the Kaplan and DeMaria (1995, 2001) wind decay models were used for inland wind estimates, and any other pertinent information. Storms and hurricanes for which the entire life cycle is available during the period of 1851 to 1885 (from genesis as a tropical storm, to peak intensity, to decay to minimal tropical storm or transformation to an extratropical storm) are so indicated in the metadata file. If this is not indicated in the metadata file, users of the data are cautioned that only a partial life cycle of the particular storm is available. Because documenting the full life cycle of tropical cyclones became somewhat more frequent starting in 1886, only those tropical cyclones that lack archival records of their full life cycle are so noted in the metadata files for the years 1886 to 1910. All of the tropical storms and hurricanes for the period of 1851 to 1910 are considered to be unnamed, although many of these storms have been recognized by various informal names. These informal names are included in the metadata file when at all possible. Here is a sample metadata file, for Storm 1, 1856:

> 1856/01: Utilized Ho's (1989) work—apparently not used in Fernández-Partagás and Díaz's (1995a) analysis—to alter the track and intensity near the US. Inland winds over SE US reduced via Kaplan and DeMaria's (1995) inland decay model. Ship with pressure measurement of 955 mb not in the hurricane's eye suggests at least 105 kt with the Gulf of Mexico wind-pressure relationship, utilize 130 kt in best track. Ho's estimate of 934 mb at landfall gives 125 kt, utilize 130 kt in best track—a major hurricane. A small RMW of 12 nmi supports slight increase of winds over suggested wind-pressure relationship. Surge value of 11–12′ provided by Ludlum

(1963) for Last Island, Louisiana. The storm is also known as the "Last Island Hurricane" after the destruction caused at that location.

For the cases where Fernández-Partagás and Díaz or the original HURDAT listed a storm, but for some reason it was not included into the revised HURDAT, an addendum to the metadata file for that year is included. For example, here is a case for 1851:

1851—Additional Notes:
1. The tropical storm listed as #5 in 1851 in Partagás and Díaz (1995a) was not included into the HURDAT because of the lack of evidence to suggest that the storm actually existed. Fernández-Partagás and Díaz had found an unsupported reference to it in Tannehill (1938), but no other information.

U.S. Landfalling Tropical Cyclones

Tables 7.8 and 7.9 summarize the continental U.S. landfalling hurricanes and tropical storms, respectively, for the years 1851 to 1910 and the states affected by these systems. In addition to the parameters also common to HURDAT (e.g. latitude, longitude, maximum sustained winds and central pressure), the U.S. landfalling hurricane compilation also includes—where available—the RMW, peak storm surge, and environmental pressure. For the period of 1851 to 1899, timings of U.S. landfalls are estimated to the nearest hour; for the later years (1900–1910), the more complete observational network allowed for an indication of U.S. landfalling hurricanes and tropical storms to the nearest 10 minutes of landfall. As was utilized in HURDAT, maximum sustained wind speeds are estimated to the nearest 10 kt for the years 1851 to 1885 and a more precise measure of 5-kt increments are used for the period of 1886 to 1910.

As mentioned previously, because of the lack of continuously populated coastal regions during this era, this record represents an incomplete listing of the frequency and intensity of tropical cyclones that have hit the United States. Based upon analysis of "settled regions" (defined as at least two inhabitants per square mile) from U.S. Census reports and other historical analyses (Department of the Interior 1895; Kagan 1966; Tanner 1995), estimated dates are provided when accurate tropical cyclone records began in specified regions of the United States (table 7.10). Prior to these dates, tropical storms or hurricanes—especially smaller systems like Andrew in 1992 and Bret in 1999—might have been missed completely or may have had their true intensity underestimated.

As an example of the intensity underestimation bias of a landfalling hurricane along a relatively uninhabited coastal region, consider the case of Storm

TABLE 7.8 Landfalling Hurricanes in the United States, 1851–1910

Number and date	Time	Latitude (N)	Longitude (W)	Maximum winds (knots)	Saffir-Simpson	RMW (nmi)	Storm surge (feet)	Central pressure (mb)	Environmental pressure (mb)	States affected
1. June 25, 1851[1]	1200	28.5	96.5	70	1			(985)		BTX1
4. August 23, 1851[1]	2100	30.1	85.7	100	3		12[5]	(960)		AFL3, GA1
1. August 22, 1852[1, 2]	1200	23.8	81.3	80	1			(977)		BFL1
1. August 26, 1852	0600	30.2	88.6	100	3	30	12[5]	961		AL3, MS3, LA2, AFL1
3. September 12, 1852[1]	0000	28.0	82.8	70	1			(985)		BFL1
5. October 9, 1852[1]	2100	29.9	84.4	90	2		7[5]	(969)		AFL2, GA1
8. October 21, 1853[2]	0600	30.9	80.9	70	1			(965)		GA1
2. September 8, 1854	2000	31.7	81.1	100	3	40		950		GA3, SC2, DFL1
3. September 18, 1854	2100	28.9	95.3	90	2			(969)		BTX2
6. September 16, 1855[1]	0300	29.2	89.5	110	3		10–15[5]	(950)		LA3, MS3
1. August 10, 1856[1]	1800	29.2	91.1	130	4	12	11–12[5]	934		LA4
5. August 31, 1856[1]	0600	30.2	85.9	90	2		6[5]	(969)		AFL2, AL1, GA1
2. September 13, 1857[3]	1100	35.2	75.7	80	1			961		NC1
3. September 16, 1858	1700	40.9	72.2	80	1	45		(976)		NY1
3. September 16, 1858	1800	41.3	72.0	70	1	45		979		CT1, RI1, MA1
5. September 16, 1859	0000	30.3	88.1	70	1			(985)		AL1
1. August 11, 1860[1]	2000	29.2	90.0	110	3		12[5]	(950)		LA3, MS3, AL2
4. September 15, 1860[1]	0400	29.3	89.6	90	2		10[5]	(969)		LA2, MS2, AL1
6. October 2, 1860[1]	1700	29.5	91.4	90	2			(969)		LA2
2. August 16, 1861[1, 2]	0000	24.2	82.0	70	1			(970)		BFL1

5. September 27, 1861	1700	34.5	77.4	70	1			(985)		NC1
8. November 2, 1861	100	34.7	76.6	70	1			(985)		NC1
4. September 13, 1865[1]	2100	29.8	93.4	90	2			(969)		LA2, CTX1
7. October 23, 1865[1]	1000	24.6	81.7	90	2			(969)		BFL2
7. October 23, 1865[1]	1400	25.4	81.1	90	2			(969)		BFL2, CFL1
1. July 15, 1866	1200	28.5	96.5	90	2			(969)		BTX2
1. June 22, 1867	1400	32.9	79.7	70	1			(985)		SC1
7. October 2, 1867[1, 4]	1500	25.4	97.1	70	1			(969)		ATX1
7. October 4, 1867[1]	1500	29.2	91.0	90	2		7[5]	(969)		LA2, CTX1
7. October 6, 1867[1]	1500	29.6	83.4	70	1			(985)		AFL1
2. August 17, 1869	0700	28.1	96.8	90	2			(969)		BTX2
5. September 5, 1869[1]	1200	29.2	90.0	70	1			(986)		LA1
6. September 8, 1869[3]	2100	41.0	71.9	80	1	30		963		NY1
6. September 8, 1869	2200	41.4	71.7	100	3	30	8[5]	965		RI3, MA3, CT1
10. October 4, 1869[3]	1900	41.3	70.5	80	1	30		(965)		MA1
10. October 4, 1869[3]	2000	41.7	70.4	80	1	30		(965)		MA1
10. October 4, 1869	2300	43.7	70.1	90	2			(968)		ME2
1. July 30, 1870	1800	30.5	88.0	70	1			(985)		AL1
6. October 10, 1870[1, 2]	0500	24.6	80.8	70	1			(970)		BFL1, CFL1
9. October 20, 1870[1]	1400	24.7	82.8	80	1			(977)		BFL1
9. October 20, 1870[1]	2000	26.0	81.6	80	1			(977)		BFL1
3. August 17, 1871[1]	0200	27.1	80.2	100	3	30		955	1016	CFL3, DFL1, AFL1
4. August 25, 1871[1]	0500	27.6	80.3	90	2			(965)		CFL2, DFL1
6. September 6, 1871[1]	1400	29.2	83.0	70	1			(985)		AFL1
3. September 19, 1873[1]	1500	29.9	84.4	70	1			(985)		AFL1

(Continued)

TABLE 7.8 Landfalling Hurricanes in the United States, 1851–1910 *(Continued)*

Number and date	Time	Latitude (N)	Longitude (W)	Maximum winds (knots)	Saffir-Simpson	RMW (nmi)	Storm surge (feet)	Central pressure (mb)	Environmental pressure (mb)	States affected
5. October 7, 1873[1]	0100	26.5	82.2	100	3	26	14[5]	959	1014	BFL3, CFL2, DFL1
6. September 28, 1874[1]	0300	29.1	82.9	70	1			(985)		AFL1
6. September 28, 1874	1800	32.8	80.0	80	1			981		SC1, NC1
3. September 16, 1875	2100	27.7	97.2	100	3		15[5]	(960)		BTX3, ATX2
2. September 17, 1876	1400	34.4	77.6	80	1			980		NC1, VA1
5. October 20, 1876[1]	0500	25.8	81.4	90	2			973		BFL2, CFL1
2. September 18, 1877[1]	1600	29.2	91.0	70	1			(985)		LA1
2. September 19, 1877[1]	2000	30.4	86.6	70	1			(985)		AFL1
4. October 3, 1877[1]	0500	30.0	85.5	100	3		12[5]	(960)		AFL3, GA1
5. September 10, 1878[1]	1100	28.6	82.6	90	2			(970)	1010	BFL2, DFL1
5. September 12, 1878	1200	32.5	80.4	80	1			(976)		NC1, SC1, GA1
11. October 23, 1878	0400	34.8	77.1	80	2		12[5]	(963)		NC2, VA1, MD1, DE1, NJ1, PA1
2. August 18, 1879	1200	34.7	76.7	100	3	16	7	971	1014	NC3, VA2
2. August 19, 1879[3]	0600	41.4	70.8	60	TS			984		(none)
3. August 23, 1879	0200	29.4	94.4	90	2			964		CTX2, LA2
4. September 1, 1879[1]	1600	29.5	91.4	110	3			(950)		LA3
2. August 13, 1880[4]	0100	25.8	97.0	110	3	12		931		ATX3
4. August 29, 1880[1]	1200	28.2	80.6	90	2			972		CFL2, DFL1
4. August 31, 1880	0400	29.7	84.8	70	1			(985)		AFL1
6. September 9, 1880	1000	34.7	77.1	70	1			987		NC1
9. October 8, 1880	1900	28.9	82.7	70	1			(985)		AFL1

5. August 28, 1881	0200	31.7	81.1	90	2	15		970	GA2, SC1
6. September 9, 1881	1600	33.9	78.1	90	2	15		975	NC2
2. September 10, 1882	0200	30.4	86.8	100	3			949	AFL3, AL1
3. September 15, 1882	0500	29.8	93.7	90	2			(969)	LA2, CTX1
6. October 11, 1882	0400	29.5	83.3	70	1			(985)	AFL1
3. September 11, 1883	1300	33.9	78.5	90	2			(965)	NC2, SC1
2. August 25, 1885	0900	32.2	80.7	100	3			(953)	SC3, NC2, GA1, DFL1
1. June 14, 1886	1600	29.6	94.2	85	2		7[5]	(973)	CTX2 LA2
2. June 21, 1886	1100	30.1	84.0	85	2			(973)	AFL2, GA1
3. June 30, 1886	2100	29.7	85.2	85	2			(973)	AFL2
4. July 19, 1886	0100	28.8	82.7	70	1			(985)	AFL1
5. August 20, 1886	1300	28.1	96.8	135	4	15	15	925	BTX4
8. September 23, 1886[4]	0700	26.0	97.2	80	1			(973)	ATX1, BTX1
10. October 12, 1886	2200	29.8	93.5	105	3		12[5]	(955)	LA3, CTX2
4. July 27, 1887	1500	30.4	86.6	75	1			(981)	AFL1
6. August 20, 1887[2]	1200	35.0	75.0	65	1			(946)	NC1
9. September 21, 1887	1700	26.1	97.2	85	2			973	ATX1
13. October 19, 1887	0200	29.1	90.4	75	1			(981)	LA1
1. June 17, 1888	0600	28.7	95.7	70	1			(985)	BTX1
3. August 16, 1888[1]	1900	25.8	80.1	110	3		14[5]	(945)	CFL3, BFL1
3. August 19, 1888	1600	29.1	90.7	95	2			(964)	LA2
6. September 26, 1888[3]	1300	41.6	69.9	55	TS			985	(none)
7. October 11, 1888	0100	29.2	83.1	95	2	11	9	970	AFL2, DFL1
6. September 23, 1889	0400	29.1	89.8	70	1			(985)	LA1
1. July 5, 1891	2200	28.8	95.5	80	1			(977)	BTX1, CTX1

(Continued)

TABLE 7.8 Landfalling Hurricanes in the United States, 1851–1910 *(Continued)*

Number and date	Time	Latitude (N)	Longitude (W)	Maximum winds (knots)	Saffir-Simpson	RMW (nmi)	Storm surge (feet)	Central pressure (mb)	Environmental pressure (mb)	States affected
3. August 24, 1891[1]	1500	25.4	80.2	70	1			(985)		CFL1
4. August 24, 1893	1200	40.6	73.9	75	1	30		986		NY1, VA1
6. August 28, 1893	0500	31.7	81.1	100	3	23	9-10	954	1010	GA3, SC3, NC1, DFL1
8. September 7, 1893	1400	29.2	91.1	85	2			973		LA2
10. October 2, 1893	0800	29.3	89.8	115	4	12		948		LA4
10. October 2, 1893	1600	30.3	88.9	95	2	17	10–12[5]	970		MS3, Al3
9. October 13, 1893	1300	33.0	79.5	105	3	15	14[5]	955		SC3, NC2, VA1
4. September 25, 1894[1]	1100	24.7	82.0	80	1			985		BFL1
4. September 25, 1894[1]	1900	26.5	82.0	90	2			(975)		BFL2, DFL1
4. September 27, 1894	0700	32.3	80.7	80	1		10[5]	(976)		SC1
4. September 29, 1894[2]	1200	37.0	75.0	70	1			(978)		VA1
5. October 9, 1894	0300	30.2	85.5	105	3			(955)		AFL3, GA1
5. October 10, 1894	1500	40.7	72.9	75	1			(978)		NY1, RI1
2. August 30, 1895[4]	0400	25.0	97.6	65	2			(973)		ATX1
1. July 7, 1896	1700	30.4	86.5	85	2			(973)		AFL2
2. September 10, 1896	1300	41.2	70.6	70	1	30		(985)		RI1, MA1
4. September 29, 1896	1100	29.2	83.1	110	3	15		960	1014	AFL3, DFL3, GA2, SC1, NC1, VA1
2. September 13, 1897	0500	29.7	93.8	75	1		6[5]	(981)		LA1, TX1
1. August 2, 1898	2300	29.7	84.8	70	1			(985)		AFL1
2. August 31, 1898	0700	32.1	80.8	75	1			(980)		GA1, SC1
7. October 2, 1898	1600	30.9	81.4	115	4	18	16	938	1010	GA4, DFL2

2. August 1, 1899	1700	29.7	84.7	85	2			979	1017	AFL2
3. August 18, 1899	0100	35.2	75.8	105	3			(945)	1012	NC3
8. October 31, 1899	0900	33.6	79.0	95	2	35	9[5]	955	1012	NC2, SC2
1. September 9, 1900	0140	29.1	95.1	125	4	14	20[5]	931	1012	CTX4
3. July 11, 1901	0720	36.0	75.8	70	1			(983)	1016	NC1
4. August 14, 1901	2110	29.3	89.6	80	1		8[5]	(973)	1013	LA1
4. August 15, 1901	1700	30.4	88.8	80	1	33	8[5]	973	1013	MS1, AL1
3. September 11, 1903	2250	26.1	80.1	75	1	43	8[5]	976	1016	CFL1
3. September 13, 1903	2330	30.1	85.6	80	1		10[5]	(977)	1016	AFL1
4. September 16, 1903	1120	39.1	74.7	70	1			990	1020	NJ1, DE1
2. September 14, 1904	1320	33.1	79.2	70	1			(985)	1017	SC1
3. October 17, 1904	0750	25.3	80.3	70	1			(985)	1016	CFL1
2. June 17, 1906	0240	24.7	81.1	70	1			(986)	1013	BFL1, CFL1
2. June 17, 1906	0750	25.2	80.7	75	1	26		979	1013	CFL1
5. September 17, 1906	2140	33.3	79.2	80	1	30		977	1018	SC1, NC1
6. September 27, 1906	1102	30.2	88.6	95	2	43	14[5]	958	1013	MS2, AL2, AFL2, LA1
8. October 18, 1906	0930	24.7	81.1	105	3	12		953	1010	BFL3, CFL3
8. October 18, 1906	1130	25.2	80.8	105	3	12		953	1010	CFL3, BFL1
2. May 29, 1908[3]	2100	35.2	75.6	55	TS			989	1015	(none)
3. July 31, 1908	1130	34.6	77.1	70	1			(985)	1017	NC1
2. June 29, 1909	1700	26.1	97.2	85	2		7[5]	972	1012	ATX2
4. July 21, 1909	1650	28.9	95.3	100	3	19	10[5]	959	1015	CTX3
6. August 27, 1909[4]	2140	23.7	97.7	65	1			(955)	1014	ATX1
8. September 21, 190[9]	0000	29.5	91.3	105	3	28	15[5]	952	1012	LA3, MS2
10. October 11, 1909[3]	1800	24.7	81.0	100	3	22		957	1009	BFL3, CFL3

(Continued)

TABLE 7.8 Landfalling Hurricanes in the United States, 1851–1910 *(Continued)*

Number and date	Time	Latitude (N)	Longitude (W)	Maximum winds (knots)	Saffir-Simpson	RMW (nmi)	Storm surge (feet)	Central pressure (mb)	Environmental pressure (mb)	States affected
3. September 14, 1910	2200	26.9	97.4	95	2			(965)	1011	ATX2
5. October 17, 1910[2]	1900	24.6	82.6	90	2	28		941	1008	BFL2
5. October 18, 1910	0600	26.5	82.0	95	2	28	15[5]	955	1008	BFL2

Notes: Date and time: Day and time when the circulation center crossed the U.S. coastline (including barrier islands). Time was estimated to the nearest hour for the period of 1851 to 1899 and to the nearest 10 minutes for the period of 1900 to 1910. Time is UTC or GMT.

Latitude and longitude: Location was estimated to the nearest 0.1° latitude and longitude (about 6 nautical miles [nmi]).

Maximum winds: Estimated maximum sustained 1-min surface (10 m) winds (in knots) to occur along the U.S. coast. Winds are estimated to the nearest 10 kt for the period of 1851 to 1885 and to the nearest 5 kt for the period of 1886 to 1910.

Saffir-Simpson: Estimated Saffir-Simpson Hurricane Scale Category at landfall, based on maximum sustained surface winds.

TS: Hurricane's center made landfall, but the hurricane-force wind remained offshore.

RMW: Radius of maximum winds at the surface (primarily for the right front quadrant of the hurricane), if available.

Storm surge: Maximum observed storm surge in feet, if available. Although a higher value may have occurred, it might not have been recorded.

Central pressure: Observed (or analyzed from peripheral pressure measurements) minimum central pressure (mb) of the hurricane at landfall. Central pressure values in parentheses indicate that the value was a simple estimation (based on a wind-pressure relationship) and not directly observed or calculated.

Environmental pressure: Sea level pressure (mb) at the outer limits of the hurricane circulation determined by moving outward from the storm center to the first anticyclonically turning isobar in four equally spaced directions and averaging the four pressures thus obtained.

States affected: Impact of the hurricane on individual U.S. states based on the Saffir-Simpson Scale (again through the estimate of the maximum sustained surface winds at each state). AFL, northwestern Florida; AL, Alabama; ATX, southern Texas; BFL, southwestern Florida; BTX, central Texas; CFL, southeastern Florida; CT, Connecticut; CTX, northern Texas; DE, Delaware; DFL, northeastern Florida; GA, Georgia; LA, Louisiana; MA, Massachusetts; MD, Maryland; ME, Maine; MS, Mississippi; NC, North Carolina; NH, New Hampshire; NJ, New Jersey; NY, New York; PA, Pennsylvania; RI, Rhode Island; SC, South Carolina; VA, Virginia.

In Texas, southern refers to the area from the Mexican border to Corpus Christi, central spans from north of Corpus Christi to Matagorda Bay, and northern refers to the region from north of Matagorda Bay to the Louisiana border. In Florida, the north-south dividing line is from Cape Canaveral (28.45N) to Tarpon Springs (28.17N). The dividing line between western and eastern Florida goes from 82.69W at the northern Florida border with Georgia, due south to Lake Okeechobee along longitude 80.85W.

[1] Hurricane may not have been reliably estimated for intensity (both central pressure and maximum sustained wind speed) because of landfall in a relatively uninhabited region. Errors in intensity are likely to be underestimates of the true intensity.

[2] Hurricane center did not make a landfall over the United States, but did produce hurricane-force winds over land. The position indicated is the point of closest approach. In this table, maximum winds refer to the strongest winds estimated to affect the United States. In this case, central pressure is given for the hurricane's point of closest approach.

[3] Hurricane center did make a direct landfall, but the strongest winds likely remained offshore. Thus the winds indicated here are lower than in HURDAT.

[4] Hurricane made landfall over Mexico, but also caused hurricane winds in Texas. The position given is that of the Mexican landfall. The strongest winds at landfall affected Mexico, while the weaker maximum sustained winds indicated here were conditions estimated to occur in Texas. Indicated central pressure given is that at Mexican landfall.

[5] Value listed is a "storm tide" observation rather than a "storm surge," which removes the astronomical tide component.

TABLE 7.9 Landfalling Tropical Storms in the United States, 1851–1910

Number and date	Time	Latitude	Longitude	Maximum winds (knots)	Landfall state
6. October 19, 1851	1500	41.1N	71.7W	50kt	NY
3. August 19, 1856	1100	34.8	76.4	50	NC
4. September 30, 1857[1]	1000	25.8	97.0	50	TX
3. September 14, 1858[1]	1500	27.6	82.7	60	FL
3. September 16, 1858[2]	0300	35.2	75.2	50	NC
7. October 17, 1859[1]	1600	26.4	80.1	60	FL
7. October 7, 1861	1200	35.3	75.3	50	NC
8. November 1, 1861[1]	0800	26.0	81.8	60	FL
8. November 3, 1861	0800	41.0	72.3	60	NY
8. November 3, 1861	0900	41.2	72.0	50	CT
6. September 18, 1863	1300	34.6	77.1	60	NC
9. September 29, 1863[1]	1200	29.3	94.8	60	TX
2. June 30, 18651	1800	26.0	97.5	50	TX
3. August 22, 1865[2]	1800	34.5	74.6	40	NC
6. September 7, 1865[1]	0000	29.7	92.0	60	LA
7. October 30, 1866	0800	39.5	74.3	60	NJ
2. August 2, 1867[2]	0100	34.9	75.0	60	NC
2. August 2, 1867[2]	2200	40.9	69.3	50	MA
2. October 4, 1868[1]	1600	29.9	85.4	60	FL
2. September 3, 1870[2]	1800	40.5	68.8	40	MA
1. June 4, 1871	0700	29.1	95.1	50	TX
2. June 9, 1871	1700	29.2	95.0	50	TX
3. August 23, 1871	0000	31.2	81.3	60	GA
7. October 5, 1871[1]	1600	30.0	83.9	60	FL
1. July 11, 1872	0500	29.1	89.1	50	LA
1. July 11, 1872	0800	30.2	89.0	50	MS
5. October 23, 1872[1]	0800	27.9	82.7	50	FL
5. October 25, 187[2]	0100	34.4	77.7	50	NC
1. June 2, 1873	1100	30.8	81.4	40	GA
4. September 23, 1873[1]	1000	27.8	82.8	50	FL
1. July 4, 1874	2000	28.5	96.2	50	TX
4. September 4, 1874[1, 4]	1200	25.0	97.6	40	TX
4. September 27, 1875[1]	1300	30.1	85.7	50	FL
2. September 16, 1876[1, 2]	1500	25.5	79.7	40	FL
7. October 26, 1877[1]	2100	29.3	83.2	40	FL
1. July 2, 1878[1]	1500	26.0	81.8	40	FL
5. September 7, 1878[1]	2100	24.7	80.9	60	FL
5. September 8, 1878[1]	0200	25.2	81.0	60	FL
8. October 10, 1878[1]	2100	29.9	85.4	50	FL
11. October 22, 1878[1, 2]	0000	25.9	79.8	50	FL

(Continued)

TABLE 7.9 Landfalling Tropical Storms in the United States, 1851–1910 *(Continued)*

Number and date	Time	Latitude	Longitude	Maximum winds (knots)	Landfall state
2. August 19, 1879[3]	0600	41.4	70.8	60	MA
5. October 7, 1879	0500	29.0	89.2	50	LA
6. October 16, 1879[1]	0800	30.4	86.6	50	FL
7. October 27, 1879[1]	2100	29.0	82.7	60	FL
1. June 24, 1880	1500	28.7	95.7	40	TX
6. September 9, 1880	1600	29.8	83.6	50	FL
11. October 23, 1880	0800	41.3	70.0	60	MA
11. October 23, 1880	1300	44.0	68.8	60	ME
1. August 3, 1881	1300	30.2	88.3	50	AL
2. August 13, 1881	2100	28.0	96.9	40	TX
4. September 22, 1882	2200	34.7	77.0	50	NC
4. September 24, 1882	0500	40.7	72.8	40	NY
3. September 11, 1884	0100	31.6	81.2	40	GA
3. August 22, 1885	2300	30.1	85.7	50	FL
4. September 21, 1885	0300	29.0	89.4	50	LA
4. September 21, 1885	1200	30.0	85.6	50	FL
4. September 23, 1885[2]	0300	41.6	69.7	50	MA
6. September 26, 1885	0400	29.6	89.0	60	LA
6. October 2, 1885[2]	1500	35.0	74.8	50	NC
8. October 11, 1885	2200	29.4	83.2	60	FL
5. August 18, 1886[1,2]	0100	23.9	81.9	55	FL
3. June 14, 1887	0700	30.2	88.7	35	MS
7. August 25, 1887[2]	0600	35.0	74.4	50	NC
16. October 30, 1887[1]	0100	26.8	82.3	40	FL
2. July 5, 1888	1600	28.8	95.6	50	TX
4. September 6, 1888[1,2]	0000	23.0	81.9	50	FL
5. September 8, 1888[1]	0000	26.7	80.0	45	FL
6. September 26, 1888[3]	1300	41.6	69.9	55	MA
7. October 11, 1888	1600	33.9	78.1	60	NC
9. November 25, 1888[2]	1800	35.3	74.2	60	NC
2. June 17, 1889	1500	29.1	82.9	45	FL
4. September 11, 1889[2]	2100	38.4	72.7	60	NJ
6. September 23, 1889	1300	30.3	87.7	60	FL
9. October 5, 1889[1]	2300	24.7	81.1	40	FL
9. October 6, 1889[1]	0100	25.2	80.9	40	FL
2. August 27, 1890	1600	29.1	90.8	50	LA
7. October 9, 1891[1]	1400	25.8	81.7	45	FL
8. October 9, 1891[1]	1400	25.8	81.7	45	FL
9. October 12, 1891[2]	0600	35.0	74.1	60	NC

Number and date	Time	Latitude	Longitude	Maximum winds (knots)	Landfall state
1. June 10, 1892[1]	2300	25.7	81.3	40	FL
4. September 12, 1892	0700	29.0	90.6	50	LA
9. October 24, 1892[1]	1900	27.6	82.8	45	FL
1. June 15, 1893	2300	29.9	83.7	60	FL
11. October 23, 1893	0300	35.2	75.6	50	NC
11. October 23, 1893	1100	38.1	75.6	45	VI
12. November 8, 18932	1800	35.6	74.6	55	NC
2. August 7, 1894	1800	30.3	87.6	50	AL
4. September 28, 1894	1200	34.7	76.7	60	NC
1. August 15, 1895	1900	29.3	89.6	50	LA
1. August 16, 1895	1300	30.2	88.8	45	MS
4. October 7, 1895	0400	29.3	94.8	35	TX
6. October 16, 18951	1300	25.7	81.3	35	FL
5. October 9, 18961	0200	26.4	82.0	50	FL
5. October 13, 18962	1200	40.0	67.2	60	RI
2. September 10, 1897[1,3]	1800	24.4	81.9	50	FL
3. September 21, 18971	0200	26.7	82.3	60	FL
3. September 23, 18973	1000	35.2	75.7	50	NC
3. September 24, 1897	1100	40.8	72.7	50	NY
3. September 24, 1897	1300	41.3	72.2	45	CT
5. October 20, 1897	2000	35.2	75.5	55	NC
6. October 25, 1897	2300	36.1	75.8	55	NC
1. August 2, 18981	0300	27.1	80.1	35	FL
5. September 20, 1898	1100	29.6	92.8	50	LA
6. September 28, 1898	0700	29.4	94.7	50	TX
8. September 26, 1898[1]	0600	25.1	80.8	40	FL
9. October 11, 1898[1,3]	1200	24.5	80.0	40	FL
1. June 27, 1899	0900	29.1	95.1	35	TX
2. July 30, 18991	1000	24.9	80.6	40	FL
3. August 13, 18992	1200	27.0	78.6	60	FL
6. October 5, 18991	1000	27.9	82.8	50	FL
3. September 13, 1900	0630	29.2	89.5	40	LA
3. September 13, 1900	1500	30.3	88.8	35	MS
6. October 12, 1900	0250	29.5	83.3	40	FL
1. June 13, 1901	2050	29.9	84.6	35	FL
2. July 10, 1901	1010	28.6	96.0	45	TX
3. July 12, 1901	2210	34.0	77.9	35	NC
4. August 10, 1901	2130	26.3	80.1	40	FL
7. September 17, 1901	1930	30.4	86.6	50	FL
9. September 28, 1901	0250	29.9	84.6	40	FL

(Continued)

TABLE 7.9 Landfalling Tropical Storms in the United States, 1851–1910 *(Continued)*

Number and date	Time	Latitude	Longitude	Maximum winds (knots)	Landfall state
1. June 14, 1902	2310	29.8	83.7	50	FL
2. June 26, 1902	2110	27.7	97.2	60	TX
4. October 10, 1902	2120	30.3	87.3	50	FL
3. October 20, 1904	1010	25.5	81.2	35	FL
5. November 3, 1904	1230	30.5	86.4	35	FL
3. September 29, 1905	0940	29.6	92.6	45	LA
5. October 9, 1905	1720	29.5	91.4	45	LA
1. June 12, 1906	2030	30.1	85.6	45	FL
8. October 21, 1906	0930	30.0	81.4	50	FL
1. June 28, 1907	2340	30.3	85.9	50	FL
2. September 21, 1907	1700	30.4	88.9	40	MS
3. September 28, 1907	2020	30.1	85.7	45	FL
2. May 29, 1908[3]	2100	35.2	75.6	55	NC
2. May 30, 1908	2250	41.3	72.0	35	CT
4. September 1, 1908	0900	34.7	76.5	45	NC
3. June 28, 1909	2010	26.0	80.1	45	FL
3. June 30, 1909	1400	30.1	84.1	35	FL
7. August 29, 1909	0900	26.4	80.1	45	FL
2. August 21, 1910[4]	0000	25.7	97.2	40	TX

Notes: Date and time: Day and time when the circulation center crossed the U.S. coastline (including barrier islands). Time was estimated to the nearest hour for the period of 1851 to 1899 and to the nearest 10 minutes for the period of 1900 to 1910. Time is UTC or GMT.

Latitude and longitude: Location was estimated to the nearest 0.1° latitude and longitude (about 6 nautical miles [nmi]).

Maximum winds: Estimated maximum sustained 1-min surface (10 m) winds (in knots) to occur along the U.S. coast. Winds are estimated to the nearest 10 kt for the period of 1851 to 1885 and to the nearest 5 kt for the period of 1886 to 1910.

Landfall states affected: AL, Alabama; CT, Connecticut; DE, Delaware; FL, Florida; GA, Georgia; LA, Louisiana; MA, Massachusetts; MD, Maryland; ME, Maine; MS, Mississippi; NC, North Carolina; NH, New Hampshire; NJ, New Jersey; NY, New York; RI, Rhode Island; SC, South Carolina; TX, Texas; VA, Virginia.

[1] Tropical storm may not have been reliably estimated for intensity (maximum sustained wind speed) because of landfall in a relatively uninhabited region. Errors in intensity are likely to be underestimates of the true intensity.

[2] Tropical storm or hurricane center did not make a landfall over the United States, but did produce tropical storm–force winds over land. The position indicated is the point of closest approach. In this table, maximum winds refer to the strongest winds estimated to affect the United States.

[3] Tropical storm or hurricane center did make a direct landfall, but the strongest winds likely remained offshore. Thus the winds indicated here are lower than in HURDAT.

[4] Tropical storm or hurricane made landfall over Mexico, but also caused tropical storm–force winds in Texas. The position given is that of the Mexican landfall. The strongest winds at landfall affected Mexico, while the weaker maximum sustained winds indicated here were conditions estimated to occur in Texas.

TABLE 7.10 Estimated Dates for Start of Accurate Tropical Cyclone Records

State[1]	Date[2]
Texas (southern)	1880
Texas (central)	1850
Texas (northern)	1860
Louisiana	1880
Mississippi	1850
Alabama	< 1851 (1830)
Florida (northwestern)	1880
Florida (southwestern)	1900
Florida (southeastern)	1900
Florida (northeastern)	1880
Georgia	< 1851 (1800)
South Carolina	< 1851 (1760)
North Carolina	< 1851 (1760)
Virginia	< 1851 (1700)
Maryland	< 1851 (1760)
Delaware	< 1851 (1700)
New Jersey	< 1851 (1760)
New York	< 1851 (1700)
Connecticut	< 1851 (1660)
Rhode Island	< 1851 (1760)
Massachusetts	< 1851 (1660)
New Hampshire	< 1851 (1660)
Maine	< 1851 (1790)

[1] Specified regions of the United States are based on U.S. census reports and other historical analyses.
[2] Years in parenthesis indicate possible starting dates for reliable records before the 1850s that may be available with additional research.

2, 1882. This tropical cyclone had been characterized by Dunn and Miller (1960) as a "minimal" storm in northwest Florida, based upon a minimum sea-level pressure measurement of 994 mb and a 50 kt (26 m s^{-1}) wind observed at Pensacola. However, only hours before landfall the barkentine *Cato* measured a central pressure of 949 mb, an observation apparently unknown to Dunn and Miller. Thus, this storm was likely a major hurricane at landfall, though the intense inner core missed making a direct strike on any populated areas. It is certain that many other landfalling storms (both in the United States and other land masses) made landfall without ships or coastal communities sampling the intense inner core, resulting in an underestimation of their intensity at landfall. Such underestimations of landfall intensity are particularly problematic for locations such as south Florida, where, for example, Miami was not incorporated until 1896. There is less uncertainty for an area like New England, which has been fairly densely populated since well before the 1850s. Despite these limitations, this analysis does allow for extending the accurate historical record back in time for several locations along the U.S. coastline.

For some U.S. landfalling hurricanes, a central pressure estimate was obtained from the work of Ho et al. (1987), Ho (1989), and other references (so noted in the metadata file for the appropriate storms). Data from these sources were then used to estimate maximum wind speeds through application of one of the new wind-pressure relationships. If no measured or analyzed (via the Ho 1989 methodology), central pressure was described in the metadata file, then the winds at landfall were determined from coastal station observations or ships immediately offshore, destruction at the coast, and/or observed storm-surge values. In general, it was extremely rare for land-based anemometers to actually measure what was suspected to be the maximum sustained surface winds. This was due to the relative sparsity of coastal stations combined with the small RMW typical of hurricanes as well as the inability of anemometers of the era to survive in extreme wind events. In the cases where there was no central pressure value directly available, the estimated winds at landfall were then used via the wind-pressure relationship to back out a reasonable central pressure. In either case, the objective was to provide both an estimate of the maximum sustained wind at landfall and a central pressure for all landfalling U.S. hurricanes.

Evaluation of the HURDAT Revision by NHC

This reanalysis effort has been done with considerable interaction with the hurricane specialists and researchers at the National Hurricane Center. The HURDAT database has been maintained and updated yearly by NHC for decades. Thus all revisions to the existing best track (or extensions back in time as is the case for the period of 1851 to 1885) have been examined and approved by the NHC Best Track Change Committee. Comments by the NHC Best Track Change Committee and the authors' replies back to the committee are also available via the HURDAT reanalysis Web page.

Future Reanalysis Work

Historical tropical cyclone reconstructions are inevitably subject to revisions whenever new archived information is uncovered. Thus, although several thousand alterations and additions to HURDAT have been completed for the years 1851 to 1910, this does not insure that there may not be further changes once new information is made available. Such an archive of historical data—especially one based upon quasi-objective interpretations of limited observations—should always be one that can be revised when more data or better interpretations of existing information becomes available.

Much more work still needs to be accomplished for the Atlantic hurricane database. One essential project is a Fernández-Partagás and Díaz–style reanalysis for the years before 1851 and for the pre-aircraft reconnaissance era of 1911 to 1943. The former may lead to a complete dataset of U.S. landfalling hurricanes for the Atlantic coast from Georgia to New England back to at least 1800, given the relatively high density of population extending that far into the past. The latter project would likely yield a much higher quality dataset for the entire Atlantic Basin—especially for frequency and intensity of tropical cyclones—given the availability of revised compilations of ship data (e.g., Comprehensive Ocean-Atmosphere Data Set [Woodruff et al. 1987]). Another possibility is to re-examine the intensity record of tropical cyclones since 1944 by utilizing the original aircraft reconnaissance data in the context of today's understanding of tropical cyclone eyewall structure and best extrapolations from flight-level winds to the surface winds (e.g., Dunion et al. 2003). Finally, efforts could be directed to extending the scope of the HURDAT database to include other parameters of interest, such as RMW and radii of gale and hurricane force winds by quadrant.

Regardless of the final direction pursued by future research into the reanalysis of Atlantic hurricanes, it is hoped that efforts detailed here have already expanded the possibilities for the utilization of the Atlantic hurricane database. Users now have access to a more complete record of Atlantic hurricanes, one that extends further back in time and one that provides more information regarding the limitations and error sources. In any planning for the future, a thorough appreciation of past events helps one prepare for possibilities to come. Atlantic hurricanes, arguably the most destructive of all natural phenomena in the Western Hemisphere, demand our attention for their understanding to better prepare society for the impacts that they bring. This reanalysis of Atlantic Basin tropical storms and hurricanes that now provide users with 150 years of record may be able to assist in such endeavors in at least a small way.

Acknowledgments

This work has been sponsored by a NOAA grant "The National Hurricane Center HURDAT File: Proposed Revision" (NA76P0369) as well as through a grant from the Insurance Friends of the National Hurricane Center. The authors wish to thank the NHC Best Track Change Committee (Jack Beven, Jim Gross, Brian Jarvinen, Richard Pasch, Ed Rappaport, and Chairman Colin McAdie) for their encouragement and detailed suggestions that have helped to control the quality of the thousands of alterations and additions to HURDAT. Special thanks for their individual contributions toward this project are also given to Sim Aberson, Auguste Boissonnade, Emery Boose, Mike Chenoweth, Hugh Cobb, Paul Hebert, Paul Hungerford, Lorne Ketch, Cary Mock, Ramon Perez Suarez, David Roth, Al Sandrik, David Vallee, and Roger Williams. The authors also thank John Kaplan, Rick Murnane and two anonymous reviewers for their constructive comments on an earlier draft of this chapter.

NOTES

1. HURDAT is freely and easily accessible at the Web site of the National Hurricane Center (http://www.nhc.noaa.gov/pastall.shtml).
2. These files, along with track maps showing all tropical storms and hurricanes for individual years, are available at the HURDAT reanalysis Web site (http://www.aoml.noaa.gov/hrd/data_sub/re_anal.html).

REFERENCES

Abraham, J., G. Parkes, and P. Bowyer. 1998. The transition of the "Saxby Gale" into an extratropical storm. In *Preprints of the 23rd Conference on Hurricanes and Tropical Meteorology*, 795–98. Boston: American Meteorological Society.

Académia de Ciencias. 1970. *Atlas nacional de Cuba*. Havana: Académia de Ciencias de Cuba.

Alexander, W. H. 1902. *Hurricanes, especially those of Puerto Rico and St. Kitts*. Bulletin 32. Washington, D.C.: Weather Bureau, Department of Agriculture.

Barnes, J. 1998a. *Florida's hurricane history*. Chapel Hill: University of North Carolina Press.

Barnes, J. 1998b. *North Carolina's hurricane history*. Chapel Hill: University of North Carolina Press.

Bender, M. A., R. E. Tuleya, and Y. Kurihara. 1985. A numerical study of the effect of a mountain range on a landfalling tropical cyclone. *Monthly Weather Review* 113:567–82.

Boose, E. R., K. E. Chamberlin, and D. R. Foster. 2001. Landscape and regional impacts of hurricanes in New England. *Ecological Monographs* 71:27–48.

Boose, E. R., M. I. Serrano, and D. R. Foster. 2004. Landscape and regional impacts of hurricanes in Puerto Rico. *Ecological Monographs* 74:335–52.

Callaghan, J., and R. K. Smith. 1998. The relationship between maximum surface wind speeds and central pressure in tropical cyclones. *Australian Meteorological Magazine* 47:191–202.

Cardone, V. J., J. G. Greenwood, and M. A. Cane. 1990. On trends in historical marine wind data. *Journal of Climate* 3:113–27.

Cline, I. M. 1926. *Tropical cyclones.* New York: Macmillan.

Coch, N. K., and B. Jarvinen. 2000. Reconstruction of the 1893 New York City hurricane from meteorological and archaeological records—Implications for the future. In *Preprints of the 24th Conference on Hurricanes and Tropical Meteorology,* 546. Boston: American Meteorological Society.

Connor, W. C. 1956. *Preliminary summary of Gulf of Mexico hurricane data.* New Orleans: New Orleans Forecast Office.

Department of the Interior. 1895. *Report on population of the United States at the eleventh census: 1890.* Part I. Washington, D.C.: Government Printing Office.

Doehring, F., I. W. Duedall, and J. M. Williams. 1994. *Florida hurricanes and tropical storms, 1871–1993, an historical survey.* Report TP-71. Gainesville: Florida Sea Grant College Program.

Dunion, J. P., C. W. Landsea, S. H. Houston, and M. D. Powell. 2003. A re-analysis of the surface winds for Hurricane Donna of 1960. *Monthly Weather Review* 131:1992–2011.

Dunn, G. E., and B. I. Miller. 1960. *Atlantic hurricanes.* Baton Rouge: Louisiana State University Press.

Dvorak, V. F. 1984. *Tropical cyclone intensity analysis using satellite data.* NOAA Technical Report, NESDIS 11. Boulder, Colo.: National Oceanic and Atmospheric Administration.

Ellis, M. J. 1988. *The hurricane almanac—1988 Texas edition.* Corpus Christi, Tex.: Hurricane.

Fergusson, S. P., and R. N. Covert. 1924. New standards of anemometry. *Monthly Weather Review* 52:216–18.

Fernández-Partagás, J., and H. F. Díaz. 1995a. *A reconstruction of historical tropical cyclone frequency in the Atlantic from documentary and other historical sources.* Part I, *1851–1870.* Boulder, Colo.: Climate Diagnostics Center, Environmental Research Laboratories, National Oceanic and Atmospheric Administration.

Fernández-Partagás, J., and H. F. Díaz. 1995b. *A reconstruction of historical tropical cyclone frequency in the Atlantic from documentary and other historical sources.* Part II, *1871–1880.* Boulder, Colo.: Climate Diagnostics Center, Environmental Research Laboratories, National Oceanic and Atmospheric Administration.

Fernández-Partagás, J., and H. F. Díaz. 1996a. Atlantic hurricanes in the second half of the Nineteenth Century. *Bulletin of the American Meteorological Society* 77:2899–906.

Fernández-Partagás, J., and H. F. Díaz. 1996b. *A reconstruction of historical tropical cyclone frequency in the Atlantic from documentary and other historical sources.* Part III, *1881–1890.* Boulder, Colo.: Climate Diagnostics Center, Environmental Research Laboratories, National Oceanic and Atmospheric Administration.

Fernández-Partagás, J., and H. F. Díaz. 1996c. *A reconstruction of historical tropical cyclone frequency in the Atlantic from documentary and other historical sources.* Part IV, *1891–1900.* Boulder, Colo.: Climate Diagnostics Center, Environmental Research Laboratories, National Oceanic and Atmospheric Administration.

Fernández-Partagás, J., and H. F. Díaz. 1997. *A reconstruction of historical tropical cyclone frequency in the Atlantic from documentary and other historical sources.* Part V, *1901–1908.* Boulder, Colo.: Climate Diagnostics Center, Environmental Research Laboratories, National Oceanic and Atmospheric Administration.

Fernández-Partagás, J., and H. F. Díaz. 1999. *A reconstruction of historical tropical cyclone frequency in the Atlantic from documentary and other historical sources.* Part VI, *1909–1910.* Boulder, Colo.: Climate Diagnostics Center, Environmental Research Laboratories, National Oceanic and Atmospheric Administration.

Fitzpatrick, P. J. 1999. *Natural disasters: Hurricanes.* Santa Barbara, Calif.: ABC-CLIO.

Garcia-Bonnelly, J. U. 1958. *Hurricanes which caused damage on the Island of Hispanola.* Final report of the Caribbean hurricane seminar, Ciudad Trujillo, Dominican Republic, February 16–25, 1956.

Garriott, E. B. 1900. *West Indian hurricanes.* Bulletin H. Washington, D.C.: Weather Bureau.

Gutierrez-Lanza, M. 1904. *Apuntes históricos acerca del Observatorio del Colegio de Belen.* Havana: Avisador Comercial.

Hebert, P. J., and C. J. McAdie. 1997. *Tropical cyclone intensity climatology of the North Atlantic Ocean, Caribbean Sea and Gulf of Mexico.* NOAA Technical Memorandum, NWS TPC Report No. 2. Miami: National Oceanic and Atmospheric Administration.

Ho, F. P. 1989. *Extreme hurricanes in the nineteenth century.* NOAA Technical Memorandum, NWS Hydro Report No. 43. Silver Spring, Md.: National Oceanic and Atmospheric Administration..

Ho, F. P., J. C. Su, K. L. Hanevich, R. J. Smith, and F. P. Richards. 1987. *Hurricane climatology for the Atlantic and Gulf Coasts of the United States.* NOAA Technical Report, NWS Report No. 38. Silver Spring, Md.: National Oceanic and Atmospheric Administration.

Holland, G. J. 1981. On the quality of the Australian tropical cyclone data base. *Australian Meteorological Magazine* 29:169–81.

Holland, G. J. 1987. Mature structure and structure changes. In *A global view of tropical cyclones*, edited by R. L. Elsberry, W. M. Frank, G. J. Holland, J. D. Jarrell, and R. L. Southern. Chicago: University of Chicago Press.

Hudgins, J. E. 2000. *Tropical cyclones affecting North Carolina since 1586—An historical perspective.* NOAA Technical Memorandum, NWS ER-92. Bohemia, N.Y.: National Oceanic and Atmospheric Administration.

Instituto Cubano de Geodesia y Cartografia. 1978. *Atlas de Cuba.* Havana: Instituto Cubano de Geodesia y Cartografia.

Jarrell, J. D., P. J. Hebert, and M. Mayfield. 1992. *Hurricane experience levels of coastal county populations from Texas to Maine.* NOAA Technical Memorandum, NWS NHC Report No. 46. Coral Gables, Fla.: National Oceanic and Atmospheric Administration.

Jarvinen, B. R., C. J. Neumann, and M. A. S. Davis. 1984. *A tropical cyclone data tape for the North Atlantic Basin, 1886–1983: Contents, limitations, and uses.* NOAA Technical Memorandum, NWS NHC Report No. 22. Coral Gables, Fla.: National Oceanic and Atmospheric Administration.

Jelesnianski, C. P. 1993. The habitation layer. In *Global guide to tropical cyclone forecasting*, edited by G. J. Holland. Technical Document WMO/TC-No. 560, Report No. TCP-31. Geneva: World Meteorological Organization.

Kadel, B.C. 1926. An interpretation of the wind velocity record at Miami Beach, Fla., September 17–18, 1926. *Monthly Weather Review* 54:414–16.

Kagan, H. H., ed. 1966. *American heritage.* New York: American Heritage.

Kaplan, J., and M. DeMaria. 1995. A simple empirical model for predicting the decay of tropical cyclone winds after landfall. *Journal of Applied Meteorology* 34:2499–512.

Kaplan, J., and M. DeMaria. 2001. On the decay of tropical cyclone winds after landfall in the New England area. *Journal of Applied Meteorology* 40:280–86.

Kinsman, B. 1969. Who put the wind speeds in Admiral Beaufort's Force Scale? *Oceans* 2:18–25.

Kraft, R. H. 1961. The hurricane's central pressure and highest wind. *Marine Weather Log* 5:155.

Landsea, C. W. 1993. A climatology of intense (or major) Atlantic hurricanes. *Monthly Weather Review* 121:1703–13.

Ludlum, D. M. 1963. *Early American hurricanes, 1492–1870.* Boston: American Meteorological Society.

Martinez-Fortun, J. A. 1942. Ciclones de Cuba. *Revista Bimestre Cubana* 50:232–49.

Mitchell, C. L. 1924. West Indian hurricanes and other tropical cyclones of the North Atlantic Ocean. *Monthly Weather Review*, Supplement 24.

Neumann, C. J. 1993. Global overview. In *Global guide to tropical cyclone forecasting*, edited by G. J. Holland, 1.1–1.56. Technical Document WMO/TC-No. 560, Report No. TCP-31. Geneva: World Meteorological Organization.

Neumann, C. J. 1994. An update to the National Hurricane Center "Track Book." In *Minutes of the 48th interdepartmental conference*, A47–A53. Miami: Office of Federal Coordinator for Meteorological Services and Supporting Research, National Oceanic and Atmospheric Administration.

Neumann, C. J., B. R. Jarvinen, C. J. McAdie, and J. D. Elms. 1993. *Tropical cyclones of the North Atlantic Ocean, 1871–1992.* NOAA Historical Climatology Series 6-2. Asheville, N.C.: National Climatic Data Center.

Neumann, C. J., B. R. Jarvinen, C. J. McAdie, and G. R. Hammer. 1999. *Tropical cyclones of the North Atlantic Ocean, 1871–1999.* NOAA Historical Climatology Series 6-2. Asheville, N.C.: National Climate Data Center.

Office of the Federal Coordinator for Meteorological Services and Supporting Research (OFCM). 2001. *National Hurricane Operations Plan (NHOP)*. FCM-P12–2001. Washington, D.C.: National Oceanic and Atmospheric Administration.

Ortiz-Hector, R. 1975. *Organismos ciclonicos tropicales extemporaneous*. Serie meteorological No. 5. Havana: Académia de Ciencias de Cuba.

Parkes, G. S., L. A. Ketch, C. T. O'Reilly, J. Shaw, and A. Ruffman. 1998. The Saxby Gale of 1869 in the Canadian Maritimes. In *Preprints of the 23rd Conference on Hurricanes and Tropical Meteorology*, 791–94. Boston: American Meteorological Society.

Perez Suarez, R., R. Vega, and M. Limia. 2000. Cronología de los ciclones tropicales de Cuba. In *Informe final del proyecto "Los ciclones tropicales de Cuba, su variabilidad y su posible vinculación con los cambios globales."* Havana: Instituto de Meteorologia.

Pielke R. A., and C. W. Landsea. 1998. Normalized hurricane damages in the hurricane United States: 1925–95. *Weather and Forecasting* 13:621–31.

Powell, M. D., and S. H. Houston. 1996. Hurricane Andrew's landfall in south Florida. Part II: Surface wind fields and potential real-time applications. *Weather and Forecasting* 11:329–49.

Rappaport, E. N., and J. Fernández-Partagás. 1995. *The deadliest Atlantic tropical cyclones, 1492–1994*. NOAA Technical Memorandum, NWS NHC Report No. 47. Coral Gables, Fla.: National Oceanic and Atmospheric Administration.

Rodriguez-Demorizi, E. 1958. *La marina de Guerra dominicana, 1874–1861*. Ciudad Trujillo: Montalvo.

Rodriguez-Ferrer, M. 1876. *Naturaleza y civilización de la grandiose Isla de Cuba*. Madrid: Noriega.

Roth, D. M. 1997a. Louisiana hurricane history. Lake Charles, Louisiana: National Weather Service, Lake Charles, La. (available at: http://www.srh.noaa.gov/lch/research/lahur.htm).

Roth, D. M. 1997b. Texas hurricane history. National Weather Service, Lake Charles, La. (available at: http://www.srh.noaa.gov/lch/research/txhur.htm).

Roth, D. M., and H. D. Cobb III. 2000. Re-analysis of the gale of '78—Storm 9 of the 1878 hurricane season. In *Preprints of the 24th Conference on Hurricanes and Tropical Meteorology*, 544–45. Boston: American Meteorological Society.

Roth, D. M., and H. D. Cobb, III. 2001. Virginia hurricane history. Hydrometeorological Prediction Center, Camp Springs, Md. (available at: http://www.hpc.ncep.noaa.gov/research/roth/vahur.htm).

Salivia, L. A. 1972. *Historia de los temporales de Puerto Rico y las Antillas (1492–1970)*. San Juan, P.R.: Edio.

Sandrik, A., and B. Jarvinen. 1999. A re-evaluation of the Georgia and northeast Florida tropical cyclone of 2 October 1898. In *Preprints of the 23rd Conference on Hurricanes and Tropical Meteorology*, 475–78. Boston: American Meteorological Society.

Sandrik, A., and C. Landsea. 2003. *Chronological listing of tropical cyclones affecting North Florida and coastal Georgia, 1565–1899.* NOAA Technical Memorandum, NWS SR-224. Fort Worth, Tex.: National Oceanic and Atmospheric Administration.

Sarasola, S. 1928. *Los huracanes en las Antillas.* Madrid: Clásica Española.

Schloemer, R. W. 1954. *Analysis and synthesis of hurricane wind patterns over Lake Okeechobee, Florida.* Hydrometeorological Report No. 31. Washington, D.C.: Weather Bureau, Department of Commerce, and Army Corps of Engineers.

Simpson, R. H., and H. Riehl. 1981. *The hurricane and its impact.* Baton Rouge: Louisiana State University Press.

Sullivan, C. L. 1986. *Hurricanes of the Mississippi Gulf Coast.* Biloxi: Gulf.

Tannehill, I. R. 1938. *Hurricanes, their nature and history.* Princeton: Princeton University Press.

Tanner, H. H., ed. 1995. *The settling of North America.* New York: Macmillan.

Tucker, T. 1982. *Beware the hurricane! The story of the cyclonic tropical storms that have struck Bermuda, 1609–1982.* Hamilton, Bermuda: Island Press.

Vickery, P. J., P. F. Skerlj, and L. A. Twisdale. 2000. Simulation of hurricane risk in the United States using an empirical storm track modeling technique. *Journal of Structural Engineering* 126:1222–37.

Viñes, B. 1877. *Apuntes relativos a los huracanes de las Antillas en Septiembre y Octubre de 1875 y 76.* Havana: Iris.

Viñes, B. 1895. *Investigaciones relatives a la circulación y translación ciclonica en los huracanes de las Antillas.* Havana: Aviasador Comercial.

Woodruff, S. D., R. J. Slutz, R. L. Jenne, and P. M. Steurer. 1987. A comprehensive ocean-atmosphere data set. *Bulletin of the American Meteorological Society* 68:1239–50.

8

Ancient Records of Typhoons in Chinese Historical Documents

Kin-sheun Louie and Kam-biu Liu

In this chapter, we survey the Chinese historical documentary record as a source of information on ancient typhoons. Records on typhoons can be found in three major types of historical documents: (1) central imperial government records and official archives, including official histories (*Zheng Shi*) and Emperor's Veritable Records (*Shi Lu*); (2) semi-official local gazettes, known as *fang zhi* in Chinese; and (3) unofficial literary works, such as personal diaries, travel logbooks, essays, and poems. The world's earliest mention and scientific description of a typhoon (*jufeng*) is contained in a book, *Nan Yue Zhi*, which was written around A.D. 470. A typhoon that struck Mizhou in Shandong in June or July 816, as recorded in the official document the *Old Tang History* (*Jiu Tang Shu*), is the earliest recorded landfall of a tropical cyclone known in the world. These voluminous documentary sources reveal mankind's earliest knowledge of typhoons. They also offer promising opportunities for quantitatively reconstructing the ancient history of typhoon landfalls in China for at least the past 1,000 years.

In China as well as in North America, the earliest instrumentally observed records of tropical cyclone activities only date to the middle to late nineteenth century (Chin 1972; J. Z. Wang 1991; Neumann et al. 1993; Fernández-Partagás and Díaz 1996; Elsner and Kara 1999; Landsea 2000). Extending this record back in time mainly depends on two sources of data—geological proxy evidence and written records. Recent developments in paleotempestology have established the use of coastal lake and marsh sediments as a proxy for reconstructing prehistoric or historic hurricane landfalls occurring during the last several millennia (Liu and Fearn 1993, 2000a, 2000b; Donnelly et al. 2001; Liu, chapter 2, and Donnelly and Webb, chapter 3, in this volume). The chronological precision of proxy records is limited by stratigraphic resolution and dating control, which typically have margins of error on the order of decades at best. For the last several centuries, at least, the proxy evidence can be supplemented by historical data extracted from written documentary records. In North America, no written record exists before the Columbian contact. Indeed, the earliest description of a possible hurricane event in the Western Hemisphere may have been written by Christopher Columbus himself nearly 500 years ago during one of his voyages to the Caribbean (Ludlam 1963). Since then, written records of hurricane strikes in the western world have been available in local newspapers, travel diaries, chronicles, and plantation records. Many of these historical records of early American hurricanes have been compiled by Ludlam (1963) and Millás (1968). In China, where written history started at least 3,500 years ago, the potential of finding much more ancient records of typhoon activities is quite promising. That potential is demonstrated by Liu, Shen, and Louie (2001) in a recent study that reconstructed a nearly 1,000-year time series of typhoon landfalls for the Guangdong Province in southern China based on Chinese historical documentary records. Similar historical archival work is being undertaken by these authors to extract a millennial history of typhoon landfalls for all other coastal provinces of China.

The Chinese historical documentary record is a tremendous archive of paleoclimatic information (Chu 1973; P. K. Wang 1979; Zhang 1980, 1991a, 1991b; Wang and Zhang 1988, 1991, 1992; Zhang and Crowley 1989). In this chapter we evaluate the potential of extracting a detailed record of typhoon activities from Chinese historical documentary evidence. Specifically, we discuss the various sources of typhoon data in three major types of Chinese historical documentary sources: (1) central imperial government records and official archives, (2) local government or semi-official gazettes, and (3) unofficial literary works. We present evidence to illustrate the ancient typhoon activities that were recorded in Chinese historical documents spanning the period from the ninth century to the end of the nineteenth century A.D. We conclude by highlighting the major milestones in the development of the scientific understanding of typhoons as a meteorological phenomenon in ancient China.

Central Imperial Government Records and Official Archives

The Chinese are famous for their special interest in the study of history and history writing. Thus, there are abundant official records written and maintained under the auspices of the central imperial government. Two kinds of official records are of particular relevance to research on natural disasters such as typhoons: official histories and Emperor's Veritable Records.

Official Histories (*Zheng Shi*)

Beginning with the Western Han Dynasty (206 B.C.–A.D. 8) (figure 8.1), it has been the custom of a succeeding dynasty to produce an official volume that records the major historical events of the previous dynasty. At times, a history of a particular dynasty might be written by a scholar in a private capacity. That work might later be recognized by the imperial government and given the status of an official history. For some dynasties, there is more than one official history. For example, for the Tang Dynasty (A.D. 618–907), there are two official histories, the *Old Tang History* (*Jiu Tang Shu*), compiled between A.D. 941 and 945, and the *New Tang History* (*Xin Tang Shu*), compiled between A.D. 1044 and 1066. These official histories (*zheng shi*) are based largely on official government records stored in the central government archives. They are written by professional historians and conform to strict historical standards. There have been 26 official histories written since the Western Han Dynasty.

The earliest unequivocal record of a typhoon landfall in China can be found in the official history of the Tang Dynasty. The *Old Tang History* contains the following entry in its "Basic Annals of Emperor Xian Zong" (*Xian Zong Ben Ji*, chapter 15 of *Old Tang History*): "On *mu-shen* day of the eighth (lunar) month (of the eleventh year of Yuanhe reign), Mizhou reported that a *jufeng* (typhoon) occurred and the seawater damaged the city wall."[1]

We have verified this particular record with the information in the *New Tang History* and inferred that the typhoon landfall should have occurred between June 29 and July 28 (of the Gregorian calendar) in A.D. 816. It is thus the earliest unequivocally recorded landfall of a tropical cyclone in the world (Louie and Liu 2003). The landfall location, Mizhou, is nowadays the coastal city of Gaomi in Shandong Province (figure 8.2). Two things are remarkable about this entry. First, the term *jufeng*, the traditional Chinese name for typhoon, was already used in this historical record of A.D. 816 (figure 8.3). This implies that observations and knowledge about typhoons as a weather phenomenon were already firmly established in China by that time. Second, the use of the word "reported" is significant, because it implies that typhoon strikes, among other natural disasters and unusual phenomena, were reported to the central government by local government officials. It is thus evident that the term *jufeng* was originally used in the local report in A.D. 816, rather than being invented by the historian who wrote the *Old Tang History* more than a century later. This implies that knowl-

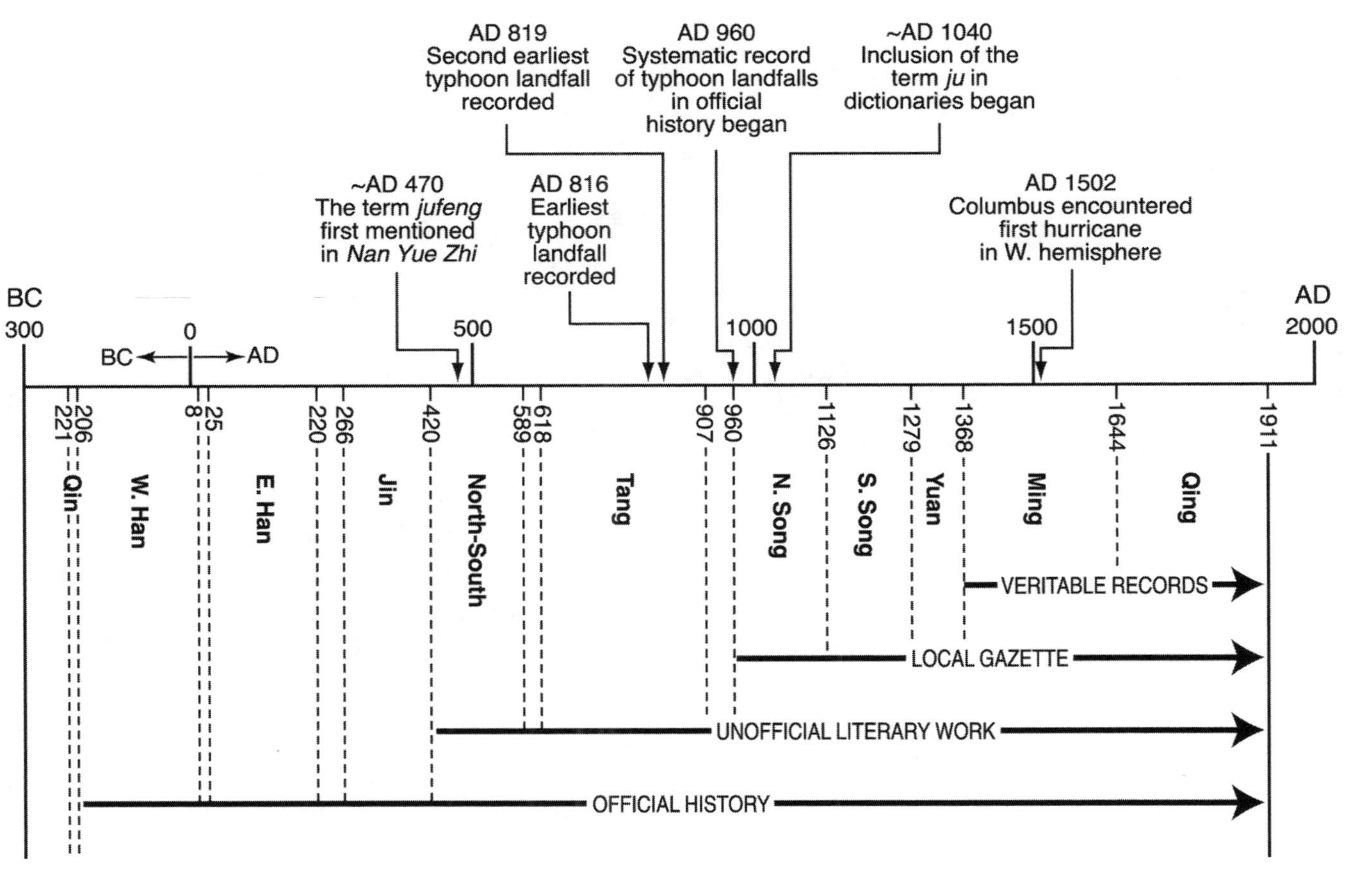

FIGURE 8.1 Timeline showing the Chinese dynasties, availability of historical documents, and major events relating to the history of typhoons in China.

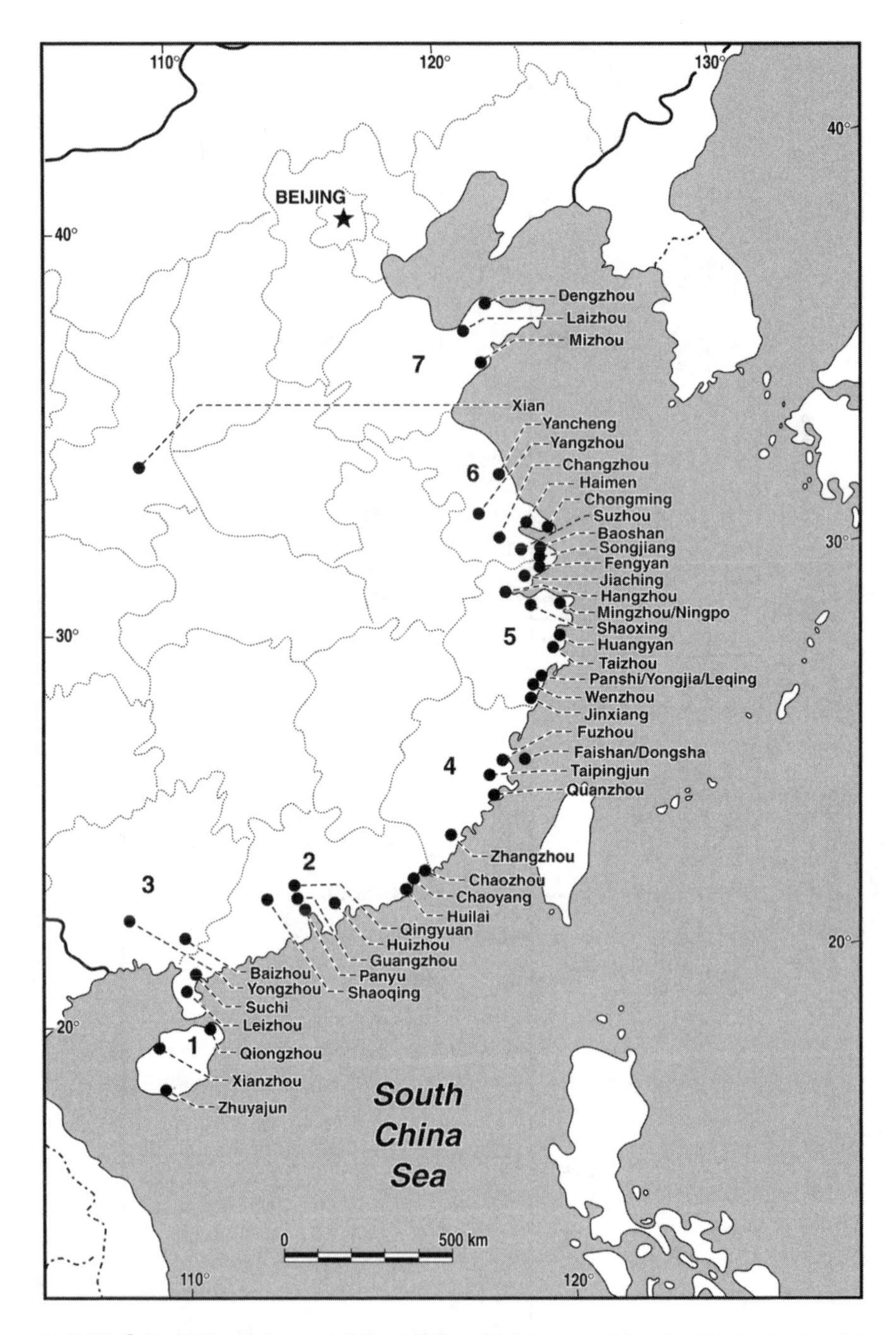

FIGURE 8.2 China, with the coastal provinces and locations mentioned in this chapter. Coastal provinces: *1*, Hainan; *2*, Guangdong; *3*, Guangxi; *4*, Fujian; *5*, Zhejiang; *6*, Jiangxu; *7*, Shandong.

FIGURE 8.3 The word *jufeng* (typhoon) in Chinese calligraphy.

edge of typhoons and typhoon strikes existed at the grass-roots level. In the third section of this chapter, we undertake a more detailed discussion of early written typhoon observations, which began in China in the fifth century A.D., and of the diffusion of related knowledge among the population.

Most records of typhoon activities in the official histories are contained in the chapters called the "Records of Five Elements" (*Wu Xing Zhi*) or "Records of Disasters" (*Zai Yi Zhi*). These records, which can be found in 16 of the 26 official histories, register unusual natural and cultural events occurring during a dynasty (S. M. Wang 1981). The systematic recording of typhoon strikes as a natural disaster in the "Records of Five Elements or Disasters" started in the Official History of Song (*Song Shi*). This practice continued in the corresponding chapters of all succeeding dynasties including the Yuan, Ming, and Qing. Clearly, a typhoon was an established category of natural event that was to be reported by local government officials to the central government during the Northern Song Dynasty (A.D. 960–1126). Fifteen typhoon strikes were registered in the "Records of Five Elements" of the Official History of Song (table 8.1).

The inventory of typhoon strikes (table 8.1) derived from the "Records of Five Elements" of the Song Dynasty illustrates the kind of data that can be extracted from the official histories. In the "Records of Five Elements" of the

TABLE 8.1 Typhoon Strikes Recorded in the Official History of Song

Year (A.D.)	Date[1]	Location (modern-day province)[2]	Description of event
975	November–December	Guangzhou (Guangdong)	A *jufeng* struck Guangzhou. Two *zhang*[3] of rain fell in a day and night. The sea rose. Boats drifted away and were lost.
982	August–September	Qiongzhou (Hainan)	A *jufeng* struck Qiongzhou. The gate of the city was damaged. Government buildings and people's houses were all destroyed.
983	October–November	Taipingjun (Fujian)	A *jufeng* felled trees in Taipingjun; 1,087 public buildings and civil residences were damaged.
983	November–December	Leizhou (Guangdong)	Leizhou was struck by a *jufeng*; 700 warehouses and residences were damaged.
984	September–October	Baizhou (Guangxi)	A *jufeng* struck Baizhou. Public buildings and civil residences were damaged.
996	September–October	Chaozhou(Guangdong)	Chaozhou was struck by a *jufeng*. Government buildings and military barracks were damaged.
1005	September–October	Fuzhou (Fujian)	A *jufeng* occurred in the sea near Fuzhou. Houses were damaged.
1076	November–December	Haiyang and Chaoyang (Guangdong)	Haiyang and Chaoyang were struck by *jufeng* and storm tides. Houses and field crops were damaged.
1081	July–August	Yongzhou (Guangxi)	A *jufeng* struck Yongzhou, causing damage to the observation tower of the city wall, government buildings, and people's residences.
1082	September–October	Zhuyajun (Hainan)	A *jufeng* struck Zhuyajun. Houses were destroyed.
1172	July 11	Huizhou (Gunagdong)	A *jufeng* struck Huizhou, damaging more than 30 naval battleships. At that time, the Ministry of Defense was deploying the naval fleet of the Guangdong Strategic Planning Unit. Three-quarters of battleships capsized. More than 130 solders were killed.
1183	September 17	Leizhou (Guangdong)	Leizhou was severely struck by a *jufeng*. People were injured by the storm surge. Crops and trees were broken or toppled.
1194	Autumn	Mingzhou (Zhejiang)	Mingzhou was struck by a *jufeng* and storm surge. Crops were damaged.
1224	Autumn	Fuzhou (Fujian)	Fuzhou was severely struck by a *jufeng*. Cultivated fields and crops were damaged.
1263	December	Fuzhou (Fujian)	A *jufeng* struck Fuzhou.

[1] Dates are converted according to the Gregorian calendar.
[2] For the locations, see figure 8.2.
[3] One *zhang* equals about 3 meters.

succeeding dynasties after Song, 7 entries of typhoon strikes were registered in the official history for the Yuan Dynasty (A.D. 1271–1368), 10 entries for the Ming Dynasty (A.D. 1368–1644), and 35 entries for the Qing Dynasty (A.D. 1644–1911). It should be noted that the official government inventories contained in the "Records of Five Elements" provide only a partial record of typhoon strikes affecting China. They probably represent those storms that caused severe societal impacts or economic losses. It is conceivable that only major, destructive events that required relief efforts from the central government were likely to be selectively reported by the local officials. Moreover, only major events in local reports were likely to be selectively retained in the official government archives of the contemporary dynasty, and only those events deemed major and significant were likely to be selectively included in the official history.

The official government records also have other biases and limitations. As the information about these typhoon strikes come mainly from official reports submitted by local government officials, reporting might be incomplete or uneven during times of political chaos or social unrest, or when the central government's power of control was weakened. There seems to be a tendency for the records of typhoon strikes to be somewhat more frequent during the early parts of the Ming and Qing dynasties when the imperial power was at its height, than during the later years when the imperial power was declining.

Emperor's Veritable Records (*Shi Lu*)

Starting around the seventh century A.D., the daily activities of each emperor were meticulously recorded and compiled into Veritable Records (*Shi Lu*) by imperial historians. Chief among these daily activities were administrative decisions made by the emperor on myriad matters reported by local government officials. These matters included reports of natural disasters such as typhoon strikes and requests for government emergency aids. These records provided the primary materials for the official histories, and, as original data reported by local governments, they usually contain a more complete record of typhoon strikes than the official histories. Most of the Veritable Records prior to the fourteenth century were lost. Only those of the last two imperial dynasties, the Ming and Qing, are preserved in whole, printed, and publicly accessible (Zhuang 1983; Qin 1994) (figure 8.1).

Records of 34 typhoon strikes found in the Veritable Records of Ming (*Ming Shi Lu*) illustrate the kind of typhoon information that can be extracted from this documentary source (table 8.2). Nine of the 34 typhoon strikes recorded in the Veritable Records of Ming were also listed in the Official His-

TABLE 8.2 Typhoon Strikes Recorded in Veritable Records of Ming

Year (A.D.)	Month or day[1]	Location[2]	Description of event
1371	August 11	Panyu and adjacent counties (Guangdong)	*Jufeng* lifted away houses; lasted for three days.
1375	February–March	Wenzhou (Zhejiang)	*Jufeng* struck severely, with thunder and rain. Seawater surged. Huts washed away; boats were damaged. Many people died.
1380	October 24	Nancheng County in Zhangzhou Prefecture (Fujian)	*Jufeng* struck; heavy rain. Tree limbs broke; houses were lifted away. Some people were killed or wounded.
1390	July 15	Haimen County in Yangzhou Prefecture (Jiangsu)	*Jufeng* struck severely. Tidewater surged. Houses and huts were damaged. Numerous people and animals were drowned.
1403	June 19	Nanhai and Panyu Counties (Guangdong)	*Jufeng* struck. Tidewater surged. Huts drifted in water. Thirty-five people drowned.
1409	September 20	Four Military Territories (Songmen, Haiman, Changguo, Taizhou) and six 1,000-household garrison bases in Chumen (Zhejiang)	*Jufeng* and heavy rain. City wall was damaged; houses drifted in water.
1410	August 26	Military Territory of Jinxiang (Zhejiang)	*Jufeng* and heavy rain. City wall and government buildings were damaged.
1411	September–October	Leizhou Prefecture (Guangdong)	*Jufeng* and torrential rain. Suchi and Haikong Counties (in Leizhou Prefecture) suffered losses of some 1,600 houses and 600 Chinese hectares of paddy rice; more than 1,600 people drowned.
1416	October 16	Yancheng County in Huai'an Prefecture (Jiangsu)	*Jufeng*. Seawater surged; more than 215 Chinese hectares of cultivated fields flooded.
1420	Summer	Haikang and Suchi Counties in Leizhou Prefecture (Guangdong)	*Jufeng* and torrential rain. Seawater surged, hurting people and damaging rice crops; 1,600 *shi*[3] of cereal grains of tax could not be levied.

1422	May 23	Guangzhou Prefecture, etc. (Guangdong)	*Jufeng* and torrential rain. Tidewater surged. More than 360 people drowned; 1,200 houses washed away; 25,300 *shi* of stockpiled crops spoiled.
1423	September–October	Qiongzhou Prefecture (Hainan)	*Jufeng* and torrential rain. Seawater surged, flooding houses and domestic animals. Fifty-two coastal residents drowned.
1424	July–August	Huangyan County in Taizhou Prefecture (Zhejiang)	*Jufeng* struck severely. Tidewater surged violently; 7,843 houses washed away; 800 people, including many old and young, drowned; more than 256 Chinese hectares and 40 Chinese acres of government-owned and private fields were inundated.
1426	June 26	Military Territory of Panshi (Zhejiang)	*Jufeng* and heavy rain, damaging government buildings and warehouses; many people were crushed to death. Tidal water surged quickly. Floodwater drifted into city and washed away money bills,[4] crops, government documents, and weapons.
	June 28 (same event)	Yongjia and Leqing Counties in Wenzhou Prefecture (Zhejiang)	*Jufeng* and intense rain raged from dawn to dusk. Floodwater rose rapidly, damaging prefecture and county offices, public buildings, warehouses, temples and ceremonial structures, and residential houses. Many people drowned or were crushed to death.
1431	July 18	Wenzhou Prefecture (Zhejiang)	*Jufeng*. Rain fell, damaging public buildings, temples, warehouses, and city wall.
1434	Autumn	Shaoqing and Leizhou Prefectures (Guangdong)	Autumn crops in the field again were inundated by storm surge driven by *Jufeng*. The harvest was destroyed; people suffered from hunger and poverty.
1436	July 10	Huangyan County (Zhejiang)	*Jufeng* and excessive rain. Tidewater surged, inundating all government buildings and warehouses. All 760 *yin* (sacks) of salt stored in warehouses drifted away and were lost.

(Continued)

TABLE 8.2 Typhoon Strikes Recorded in Veritable Records of Ming *(Continued)*

Year (A.D.)	Month or day[1]	Location[2]	Description of event
1436	October 29	Zidi Yangzhou, Suzhou, and Changzhou Prefectures (Jiangsu)	*Jufeng* struck severely. Tidewater surged. People from several hundred families in each *zhou* and county were drifting in the floodwater.
1495	October 2	Chaozhou Prefecture, etc. (Guangdong)	*Jufeng* and violent rain, damaging city walls and houses, drowning people and domesticated animals.
1503	October 8	Chongming County in Zidi Suzhou Prefecture (Jiangsu)	*Jufeng* struck severely. Sea surged and brought disaster. Government decree to distribute relief: one *shi* of rice to people who had lost houses or domesticated animals to flooding, and two *shi* of rice to people who had lost family members; autumn taxes were reduced by 19,560 *shi* of cereals and 23,190 bundles of hay.
1555	August 7	Mazhi, Matu, Baoshan, etc. (Shanghai)	A *jufeng* struck severely; most of the pirate ships capsized, and many of our naval ships were also damaged.[5] The next day, the pirates in Fulin also set sail to sea. They were attacked by our battleships and blown away by a strong sea wind. More than 20 pirate ships were sunk.
1568	August 21	Taizhou Prefecture (Zhejiang)	*Jufeng* struck severely. Tidewater surged quickly. The rivers from Tientai Mountain swelled and converged rapidly. The floodwater intruded into the city limit of Taizhou Prefecture, not receding until three days later. More than 30,000 people drowned, more than 115,000 Chinese acres of cultivated fields were damaged, and more than 50,000 houses were inundated or destroyed.
1584	August 9	(Zhejiang)	*Jufeng* and flood.

1587	August 2	Suzhou, Songjiang, Changzhou, and Zhenjiang (Jiangsu)	[Counties in these four prefectures] were struck by *jufeng* and heavy rain that lasted for several days. Floodwater rose violently, inundating numerous residential houses.
	(End of seventh lunar month, same event)	Zidi Yingtien Prefecture, etc. (Jiangsu)	*Jufeng* struck severely. Excessive flooding turned an area of several hundred square *li*[6] into a lake. Taiping, situated at the lowest topographic spot, suffered most severely.
1589	July 20	Hangzhou, Jiaching, Ningpo, Shaoxing, and Taizhou (Zhejiang)	*Jufeng* struck severely. Seawater surged as if it were boiling. [Counties in these several prefectures] had hundreds of houses and public buildings collapsed and destroyed. Some government and civilian vessels as well as battleships were damaged. More than 200 people drowned or were crushed to death. Crops and trees were destroyed by floodwater. The senior citizens remarked that this was another disaster following the one in the fifteenth year of Wanli reign [A.D. 1587].
1591	September 4	Songjiang, Taiyin, etc. (Shanghai)	*Jufeng* and sea surge caused houses to collapse and inflicted casualties.
1603	September 8	Quanzhou, etc.(Fujian)	Heavy rain. Seawater rose rapidly. *Jufeng* struck suddenly. More than 10,000 people drowned. Numerous residential houses, household belongings, and domesticated animals drifted in the water.
1611	July-August	South Zidi Tolin in Fengyan County (Shanghai)	Three Japanese aliens were captured near the military camp in South Zidi Tolin. On their way to Indochina, they ended up drifting to here after their vessel was capsized by a *jufeng*.
1615	May 3	Yi County in Laizhou Prefecture (Shandong)	*Jufeng* was howling fiercely.

(Continued)

TABLE 8.2 Typhoon Strikes Recorded in Veritable Records of Ming *(Continued)*

Year (A.D.)	Month or day[1]	Location[2]	Description of event
1617	June–July	Faishan, etc. (Fujian)	In the fifth lunar month . . . our naval soldiers were battling again with the pirate forces of Yu Qian near Dongxiji in Haimen.[7] We killed 16 and captured 11 enemy soldiers. Yu Qian then grabbed one large vessel and sailed along Faishan, Niulanji, Nanji, and Baichuan'ou, kidnapping fishermen's families and looting back and forth. After being struck by a *jufeng*, they had to build temporary lodges to repair the vessel. They were thus captured by our forces.
		Dongsha (Fujian)	In the middle of the fifth lunar month of the forty-fifth year of Wanli reign (June–July A.D. 1617), a vessel of Japanese pirates was drifting near Dongsha, Fujian. We captured 67 Japanese pirates alive. Execution was carried out, and a second-class military merit was granted. The vessel was forced into our waters by a *jufeng*.
1617	November–December	Laizhou and Dengzhou (Shandong)	Shandong was hit by natural disasters. Relief crops brought from Dengzhou and Laizhou were inundated and ruined by a *jufeng*.
1618	September 22	Chaoyang, Cheyang, Shaoping, Huilai, Punning, and Chenghai Counties in Chaozhou Prefecture (Guangdong)	*Jufeng* struck severely. A huge fire lit up the sky amid torrential rain. Seawater surged by several *zhang*.[8] Several tens of thousands of people [in these counties] were drifting and lost. Numerous government buildings, city walls, levees, and farmlands were damaged and collapsed.

			On the fourth day of the eighth lunar month of last year (September 22, 1618),[9] an extraordinary hydrological disaster occurred. Fire, thunder, sea surge, and *jufeng* struck concurrently. More than 12,530 men and women were drowned, and 31,869 houses collapsed. More than 5,000 Chinese hectares of cultivated fields and salt ponds were flooded. More than 1,270 *zhang* of levees were breached. Virtually all the houses, city walls, and government buildings in all the affected counties were destroyed. Civilians suffered losses even more severely.
1620	September 24	Dengzhou and Laizhou (Shandong)	*Jufeng* disrupted the cargo shipping at sea . . . 85 cargo ships belonging to Dengzhou and 16 vessels belonging to Laizhou were damaged. More than 44,900 *shi* of cereal grains were lost.

[1] Dates are converted according to the Gregorian calendar.
[2] In the Ming Dynasty, there were basically three local administrative levels: *sheng* (province), *fu* (prefecture), and *xian* (county). The Chinese word *zhou* in place-names (e.g., Wenzhou) usually denotes a status of prefecture. In this table, the name of the province is put in parentheses. For the locations, see figure 8.2.
[3] One *shi* equals about 100 liters.
[4] Money bills were paper notes printed by the government and issued as currency.
[5] This entry describes a naval battle between the Chinese navy and the Guo pirate ships from Japan.
[6] One *li* equals about 500 meters.
[7] This entry describes government naval soldiers sailing to sea to attack pirates.
[8] One *zhang* equals about 3 meters.
[9] This is an entry of A.D. 1619.

tory of Ming (*Ming Shi*). Indeed, of the 10 entries for typhoon strikes registered in the Official History of Ming, only one (occurring in the fourteenth year of Hongzhi Reign, A.D. 1501) was absent in the Veritable Records. This comparison confirms that the records in the Veritable Records were a primary source used by the imperial historians for writing the official history, and it suggests that the imperial historian relied on other sources to provide supplementary information. Of the remaining 25 typhoon strikes recorded in the Veritable Records, 7 were also listed in the official history but were characterized as "storm surge" or "catastrophic tidal surge" events without explicitly using the term *jufeng*. We judged that these were most likely typhoon strikes as correctly described and reported by the local government officials, but the terms "storm surge" and "catastrophic tidal surge" were substituted for *jufeng* by imperial historians years later. In all, 18 typhoon strikes recorded in the Veritable Records were missing from the Official History of Ming. Their omission from the official history may suggest that they were deemed to be minor or nondisastrous events by the imperial historians. Among these 18 typhoon strikes, only 5 (in A.D. 1368, 1380, 1391, 1431, and 1503) clearly involved casualties. Records of the other 13 strikes contain descriptions of damages to crops and premises but do not explicitly mention casualties. It may be inferred, then, that typhoon strikes selected for inclusion in the official history were those that were considered to be major or disastrous events.

In addition to these 34 entries in which the event was explicitly characterized as a *jufeng* strike, there are another 25 events recorded in the Veritable Records of Ming that we inferred to be probable typhoon strikes, although the term *jufeng* was not explicitly used. Most of these events were characterized variously as storms, catastrophic winds, strong winds and severe rains, storm surge, sea surge, tidal surge, strong winds and floods, or as some combination of these terms. Twelve of these inferred typhoon events were also recorded in the Official History of Ming. The 25 events that we inferred to be probable typhoon strikes all occurred during the normal typhoon landfall period—from late spring to late autumn—and their descriptions in the Veritable Records all invoked elements of rain, storm, and sea surge.

The Veritable Records of Qing (*Qing Shi Lu*) is a large publication, containing 60 volumes and more than 60,000 pages. It contains abundant information on events of natural disasters in the Qing Dynasty but up to now no systematic endeavor to extract the relevant data has been undertaken. We are currently working to retrieve from the Veritable Records of Qing information on typhoon landfalls.

Semi-official Local Gazettes (*fang zhi* or *di fang zhi*)

The local gazettes, or *fang zhi* (also known as *di fang zhi*), are probably the most important and most voluminous source of typhoon data in China. As early as in the early Sui Dynasty (A.D. 581–618), the Emperor Wen Di had ordered every province (*jun*, typically comprising several counties) to compile and submit to the central government a volume comprising all facts relating to the district's culture, history, and geography, including maps. This national initiative marks the beginning of government-commissioned local gazettes in China. After the middle of the Tang Dynasty (ca. A.D. 780), and especially during the Song Dynasty (ca. A.D. 960), the writing and updating of local gazettes were progressively institutionalized (figure 8.1). During the ensuing Ming and Qing dynasties, the central government issued guidelines and standards for local authorities to follow in their writing or compilation of the local gazettes (Zhou 1994; Lai 1995). These local gazettes typically include chapters on unusual phenological events and natural disasters and meteorological events such as floods, droughts, hailstorms, snowstorms, deep freeze, earthquakes, and typhoons. These local gazettes are semi-official in nature, because they were written by local scholars under the auspices of the local government. They were not subject to the same strict editorial standard that was applied to the writing of the official histories. Thus their quality varies from one county to another, and sometimes from one edition to another.

Despite the general lack of editorial uniformity and quality control, the local gazettes constitute a tremendous data source for typhoon activities because of their enormous quantities and spatial coverage. According to a bibliography compiled by the Beijing Observatory (1985), more than 8,200 volumes of local gazettes currently exist in mainland China. Among these, about 40 were written during the time of the Song and Yuan dynasties, 900 during the Ming Dynasty, 5,000 during the Qing Dynasty, and 2,000 during the Republican Period from A.D. 1912 to 1949. Many of these volumes are scattered in various local libraries in China, and some are original or single copies. Nevertheless, most counties in China have at least one volume of *fang zhi* that are accessible in a public library in China or overseas. These tremendous volumes of historical and geographical documents contain a much more complete record of typhoon activities than that provided by the official histories compiled by the central government. For example, in a survey of 83 local gazettes for the Guangdong Province of China, Lee and Hsu (1989) found records of 571 typhoon strikes since A.D. 971. This compares with only 28 records of typhoon strikes in Guangdong contained in the official histories of the Song, Yuan, Ming, and Qing dynasties covering the same time period.

The fact that the majority of the local gazettes in China were compiled during the Ming and Qing dynasties (ca. fourteenth to the twentieth centuries) suggests that these documents contain, on a county-by-county basis, fairly detailed and reliable records of typhoon activities for at least the past five to six centuries. After 1884, instrumental records of typhoon activities in the western Pacific became available (Chin 1972; J. Z. Wang 1991), thereby providing an objective, scientific, and more complete database that replaces the importance of the documentary record as the primary source of typhoon data in China. The co-existence of documentary and instrumental data, especially during the early years of the instrumental period, offers an extraordinary opportunity to test the reliability and completeness of the documentary typhoon record extracted from the local gazettes.

In a pioneer study along this line, Liu, Shen, and Louie (2001) compared the historical typhoon landfall record for Guangdong Province, extracted from the local gazettes, with that from the early instrumental dataset for the 26-year period (1884 to 1909) during which the two datasets overlap. They found records of 41 typhoon landfalls in the documentary dataset, compared with 126 strikes by tropical cyclones of all intensities (tropical depressions, tropical storms, and typhoons) in the instrumental dataset. Thus the documentary record under-represents the total number of tropical cyclone strikes by a factor of three. However, all of the storm events in the documentary dataset that can be matched with an intensity designation in the instrumental dataset were classified as typhoons. This implies that the strikes recorded in the local gazettes are most likely to be caused by typhoons with destructive impacts rather than by the more benign tropical storms and depressions. More important, they found that trends of the two time series parallel each other remarkably well, with a correlation coefficient of 0.71 (Liu, Shen, and Louie 2001). These results confirm that, at least for the late-Qing Dynasty represented in this calibration study, the documentary dataset compiled from the local gazettes provides a reasonably accurate record of the temporal trends and variability of typhoon activities in China. There is no reason to suspect that the quality or reliability of the local gazette dataset had been different or lower for the earlier parts of the Qing Dynasty.

An example of the enormous potential of the Chinese local gazettes as a source for paleoclimatic information is a recent study that produced a 935-year record of typhoon landfalls for the Guangdong Province. Liu, Shen, and Louie (2001) reconstructed 571 typhoon strikes for the period A.D. 975 to 1909 from an

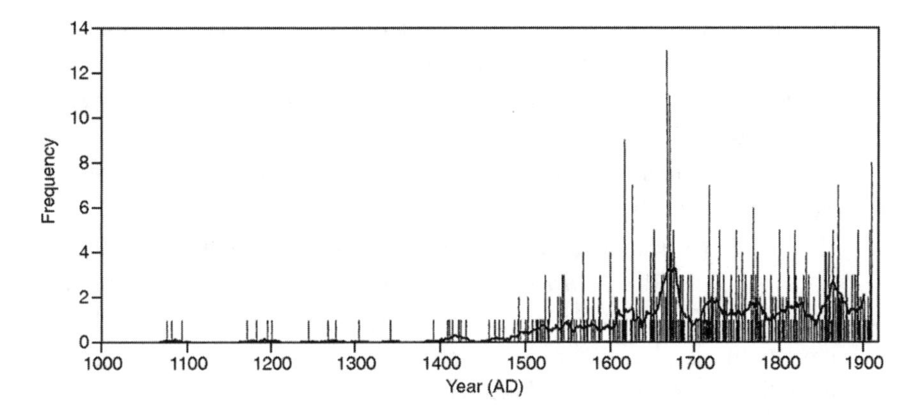

FIGURE 8.4 A 935-year time series of typhoon strikes for the Guangdong Province during the period A.D. 975 to 1909 reconstructed from Chinese *fang zhi* (local gazettes) records. The continuous curve is the 21-year moving averages derived from the annual frequencies (after Liu, Shen, and Louie 2001).

inventory compiled principally from 83 volumes of local gazettes representing 60 counties or districts in Guangdong (Lee and Hsu 1989) (figure 8.4). This 935-year time series—the longest record of tropical cyclone activity in the world—displays a marked periodicity of approximately 50 years. Two prominent peaks in typhoon landfalls in Guangdong, ca. A.D. 1660 to 1680 and A.D. 1850 to 1880, seem to coincide with two of the coldest and driest episodes in northern and central China during the Little Ice Age (Liu, Shen, and Louie 2001). This may have been caused by a displacement of the storm tracks to the south associated with an intensification of the westerlies and a southward shift of the subtropical anticyclone during the colder periods, resulting in more typhoons striking Guangdong than places further north (Liu, Shen, and Louie 2001).

Time-series data of historical typhoon strikes can be analyzed quantitatively. For example, Chan and Shi (2000) used wavelet analysis to study the periodicity inherent in a smaller dataset compiled from the local gazettes of 22 cities or counties in central Guangdong (Qiao and Tang 1993).

Unofficial Literary Works and Early Observations of Typhoons

This category includes a variety of personal diaries, travel logs, essays, chronicles, and poems written by individual scholars in a private capacity. Although storms were not the specific subject of any book in ancient China, knowledge of typhoons, or *jufeng*, can be found in many unofficial publications since the fifth century A.D. These unofficial records of typhoons have not been systematically studied. For example, there is no discussion of typhoons in Needham's (1959) *Science and Civilization in China*, the most authoritative work on the history of scientific thoughts and technological developments in ancient China.

The world's earliest mention and scientific description of typhoons is contained in a book entitled *Nan Yue Zhi* (*Book of the Southern Yue Region*) written around A.D. 470 (Louie and Liu 2003) (figure 8.1). The author, Shen Huaiyuan, wrote this book to record his observations of the culture and geography of Guangdong (known as Southern Yue Region at that time) after he was exiled to Guangzhou in A.D. 453. The following excerpt contains the relevant information:

> Many *jufeng* [typhoons] occur around Xi'an County. *Ju* is a wind [or storms] that comes in all four directions. Another meaning for *jufeng* is that it is a scary wind. It frequently occurs in the sixth and seventh month. Before it comes, roosters and dogs are silent for three days. Major ones (*storms*) may last up to seven days. Minor ones last one or two days. These are called *heifeng* [black storms/winds] in foreign countries.

This is the earliest appearance of the word *ju* or *jufeng* in Chinese literature (figure 8.5). This term, and the concept behind it, was probably widely used and understood in Guangdong before the mid-fifth century A.D. Shen Huaiyuan, a native of central China and born in today's Zhejiang Province, must have learned about this novel term during his exile to Guangdong and felt it necessary to record and explain it. Most remarkably, he gave the first scientific definition of typhoon as "a wind that comes in all four directions." This brief, yet elegant, definition underscores the essence of typhoon as a distinct meteorological phenomenon—the cyclonic circulation around the eye. The paragraph also provides an etymological explanation concerning the origin of the word *ju*. The Chinese words for *ju* and "scary" are homophones, so the Chinese word for *ju* might have been derived phonetically from the vivid and popular description of a scary wind. The word *ju* can be used independently or together with the word *feng*, the latter being a generic term meaning "wind" in the Chinese language.

In several ways, the paragraph defining *jufeng* is a condensed summary, reflecting the state-of-the-art understanding of typhoon as a meteorological phenomenon in China during the mid-fifth century. Xi'an was a coastal county in

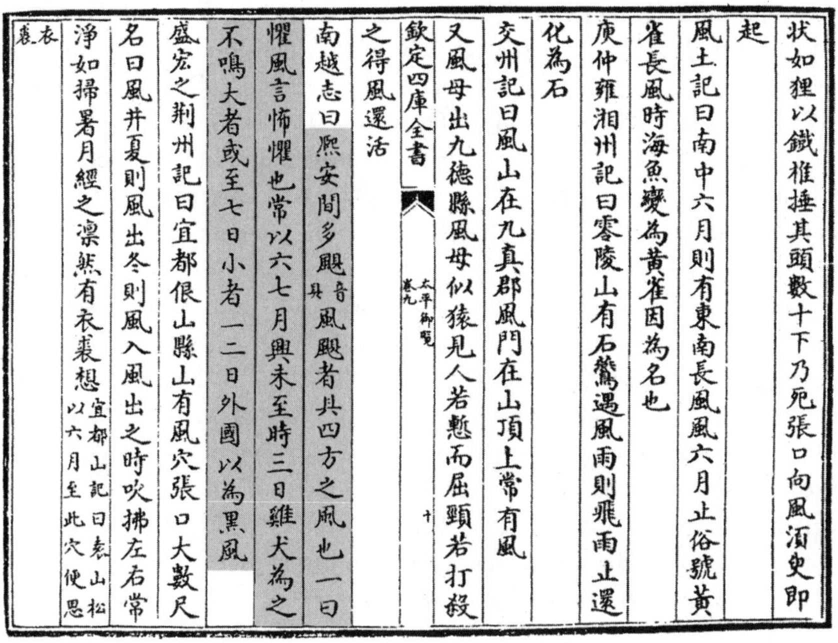

狀如狸以鐵椎摧其頭數十下乃死張口向風須臾即
起
風土記曰南中六月則有東南長風風六月止俗號黃
雀長風時海魚變為黃雀因為名也
庚仲雍湘州記曰零陵山有石鷰遇風雨則飛雨止還
化為石
交州記曰風山在九真郡風門在山頂上常有風
又風母出九德縣風母似猿見人若慙而屈頸若打殺
欽定四庫全書　太平御覽 卷九　十
之得風還活
南越志曰熙安間多颶(音具)風颶者具四方之風也一曰
懼風言怖懼也常以六七月興未至時三日雞犬為之
不鳴大者或至七日小者一二日外國以為黑風
盛宏之荊州記曰宜都佷山縣山有風穴張口大數尺
名曰風井夏則風出冬則風入風出之時吹拂左右常
淨如掃暑月經之凜然有衣裘想(宜都山記曰袁山松以六月至此穴便思
衣裘)

FIGURE 8.5 Excerpt from *Nan Yue Zhi* containing the earliest mention and scientific description of *jufeng* (typhoon) in the world (highlighted paragraph).

the vicinity of modern Guangzhou City, a region highly vulnerable to typhoon strikes. The sixth and seventh months of the Chinese lunar calendar are roughly equivalent to July and August according to the Gregorian calendar—the season of frequent typhoon landfalls in Guangdong. The paragraph also correctly identifies the duration of a typhoon strike event as a few days to a week. In a typhoon-prone region like Guangdong, people must have strived to relate the impending arrival of a typhoon to observations of anomalous biological activities, such as the behavior of domesticated animals. If so, then attempts at forecasting, however crude and rudimentary, were made as the understanding of typhoons as a weather phenomenon grew.

The last statement in the excerpt—"These are called *heifeng* [black winds/storms] in foreign countries"—deserves special attention. It implies that observations or understanding about this tropical weather system also existed, at least contemporaneously, in other places outside of China. "Foreign countries" most likely refers to neighboring countries in Southeast and South Asia that are also vulnerable to tropical cyclone strikes. Strong corroborative evidence for this inference can be found in a book entitled *Fo Guo Ji* (*Record of the Nation of Buddha*) written in A.D. 416, about a half-century before the writing of *Nan Yue Zhi*. The book's author, Fa Xian, was a Buddhist monk who traveled from today's Xi'an City (in west-central China) by land through Central Asia to India. After spending 14 years on the road and in India, he returned to China by sea in A.D. 413. His return voyage followed a route that included stops in today's Sri Lanka and Indonesia before reaching China. In *Fo Guo Ji*, essentially a log book of his travel, Fa Xian recorded two storms that he and his entourage of 200 passengers encountered at sea during their return voyage. The first storm occurred two days after their large vessel set sail: "encountered *dafeng* [strong wind] . . . water leaked into the ship." Their ordeal with the storm lasted for thirteen days and nights, until they arrived at an unknown island. The second storm occurred about a month after they resumed their voyage from Indonesia. They encountered "*heifeng baoyu*" (black winds and violent rainstorm) at a location inferred to be in the South China Sea near Guangdong. They eventually landed in the southern part of the Shandong Peninsula after their ship drifted for days in the sea. The *heifeng*, or black winds, that Fa Xian described was probably a typhoon.

This passage confirms that by the early part of the fifth century A.D., the term *heifeng* had already been used by the people of Southeast Asia and southern China to describe the weather phenomena associated with typhoons or tropical cyclones. Notably, a different term—*dafeng*, or strong wind—was used by Fa Xian to describe the first storm, which, judging from its longer duration, may be associated with the summer monsoon or other tropical disturbances instead of a typhoon. This implies that by that time typhoons had already been recognized as a distinct meteorological phenomenon apart from other kinds of storm system (Louie and Liu 2003).

After the term *ju* or *jufeng* first appeared around A.D. 470, its use in Chinese literature proliferated during the Tang Dynasty, especially in the ninth century A.D. At about the same time, the first-ever unequivocal typhoon strike was officially reported to the imperial government in A.D. 816 (previously discussed), the term *ju* or *jufeng* was mentioned in many poems or essays written by the intelligentsia. Han Yu, a famous and prolific poet as well as a leading scholar and senior government official of the Tang Dynasty who was twice exiled to Guangdong, mentioned *jufeng* in five of his poems written between A.D. 804 and 819. These poems describe the culture and environment of Guangdong—at that time considered to be a land of barbarians within the territorial frontiers of China. *Jufeng* was mentioned together with other disadvantages of this terri-

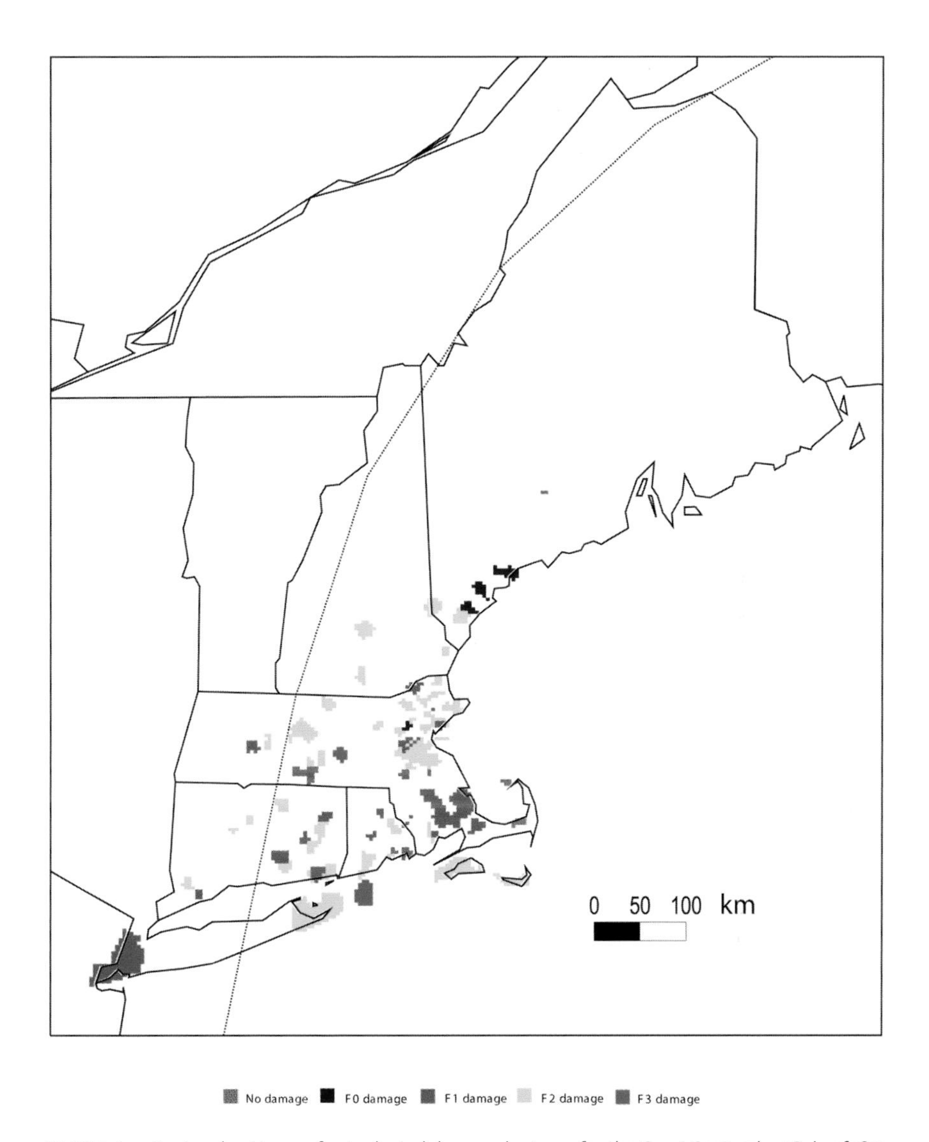

PLATE 1 Regional patterns of actual wind damage by town for the Great September Gale of 1815 in New England (adapted from Boose et al. 2001).

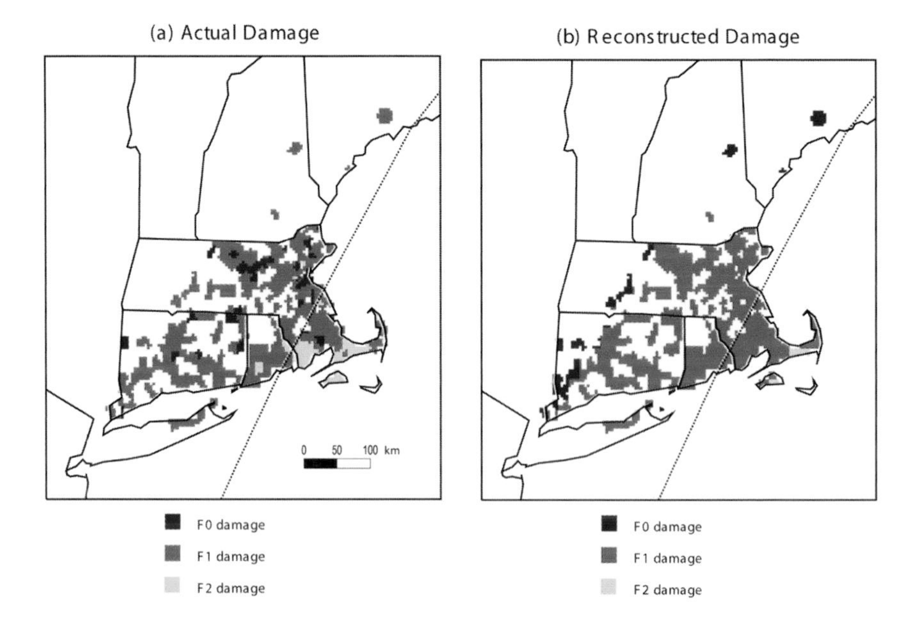

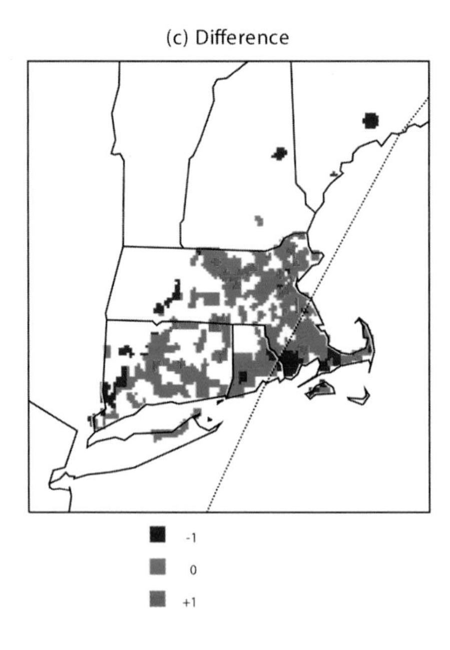

PLATE 2 Comparison of actual and reconstructed wind damage on the Fujita scale for Hurricane Bob (1991) in New England: (*a*) actual damage by town; (*b*) reconstructed damage for the same towns at 10-km resolution using the HURRECON model; and (*c*) difference map showing reconstructed damage minus actual damage (adapted from Boose et al. 2001).

a) North-facing Slopes

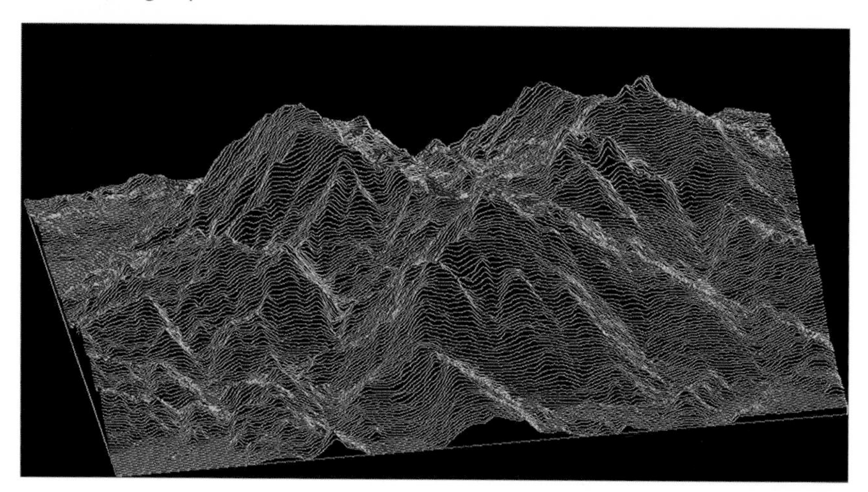

b) South-facing Slopes

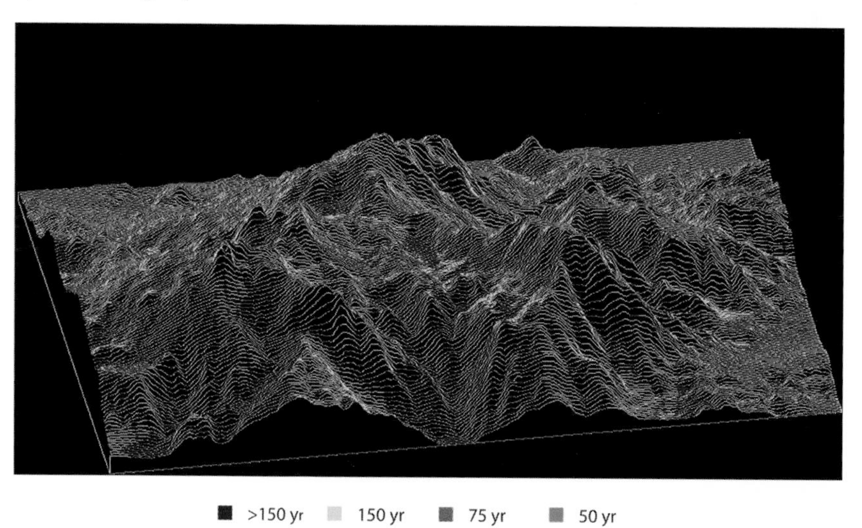

■ >150 yr ■ 150 yr ■ 75 yr ■ 50 yr

PLATE 3 Landscape gradients in reconstructed F3 damage for the Luquillo Experimental Forest, northeastern Puerto Rico (1851–1997) using the EXPOS model, showing average return intervals for (*a*) north-facing slopes and (*b*) south-facing slopes (adapted from Boose et al. 2004).

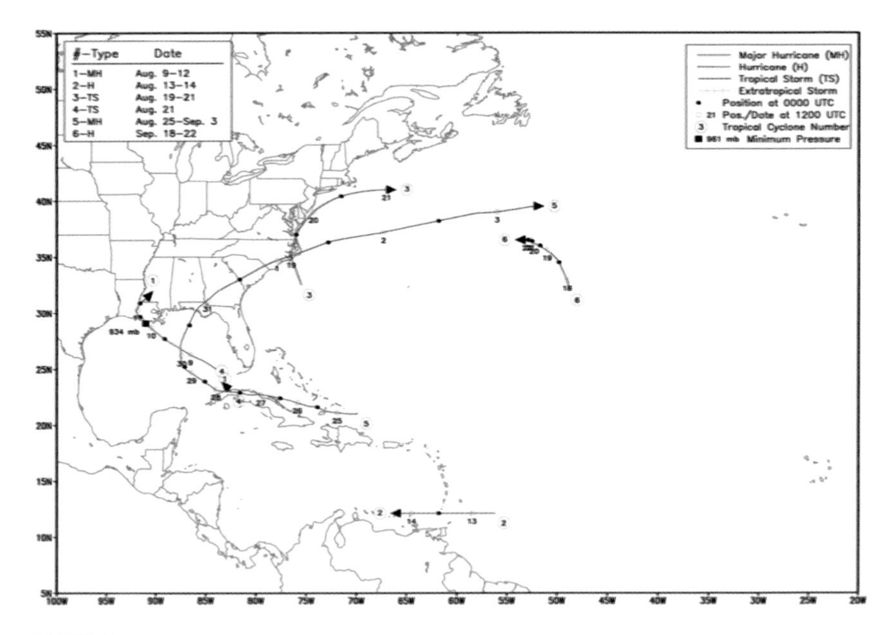

PLATE 4

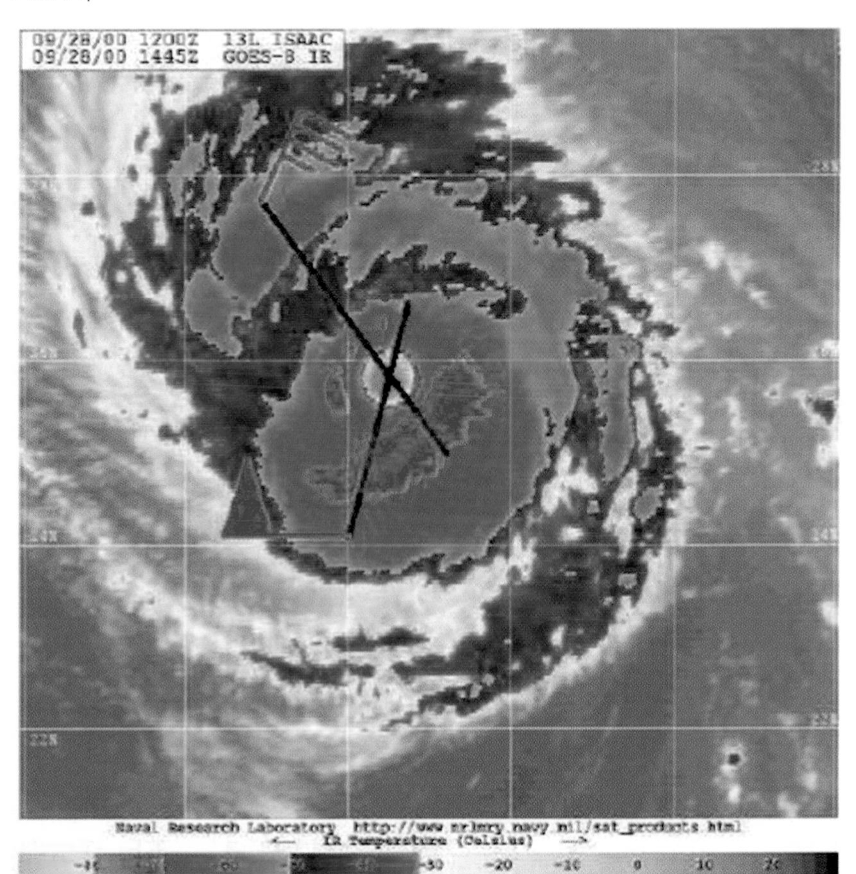

PLATE 5

PLATE 4 Reconstructed Atlantic tropical cyclone tracks and intensities for 1856.

PLATE 5 An idealized representation for finding the center of a tropical cyclone based on peripheral wind observations. Two ship observations (indicated by the red wind barbs) roughly indicate the tropical cyclone center (where the two black lines cross), assuming cyclonic flow with a 20° inflow angle.

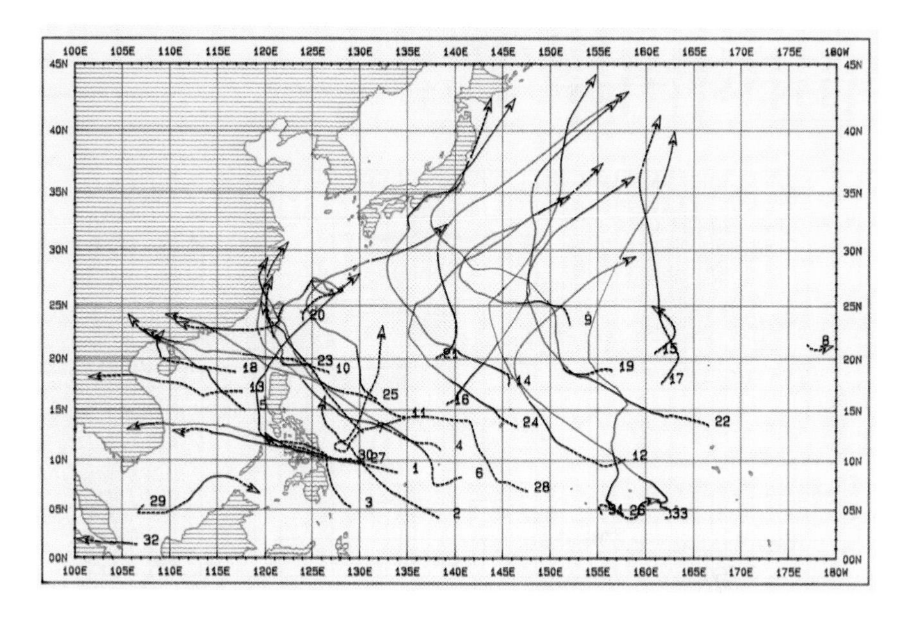

PLATE 6 Tracks and intensity of all tropical cyclones in the western North Pacific during 2001. Note, in particular, the track for Typhoon Lingling (TC-27W), which is discussed in the text. The maximum sustained wind speed for each storm is color coded: less than 34 knots, dashed black; 34 to 63 knots, solid black; 64 to 99 knots, green; more than 100 knots, red. All wind speeds are for 1-minute average sustained winds.

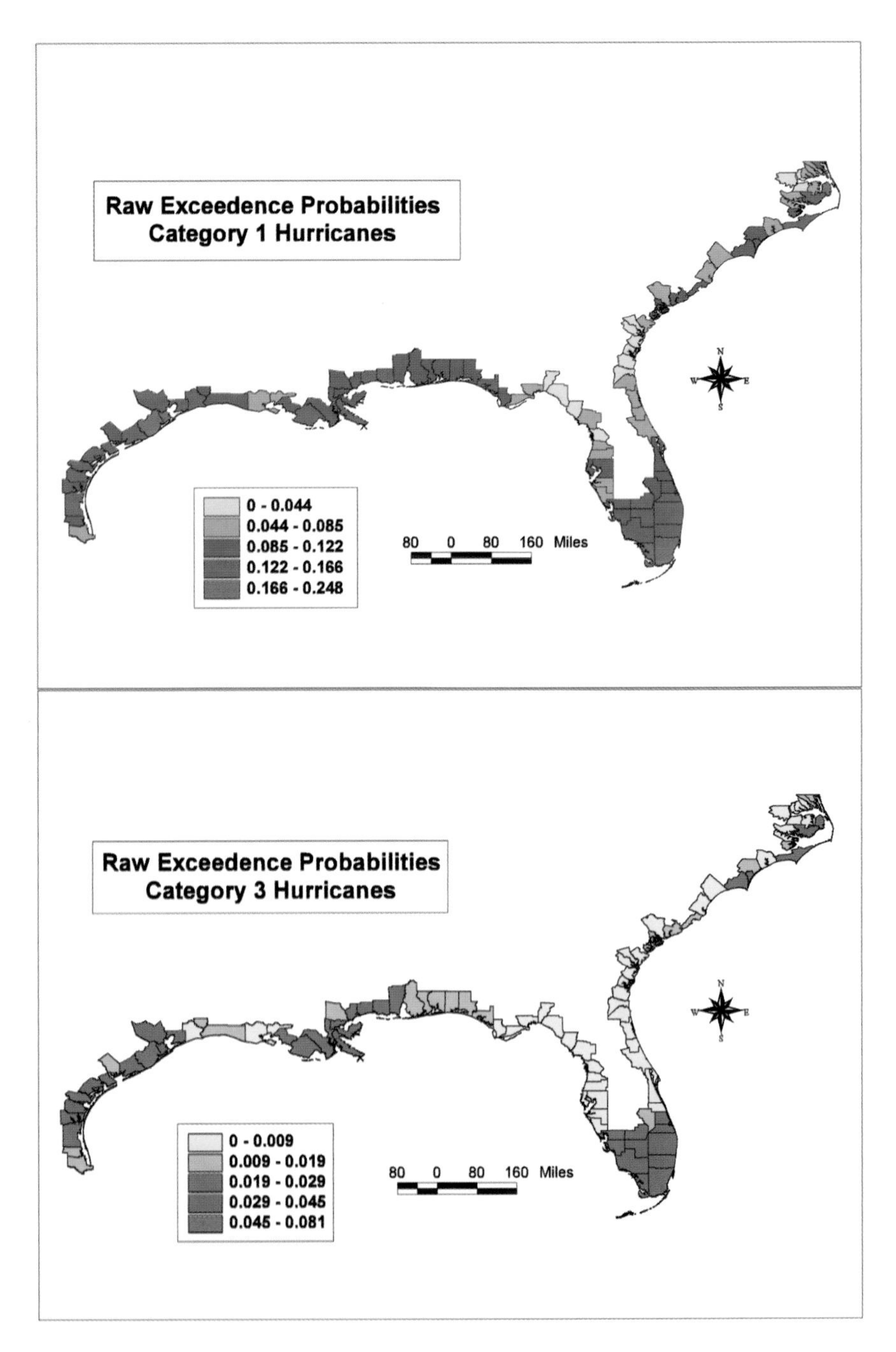

PLATE 7 Annual exceedence probabilities for Category 1 (*top*) and Category 3 (*bottom*) hurricane winds in coastal counties from Texas to North Carolina using the dynamic probability model run in the raw climatological mode. Exceedence refers to wind speeds greater than or equal to the minimum categorical value.

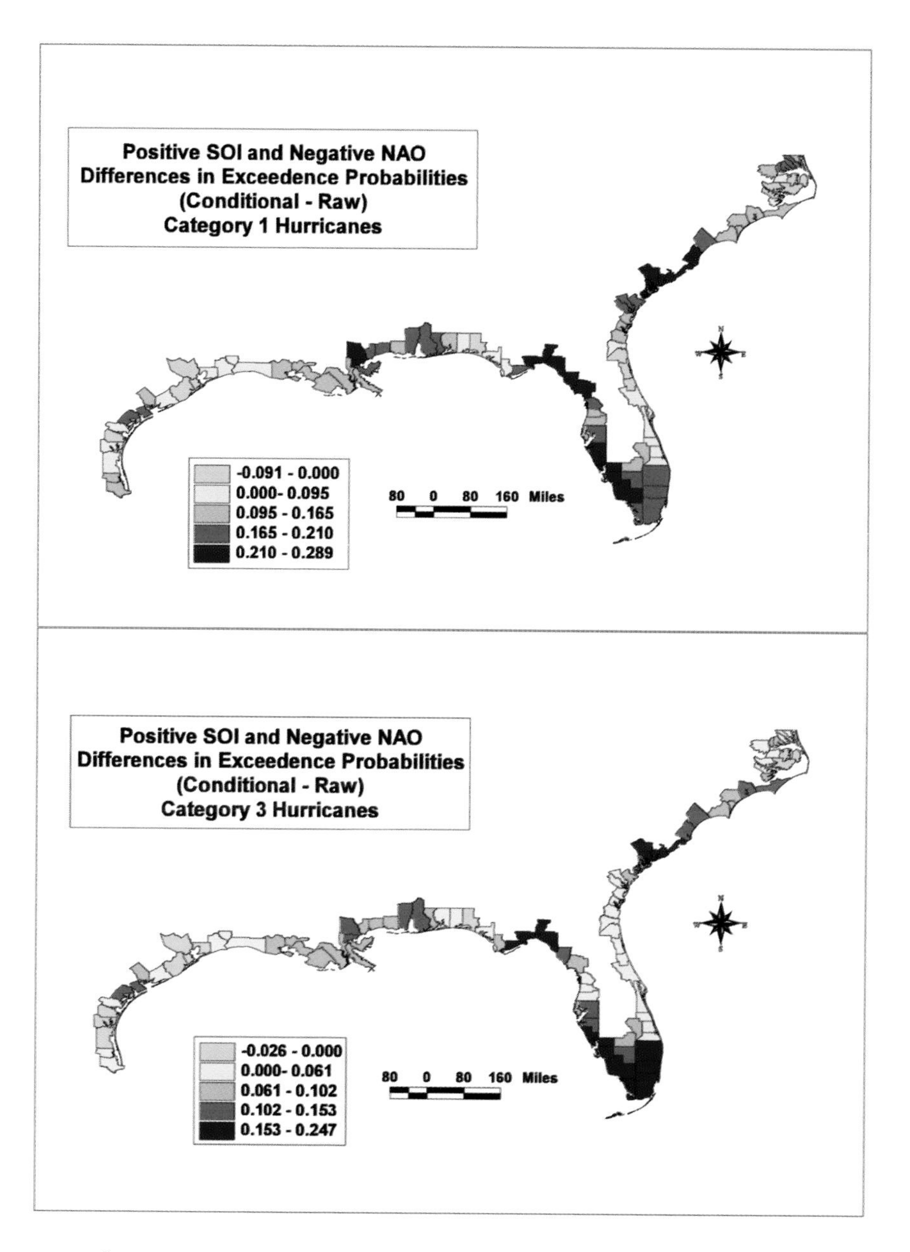

PLATE 8 Differences in annual exceedence probabilities (conditional minus raw) for Category 1 (*top*) and Category 3 (*bottom*) hurricane winds in coastal counties from Texas to North Carolina. The probabilities are based on an SOI value of +5.2 s.d. and an NAO value of −3.1 s.d. The differences in probabilities are spatially smoothed using a triangle kernel-type smoother with a bandwidth of three counties.

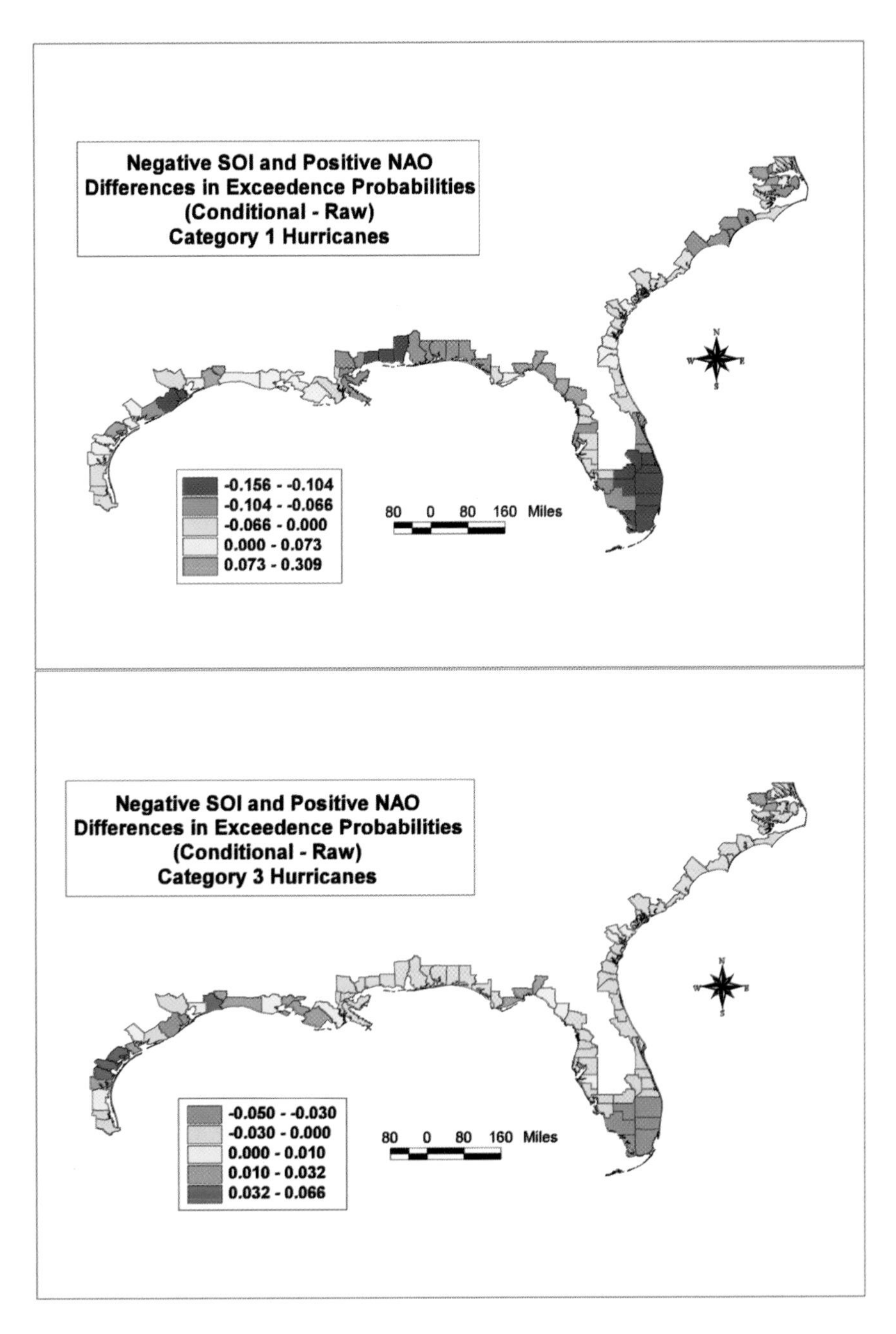

PLATE 9 Differences in annual exceedence probabilities (conditional minus raw) for Category 1 (*top*) and Category 3 (*bottom*) hurricane winds in coastal counties from Texas to North Carolina. The probabilities are based on an SOI value of −5.2 s.d. and an NAO value of +3.1 s.d. The differences in probabilities are spatially smoothed using a triangle kernel-type smoother with a bandwidth of three counties.

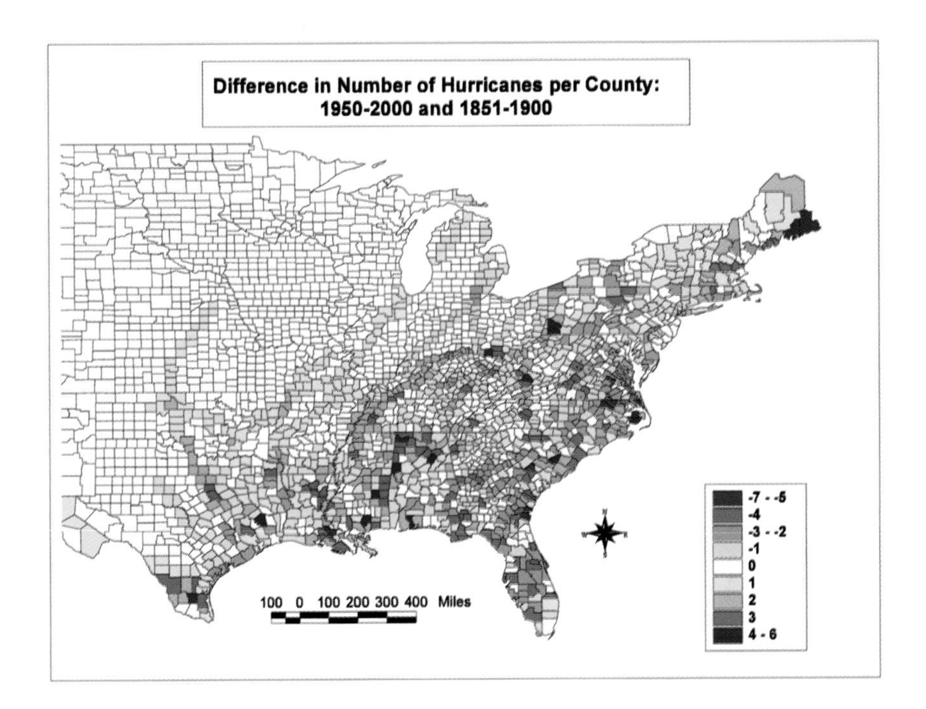

PLATE 10 Differences in the number of hurricanes per county over two 50-year periods. Negative values (blue) indicate that there were more hurricanes in the county during the period 1851 to 1900 than during the period 1950 to 2000. A hurricane is defined as the center of circulation of a tropical cyclone passing through the county borders based on 6-hour positions given in the HURDAT reanalysis dataset. The map considers only tropical cyclones designated as hurricanes at landfall. For inland counties, tropical cyclones have typical wind speeds well below the hurricane threshold.

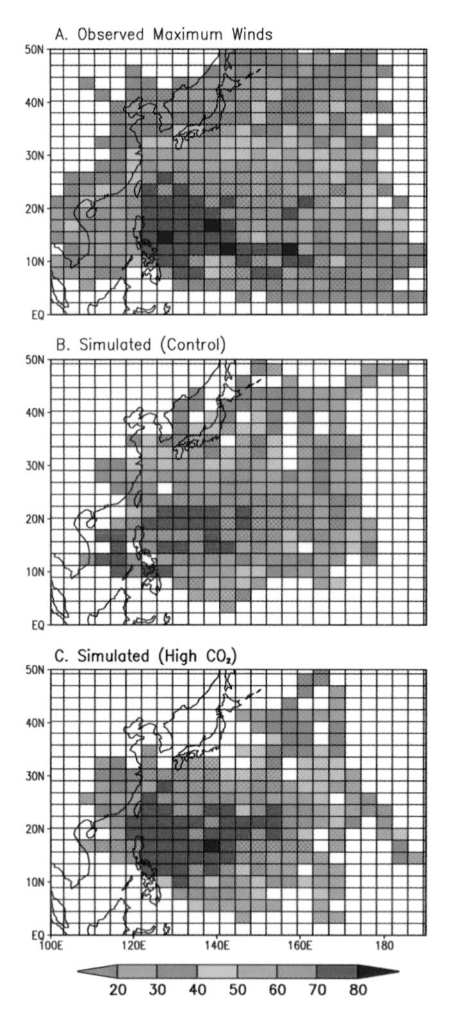

PLATE 11 Geographical distribution of the maximum surface wind speeds (m s^{-1}) observed during Northwest Pacific Basin tropical cyclones: (*A*) 1971 to 1992, and (*B* and *C*) simulated. Observations are from the Joint Typhoon Warning Center (Guam), as compiled by Neumann as of 1993, and are available from the National Center for Atmospheric Research (http://dss.ucar.edu/datasets/ds824.1). The simulated distributions are based on 71 case studies each from (*B*) control and (*C*) high-CO_2 conditions; results from 20 preliminary cases each were included in order to increase spatial coverage. Blank (white) regions denote areas where no tropical storms were reported during the 1971 to 1992 period or where none occurred in the case studies.

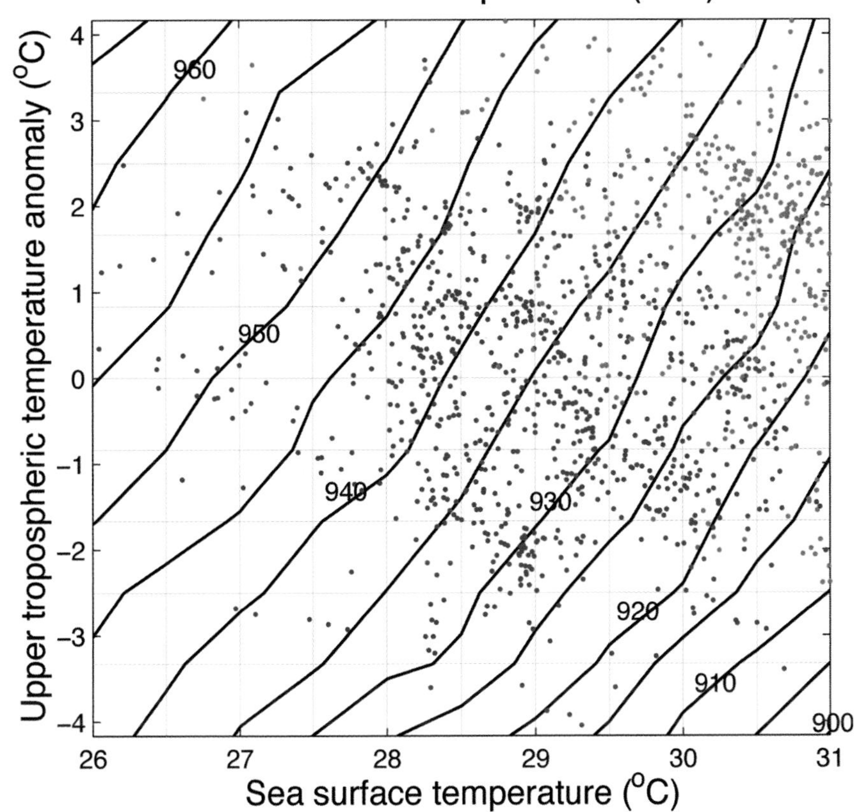

PLATE 12 Maximum tropical cyclone intensities (thick lines) as a function of sea-surface temperature and environmental temperature lapse rate for a suite of idealized GFDL hurricane model simulations. The maximum intensity is estimated as the minimum central surface pressure averaged over the final 24 hours of each 72-hour idealized model integration. The upper-tropospheric temperature anomaly denotes the maximum specified initial temperature anomaly in the upper troposphere relative to the surface. Positive anomalies denote stabilization. Note the higher intensities (lower central pressures) for higher sea-surface temperatures and cooler upper-tropospheric anomalies. Experiments are conducted at intervals of 0.5°C for SST and 0.83°C for upper-tropospheric temperature anomalies, denoted by the intersection points of the fine grid on the diagram, and the resulting two-dimensional maximum intensity distribution is contoured. The blue dots denote SST/upper-tropospheric temperature anomaly combinations as simulated in a control experiment of the GFDL global climate model. The red dots denote those conditions for a high-CO_2 experiment. The upper-tropospheric temperature anomalies denoted by the dots are the deviations of $T_{320mb}-T_{1000mb}$ from the mean value in the Northern Hemisphere Summer control climate. The present-day climate is represented by an SST value of 28.5°C.

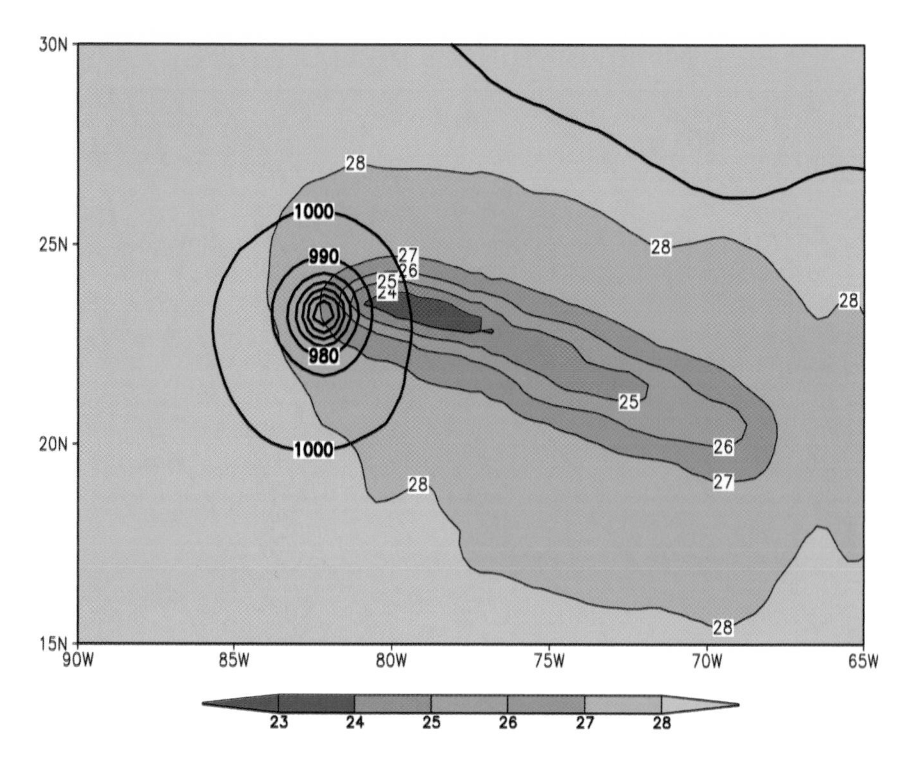

PLATE 13 A cool wake generated by a propagating hurricane, as simulated in a hurricane model coupled to a high-resolution ocean model. The SSTs (light contours; in °C) and sea level pressure (dark contours; in mb) are shown at hour 72 for one of the idealized coupled hurricane model–ocean model cases (using the Northwest Atlantic highly stratified ocean temperature vertical profile). The cool SST wake induced by the hurricane is indicated by the lower SSTs to the east-southeast of the storm. The storm motion is toward the west-northwest.

tory, such as misty swamps, excessive summer heat, fierce crocodiles, epidemic diseases, and loud thunder. Of particular interest is a poem written in A.D. 819, in which he described his personal experience with a typhoon strike while he, on his way to his exile destination of Chaozhou in eastern Guangdong, arrived at a place called Shashan, a locality in today's Qingyuan County in central Guangdong, a place some 70 kilometers from the sea (figure 8.2):

> Encountering *jufeng* at Shashan
> thunder and lightning struck hard;
> I then arrived Fuxu among strong tidal waves[2]
> the scene was terrible on the shore;
> although the two bluffs looked sturdy
> wood and rock debris were flying around;
> although Tunmen seemed high[3]
> it was submerged in the choppy water.

This event, inferred to have occurred on or around the twenty-fifth day of the third lunar month in A.D. 819 (around May 22), should be the second earliest typhoon landfall recorded in Chinese historical documents (figure 8.1).

The use of the term *jufeng* continued to proliferate during the ensuing Northern Song Dynasty (A.D. 960–1126), reflecting the diffusion of knowledge about typhoons as a meteorological phenomenon and as a natural hazard across a wide spectrum of society. Records of and references to *jufeng* appeared in many literary works as well as local gazettes throughout the nation, and were no longer restricted to travelers' reports from Guangdong. These writings contain a lot of ethnographic and anecdotal information about typhoons as a meteorological phenomenon, as well as information about the response of the ancient Chinese societies to this natural hazard. This information reveals the development of early popular and scientific thinking of typhoons in China, which most likely predates that in other parts of the world.

A remarkable development during the Northern Song Dynasty is the inclusion of the word *ju* in the important dictionaries and encyclopedias. For example, in a dictionary entitled *Lei Bian* (*Book of Classifications*) compiled during A.D. 1039 and 1076 by Ding Du and Sima Guang (the latter being a leading scholar in Chinese history and one-time prime minister of the Northern Song Dynasty), an entry for *ju* is found in chapter 38—"a word used by the people of Yue to refer to winds coming from all four directions" (figure 8.1). And, in a phonetic dictionary entitled *Ji Yun* (*A Collection of Phonemes*), also compiled by Sima Guang, there is an entry for *ju* with an explanation similar to that in *Lei Bian*. Since then, the word *ju*, and the concept of typhoon that it implies, have become integral parts of the Chinese vocabulary and the Chinese system of scientific knowledge.

Conclusions

Numerous studies demonstrate that the Chinese historical documentary record is a major source of paleoclimatic information (e.g., Chu 1973; Wang 1979, 1980; Zhang 1980, 1984, 1991a, 1991b; Wang and Zhao 1981; Zhang and Liu 1987; Wang and Zhang 1988, 1991, 1992; Zhang and Crowley 1989; Zhang and Wang 1989, 1991; Gong and Hameed 1991; Wang et al. 1992). In this chapter, we have shown that the Chinese documentary record is a valuable source of historical information on ancient tropical cyclones in the world prior to the Columbian contact. Its value is enhanced by the fact that there are no known pre-Columbian written records of tropical cyclones in other regions affected by tropical cyclones: North America and the Caribbean, Australia, and South and Southeast Asia. The pre-Columbian record provides a window that allows us to understand how popular ideas and scientific knowledge of typhoons as a meteorological phenomenon have evolved through time. More importantly, the sheer volume and consistency of information offered by the Chinese documentary records make it possible at least to qualitatively reconstruct the ancient history of typhoon activities over the past millennium. And, for the past five centuries, a quantitative reconstruction can even be attempted.

In summary, information derived from Chinese historical documentary sources contributes significantly to our understanding of the history of ancient typhoon activities. As early as in the fifth century A.D., tropical cyclones, and typhoons in particular, were recognized by the people of southern China as a distinct meteorological phenomenon separate from other types of storms. The term *jufeng*, the predecessor of the modern Chinese term *taifeng* (typhoon), was in common usage by about A.D. 470. The earliest definition of *jufeng*—"a wind that comes in all four directions"—accurately captures the essential feature of the cyclonic circulation around the eye.

A typhoon that struck the coastal city of Mizhou (today's Gaomi City) in the Shandong Peninsula of northern China in A.D. 816 is the earliest known record of tropical cyclone landfall in the world. That early record is followed by another, which described a typhoon that struck Qingyuan County in central Guangdong in A.D. 819.

Jufeng as a weather phenomenon was frequently mentioned, described, and discussed in many literary works, diaries, and poems of the Tang Dynasty, beginning in the ninth century A.D., reflecting the diffusion of knowledge about typhoons across various sectors of society and across the nation geographically. By the time of the Northern Song Dynasty (A.D. 960–1126), a well-structured system was established to regularly report typhoon strikes as a local disaster to the central government.

The large quantities and fairly complete geographical coverage of *fang zhi* or local semi-official gazettes in China provide an enormous archive of historical data on typhoon strikes going back to at least a millennium before present. A recent study based on the *fang zhi* records of one province (Guangdong) alone developed a 935-year time series consisting of 571 typhoon landfalls in China during the period from A.D. 975 to 1909 (Liu, Shen, and Louie 2001). Similar documentary records from all coastal provinces of China, when compiled and integrated into a geographic information system, offer the promise of quantitatively reconstructing the earliest and longest history of tropical cyclone activity in the world.

Written records of ancient typhoon activities in China can be found in three major types of historical documents: (1) official histories or *Zheng Shi*, including in particular the chapters of "Records of Five Elements" (*Wu Xing Zhi*), and the Emperor's Veritable Records (*Shi Lu*); (2) semi-official local gazettes (*fang zhi*); and (3) unofficial literary works. Information derived from these three major documentary sources can be cross-checked and verified with each other. Collectively they represent an enormous archive of paleotempestological data that are vital for the historical, sociological, and climatological studies of typhoons as a meteorological phenomenon and a natural hazard of great societal concern.

Acknowledgments

This research was supported by grants from the U.S. National Science Foundation and the Risk Prediction Initiative of the Bermuda Biological Station for Research. Additional funding was provided by the City University of Hong Kong. We thank Cliff Duplechin and Caiming Shen for cartographic and technical assistance.

NOTES

1. Day *mu-shen* was the fifteenth day of the eighth lunar month in that particular year.
2. Fuxu was a locality in the vicinity of modern Guangzhou City.
3. Tunmen was a military base near Guangzhou.

REFERENCES

Beijing Observatory, Chinese Academy of Sciences. 1985. *Combined catalog of Chinese local gazettes* (in Chinese). Beijing: Zhong Hua Bookstore.

Chan, J. C. L., and J. E. Shi. 2000. Frequency of typhoon landfall over Guangdong Province of China during the period 1470–1931. *International Journal of Climatology* 20:183–90.

Chin, P. C. 1972. *Tropical cyclone climatology for the China seas and western Pacific from 1884 to 1970*. Vol. 1, *Basic data*. Technical Memoir No. 11. Hong Kong: Royal Observatory.

Chu, K. C. 1973. A preliminary study on the climate fluctuations during the last 5,000 years in China (in Chinese). *Scientia Sinica* 16:169–89.

Donnelly, J. P., S. Roll, M. Wengren, J. Butler, R. Lederer, and T. Webb III. 2001. Sedimentary evidence of intense hurricane strikes from New Jersey. *Geology* 29:615–18.

Elsner, J. B., and A. B. Kara. 1999. *Hurricanes of the North Atlantic: Climate and society*. New York: Oxford University Press.

Fernández-Partagás, J., and H. F. Díaz. 1996. Atlantic hurricanes in the second half of the nineteenth century. *Bulletin of the American Meteorological Society* 77:2899–906.

Gong, G. F., and S. Hameed. 1991. The variation of moisture conditions in China during the last 2000 years. *International Journal of Climatology* 11:271–83.

Lai, X. 1995. *The Chinese local gazettes* (in Chinese). Taipei: Commercial Press.

Landsea, C. W. 2000. Climate variability of tropical cyclones: Past, present and future. In *Storms*, edited by R. A. Pielke, Sr., and R. A. Pielke, Jr., 220–41. New York: Routledge.

Lee, K., and S. I. Hsu. 1989. *Typhoon records from ancient chronicles of Guangdong Province* (in Chinese). Occasional Paper 98. Hong Kong: Department of Geography, Chinese University of Hong Kong.

Liu, K.-b., and M. L. Fearn. 1993. Lake-sediment record of late Holocene hurricane activities from coastal Alabama. *Geology* 21:793–96.

Liu, K.-b., and M. L. Fearn. 2000a. Holocene history of catastrophic hurricane landfalls along the Gulf of Mexico coast reconstructed from coastal lake and marsh sediments. In *Current stresses and potential vulnerabilities: Implications of global change for the Gulf Coast region of the United States*, edited by Z. H. Ning and K. Abdollahi, 38–47. Baton Rouge: Franklin Press.

Liu, K.-b., and M. L. Fearn. 2000b. Reconstruction of prehistoric landfall frequencies of catastrophic hurricanes in northwestern Florida from lake sediment records. *Quaternary Research* 54:238–45.

Liu, K.-b., C. Shen, and K.-s. Louie. 2001. A 1,000-year history of typhoon landfalls in Guangdong, southern China, reconstructed from Chinese historical documentary records. *Annals of the Association of American Geographers* 91:453–64.

Louie, K.-s., and K.-b. Liu. 2001. Earliest historical records of typhoon in ancient China. *Abstracts*, Annual Meeting of the Association of American Geographers, New York, February 28–March 3.

Louie, K.-s., and Liu, K.-b. 2003. Earliest historical records of typhoons in China. *Journal of Historical Geography* 29:299–316.

Ludlam, D. M. 1963. *Early American hurricanes, 1492–1870.* Boston: American Meteorological Society.

Millás, J. C. 1968. *Hurricanes of the Caribbean and adjacent regions, 1492–1800.* Miami: Academy of the Arts and Sciences of the Americas.

Needham, J. 1959. *Science and civilization in China.* Cambridge: Cambridge University Press.

Neumann, C. J., B. R. Jarvinen, C. J. McAdie, and J. D. Elms. 1993. *Tropical cyclones of the North Atlantic Ocean, 1871–1992.* NOAA Historical Climatology Series 6-2. Ashville, N.C.: National Climatic Data Center.

Qiao, S. X., and W. Y. Tang. 1993. *Collection and research of climate data from historical records in the Guangzhou area* (in Chinese). Guangdong: Guangdong People's Press.

Qin, G. 1994. *A guide to the precious archives of the Ming and Qing dynasties* (in Chinese). Beijing: People's Press.

Wang, J. Z. 1991. *Typhoon activity in Western North Pacific for the last hundred years* (in Chinese). Beijing: China Ocean Press.

Wang, P. K. 1979. Meteorological records from ancient chronicles of China. *Bulletin of the American Meteorological Society* 60:313–18.

Wang, P. K. 1980. On the relationship between winter thunder and the climatic change in China in the past 2200 years. *Climatic Change* 3:37–46.

Wang, P. K., and D. Zhang. 1988. An introduction to some historical governmental weather records of China. *Bulletin of the American Meteorological Society* 69:753–58.

Wang, P. K., and D. Zhang. 1991. Reconstruction of the 18th century precipitation of Nanjing, Suzhou and Hangzhou using the Clear and Rain Records. In *Climate Since A.D. 1500*, edited by R. S. Bradley and P. D. Jones, 184–209. London: Routledge.

Wang, P. K., and D. Zhang. 1992. Recent studies of the reconstruction of East Asian monsoon climate in the past using historical literature of China. *Journal of the Meteorological Society of Japan* 70:423–45.

Wang, S. M. 1981. *Introduction to major Chinese history works* (in Chinese). Beijing: Zhong Hua Bookstore.

Wang, S. W., and Z. Zhao. 1981. Droughts and floods in China, 1470–1979. In *Climate and history*, edited by T. M. L. Wigley, M. J. Ingram, and G. Farmer, 271–88. Cambridge: Cambridge University Press.

Wang, W. C., D. Portman, G. Gong, P. Zhang, and T. Karl. 1992. Beijing summer temperatures since 1724. In *Climate Since A.D. 1500*, edited by R. S. Bradley and P. D. Jones, 210–23. London: Routledge.

Zhang, D. 1980. Winter temperature variation during the last 500 years in southern China. *Kexue Tongbao* 25:497–500.

Zhang, D. 1984. Synoptic-climatic studies of dust fall in China since historic times. *Scientia Sinica* B 27:825–36.

Zhang, D. 1991a. Historical records of climate change in China. *Quaternary Science Reviews* 10:551–54.

Zhang, D. 1991b. The Little Ice Age climate of China and its relation with the global change (in Chinese). *Quaternary Science* 2:104–12.

Zhang, D., and C. Liu. 1987. Reconstruction of summer temperature series (1724–1903) in Beijing. *Kexue Tongbao* 32:1046–49.

Zhang, D., and P. K. Wang. 1989. Reconstruction of the 18th century summer monthly precipitation series of Nanjing, Suzhou, and Hangzhou using the Clear and Rain Records of Ching Dynasty. *Acta Meteorologica Sinica* 3:261–78.

Zhang, D., and P. K. Wang. 1991. A study on the reconstruction of the 18th-century Meiyu activity of Lower Yangtze region of China. *Scientia Sinica* B 34:1237–45.

Zhang, J., and T. J. Crowley. 1989. Historical climate records in China and reconstruction of past climates. *Journal of Climate* 2:833–49.

Zhou, X. 1994. *The local gazettes of China* (in Chinese). Taipei: Commercial Press.

Zhuang, J. 1983. *An introduction to the palace archives* (in Chinese). Taipei: National Palace Museum.

9

The Importance of Best-Track Data for Understanding the Past, Present, and Future of Hurricanes and Typhoons

Richard J. Murnane

The latitude, longitude, central pressure, and maximum wind speed of a hurricane or typhoon is observed or estimated at 6-hour intervals and archived in "best-track" data sets. This information is used in studies of tropical cyclone climatology, for the purposes of assessing the quality of model output and forecasts, developing building codes, pricing insurance, and managing responses to hurricane landfall. The data will also be used to assess potential future changes in tropical cyclone activity. The range of users and the scientific and social value of the data are consistent with the term "best-track" data, but the quality of the data and the effort required to produce them are not generally appreciated. This chapter discusses scientific (e.g., wind speed averaging times) and political (e.g., the classification and naming of storms) factors involved in the development of tropical cyclone best-track data. The current state of best-track data is also considered and a case is made for collecting and archiving an enhanced set of best-track data.

Hurricanes and typhoons inspire and frighten individuals and society. This love-hate relationship is one motivation for many societal activities including emergency management plans, regional building codes, and insurance regulation. Most of these activities implicitly account for short-term natural variability in the occurrence of hurricanes and typhoons. But longer-term natural climate variability and anthropogenic climate change potentially could induce significant changes in the frequency, intensity, or location of where hurricanes and typhoons strike. Such changes would have important implications for many sectors of society. Unfortunately, identifying if and when significant changes occur will be difficult given current knowledge and observational practices. A prerequisite for detecting future changes in hurricane and typhoon activity is knowledge of their occurrence and variability in the past and improvements in current observational practices and record keeping.

This chapter provides an overview of the status of modern records of hurricane and typhoon winds as recorded in the "best-track" data for a hurricane or typhoon. An understanding of best-track data permits a more rigorous evaluation of past and current tropical cyclone variability and of the potential for future changes. Best-track data typically give the location, central pressure, and maximum sustained wind speed of a storm at 6-hour intervals. The first best-track data set involved North Atlantic hurricanes and was developed in the 1960s by Hope and Neumann (1968) to estimate the risk of hurricane winds near Cape Canaveral for use by the National Aeronautics and Space Administration (NASA). Since then the uses and users of best-track data have grown far beyond what was initially envisioned and planned, and best-track data for storms in other ocean basins have been developed.

A wide range of users in the private and public sectors (e.g., insurers, the military, forecasters, offshore drilling companies, agricultural interests, etc.) base decisions on information derived from best-track data. Unfortunately, the best-track data are often not as reliable as their name implies. The goal of this chapter is to raise awareness of the issues involved with developing the best-track data sets as one step toward creating more beneficial and valuable tools for understanding hurricanes, typhoons, and tropical cyclones in general.

Hurricane and Typhoon Winds

Naming and Classifying Storms

A tropical cyclone is given an individual name when its maximum sustained wind at 10 meters above the surface exceeds 63 km per hour (34 knots or 39 miles per hour). In addition to being named, the tropical cyclone's strength is classified using terms such as "tropical storm," "tropical storm with gale force winds," and "cyclonic storm." A storm's classification changes again once tropical cyclone winds exceed 118 km per hour (64 knots or 74 miles per hour) and is denoted by terms such as "hurricane," "typhoon," and "severe cyclonic storm." Tropical cyclones with maximum sustained winds less than 63 km per hour are called tropical depressions. Intense tropical cyclones have been inferred to have winds of up to 306 km per hour (165 knots or 195 miles per hour) (Dunnavan and Diercks 1980) and are some of the most destructive and deadly phenomena in nature.[1]

Many people think that the classification and naming of tropical cyclones, typhoons, and hurricanes is done with great precision and that best-track data reflect the best knowledge of a hurricane's or typhoon's track and intensity. In fact, the process of classifying and naming is complex and a function of science, custom, and politics. Decisions to name storms involve scientific uncertainty, regional practices and histories, international negotiation, and, at times, international disagreement. Forecasters working in national meteorological agencies decide on when a tropical cyclone should be named and the classification of existing storms and issue warnings once a tropical cyclone forms in or enters their territorial waters. The same forecasters often compile the best-track data. But, before discussing the best-track data and the "best" records contained in them, it is worth examining some of the science, customs, and politics involved in classifying and naming storms.

"Cyclone" is a generic meteorological term sometimes used to describe a closed cyclonic circulation in the atmosphere.[2] The most general definition of "tropical cyclone" is "a cyclone that originates over the tropical oceans" (Glickman 2000). This definition can include a wide variety of phenomena ranging from midget tropical cyclones whose diameter of outer closed isobars is on the order of 50 km, to monsoon troughs with a size on the order of 1,000 km. Holland (1993) provides a more specific definition for a tropical cyclone: "A synoptic-scale to meso-scale low pressure system which derives its energy primarily from: (1) evaporation from the sea in the presence of high winds and low surface pressure; and (2) condensation in convective clouds concentrated near its centre."

Given the range of phenomena that fit the definition of a tropical cyclone, it is not surprising that forecasters often debate whether an atmospheric feature should be named. These debates are more than esoteric discussions because naming a storm initiates responses from a wide range of organizations including the military, the private sector, and emergency managers. The major problem with naming and classifying tropical cyclones is that forecasters rarely have direct observations of their maximum sustained winds.

The names used to identify individual storms and terms used to classify storm strength differ by region and can reflect local customs. The regions are defined by an international agreement through the World Meteorological Organization (WMO) that divides the world into five Regional Associations (RAs) on the basis of geographic location. Individual storms within a region are assigned names from local languages, or numbered sequentially, and set through international agreement (WMO 2000). Neumann (1993) provides an overview of storm classification in different regions. A complication in best-track data occurs when named storms leave one RA and enter another. For example, a storm might acquire a different name or change classification. Many times such changes or discrepancies in classification arise from differences in the averaging time used to determine the maximum sustained winds in a tropical cyclone.

Gusts, Sustained Winds, Height, and Exposure

The averaging time used to determine the sustained wind speed is an important, but at times overlooked, factor related to classifying tropical cyclones. Surface winds can change greatly over short periods of time. Therefore wind speeds are averaged over a period of time to determine a reproducible sustained speed. WMO guidelines suggest that a 10-minute average be used to determine sustained wind speeds, but not all countries follow this convention. All RAs but the U.S. National Hurricane Center (NHC) and the Joint Typhoon Warning Center (JTWC) follow the WMO recommendation and use 10-minute average winds. The NHC and the JTWC use 1-minute averages. In all cases, wind speeds should be adjusted to a standard observational height and surface roughness, which internationally is taken to be 10 m above flat, unobstructed terrain. Ironically, over rough, open ocean, where tropical cyclones can only naturally persist, this standard height is almost impossible to reproduce due to the influence of waves. In practice, many wind speeds in the official records are not appropriately adjusted to this standard.

The averaging period for measuring sustained winds is a function of an individual country's custom and can be different from that used by the RA associated with a country. For example, some countries use 3-minute averages for estimating sustained wind speeds despite the 10-minute average sanctioned by the WMO. The longer the averaging period, the slower the average wind speed. One result of the differences in averaging times for wind speeds is that a tropical cyclone can be named by one country but not by another that uses a longer time for averaging. A political solution for unifying the different wind averaging periods has yet to be found.

The different averaging times used to characterize maximum sustained winds complicate efforts to compile best-track data. Maximum sustained wind speeds based on different averaging times can be compared using conversion factors (Simiu and Scanlon 1978). The conversion factors are based on empirical relationships that vary in detail as a function of surface roughness and atmospheric conditions such as the stability of the surface boundary layer. Significant differences can occur between the maximum sustained wind estimates from different agencies, particularly at high wind speeds. The JTWC uses a value of 0.88 to convert from 1-minute to 10-minute sustained winds; other agencies use values that range from 0.9 to 0.8.

Although maximum sustained winds are important, the greatest damage in a storm is often caused by wind gusts.[3] Expected gusts are linearly related to sustained wind through a mean-gust factor (Atkinson 1974). For example, consider a wind record of time duration, T, and average wind speed, U. Divide T into m equal segments of length t = T/m and measure the wind speed for each segment u. The largest value of u is called the maximum gust, u_g. The mean gust factor, g, is average of u_g/U determined from multiple observations of duration T. The mean gust factor can be used to find the expected gust of duration t associated with a sustained wind over a period T. For example, if t is one minute, T is one hour, g is 1.5, and U is 50 knots, then, on average, over a 1-hour period there should be a single minute with sustained winds of 75 knots. The gust factor can vary as a function of wind speed (particularly for extreme winds associated with typhoons and hurricanes) and surrounding terrain (Durst 1960; Cook 1985; Krayer and Marshall 1992). Exceedance statistics are often used as an alternative to gust factors in engineering studies because they can be more easily used in mathematical models of the dynamic response of structures (Davenport 1964).

When developing best-track data the details of converting from one convention to another and the uncertainty associated with the conversion must be considered. Neumann (1993) provides a sense of the importance of the conversion factor by comparing tropical cyclone frequency and hurricane and typhoon frequency in the Atlantic, western North Pacific, and eastern North Pacific. Neumann finds that tropical cyclone frequencies based on 1-minute and 10-minute maximum sustained wind differ by less than 4% but that the frequency for hurricanes and typhoons based on the two averaging periods varies by 17%. Although the error associated with converting between records based on 10-minute and 1-minute sustained winds is significant for severe storms, a potentially even more significant error in estimating winds is the paucity, or complete lack, of in-situ observations of maximum winds in a hurricane or typhoon.

Wind Observations

Direct measurement of maximum sustained winds in a hurricane or typhoon is uncommon for several reasons. Hurricanes and typhoons form over the ocean where there are few observational platforms, and only a fraction of all storms approach or reach land where they are more likely to cross observational stations. Ships at sea, aircraft (except for special purpose research or reconnaissance flights), and people generally try to avoid these intense storms. In addition, maximum winds occur in a relatively small area relative to the total size of a storm. An anemometer provides the most direct measure of maximum sustained winds. When the maximum winds in a hurricane or typhoon cross an anemometer, however, it often fails due to a loss of power or damage from flying debris; or it provides measurements that are controversial because of calibration issues.

Direct measurements of a storm's minimum central pressure are more common and less subject to error. Minimum central pressures have been recorded back to 1899 for many U.S. landfalling hurricanes (Jarrell, Hebert, and Mayfield 1992). The relationship between minimum central pressure and maximum sustained wind speed is a function of several variables such as storm size, surrounding environmental pressures, and forward motion.

Many historical records of wind speed are based on qualitative measures such as the Beaufort scale or Fujita scales.[4] Wind speeds derived from these scales are based on empirical relationships between wind speed and observed phenomena. Historical reconstructions of hurricane and typhoon wind speeds prior to development of the Beaufort scale may involve translating descriptive terms such as "fierce gale" into modern terms (see, e.g., Boose, chapter 4; Mock, chapter 5; García Herrera et al., chapter 6; and Landsea et al., chapter 7 in this volume). Reconstructions based on the Fujita scale must account for differences in construction practices, materials, and building styles through time. In most cases historical records provide anecdotal accounts from areas away from the maximum sustained winds. Empirical relationships that link storm characteristics such as central pressures, the radius of maximum winds, and storm size must be used to infer the maximum sustained wind (e.g., Emanuel 1986; Holland 1997; Landsea et al., chapter 7 in this volume). The wind-pressure relationship can have a significant impact on the estimated maximum sustained wind and minimum central pressure estimates that are based on observations that are far from a storm's center.

Technological advances have made a wider variety of wind-speed measurements feasible. One of the first significant observational advances was the initiation of routine aircraft reconnaissance in the Atlantic Basin in 1944 (Neumann et al. 1993). Some wind speeds reported from early flights were based on the Beaufort scale. In addition, the maximum sustained winds in intense hurricanes were apparently overestimated and this bias has been removed for data from the 1940s through the 1960s (Landsea et al. 1999).

The advent of satellite remote sensing in the mid-1960s provided the first complete global tracking of tropical cyclones. It is likely that most best-track data sets lack a complete account of storm occurrence prior to satellite observations. The development of satellite remote-sensing data analysis methods, particularly the Dvorak technique (Erickson 1972; Dvorak 1975, 1984), enabled forecasters to estimate the intensity and maximum sustained winds of a tropical cyclone by comparing satellite imagery with specified patterns. An objective Dvorak technique has also been developed that automates the analysis of satellite imagery and provides tropical cyclone intensity estimates (Velden, Olander, and Zehr 1998).

More recently, mobile Doppler radar and Global Positioning System (GPS) dropsondes (Hock and Franklin 1999) offer direct measurements of wind speed in hurricanes and typhoons. Doppler radar provides information on wind fields with unprecedented levels of detail and GPS dropsondes provide unique information about the vertical variation of a storm's winds. Although GPS dropsondes do not measure sustained winds, they can be used to scale maximum sustained winds from flight levels to the surface.

Each observational technique has advantages and disadvantages and a major challenge is to develop algorithms that will produce from the wide range of available information an optimum estimate of maximum sustained wind speeds at 10 meters above the surface. On a real-time basis the H*WIND system (Powell et al. 1998) developed by the Hurricane Research Division is one of the most complete efforts at developing wind fields associated with hurricanes. The H*WIND system melds essentially all observational data into a unified wind field that is consistent with 1-minute sustained winds at 10 meters height over water (Powell, Houston, and Reinhold 1996). A spline function is used to fit the wind-field model to the observations in a least-squares sense. On a retrospective basis a reanalysis of all available data can provide an improved best-track record (Landsea et al., chapter 7 in this volume). A rather dramatic example of this is the recent reclassification of Hurricane Andrew (1992) from a Category 4 to a Category 5 hurricane.

A comparison of flight level winds and surface winds derived from GPS dropsondes provides an example of some of the factors that must be considered when reconciling different observations. Aircraft observations of flight level winds must be extrapolated first to the top of the boundary layer (the layer of the atmosphere that responds to frictional effects of the surface), and then through the boundary layer to a height of 10 meters. Theoretical considerations and empirical relationships are used when making these extrapolations. Powell and Black (1990) analyzed flight level and ocean buoy measurements and found that surface winds are 63 to 73% of flight level winds. Most of the observations used in the Powell and Black (1990) study were taken away from the location of maximum winds in a storm. The operational practice of the NHC is to assume that maximum sustained winds (1-minute average) at 10 meters are 90% of flight level winds (J. L. Franklin, Black, and Valde 2000). Franklin and colleagues suggest that the higher value is more appropriate for estimating maximum sustained winds near the radius of maximum wind and that Powell and Black (1990) found lower values because their data mainly came from locations distant from the maximum winds.

When a storm is at sea and there is no aircraft reconnaissance, the typical situation for all ocean basins except that of the North Atlantic, the major source of information on maximum sustained wind speed is through the analysis of satellite information using the Dvorak technique or objective Dvorak technique. The empirical relationship between the current intensity estimate of a storm based on the Dvorak analysis and the maximum wind and central pressure of a storm differs between the Atlantic and Northwest Pacific. Harper (2002) offers a thorough analysis of the historical development of different pressure–wind relationships and a possible explanation for the different relations in the two ocean basins.

This brief discussion suggests that significant efforts are required when estimating maximum sustained winds and wind fields for hurricanes and typhoons from a variety of observational platforms. A forecaster does not, however, always have access to all, or the best, data, especially for regions with multiple jurisdictions such as the western North Pacific. Best-track data for the storm are assembled several weeks after a storm decays, or after the storm season ends. Because of time, personnel, and monetary constraints, the data used to develop the best track for a storm are not necessarily the "best"; they are the data on hand at the time that the best tracks are developed.

Best-Track Data

Best Tracks Are Not True Tracks

Synoptic scale (on the order of 10^2 to 10^3 miles) atmospheric circulation patterns determine the track of a tropical cyclone. The innermost portions of the storm (the eye and eyewall) that are used to "fix" the location of the storms exhibit additional small-scale oscillatory movements (usually less than 25 miles) referred to as trochoidal motion (Neumann 1979; Jarvinen, Neumann, and Davis 1984). These small-scale movements are sometimes not representative of the larger-scale motion of the entire storm (Lawrence and Mayfield 1977; Jarvinen, Neumann, and Davis 1984).

Continuous estimates of storm motion and location are provided to a forecaster in an operational forecast framework. The trochoidal motions complicate the development of a forecast because a major factor in the forecast is the location of a tropical cyclone's eye. A sudden deceleration, acceleration, turn to the left, or turn to the right of the eye might be merely a transitory trochoidal shift and not representative of the more conservative motion of the entire storm. It is not until later analysis that this becomes known. A storm's best track is intended to reflect the synoptic scale of motion and specifically excludes smaller-scale shifts such as trochoidal motion.

The most extensive best-track records are for storms in the North Atlantic, but only since 1956 have storm positions been recorded at 6-hour intervals. Between 1931 and 1956 storm positions were determined four times a day, and only the 0000 and 1200 GMT positions were recorded. Prior to 1931, storm positions, when known, were recorded once per day at 1200 GMT. Storm positions at 6-hour intervals are inferred from recorded observations using a polynomial interpolation scheme. Wind speeds are rounded to the nearest 5-knot value and assumed to represent 1-minute sustained winds at 10-m height over a flat, unobstructed surface. A summary of the different factors that must be considered when developing best-track data is given by Jarvinen, Neumann, and Davis (1984) and Landsea et al. (chapter 7 in this volume).

Best-track data for other regions of the world are more limited. For example, Holland (1981) provides an overview of the status of the Australian best-track database. Generally only storm data recorded after 1959 or 1960 are deemed reliable.

Current Status

The Regional Specialized Meteorological Center (RSMC) is responsible for issuing a best-track record for storms in its area of responsibility. The JTWC also develops and maintains its own set of best-track data for the Pacific and Indian Oceans. The starting years for best-track data differ among basins. Best-track data for the Atlantic Basin starts in 1851 (Landsea et al., chapter 7 in this volume); best-track data in other basins start at later dates. For example, the JTWC best-track data for the Northwest Pacific start in 1950, for the northern Indian Ocean the data start in 1971, and in the Southern Hemisphere the data start in 1985. In contrast, the Australian Bureau of Meteorology provides records of storms near Australia starting in 1907. The quality and completeness of all records improve over time as aircraft and satellite observations increase in frequency and with instrumentation advances.

A good indication of the typical location and track of tropical cyclones is shown in figure 9.1. These tracks are based on JTWC best-track data for the period from 1992 to 2001. The track characteristics are not surprising. In the North Atlantic and Pacific Oceans the storms initially tend to follow westward path and then recurve toward the pole of each hemisphere. In addition, all ocean basins except the South Atlantic experience tropical cyclones. What is not evident in compilations and plots such as this is the uncertainty in storm track and strength and in the formation and end points of a storm. An examination of the maximum sustained wind speeds issued by different meteorological agencies for Typhoon Lingling in November 2001, illustrates the subjective nature of the best-track data.[5]

The tropical depression that was the source for Typhoon Lingling was first noted on November 5. It began to the east of the Philippines and maintained a generally westward track until it dissipated over Vietnam (plate 6). The Philippine Atmospheric, Geophysical, and Astronomical Services Administration (PAGASA) was the first meteorological agency to upgrade the tropical depression to a tropical storm and it gave the storm the Filipino nickname Nanang on the 0600 UTC warning on November 6. The Japanese Meteorological Agency (JMA) upgraded the tropical depression to Tropical Storm Lingling at 1800 UTC on the same day. The JTWC and the National Meteorological Center of China (NMCC) classified the storm as Tropical Storm Lingling for the 0000 UTC warning on November 7.

The NMCC estimated that Typhoon Lingling had peak 10-minute average maximum sustained winds of 100 knots and that these winds were maintained for 24 hours. This peak wind estimate agrees well with the JTWC estimate of a 115 knot 1-minute average sustained wind (115*0.88 = 101, where 0.88 is the conversion factor between a 1-minute and 10-minute sustained wind) and is close to the Hong Kong Observatory (HKO) (10-minute average) estimate of 95 kts. The HKO peak winds were reported for only one hour, and the JMA reported that 10-minute average sustained winds of 85 knots were maintained for 24 hours. These maximum-wind estimates are of particular interest because they occurred just over 6 hours prior to landfall on the coast of Vietnam.

On November 11, 1800 UTC, the final warning time prior to Lingling striking Vietnam's coast, there was a wide range in estimates of maximum sustained winds. The JTWC reported 1-minute sustained peak winds of 95 kt (equivalent to 10-minute sustained winds of 84 kt). This estimate was based on a range of winds (77 kt to 102 kt) estimated from Dvorak analyses. The JTWC estimate (95*0.88 = 84) was close to the NMCC estimate of 80 kt, but greater than the

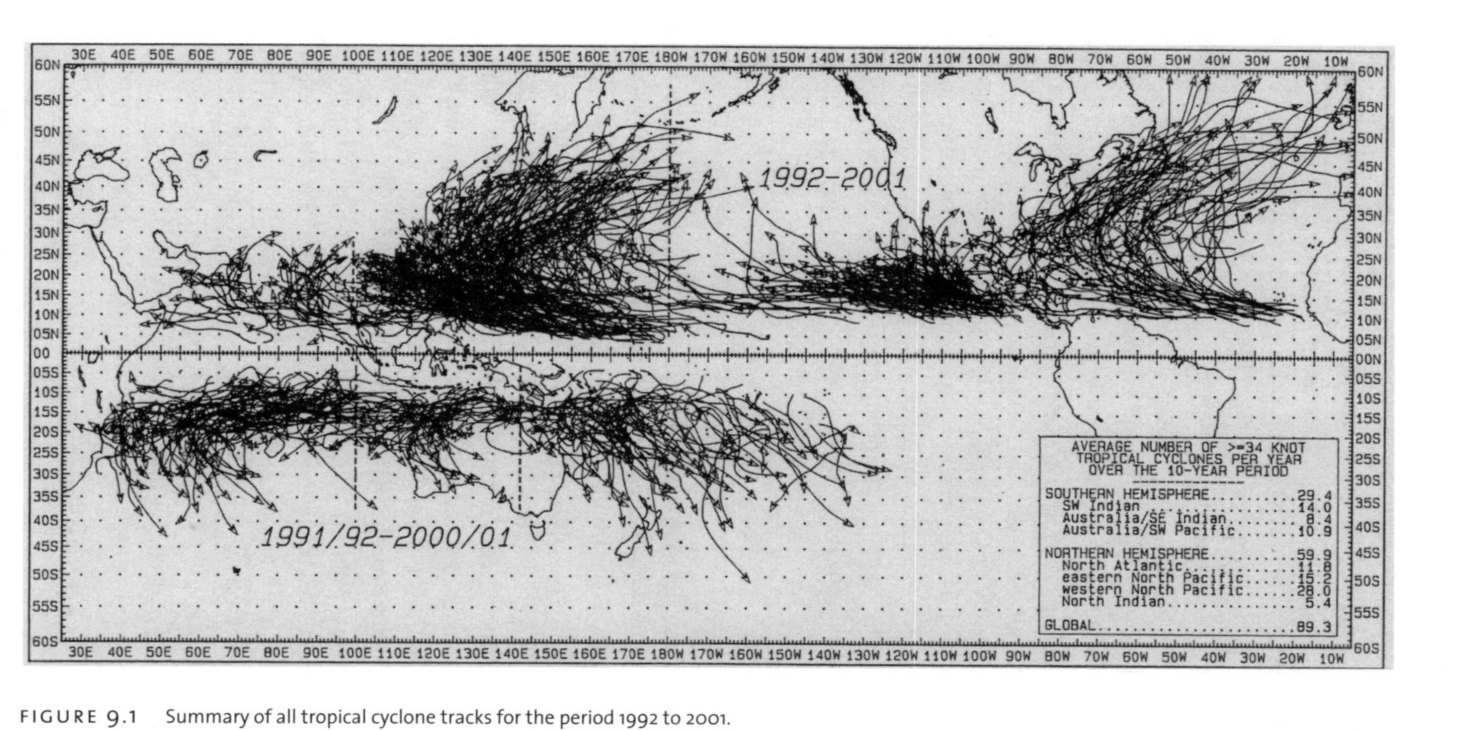

FIGURE 9.1 Summary of all tropical cyclone tracks for the period 1992 to 2001.

HKO estimate of 65 kt, and much greater than the JMA estimate of 50 kt (all but the JTWC winds are 10-minute averages). Thus, there was at least a 30-knot range in estimates of maximum sustained winds prior to landfall. These estimates were based on analyses of satellite imagery, so variability can be ascribed to differences in the interpretation of satellite data. The estimates do not, however, account for hourly surface observations of the storm made by Vietnam. According to Vietnamese observations Typhoon Lingling made landfall on November 11 around 2100 UTC and the maximum sustained winds were 62 kt (this is probably based on a 10-minute average). Typhoon Lingling quickly dissipated after landfall and JTWC issued its last tropical storm warning on 0600 UTC, November 12.

This example includes a range in wind-speed estimates that may be larger than average, but it is not a worst-case scenario. It illustrates the wide range of feasible wind speed estimates for a tropical cyclone even when using satellite observations in the modern era and it provides a sense of the possible uncertainty in earlier best-track wind speed estimates.

Enhanced Best-Track Data

Current best-track data sets omit information of great value to users ranging from scientists studying tropical cyclones and climate variability to businessmen trying to price catastrophe bonds for typhoon winds in Japan. Among the statistics of interest are radius of maximum winds in a storm, radial distance to gale force winds in different storm quadrants, and an estimate of data quality. An example of the perceived value of enhancing best-track data is the support from a group of insurance and reinsurance companies for digitizing a set of data on the radius of maximum and other winds for tropical storms and hurricanes in the North Atlantic for the period from 1988 to 1997.[6]

Forecasters and scientists are aware of the need for this additional information. The official WMO format for tracking tropical cyclones was recently upgraded to include these and other data. The items included in the WMO Global Tropical Cyclone Track and Intensity Data Report are listed in table 9.1. In addition, the U.S. National Climatic Data Center (NCDC) will archive this information for each storm so that it is readily available.

Discussion and Summary

Best-track data for tropical cyclones are used in a majority of studies of tropical cyclone variability. For example, every chapter in this book directly uses or

TABLE 9.1 Data Fields in the WMO Global Tropical Cyclone Track and Intensity Data Report

Field	Description
1–9	Cyclone identification code composed (2 digits) in order within the cyclone season, area code (3 digits) and year code (4 digits)
10–19	Storm name
20–23	Year
24–25	Month (01–12)
26–27	Day (01–31)
28–29	Hour: universal time (at least every 6 hourly position –00Z,06Z,12Z, and 18Z)
30	Latitude indicator: 1 = North latitude; 2 = South latitude
31–33	Latitude (degrees and tenths)
34–35	Check sum (sum of all digits in the latitude)
36	Longitude indicator: 1 = West longitude; 2 = East longitude
37–40	Longitude (degrees and tenths)
41–42	Check sum (sum of all digits in the longitude)
43	Position confidence: 1 = good (<30nm; <55km); 2 = fair (30–60nm; 55–110 km); 3 = poor (>60nm; >110km); 9 = unknown
44–45	Dvorak T-number (99 for no report)
46–47	Dvorak CI-number (99 for no report)
48–50	Maximum average wind speed (whole values) (999 for no report)
51	Units: 1 = kt; 2 = m/s; 3 = km per hour
52–53	Time interval for averaging wind speed (minutes for measured or derived wind speed, 99 if unknown or estimated)
54–56	Maximum wind gust (999 for no report)
57	Gust period (seconds, 9 for unknown)
58	Quality code for wind reports: 1 = aircraft or dropsonde observation; 2 = over-water observation (e.g. buoy); 3 = over-land observation; 4 = Dvorak estimate; 5 = other
59–62	Central pressure (nearest hectopascal) (9999 if unknown or unavailable)
63	Quality code for pressure report (same code as for winds)
64	Units of length: 1 = nm; 2 = km
65–67	Radius of maximum winds (999 for no report)
68	Quality code for RMW: 1 = aircraft observation; 2 = radar with well-defined eye; 3 = satellite with well-defined eye; 4 = radar or satellite, poorly defined eye; 5 = other estimate
69–71	Threshold value for wind speed (gale force preferred, 999 for no report)
72–75	Radius in Sector 1: 3150–450
76–79	Radius in Sector 2: 450–1350
80–83	Radius in Sector 3: 1350–2250
84–87	Radius in Sector 4: 2250–3150

(Continued)

TABLE 9.1 Data Fields in the WMO Global Tropical Cyclone Track and Intensity Data Report *(Continued)*

Field	Description
88	Quality code for wind threshold: 1 = aircraft observations; 2 = surface observations; 3 = estimate from outer closed isobar; 4 = other estimate
89–91	Second threshold value for wind speed (999 for no report)
92–95	Radius in Sector 1: 3150–450
96–99	Radius in Sector 2: 450–1350
100–103	Radius in Sector 3: 1350–2250
104–107	Radius in Sector 4: 2250–3150
108	Quality code for wind threshold (code as for row 88)
109–110	Cyclone type: 01 = tropics; disturbance (no closed isobars); 02 = <34 knot winds, <17m/s winds, and at least one closed isobar; 03 = 34–63 knots, 17–32m/s; 04 = >63 knots, >32m/s; 05 = extratropical; 06 = dissipating; 07 = subtropical cyclone (nonfrontal, low-pressure system that comprises initially baroclinic circulation developing over subtropical water); 08 = overland; 09 = unknown
111–112	Source code (2-digit code to represent the country or organization that provided the data to NCDC USA; WMO Secretariat is authorized to assign number to additional participating centers, organizations): 01 RSMC Miami-Hurricane Center; 02 RSMC Tokyo-Typhoon Center; 03 RSMC-tropical cyclones New Delhi; 04 RSMC La Reunion-Tropical Cyclone Centre; 05 Australian Bureau of Meteorology; 06 Meteorological Service of New Zealand Ltd.; 07 RSMC Nadi-Tropical Cyclone Centre; 08–11 no longer used; 12 Central Pacific Hurricane Center, Honolulu

refers to best-track data sets, and future efforts to determine if tropical cyclone activity responds to global climate change will be based on a comparison with existing best-track data. In addition, the best-track data are important to a range of users in the private sector. Users, however, do not generally consider the quality of best-track data.

A variety of factors detract from the quality of best-track data. For modern storms these factors range from simple differences in the time span used to determine sustained wind speeds to complicated details associated with estimating maximum sustained winds from satellite observations. Factors affecting the quality of best-track data for earlier storms range from the paucity of observations, to changes in descriptive terms, to differences in buildings and their structural characteristics. Nevertheless, best-track data are used widely despite these known limitations. Moreover, the data will continue to be used whether or not their quality improves simply because there are no ready alternatives.[7]

Forecasters and scientists recognize the value of best-track data and through the WMO agreed to enhance the official records of tropical cyclones (table 9.1), although it is not uncommon for many of the fields to be left unfilled. Still, there is a value gained through the inclusion of any additional information beyond the standard latitude, longitude, central pressure, and maximum sustained wind speed that comprise current best-track data sets.

An important challenge is to ensure that best-track data contain the best available data. In many cases the official record for a storm needs to be completed within a few weeks of a storm's demise. The person completing the best track often cannot devote sufficient time to collect and assess data from every observational platform and country. As a result, the best-track data do not always live up to their name. The diverse uses for, and the economic value of decisions based on, the best-track data suggest that it is worth devoting significant effort to compiling and maintaining the best possible data set. An awareness of the issues associated with existing best-track data should be one of the first steps toward acquiring the money and personnel needed to develop and maintain best-track databases.

Acknowledgments

Comments and suggestions from C. Neumann and B. Harper are greatly appreciated and improved the manuscript. This is contribution 1642 from the Bermuda Biological Station for Research, Inc. This work was supported by the sponsors of the Risk Prediction Initiative.

Notes

1. The 306 km per hour estimate is for maximum winds associated with Typhoon Tip in the western North Pacific in 1979. In the spirit of this chapter, it should be noted that Tip's maximum wind speed appears to be derived from a recorded sea level pressure of 870 hPa and the Atkinson and Holliday (1975, 1977) equation relating sea level pressure to maximum sustained wind speed in the western North Pacific; it is not based on a direct measurement. For a discussion on wind speeds and pressures in Tip and other storms, see Appendix C in Harper (2002).
2. Cyclonic circulation is counterclockwise in the northern hemisphere and clockwise in the southern hemisphere.
3. A brief increase in wind is a gust and a brief decrease is a lull.

4. The Beaufort scale was invented by Admiral Beaufort and rates the *force* of the wind on a scale of 1 to 12. Originally, scale values 0 through 4 described the wind in terms of a ship's speed and 5 through 12 in terms of the ship's sail carrying ability. The conversion between numbers in the Beaufort scale and wind speed was standardized in 1926. B12 on the Beaufort scale is equivalent to hurricane-force winds. The Fujita scale was developed in 1971 by T. Fujita and classifies wind speeds on a scale of 0 to 6 based on damage to buildings. F1 on the Fujita scale is equivalent to B12 on the Beaufort scale. The wind-speed equivalents of the Fujita scale are only estimates, however, and have not been experimentally verified.
5. This example is derived from monthly tropical storm summaries issued by Gary Padgett. The summaries are available at http://australiasevereweather.com/cyclones/. Lingling, a Chinese nickname for young girls, is one of Hong Kong's contributions to the WMO list of typhoon names for the western North Pacific.
6. This data set was developed by J. Pennington, M. DeMaria, and K. Williams and is available at http://www.bbsr.edu/rpi/research/demaria/demaria4.html.
7. Interestingly, there are efforts to develop proprietary libraries of "synthetic" best tracks using statistical and dynamical models as a way to overcome the limitations associated with existing best-track data; however, these synthetic tracks are, at least in part, based on analyses of, and statistics derived from, the best-track data.

REFERENCES

Atkinson, G. D. 1974. *Investigation of gust factors in tropical cyclones*. FLEWEACEN Technical Note, JTWC 74-1. Guam: U.S. Fleet Weather Center.

Atkinson, G. D., and C. R. Holliday. 1975. *Tropical cyclone minimum sea level pressure—Maximum sustained wind relationship for western North Pacific*. FLEWACEN Technical Note, JTWC 75-1. Guam: U.S. Fleet Weather Center.

Atkinson, G. D., and C. R. Holliday. 1977. Tropical cyclone minimum sea level pressure/maximum sustained wind relationship for the western North Pacific. *Monthly Weather Review* 105:421–27.

Cook, N. J. 1985. *The designer's guide to wind loading of building structures*. Part 1, *Background, damage survey, wind data and structural classification*. London: Butterworths.

Davenport, A. G. 1964. Note on the distribution of the largest value of a random function with applications to gust loading. *Proceedings of the Institute of Civil Engineers* 28: 187–96.

Dunnavan, G. M., and J. W. Diercks. 1980. An analysis of Supertyphoon Tip (October 1979). *Monthly Weather Review* 180:1915–23.

Durst, C. S. 1960. Wind speeds over short periods of time. *Meteorological Magazine* 89:181–87.

Dvorak, V. F. 1975. Tropical cyclone intensity analysis and forecasting from satellite imagery. *Monthly Weather Review* 103:420–30.

Dvorak, V. F. 1984. *Tropical cyclone intensity analysis using satellite data*. NOAA

Technical Report No. 11. Washington, D.C.: National Oceanic and Atmospheric Administration.

Emanuel, K. A. 1986. An air-sea interaction theory for tropical cyclones. Part I: Steady-state maintenance. *Journal of Atmospheric Sciences* 43:485–604.

Erickson, C. O. 1972. *Evaluation of a technique for the analysis and forecasting of tropical cyclone intensities from satellite pictures*. NOAA Technical Memorandum, Report No. NESS 42. Washington, D.C.: National Oceanic and Atmospheric Administration.

Franklin, J. L., M. L. Black, and K. Valde. 2000. Eyewall wind profiles in hurricanes determined by GPS dropwind sondes (available at: http://www.nhc.noaa.gov/aboutwindprofile_text.html).

Glickman, T. S. 2000. *Glossary of meteorology*. Boston: American Meteorological Society.

Harper, B. 2002. *Tropical cyclone parameter estimation in the Australian region: Wind-pressure relationships and related issues for engineering planning and deesign*. Report No. J0106-PR003E. Bridgeman Downs, Queensland, Australia: Systems Engineering Australia.

Hock, T. F., and J. L. Franklin. 1999. The NCAR GPS dropwindsonde. *Bulletin of the American Meteorological Society* 80:407–20.

Holland, G. J. 1981. On the quality of the Australian tropical cyclone data base. *Australian Meteroological Magazine* 29:169–81.

Holland, G. J. 1993. Ready reconer. In *Global guide to tropical cyclone forecasting*, edited by G. J. Holland. Technical Document WMO/TD-No. 560, Report No. TCP-31. Geneva: World Meteorological Organization.

Holland, G. J. 1997. The maximum potential intensity of tropical cyclones. *Journal of Atmospheric Sciences* 54:2519–41.

Hope, J. R., and C. J. Neumann. 1968. *Probability of tropical cyclone induced winds at Cape Kennedy*. Weather Bureau Technical Memorandum SOS-1. Silver Spring, Md.: National Oceanic and Atmospheric Administration.

Jarrell, J. D., P. J. Hebert, and M. Mayfield. 1992. *Hurricane experience levels of coastal county populations from Texas to Maine*. NOAA Technical Memorandum, NWS NHC Report No. 46. Coral Gables, Fla.: National Oceanic and Atmospheric Administration.

Jarvinen, B. R., C. J. Neumann, and M. A. S. Davis. 1984. A *tropical cyclone data tape for the North Atlantic Basin, 1886–1983: Contents, limitations, and uses*. NOAA Technical Memorandum, NWS NHC Report No. 22. Coral Gables, Fla.: National Oceanic and Atmospheric Administration.

Krayer, W. R., and R. D. Marshall. 1992. Gust factors applied to hurricane winds. *Bulletin of the American Meteorological Society* 73:613–17.

Landsea, C. W., R. A. Pielke, Jr., A. M. Mestas-Nuñez, and J. A. Knaff. 1999. Atlantic Basin hurricanes: Indices of climatic changes. *Climatic Change* 42:89–129.

Lawrence, M. B., and B. M. Mayfield. 1977. Satellite observations of trochoidal motion during Hurricane Belle, 1976. *Monthly Weather Review* 105:1458–61.

Neumann, C. J. 1979. *Operational techniques for forecasting tropical cyclone intensity and movement*. WMO-No. 528 World Weather Watch, TCP Sub-project No. 6. Geneva: World Meteorological Organization.

Neumann, C. J. 1993. Global overview. In *Global guide to tropical cyclone forecasting*, edited by G. J. Holland. Technical Document WMO/TD-No. 560, Report No. TCP-31. Geneva: World Meteorological Organization.

Neumann, C. J., B. R. Jarvinen, C. J. McAdie, and J. D. Elms. 1993. *Tropical cyclones of the North Atlantic Ocean, 1871–1992*. NOAA Historical Climatology Series 6-2. Asheville, N.C. National Climatic Data Center.

Powell, M. D., and P. G. Black. 1990. The relationship of hurricane reconnaissance flight-level wind measurements to winds measured by NOAA's oceanic platforms. *Journal of Wind Engineering and Industrial Aerodynamics* 36:381–92.

Powell, M. D., S. H. Houston, L. R. Amat, and N. Morisseau-Leroy. 1998. The HRD real-time hurricane wind analysis system. *Journal of Wind Engineering and Industrial Aerodynamics* 77–78:53–64.

Powell, M. D., S. H. Houston, and T. A. Reinhold. 1996. Hurricane Andrew's landfall in south Florida. Part I: Standardizing measurements for documentation of surface wind fields. *Weather and Forecasting* 11:329–49.

Simiu, E., and R. H. Scanlon. 1978. *Wind effects on structures: An introduction to wind engineering*. New York: Wiley-Interscience.

Velden, C. S., T. L. Olander, and R. M. Zehr. 1998. Development of an objective scheme to estimate tropical cyclone intensity from digital geostationary infrared imagery. *Weather and Forecasting* 13:172–86.

World Meteorological Organization (WMO). 2000. *Tropical cyclone names*. Fact Sheet No. 15B: 4. Geneva: World Meteorological Organization.

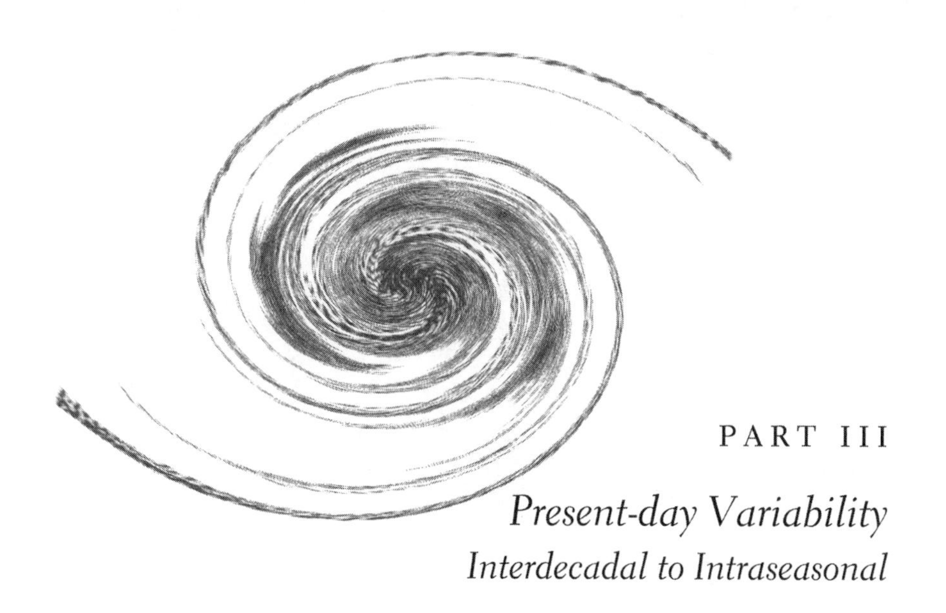

PART III

Present-day Variability

Interdecadal to Intraseasonal

10

Variations in the Activity of Tropical Cyclones over the Western North Pacific:

From Interdecadal to Intraseasonal

Johnny C. L. Chan

This chapter describes variations in tropical cyclone activity over the western North Pacific on time scales from interdecadal to intraseasonal. Although the physical mechanisms that cause the interdecadal variations are still unknown, those on interannual and seasonal scales can be largely explained by changes in the planetary-scale flow patterns. In particular, modifications of the flow conditions due to the occurrence of the El Niño–Southern Oscillation phenomenon as well as the stratospheric Quasi-Biennial Oscillation are found to contribute significantly toward the variations in the number, formation locations, life span and tracks of tropical cyclones. None of the studies demonstrate any contribution from local sea-surface temperature variations. Intraseasonal variations in tropical cyclone activity are mostly related to the eastward-propagating Madden-Julian Oscillation, as well as the 10- to 25-day and synoptic-scale (2–8-day) oscillations that propagate northwestward from the equatorial region of the western North Pacific. On the basis of some of these results, real-time predictions of seasonal or annual tropical cyclone activity have been made with considerable success.

The western North Pacific is the only ocean basin in the world in which tropical cyclones can form throughout the year. The monthly frequency is at its lowest between January and March (figure 10.1) and rises to a peak in August before decreasing again in the boreal autumn and winter. The average number of tropical cyclones per year is around 30 (the exact number depends on the period of average because of an appreciable interdecadal variability). This number is the highest among all ocean basins and exceeds 30% of the global total of around 80. This intense activity results primarily from the frequent presence of favorable thermodynamic and dynamic conditions for tropical cyclone development (Gray 1979). Probably because of this fact, low-

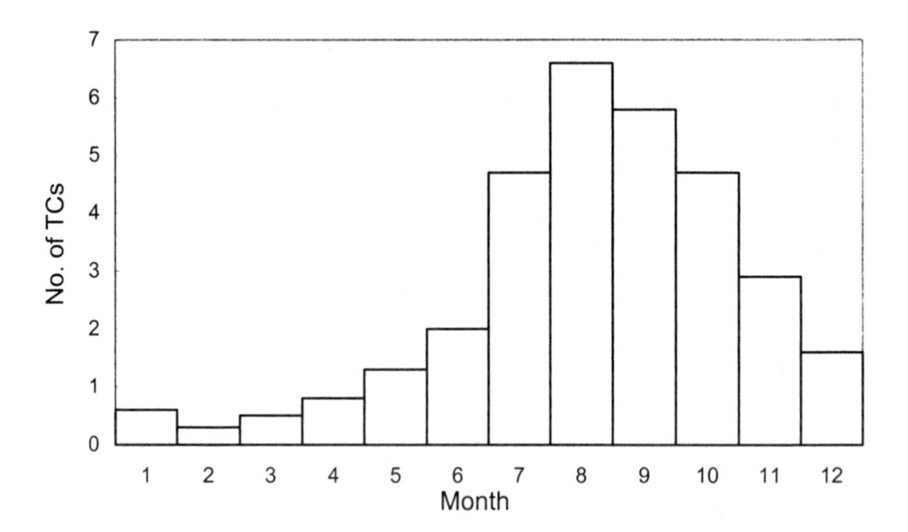

FIGURE 10.1 Monthly mean number of tropical cyclones (TCs) over the western North Pacific based on the 1959 to 2001 data from the Joint Typhoon Warning Center. Month 1 is January, 2 is February, and so on.

frequency variations of tropical cyclone activity in this ocean basin did not receive much attention until the 1980s.

Since then, a number of studies have related variations in tropical cyclone activity on intraseasonal to interdecadal time scales to known atmospheric oscillations such as the Madden-Julian Oscillation (MJO), the El Niño–Southern Oscillation (ENSO), and the stratospheric Quasi-Biennial Oscillation (QBO). This chapter discusses variations in tropical cyclone activity on intraseasonal to interdecadal time scales as well as the possible physical processes involved with these variations. In addition, an application of these results in real-time seasonal forecasting of tropical cyclone activity over the western North Pacific is described.

General Conditions over the Western North Pacific

Before discussing the different variations in tropical cyclone activity, it would be useful to review the general atmospheric and oceanographic conditions over the western North Pacific during the summer months when most of the tropical cyclones form (figure 10.1). During the peak tropical cyclone season (July to September), the low-level flow over the western North Pacific tropics is characterized by a band of cyclonic flow that extends from the South China Sea to about 150°E (figure 10.2). This flow is labeled as either the intertropical convergence zone or the monsoon trough. The former name refers to the convergence between the cross-equatorial flow from the Southern Hemisphere with the flow associated with the subtropical high over the western North Pacific. As Ramage (1971) pointed out, however, a more appropriate name for this feature is "monsoon trough," because it is associated with the flow of the South Asia summer monsoon over the Indian subcontinent, the East Asia summer monsoon over the South China Sea and the western Pacific summer monsoon over the western North Pacific. This band of cyclonic flow provides the background vorticity necessary for tropical cyclone genesis. In the vicinity of the trough axis, the vertical wind shear is minimal (figure 10.3), which would allow deep convection to be sustained. Most of the tropical cyclones form along the monsoon trough (Xue and Neumann 1984).

Over the western North Pacific tropics, the sea-surface temperature is above 28°C throughout most of the year (figure 10.4). Together with the high relative humidity in the low-mid troposphere (figure 10.2b), atmospheric conditions are conducive to the development of deep convection. In other words, the thermodynamic conditions necessary for tropical cyclone formation are satisfied throughout much of the year. Therefore, whether tropical cyclones form at certain locations very much depends on the dynamic conditions. The local sea-surface temperature distribution bears no relationship with the variations in tropical cyclone activity in the western North Pacific.

Landfalling Tropical Cyclones Along the South China Coast

Prior to satellite observations, some tropical cyclones might never have been detected if they did not make landfall or cross the path of a ship. Thus, studies of long-term variations of tropical cyclone activity are often limited to the last few decades. Historically, however, records of tropical cyclone–landfall events should have been recorded by various government agencies, making it possible to study landfall variations on much longer time scales.

One such study was carried out by Chan and Shi (2000; see also Louie and Liu, chapter 8 in this volume). Chan and Shi (2000) identified typhoons that occurred during the last 500 years in Guangdong Province along the South China coast using records of weather-related disasters compiled by Qiao and Tang (1993). Each record contains descriptions of damage from severe winds, heavy rain, and storm surges that lasted for at least a day. Although records date back more than 1,000 years, Chan and Shi (2000) only considered the more

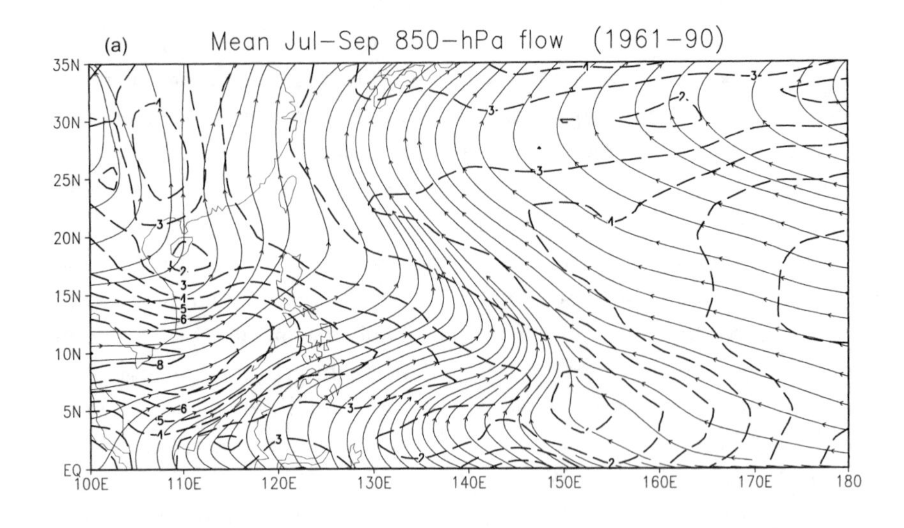

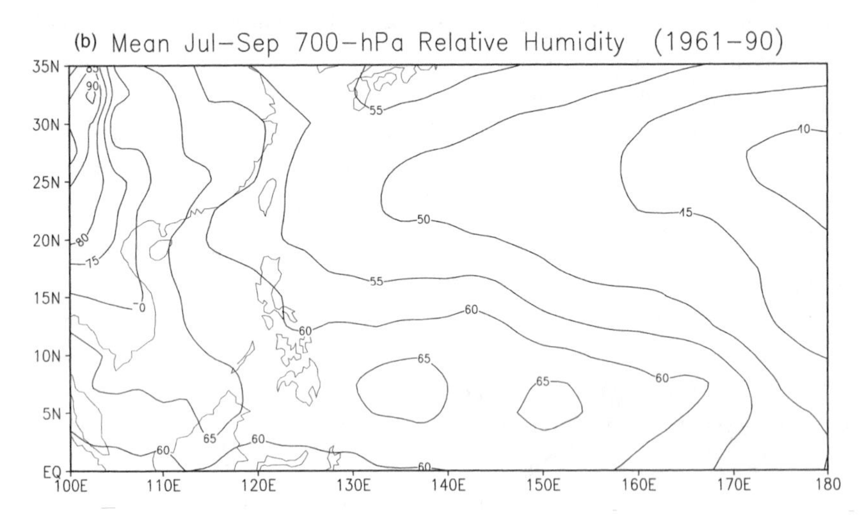

FIGURE 10.2 Average July to September conditions over the western North Pacific for the period from 1961 to 1990, constructed from National Centers for Environmental Prediction monthly reanalysis data: (*a*) 850-hPa flow (dashed lines: isotachs in intervals of 1 m s^{-1}) and (*b*) 700-hPa relative humidity (percent, contour interval: 5%). A pressure of 850 hPa is ~ 1.5 km in altitude and that of 700 hPa ~ 3 km. The arrows on the streamlines give the direction of flow. Isotachs are contours of constant wind speed.

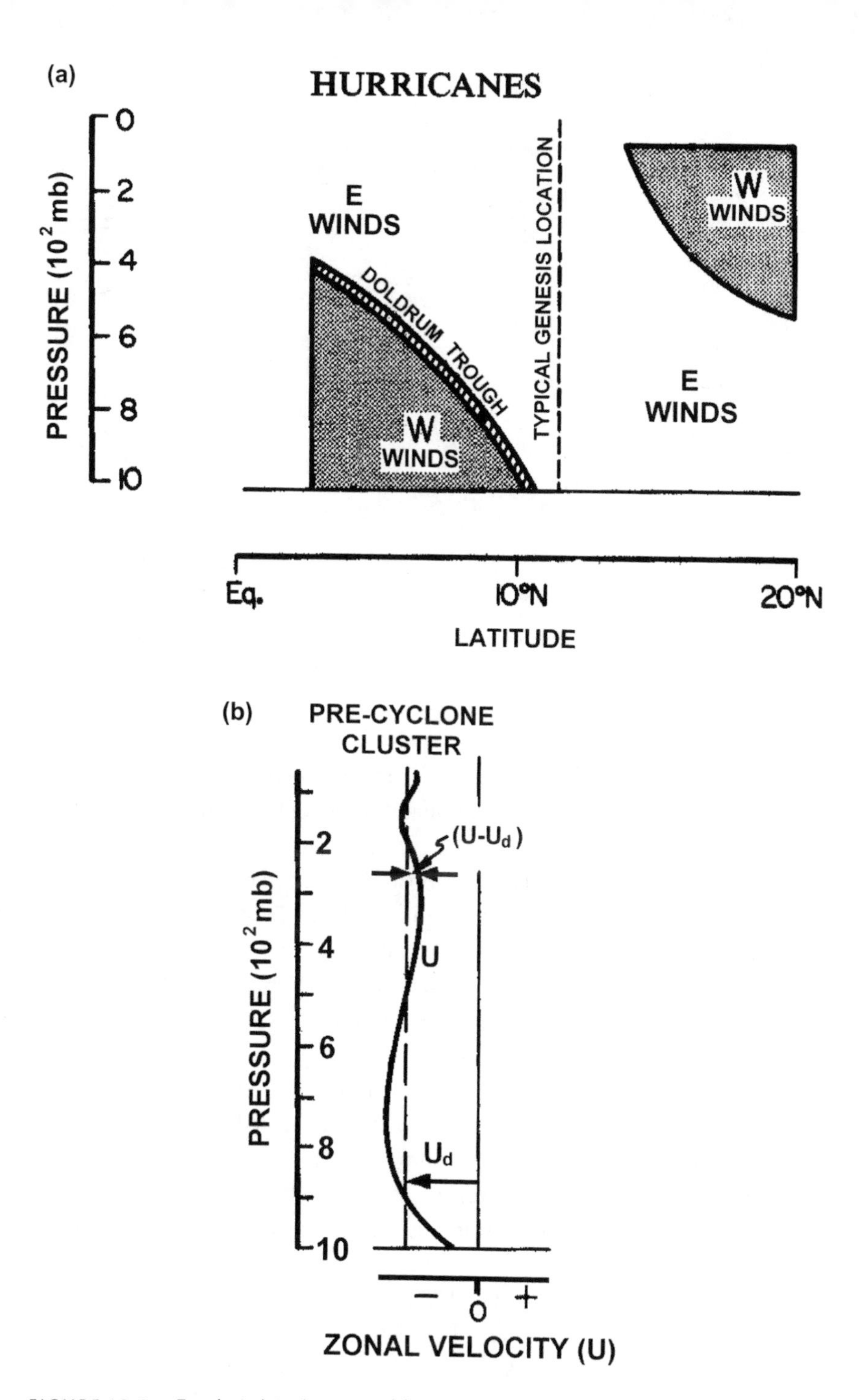

FIGURE 10.3 Zonal winds in the tropics: (*a*) typical meridional-vertical distribution (W: westerly, E: easterly) associated with the monsoon (doldrum) trough; (*b*) typical vertical distribution of zonal winds U at the typical genesis location indicated in (*a*). The zonal velocity of the weather disturbance that eventually became a tropical cyclone is U_d. (adapted from Gray 1979, by permission of the Royal Meteorological Society).

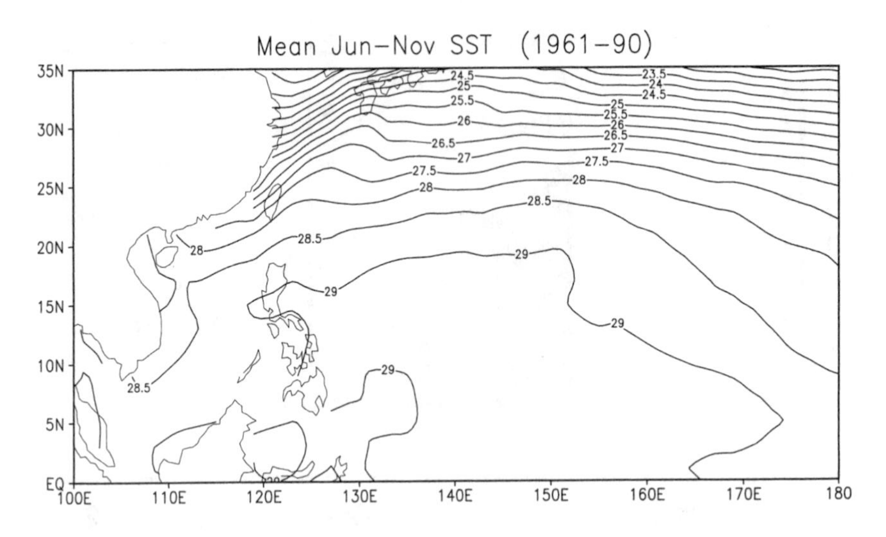

FIGURE 10.4 Average June to November sea-surface temperature over the western North Pacific for the period 1961 to 1990, constructed from NCEP monthly reanalysis data. Unit: °C; contour interval: 0.5°C.

recent events (from 1470) because earlier records were rather sporadic. Based on these records, they counted the annual number of typhoons that made landfall along the coast of South China. Presumably, these were systems that inflicted significant damage. To study the long-term variations, they also totaled the annual number over 5- and 10-year intervals. In addition to a linear trend, some possible periodicities are evident in all three series (figure 10.5). The linear trend should not simply be attributed to the increasing number of records in more recent times, because low landfall frequencies are also observed in latter years. Indeed, a variance analysis (for a description of the technique, see Chan and Shi 1996) suggests oscillations with time scales from decadal to centennial (table 10.1). Chan and Shi (2000) also performed other statistical tests to demonstrate the general validity of these results.

Thus, at least for the South China coast, landfall frequencies appear to have a variety of long-term variations. It is reasonable to assume that such variations likely exist at other locales; similar studies should therefore be carried out. Although these results may not be of immediate use by government planners, they can be used to test the ability of climate models to simulate realistic tropical cyclone landfall statistics. A reproduction of similar periodicities would not only help in estimating future landfall frequency, but would also contribute toward a better understanding of the physical mechanisms of such variations. Analyses of this type can also be combined with other proxy data (Liu, chapter

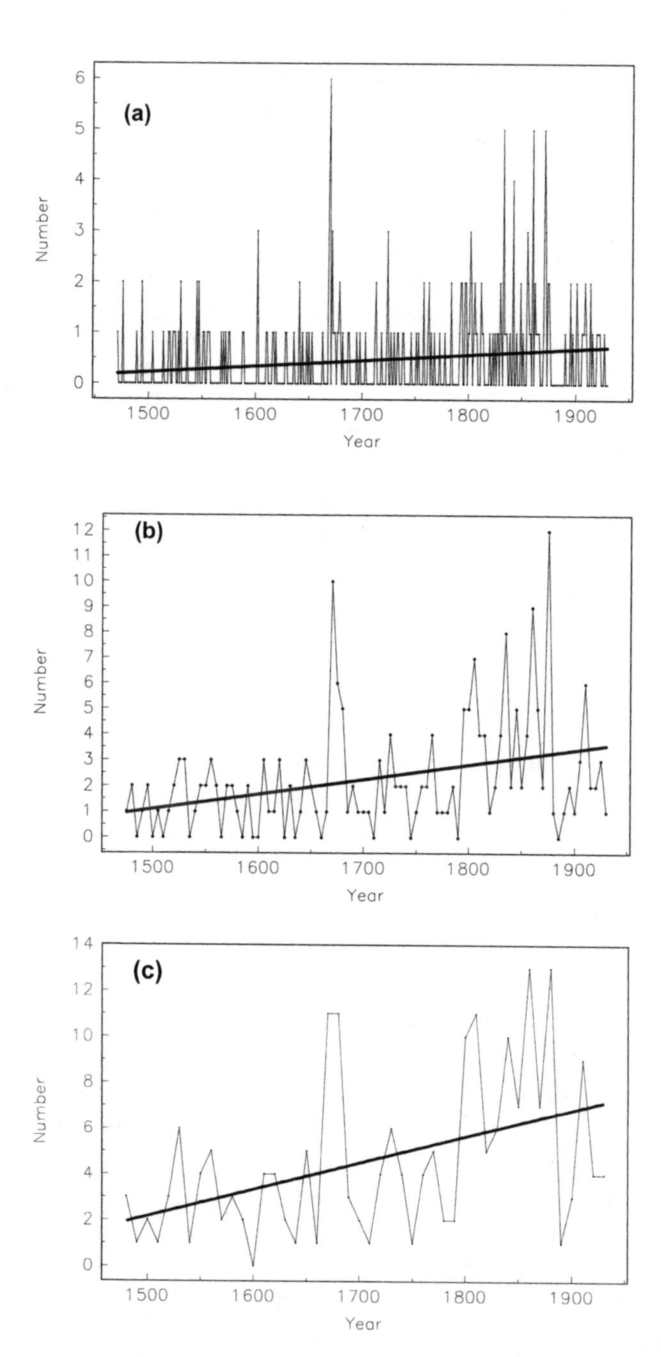

FIGURE 10.5 Time series of the number of typhoons making landfall over Guangdong Province between 1470 and 1931: (*a*) every year, (*b*) 5-year total, and (*c*) 10-year total (from Chan and Shi 2000, by permission of the Royal Meteorological Society).

TABLE 10.1 Major Periods for Typhoons Making Landfall over Guangdong Province, 1470–1931

Time series	Major periods (years)
Y_1	201, 142, 94, 59, 26
Y_5	205, 185, 130, 55, 15
Y_{10}	190, 150, 40

Note: Periods were identified from the variance analysis method. Y1, every year; Y5, 5-year total; Y10, 10-year total. Note that for the series Y5 (Y10), the actual period obtained is 1/5 (1/10) of the period shown.

2, and Donnelly and Webb, chapter 3 in this volume) so that a long historical record may be established (Boose, chapter 4; Mock, chapter 5; García Herrera et al., chapter 6; and Louie and Liu, chapter 8 in this volume).

Interdecadal Variability

Subsequent to Landsea et al.'s (1996) findings that since the 1970s there has been a downward trend in the number of intense hurricanes in the Atlantic, Chan and Shi (1996) also found an interdecadal variation in the annual number of tropical storms and typhoons in the western North Pacific (figure 10.6). The number decreased from a high of about 35 in the 1960s to a minimum of about 23 in the mid- to late 1970s. The number of storms then increased to a peak of around 35 in the mid-1990s. Chan and Shi (1996) found that a second-order polynomial gave the best fit to the data.

In an earlier study of the variation of annual tropical cyclone frequencies during the period from 1884 to 1988, Zhang, Zhang, and Wei (1994) found oscillations with periods of 31, 21, 15, and 6 years although no physical explanation was given. They also identified significant changes in the pattern in 1931, 1959, and 1977.

Despite the rising trend in tropical cyclone activity in the 1990s, the year 1998 saw the smallest number of tropical storms and typhoons (18) for the 1965 to 2001 period (figure 10.7b), and the lowest (9) number of typhoons for that same period (figure 10.7c). When tropical depressions are also included, however, the tropical cyclone activity in 1998 was not unusually low (total number of events = 27) (figure 10.7a). Tropical cyclone activity in the subsequent three years remained below those in the 1990s, but the number of typhoons appeared to be increasing again. Various polynomial fits have been tested and a fourth-order polynomial appears to give the "optimal" fit for all the different categories

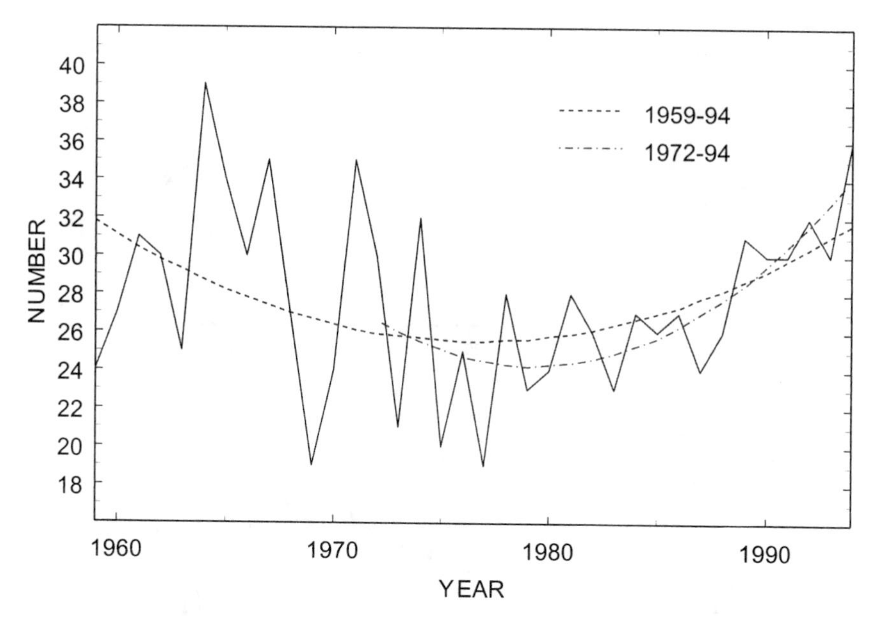

FIGURE 10.6 Annual number of tropical storms and typhoons over the western North Pacific from 1959 to 1994. The dotted and dot-dashed lines indicate second-order polynomial fits to the time series for the two periods shown (from Chan and Shi 1996, reproduced with modification by permission of the American Geophysical Union).

of tropical cyclones. The fitted curves suggest a general downward trend (figure 10.7). At the same time, the trend in the number of tropical cyclones in the Atlantic appears to have turned upward recently (Goldenberg et al. 2001). Therefore, it seems that perhaps on a global basis, the total tropical cyclone activity might be relatively constant.

The physical mechanisms for such interdecadal variations are still not clear. Because tropical cyclone activity is very much related to the ENSO, which has been found to go through interdecadal changes (Wang 1995), it is likely that interdecadal variability in ENSO might help to explain similar changes in tropical cyclone activity.

ENSO–Tropical Cyclone Relationships

In addition to interdecadal variability, Chan and Shi (1996) found interannual oscillations in the frequency of tropical cyclones, with major periods of two and seven years. Chan (1985, 1995) has identified the two-year period and related it to the stratospheric QBO. The 7-year period could be related to ENSO,

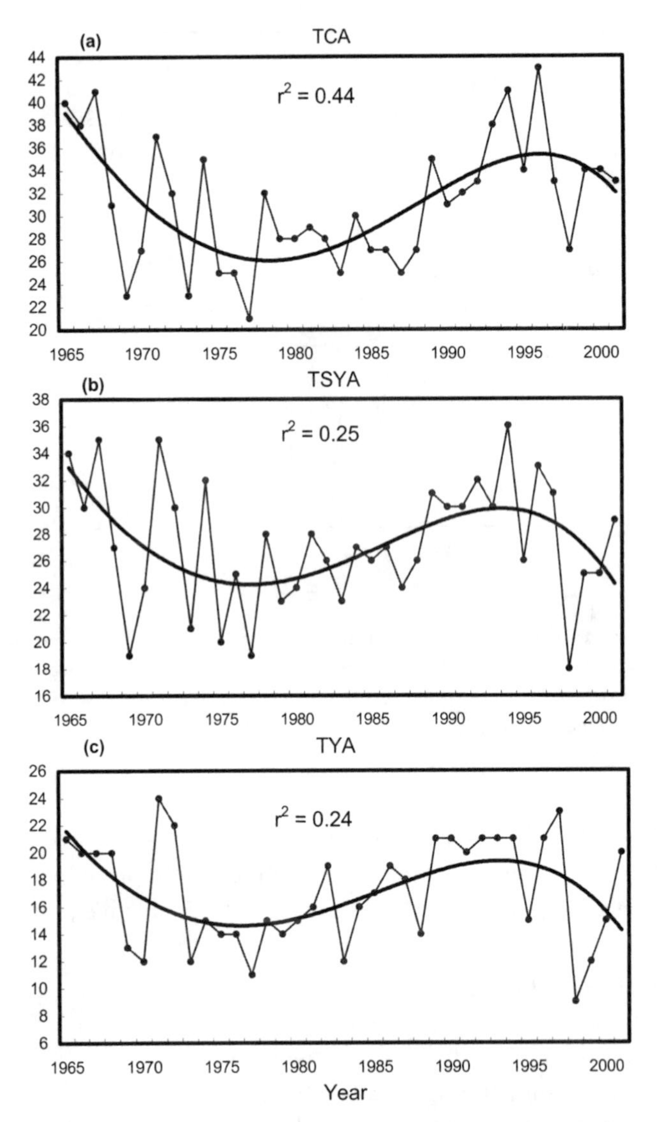

FIGURE 10.7 Annual number of (*a*) all tropical cyclones, (*b*) tropical storms and typhoons, and (*c*) typhoons over the western North Pacific from 1965 to 2001. The curved line in each figure indicates a fourth-order polynomial fit to the time series. The value of r^2 in each figure is the percent of variance explained by the polynomial.

which has been shown by a number of authors to affect interannual variations in tropical cyclone activity in various ocean basins: the Atlantic (Landsea 2000), the central North Pacific (Chu and Wang 1997; Chu and Clark 1999), the Australian region (Nicholls 1984), and the western North Pacific (Chan 1985, 2000; Dong 1988; Dong and Holland 1994; Lander 1994; Chen et al. 1998; Wang and Chan 2002). For a summary of the relationship between ENSO and tropical cyclone activity in various ocean basins, see also Chu (chapter 8 in this volume). Because extensive studies have been made on the relationship between ENSO and tropical cyclone activity in the western North Pacific, it is worthwhile to review such a relationship in greater detail, from both statistical and physical points of view.

Interannual Variations

Chan (1985) first noted that the annual number of typhoons over the western North Pacific correlates significantly with the Southern Oscillation Index (SOI) at the 3- to 3.5-year period. The SOI is found to lead by about one year. In other words, a decrease in SOI (corresponding to an El Niño event) is likely to be followed by a decrease in tropical cyclone activity in the next year. Further, the number of tropical cyclones that form east of 150°E during an El Niño year tends to be above normal. Chan (1985) attributed these relationships to changes in the Walker circulation associated with the ENSO phenomenon. Lander (1994) identified two El Niño events during which tropical cyclone activity did not follow these general relationships and disputed the existence of an ENSO signal in tropical cyclone activity. He nevertheless concurred with Chan's (1985) finding that more tropical cyclones tend to form in the eastern part of the western North Pacific during an El Niño event. Dong and Holland also drew a similar conclusion (Dong 1988; Dong and Holland 1994). Using a coarse-grid general circulation model, Wu and Lau (1992) were able to simulate shifts in the location of tropical cyclone formation in association with the migration of the sea-surface temperature anomalies in the equatorial Pacific. (The migrating anomalies can be used as proxies for the occurrence of an ENSO event).

Chan (2000) extended his 1985 study to examine tropical cyclone activities in the year prior to, during and after an El Niño event and a La Niña event. He found that during the year before an El Niño, the number of tropical cyclones to the southeast of Japan tends to be below normal (figure 10.8a). The activities during and after an El Niño event (figure 10.8b and c) are consistent with the

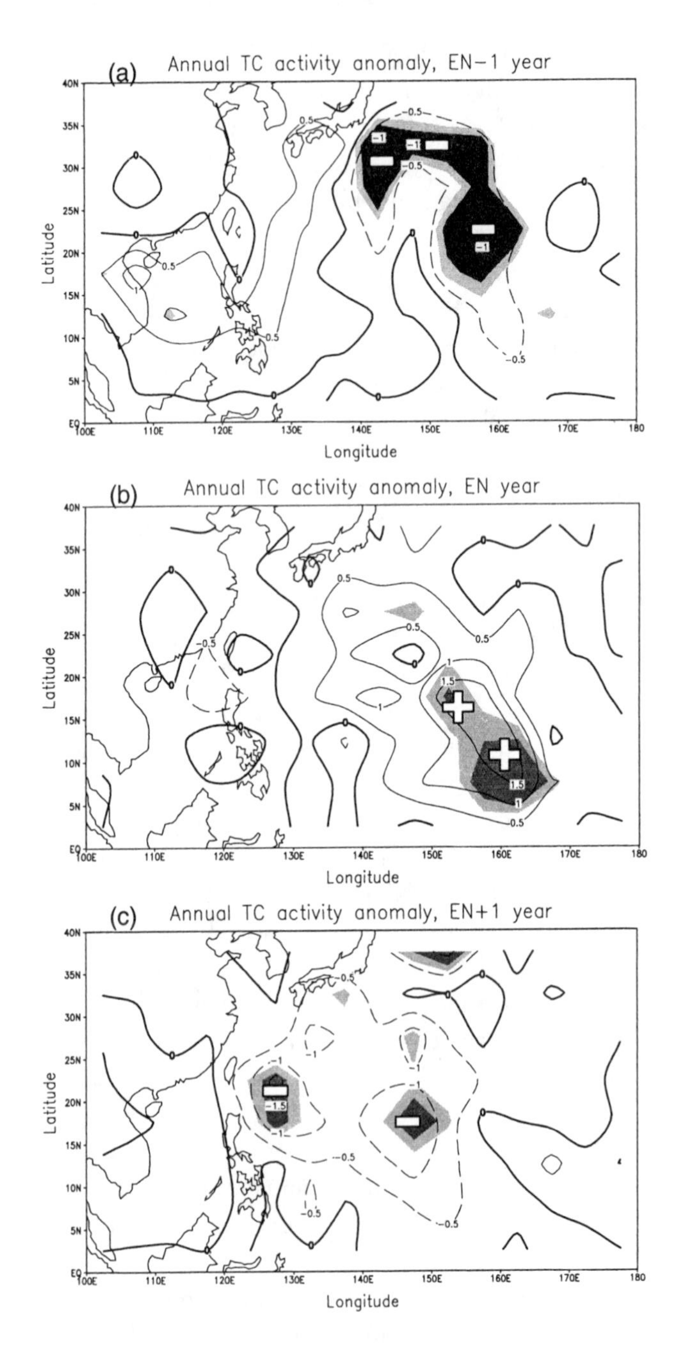

FIGURE 10.8 Composite anomalies of the annual tropical cyclone activity during (*a*) El Niño −1, (*b*) El Niño, (*c*) El Niño + 1, (*d*) La Niña −1, (*e*) La Niña, and (*f*) La Niña + 1 years over the western North Pacific. Solid (dashed) lines indicate positive (negative) anomalies, with the plus (minus) signs indicating the approximate locations of the maximum (negative maximum) values. Contour interval: 0.5. Light and dark shades indicate areas where the *t*-test is significant at the 90% and 95% level respectively (from Chan 2000, by permission of the American Meteorological Society).

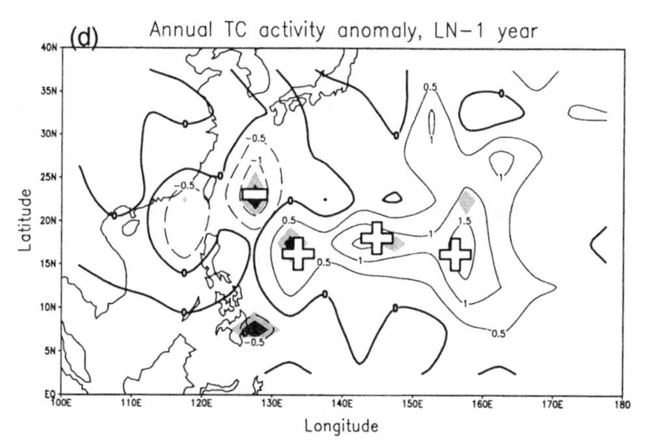

(d)
Annual TC activity anomaly, LN−1 year
Latitude
40N
35N
30N
25N
20N
15N
10N
5N
EQ
100E
110E
120E
130E
140E
150E
160E
170E
180
Longitude

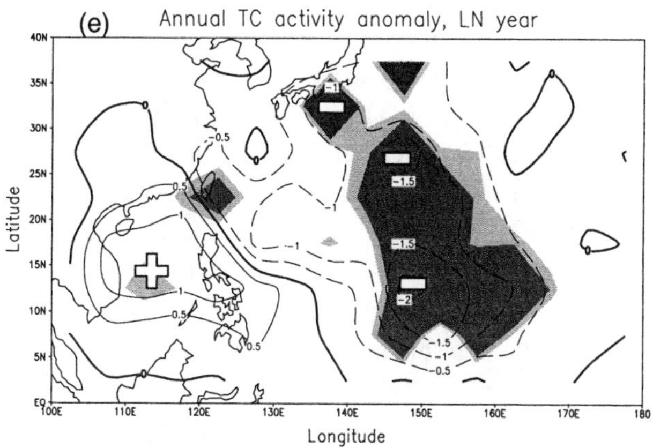

(e)
Annual TC activity anomaly, LN year
Latitude
40N
35N
30N
25N
20N
15N
10N
5N
EQ
100E
110E
120E
130E
140E
150E
160E
170E
180
Longitude

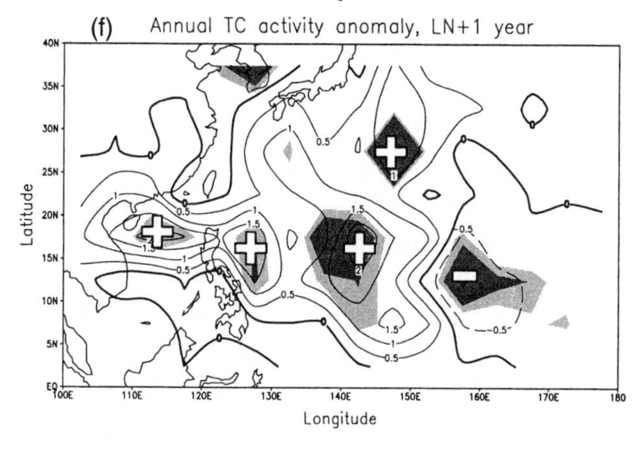

(f)
Annual TC activity anomaly, LN+1 year
Latitude
40N
35N
30N
25N
20N
15N
10N
5N
EQ
100E
110E
120E
130E
140E
150E
160E
170E
180
Longitude

results obtained by Chan (1985). During the year prior to a La Niña event, the area east of the Philippines tends to have above-normal tropical cyclone activity (figure 10.8d). These anomalies do not pass a significance test. In a La Niña year, the distribution of tropical cyclone activity is generally opposite to that in an El Niño year, with fewer tropical cyclones over much, especially the eastern part, of the western North Pacific, but more over the South China Sea (figure 10.8e). The entire western North Pacific generally has above-normal tropical cyclone activity during the year after a La Niña event (figure 10.8f).

Seasonal and Monthly Variations

In examining the anomalies in tropical cyclone activity during El Niño and La Niña events, Chan (2000) found that these anomalies do not occur throughout the year, but only in certain months. During these months, the circulations that govern either tropical cyclone genesis and development (850 hPa) or motion (500 hPa) have significant deviations from climatology, in association with the different phases of the ENSO cycle. In fact, all of the tropical cyclone anomalies can be explained by such deviations. Changes in the planetary-scale circulation from before to after the occurrence of an ENSO event modify the locations where tropical cyclones form and intensify, as well as the tropical cyclone tracks.

For example, during the late season in an El Niño year, low-level anomalous westerlies are found in the equatorial regions in the eastern part of the western North Pacific (figure 10.9). This creates positive relative vorticity anomalies, which therefore provides a favorable environment for tropical cyclone genesis so that more tropical cyclones form in the latter part of an El Niño year in the southeastern part of the western North Pacific (see also Wang and Chan 2002). Another example is the change in the steering flow. During the late season of a La Niña year, an anomalous anticyclone is found at 500 hPa over the East China Sea with significant easterly anomalies to its south (figure 10.10). As a result, tropical cyclones that form to the east of the Philippines have a tendency to enter the South China Sea (figure 10.11). Other examples of the association between tropical cyclone activity anomalies and large-scale atmospheric circulation can be found in Chan (2000) and Wang and Chan (2002). All these results therefore reinforce the existence of a relationship between ENSO and tropical cyclone activity.

Chen et al. (1998) found that the locations of tropical cyclone genesis during June to August differed between years when the sea-surface temperature in

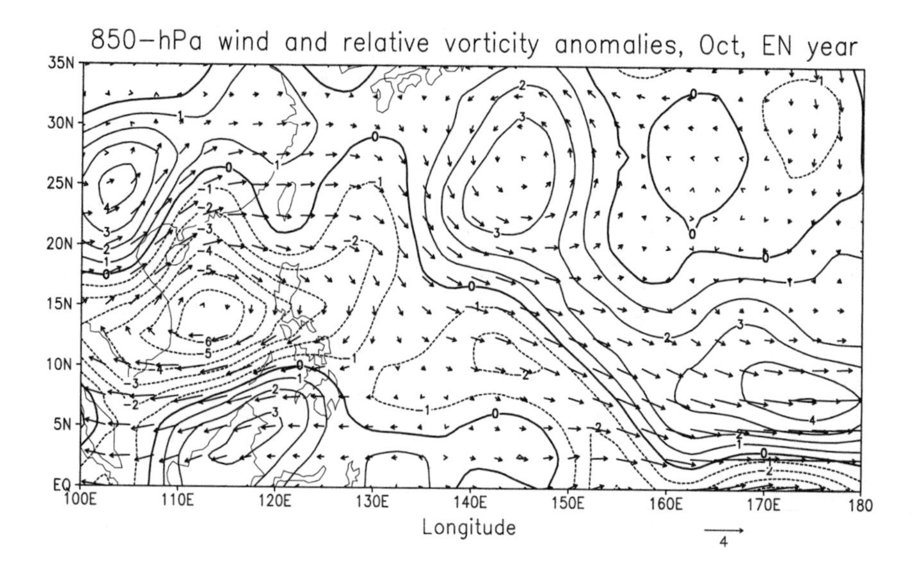

FIGURE 10.9 Anomalies in 850-hPa wind (vector, scale indicated on lower right of diagram; unit: m s^{-1}) and relative vorticity (contour line, unit: 10^{-5} s^{-1}; contour interval: 1×10^{-5} s^{-1}) during October in an El Niño year (adapted from Chan 2000, by permission of the American Meteorological Society).

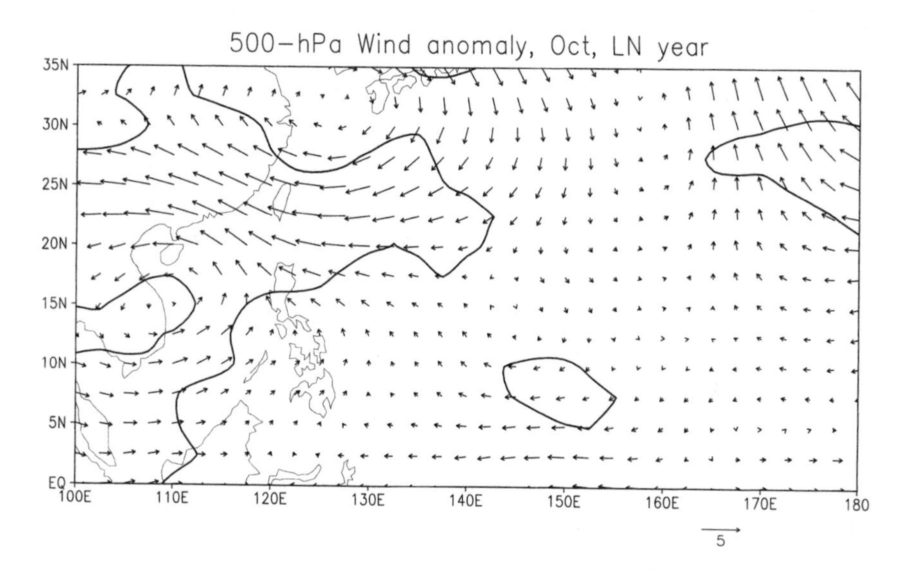

FIGURE 10.10 Anomalous 500-hPa wind vectors (scale indicated on lower right of diagram; unit: m s^{-1}) during October of a La Niña year. Contour lines indicate 80% wind steadiness of the composite (adapted from Chan 2000, by permission of the American Meteorological Society).

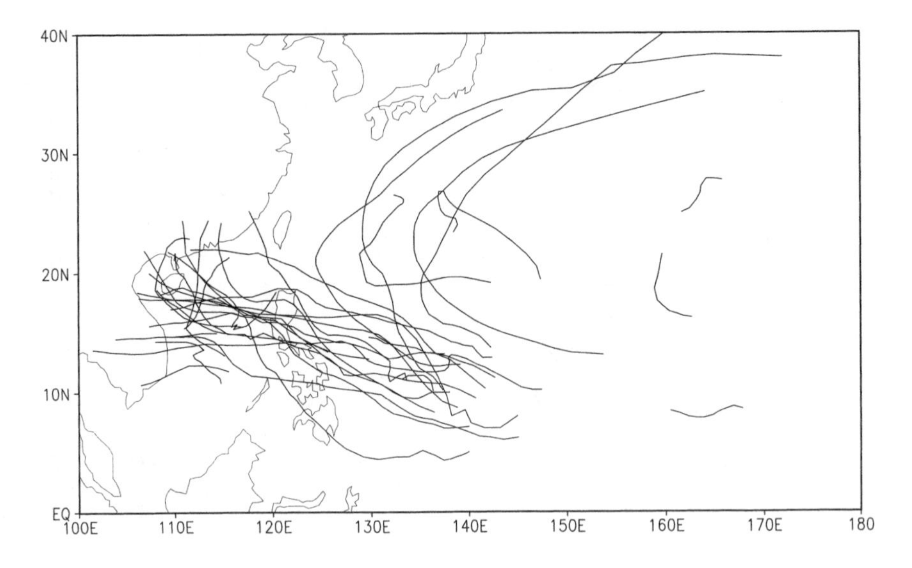

FIGURE 10.11 Tracks of tropical cyclones in October of all the La Niña years between 1959 and 1997 (from Chan 2000, by permission of the American Meteorological Society).

the Niño 3 region (4°N–4°S, 150–90°W) was significantly above normal (an El Niño event) and those when the sea-surface temperature in this region was below normal (a La Niña event). Specifically, when the sea-surface temperature was above normal (i.e., a warm event), tropical cyclones tended to form at lower latitudes and farther east (figure 10.12). During a cold event (sea-surface temperature below normal), more tropical cyclones occurred at higher latitudes and farther west. Wang and Chan (2002), who extended Chen et al.'s study to include more recent events, also identified such a southeast-northwest difference between warm and cold years. In addition, Chen et al. (1998) found similar, but less pronounced, east-west differences for the fall season (September to November).

Wang and Chan (2002) further noted that because tropical cyclones form further to the southeast during warm years, they tend to last longer, with a mean life span of seven days versus only four days for tropical cyclones in cold years. This also leads to a greater number of days of tropical storm occurrence in warm years (159), almost twice that of cold years (84). The tracks also differ between warm and cold years. During the peak season (July to September), more tropical cyclones take on recurving tracks in warm years, but those in cold years tend to move more northward after forming at locations farther north than

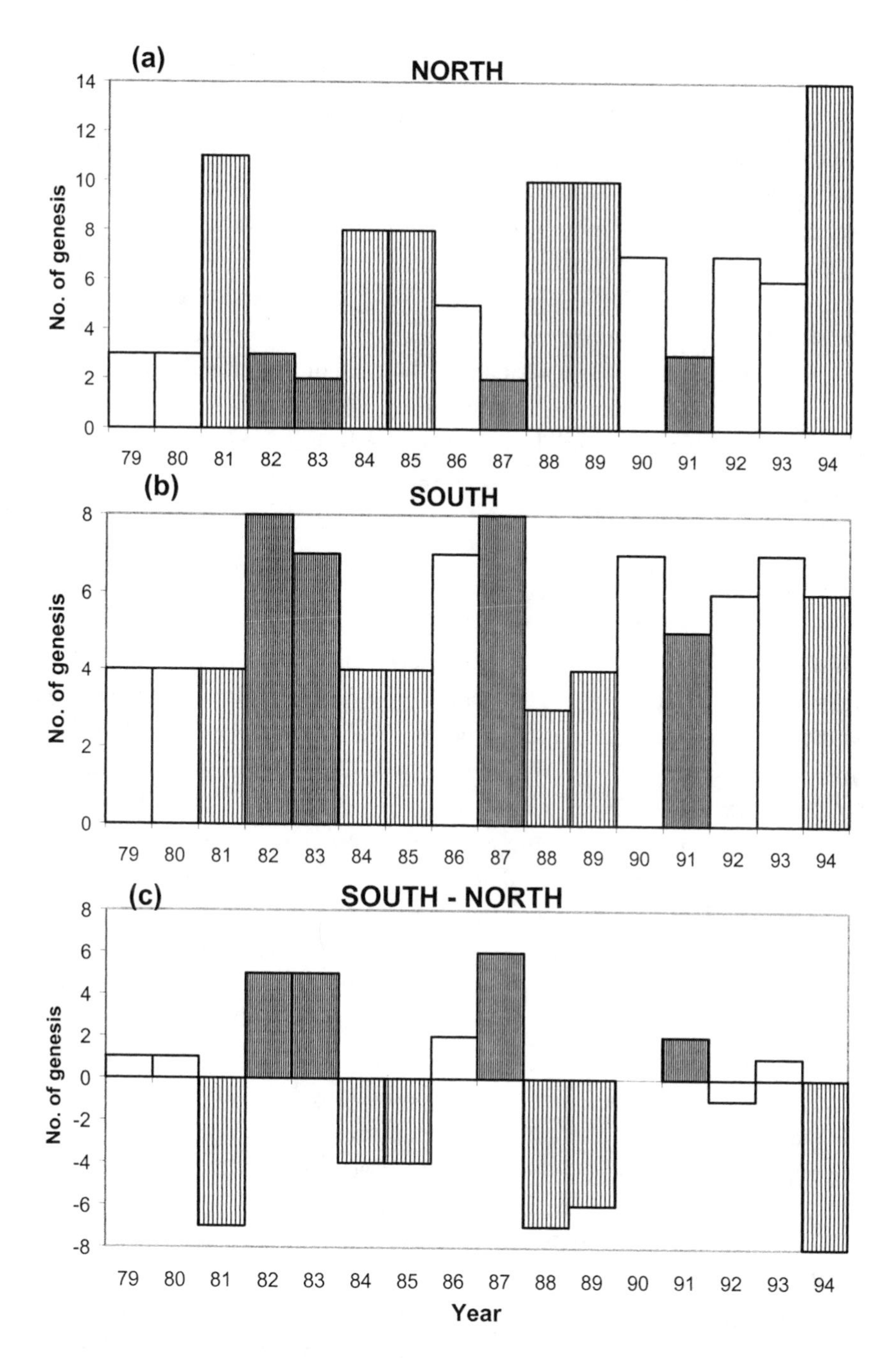

FIGURE 10.12 Tropical cyclone genesis frequency of every summer (June–August) during the period from 1979 to 1994 over (*a*) 15°–30°N, 120°E–180° (the north region) and (*b*) 0–15°N, 120°E–180° (the south region). The difference between south and north regions (south minus north) is shown in (*c*). Light striped bars indicate summers with Niño 3 sea-surface temperature anomalies below normal; heavy bars indicate summers with Niño 3 sea-surface temperature anomalies above normal (redrawn from Chen et al. 1998, by permission of the American Meteorological Society). Niño 3 is the region in the equatorial Pacific between 5°S and 5°N and 150°W and 90°W.

normal (figure 10.13). In the late season (October to December), recurving tropical cyclones continue to dominate in warm years, but those in cold years have generally westward-moving tracks. As a result, in cold years the South China Sea experiences more tropical cyclones than in warm years, which is consistent with the results of Chan (2000).

It is important to note a conclusion of Wang and Chan (2002) that no relationship has been found between the local sea-surface temperature and the locations of tropical cyclone formation. This is consistent with the result of Gray (1979) and the earlier discussion in this chapter that thermodynamic conditions over the western North Pacific are generally suitable for tropical cyclone formation. Evans (1993) also found no relationship between tropical cyclone intensity and sea-surface temperature.

Physical Processes

Most of the studies on the relationship between ENSO and tropical cyclone activity show consistent results that can be physically explained by the changes in the planetary-scale circulation associated with ENSO. The following summarizes the explanations of various researchers (Chan 1985, 2000; Wu and Lau 1992; Lander 1994; Chen et al. 1998; Wang and Chan 2002). Before the mature stage of the El Niño, strong westerly anomalies in the western equatorial Pacific have extended to the dateline and beyond. These anomalies, when coupled with the trade winds, enhance the cyclonic shear so that the monsoon trough is located at lower latitudes and extends farther eastward, which results in more tropical cyclones forming in the southeastern region of the western North Pacific. This situation begins to occur from the early (April to June) to peak season (depending on the onset time of the warm event [Xu and Chan 2001]) of an El Niño year (defined as the year in which sea-surface temperature anomalies in the eastern equatorial Pacific first reach a maximum). Because the tropical cyclones form farther to the southeast, they tend to last longer when moving westward. At the same time, the subtropical high tends to be not as strong so that westerly troughs are more likely to penetrate southward, which results in more recurving tropical cyclones.

Wang, Wu, and Fu (2000) suggest that during the year after the mature stage of a warm event, an anomalous anticyclone tends to form just to the east of the Philippines, where tropical cyclone formation is climatologically a maximum (Xue and Neumann 1984). This anticyclonic flow suppresses the formation and development of tropical cyclones. As a result, in the year after a warm event, tropical cyclone numbers tend to be below normal.

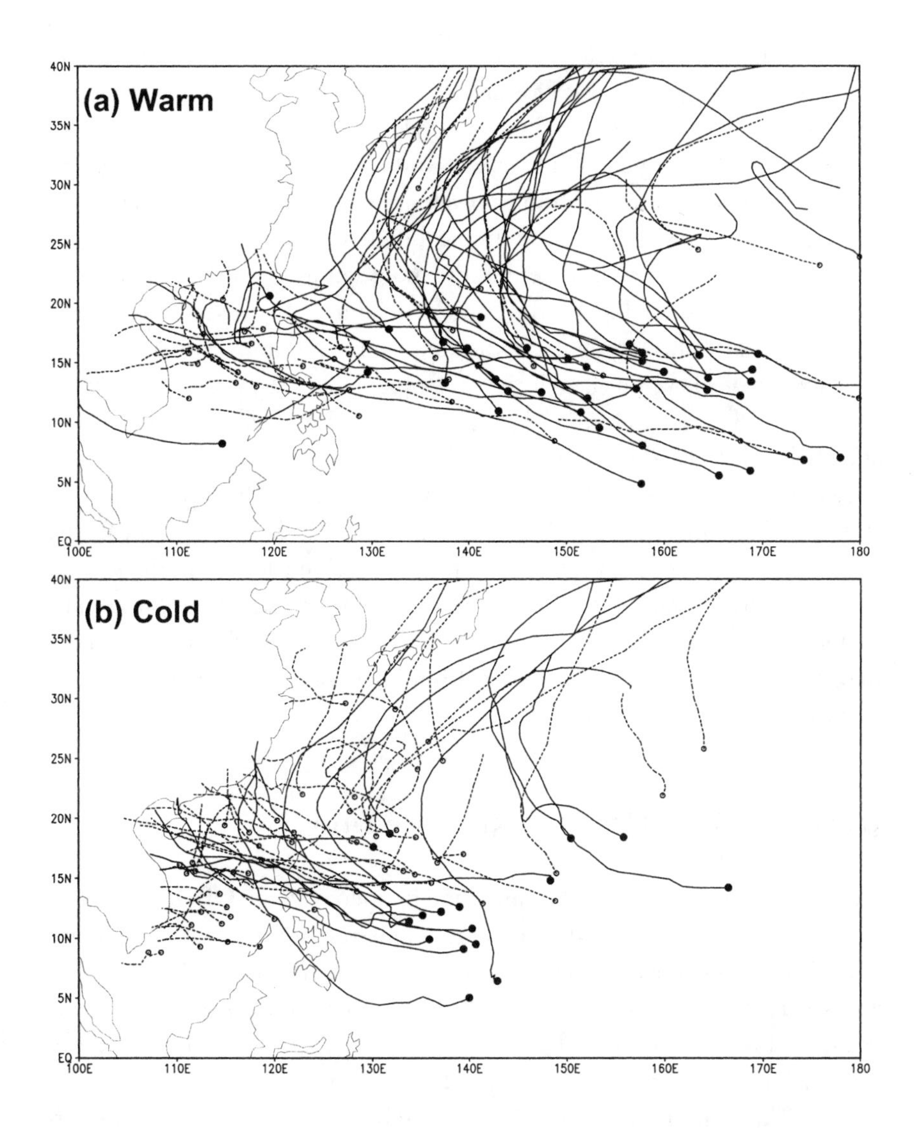

FIGURE 10.13 September to November tropical cyclone tracks: (*a*) during the six strongest warm years examined for this study (1965, 1972, 1982, 1987, 1991, 1997); (*b*) during the six strongest cold years (1970, 1973, 1975, 1988, 1998, 1999). Genesis locations (tracks) of the long-lived tropical storms are marked by heavy solid dots (solid lines). A long-lived tropical storm has a life span exceeding seven days. Genesis locations (tracks) of other storms are denoted by open circles (dashed lines) (adapted from Wang and Chan 2002, by permission of the American Meteorological Society).

If a cold, La Niña event develops, easterly anomalies are present near the dateline. This is especially the case after the La Niña event has fully developed. The easterly anomalies then reduce the cyclonic shear. As a result, conditions in the southeastern region of the western North Pacific become unfavorable for tropical cyclone genesis and tropical cyclones can only form much farther to the west. At the same time, the subtropical high tends to be enhanced, which suggests a northward displacement of the monsoon trough so that more tropical cyclones form in the northwestward quadrant. Because these tropical cyclones are already closer to land, they tend to have a shorter life span.

Thus, tropical cyclone activity is being modulated to a certain extent by the occurrence of an ENSO event. Using a limited sample, Zhang, Drosdowsky, and Nicholls (1990) showed that the SOI could explain about 40% of the variance of tropical cyclone activity in the western North Pacific. Prediction studies (e.g., Chan, Shi, and Lam 1998; Chan, Shi, and Liu 2001) also indicate that ENSO is only one of the predictors of tropical cyclone activity in the western North Pacific. Other factors also contribute toward variations in tropical cyclone activity, which include changes in the planetary-scale circulation that may not be related to ENSO. Physical links between these other factors and tropical cyclone activity have yet to be established.

QBO–Tropical Cyclone Relationship

Gray (1984) has linked Atlantic hurricane frequency to the QBO. During the westerly phase of the QBO, the frequency of intense hurricane development is almost three times the frequency seen during the easterly phase (Gray, Sheaffer, and Knaff 1992). The proposed explanation is that when the zonal winds are westerly in the lower stratosphere, the vertical wind shear in the upper troposphere off the equator tends to be smaller, therefore favoring tropical cyclone formation.

By examining the annual number of tropical cyclones over the western North Pacific during the period from 1884 to 1988, Zhang, Zhang, and Wei (1994) also found that tropical cyclone activity tends to be enhanced during the westerly phase of the zonal winds in the lower stratosphere. In addition, Chan (1995) identified a significant correlation between time series of tropical cyclone activity in the western North Pacific and zonal winds at 30 and 50 hPa at the QBO frequency. The phase difference between the two time series is consistent with the explanations of Gray, Sheaffer, and Knaff (1992) and Zhang, Zhang, and Wei (1994) and implies that when lower stratospheric winds begin to strengthen from the west, tropical cyclone activity over the western North Pacific is likely to increase. Chan (1995) also found that this correlation sometimes weakened during ENSO years, which suggests again that more than one factor must be considered in understanding the low-frequency variability of tropical cyclone activity.

Intraseasonal Oscillations in Tropical Cyclone Activity

Gray (1979), and later Harr and Elsberry (1991), noted that the formation of tropical cyclones over the western North Pacific tends to cluster in time with periods of one to two weeks. Further investigations by Harr and Elsberry (1995a, 1995b) suggested that over the western North Pacific, specific large-scale circulation characteristics are favorable for periods of tropical cyclone activity and inactivity. These circulation characteristics tend to alternate with unfavorable conditions occurring on time scales of 10 to 30 days.

A recent study by Harr, Elsberry, and Chan (2000) indicates that variability in the monsoon trough, which is the breeding ground for about 80% of western North Pacific tropical cyclones (Gray 1979), can be explained by a combination of circulation patterns with 30- to 60-day, 10- to 25-day, and 2- to 8-day periods. Therefore, the combination of the dominant circulation characteristics over each of the three time scales may define a separate time scale of alternating active and inactive tropical cyclone periods. Harr, Elsberry, and Chan (2000) indicate that the spectral power of the three dominant periods decreases with increasing frequency. That is, the dominant forcing apparently comes from the 30- to 60-day period.

Over the western North Pacific, the 30- to 60-day period is dominated by the MJO (Madden and Julian 1994). The dominant characteristic of the MJO is a large eastward-propagating convective region that is best defined in the equatorial Indian Ocean and the western North Pacific. The associated circulation is mainly in the equatorial zonal wind, but low-level cyclonic circulations during the convective phase of the MJO are present just off the equator and contribute to surface convergence, positive relative vorticity, enhanced convection, and other conditions that would favor the formation of tropical cyclones. During the reduced convective phase of the MJO, anticyclonic circulations contribute to low-level divergence and unfavorable conditions for tropical cyclone formation.

Liebmann, Hendon, and Glick (1994) examined the relationship between the MJO and tropical cyclone activity over the western North Pacific. Although the numbers of tropical cyclones increased during the enhanced convective phase of the MJO, they found that the ratio of storms and typhoons that formed per depression was similar for the active and inactive convective phases of the MJO. The increased numbers of tropical cyclones during the active MJO is primarily due to an increased number of depressions, which are not any more likely to intensify to typhoon strength than a depression during the inactive phase of the MJO. They also pointed out that the increase in tropical cyclone activity with a 30- to 60-day period was not unique, because increased tropical cyclone activity was also identified during arbitrarily chosen periods other than the MJO. The MJO should still be considered to have a significant influence on tropical cyclone activity because it explains a large percentage of the variance in low-frequency tropical variability.

The variation in the influence of the MJO on tropical cyclone activity is most likely due to circulation characteristics on other scales of motion as defined by Harr, Elsberry, and Chan (2000). Hartmann, Michelsen, and Klein (1992) identified circulation characteristics over the western North Pacific that were associated with a 20- to 25-day period. These disturbances propagate from equatorial regions east of New Guinea northwestward toward eastern Asia. These circulations have characteristics similar to those simulated by Wang and Xie (1997) with respect to moist Rossby waves emanating from enhanced equatorial convection. Hartmann, Michelsen, and Klein (1992) found some modulation of tropical cyclone activity over the Philippine Sea during periods of active 20- to 25-day circulations.

Several studies have examined synoptic-scale (2–8 days) circulations over the western North Pacific (Lau and Lau 1990; Takayabu and Nitta 1993; Chang et al. 1996) that also propagate northwestward over the Philippine Sea. Chang et al. (1996) identified one period of enhanced northwestward-moving wave activity with enhanced tropical cyclone activity over the western North Pacific.

Although tropical cyclone activity has been related to circulation variability over a range of periods, it seems likely that a combination of circulations act to influence intraseasonal tropical cyclone activity. Maloney and Hartmann (2001) and Hartmann and Maloney (2001) examined the relationship between MJO activity and eddy kinetic energy over the Philippine Sea. They found that the increased eddy kinetic energy was related to the enhanced convective phase of the MJO. Barotropic energy conversion from the mean atmospheric flow forced by the enhanced MJO phase increased the eddy kinetic energy, and the eddy circulations then moved northwestward over the Philippine Sea. Although no direct association was made between increased eddy kinetic energy and tropical cyclone activity, their studies provide evidence of the potential influence of low-frequency circulation variability on higher-frequency circulations over regions where tropical cyclones form. The increased eddy activity they identified during the active MJO phase seems to be related to the results of Liebmann, Hendon, and Glick (1994) who identified increased numbers of depressions during the active MJO period.

Although a precursor disturbance is an important component to the formation of tropical cyclones, other physical mechanisms also contribute to their genesis. Dependence on other mechanisms is likely responsible for the observation by Liebmann, Hendon, and Glick (1994) that the proportion of typhoons to depressions during the active MJO phase is not different from the inactive phase, or that the MJO does not uniquely control tropical cyclone activity. Further research is required to identify the combination of factors that control tropical cyclones over intraseasonal time scales. Existing studies indicate that these factors have variations over a broader frequency range than just the MJO.

Seasonal Prediction of Tropical Cyclone Activity over the Western North Pacific

Because variations in tropical cyclone activity appear to be related to changes in the large-scale circulations, it should be possible to predict such variations if the circulation changes can be identified in advance. This consideration has led to the development of statistical schemes for predicting the annual tropical cyclone activity in various ocean basins (see the review in Cheung and Kyle 2000). Chan, Shi, and Lam (1998) introduced the first real-time seasonal prediction scheme for the western North Pacific. They applied the projection pursuit regression technique (which is similar to a multiple regression except that it is more robust) and derived equations to predict the annual number of tropical cyclones, tropical storms, and typhoons for the western North Pacific and the South China Sea. Predictors include monthly values of indices that represent the ENSO phenomenon and environmental conditions over East Asia and the western North Pacific from April of the previous year to March of the current year. Trends and short-term oscillations of tropical cyclone activity are also used as predictors. Although the real-time predictions for tropical cyclone activity over the western North Pacific for 1997 made by Chan, Shi, and Lam (1998) appeared to be satisfactory, those for the South China Sea were greater than observed. Chan, Shi, and Lam (1998) attributed the over-prediction to the effect of the strong 1997 El Niño event that apparently was not anticipated early enough in their prediction scheme.

In an attempt to improve the accuracy of the forecasts, Chan, Shi, and Liu (2001) re-derived their prediction equations by incorporating the results of Chan and Xu (2000) and Xu and Chan (2001) who identified precursors of El Niño and La Niña events in the large-scale flow patterns over the entire western Pacific. Chan, Shi, and Liu (2001) introduced an update scheme that generates the forecasts based on data up to May of each year (as opposed to the original forecasts that use data only up to March). Hindcasts of the 1998 and 1999 seasons produced using the improved prediction scheme gave better predictions than those from Chan, Shi, and Lam's (1998) original scheme. Because of these encouraging results, Chan, Shi, and Liu (2001) made real-time predictions for 2000, all of which turned out to be correct within the standard errors of the scheme. Similar success was also achieved for the 2001 forecasts.

Encouraged by such success, Liu and Chan (2003) attempted to develop a scheme for predicting the number of landfalling tropical cyclones along the South China coast using the same methodology. They found the most significant predictors to be indices that relate to the occurrence or non-occurrence of the ENSO event. Other indices representing the planetary-scale circulation are also used. A real-time prediction for 2001 gave reasonable results.

Thus, seasonal predictions of tropical cyclone activity over the western North Pacific appear to be possible on an operational basis. Similar schemes could likely be implemented for other locations where landfalling tropical cyclones are frequent, such as the Philippines, the island of Taiwan, and Japan.[1] Extensions of such statistical schemes to predict intraseasonal variations may also be explored.

Summary

The formation of tropical cyclones and their development in ocean basins are controlled by the large-scale flow, the thermodynamic structure of the atmosphere, and the underlying ocean conditions. Over the western North Pacific, the latter two factors are generally favorable for tropical cyclone activity (Gray 1979). Therefore, changes in the large-scale flow patterns result in concomitant changes in the location of tropical cyclone formation, the extent of cyclone development, and the track of tropical cyclones. Such changes occur over different time scales so that variations in tropical cyclone activity are observed from intraseasonal to interdecadal, or even centennial scales. Although some understanding of the physical processes involved has been achieved, much more research is necessary, especially on variations over very long time scales. Although some success has been achieved in the prediction of annual tropical cyclone activity, an improved understanding could lead to better predictions.

Acknowledgments

The author would like to dedicate this chapter to William M Gray of Colorado State University, whose original ideas on interannual variations of tropical cyclone activity in the Atlantic inspired the author to pursue similar studies for the western North Pacific. Professor Gray was also instrumental in urging the author to develop seasonal prediction schemes for this ocean basin. In addition, the author would like to thank Dr. Patrick Harr of the Naval Postgraduate School (NPS), who contributed substantially to this chapter's section on intraseasonal variations. The author benefited over the years from many discussions with him, Prof. Russell Elsberry of NPS, J.-E. Shi of the Beijing Meteorological College, and K. S. Liu of City University of Hong Kong.

The author also gratefully acknowledges the following organizations for their support of the author's research efforts in the low-frequency variations of tropical-cyclone activity in the western North Pacific through the years: the Research Grants Council of the Hong Kong Special Administrative Region Government of China, City University of Hong Kong, and the Risk Prediction Initiative of the Bermuda Biological Station for Research.

Note

1. For an example of a forecast of typhoon landfall in Japan, see <http://forecast.mssl.ucl.ac.uk/for_typh.html>.

References

Chan, J. C. L. 1985. Tropical cyclone activity in the northwest Pacific in relation to the El Niño/Southern Oscillation phenomenon. *Monthly Weather Review* 113:599–606.

Chan, J. C. L. 1995. Tropical cyclone activity in the western North Pacific in relation to the stratospheric Quasi-Biennial Oscillation. *Monthly Weather Review* 123:2567–71.

Chan, J. C. L. 2000. Tropical cyclone activity over the western North Pacific associated with El Niño and La Niña events. *Journal of Climate* 13:2960–72.

Chan, J. C. L., and J. E. Shi. 1996. Long-term trends and interannual variability in tropical cyclone activity over the western North Pacific. *Geophysical Research Letters* 23:2765–67.

Chan, J. C. L., and J. E. Shi. 2000. Frequency of typhoon landfall over Guandong Province of China during the period 1470–1931. *International Journal of Climatology* 20:183–90.

Chan, J. C. L., J. E. Shi, and C. M. Lam. 1998. Seasonal forecasting of tropical cyclone activity over the western North Pacific and the South China Sea. *Weather and Forecasting* 13:997–1004.

Chan, J. C. L., J. E. Shi, and K. S. Liu. 2001. Improvements in the seasonal forecasting of tropical cyclone activity over the western North Pacific. *Weather and Forecasting* 16:491–98.

Chan, J. C. L., and J. Xu. 2000. Physical mechanisms responsible for the transition from a warm to a cold state of the El Niño/Southern Oscillation. *Journal of Climate* 13:2056–71.

Chang, C.-P., J.-M. Chen, P. A. Harr, and L. E. Carr. 1996. Northwestward propagating wave patterns over the tropical western North Pacific during summer. *Monthly Weather Review* 124:2245–66.

Chen, T.-C., S.-P. Weng, N. Yamazaki, and S. Kiehne. 1998. Interannual variation in the tropical cyclone formation over the western North Pacific. *Monthly Weather Review* 126:1080–90.

Cheung, N. K. W., and W. J. Kyle. 2000. Trends in seasonal forecasting of tropical cyclone activity. *Australian Meteorological Magazine* 49:201–21.

Chu, P.-S., and J. D. Clark. 1999. Decadal variations of tropical cyclone activity over the central North Pacific. *Bulletin of the American Meteorological Society* 80:1875–81.

Chu, P.-S., and J. Wang. 1997. Tropical cyclone occurrences in the vicinity of Hawaii: Are the differences between El Niño and Non-El Niño years significant? *Journal of Climate* 10:2683–89.

Dong, K. 1988. El Niño and tropical cyclone frequency in the Australian region and the northwest Pacific. *Australian Meteorological Magazine* 36:219–25.

Dong, K., and G. J. Holland. 1994. A global view of the relationship between ENSO and tropical cyclone frequencies. *Acta Meteorologica Sinica* 8:19–29.

Evans, J. L. 1993. Sensitivity of tropical cyclone intensity to sea surface temperature. *Journal of Climate* 6:1133–40.

Goldenberg, S. B., C. W. Landsea, A. M. Mestas-Nuñez, and W. M. Gray, 2001. The recent increase in Atlantic hurricane activity: Causes and implications. *Science* 293:474–79.

Gray, W. M. 1979. Hurricanes: Their formation, structure and likely role in the tropical circulation. In *Meteorology over the tropical oceans*, edited by D. B. Shaw, 155–218. London: Royal Meteorological Society.

Gray, W. M. 1984. Atlantic seasonal hurricane frequency. Part I: El Niño and 30 mb Quasi-Biennial Oscillation influences. *Monthly Weather Review* 112:1649–68.

Gray, W. M., J. D. Sheaffer, and J. A. Knaff. 1992. Influence of stratospheric QBO on ENSO variability. *Journal of the Meteorological Society of Japan* 70:975–94.

Harr, P. A., and R. L. Elsberry. 1991. Tropical cyclone track characteristics as a function of large-scale circulation anomalies. *Monthly Weather Review* 119:1448–68.

Harr, P. A., and R. L. Elsberry. 1995a. Large-scale circulation variability over the tropical western North Pacific. Part I: Spatial patterns and tropical cyclone characteristics. *Monthly Weather Review* 123:1225–46.

Harr, P. A., and R. L. Elsberry. 1995b. Large-scale circulation variability over the tropical western North Pacific. Part II: Persistence and transition characteristics. *Monthly Weather Review* 123:1247–68.

Harr, P. A., R. L. Elsberry, and J. C. L. Chan. 2000. Forecasts of intraseasonal periods of tropical cyclone inactivity over the tropical western North Pacific. In *Preprints of the 24th Conference on Hurricanes and Tropical Meteorology*, 51–52. Boston: American Meteorological Society.

Hartmann, D. L., and E. D. Maloney. 2001. The Madden-Julian oscillation, barotropic dynamics, and North Pacific tropical cyclone formation. Part II: Stochastic barotropic modeling. *Journal of the Atmospheric Sciences* 58:2259–572.

Hartmann, D. L., M. L. Michelsen, and S. A. Klein. 1992. Seasonal variations of tropical intraseasonal oscillations: A 20–25 day oscillation in the western Pacific. *Journal of the Atmospheric Sciences* 49:1277–89.

Lander, M. A. 1994. An exploratory analysis of the relationship between tropical storm formation in the western North Pacific and ENSO. *Monthly Weather Review* 122:636–51.

Landsea, C. W. 2000. El Niño–Southern Oscillation and the seasonal predictability of tropical cyclones. In *El Niño and the Southern Oscillation: Multiscale variability and global and regional impacts*, edited by H. F. Díaz and V. Markgraf, 149–81. Cambridge: Cambridge University Press.

Landsea, C. W., N. Nicholls, W. M. Gray, and L. A. Avilla. 1996. Quiet early 1990s continues trend of fewer intense Atlantic hurricanes. *Geophysical Research Letters* 23:1697–700.

Lau, K.-H., and N.-C. Lau. 1990. Observed structure and propagation characteristics of tropical summertime synoptic-scale disturbances. *Monthly Weather Review* 118:1888–913.

Liebmann, B., H. Hendon, and J. D. Glick. 1994. The relationship between tropical cyclones of the western Pacific and Indian Oceans and the Madden-Julian Oscillation. *Journal of the Meteorological Society of Japan* 72:401–11.

Liu, K. S., and J. C. L. Chan. 2003. Climatological characteristics and seasonal forecasting of tropical cyclones making landfall along the South China coast. *Monthly Weather Review* 131:1650–62.

Madden, R. A., and P. R. Julian. 1994. Observations of the 40–50 day tropical oscillation—A review. *Monthly Weather Review* 122:814–37.

Maloney, E. D., and D. L. Hartmann. 2001. The Madden-Julian oscillation, barotropic dynamics, and North Pacific tropical cyclone formation. Part I: Observations. *Journal of the Atmospheric Sciences* 58:2545–71.

Nicholls, N. 1984. The Southern Oscillation, sea-surface temperature and interannual fluctuations in Australian tropical cyclone activity. *Journal of Climatology* 4:661–70.

Qiao, S. X., and W. Y. Tang. 1993. *Collection and research of climate data from historical records in the Guangzhou area* (in Chinese). Guangdong: Guangdong People's Press.

Ramage, C. S. 1971. *Monsoon meteorology*. New York: Academic Press.

Takayabu, Y. N., and T. Nitta. 1993. 3–5 day disturbances coupled with convection over the tropical Pacific Ocean. *Journal of the Meteorological Society of Japan* 71:221–46.

Wang, B. 1995. Interdecadal changes in El Niño onset in the last four decades. *Journal of Climate* 8:267–85.

Wang, B., and J. C. L. Chan. 2002. How does ENSO regulate tropical storm activity over the western North Pacific? *Journal of Climate* 15:1643–58.

Wang, B., R. Wu, and X. Fu. 2000. Pacific-East Asian teleconnection: How does ENSO affect East Asian climate? *Journal of Climate* 13:1517–36.

Wang, B., and X. Xie. 1997. A model for the boreal summer intraseasonal oscillation. *Journal of the Atmospheric Sciences* 54:72–86.

Wu, G., and N. C. Lau. 1992. A GCM simulation of the relationship between tropical storm formation and ENSO. *Monthly Weather Review* 120:958–77.

Xu, J., and J. C. L. Chan. 2001. The role of the Asian/Australian monsoon system in the onset time of El Niño events. *Journal of Climate* 14:418–33.

Xue, Z., and C. J. Neumann. 1984. *Frequency and motion of western North Pacific tropical cyclones.* Technical Memorandum, NWS NHC Report No. 23. Washington, D.C.: National Oceanic and Atmospheric Administration.

Zhang, G. Z., W. Drosdowsky, and N. Nicholls. 1990. Environmental influences on northwest Pacific tropical cyclone numbers. *Acta Meteorologica Sinica* 4:180–88.

Zhang, G., X. Zhang, and F. Wei. 1994. A study on the variations of annual frequency of tropical cyclone in northwest Pacific during the last hundred years. *Journal of Tropical Meteorology* 11:315–23.

11

ENSO and Tropical Cyclone Activity

Pao-Shin Chu

The present state of knowledge regarding tropical cyclone activity in various ocean basins and the El Niño–Southern Oscillation phenomenon is reviewed in this chapter. The ocean basins include the western North Pacific, the eastern and central North Pacific, the southwestern Pacific, the southeastern Pacific, and the North Atlantic. Following a description of the ENSO phenomenon, tropical cyclone activity in each basin is discussed in the context of frequency, genesis location, track, life span, and intensity.

For the western North Pacific, the pronounced change in tropical cyclone activity due to warm ENSO is the eastward and equatorial shift in genesis location, longer life span, and more recurvature of tropical cyclone tracks (Chan, chapter 10 in this volume). There is also a notable decrease in tropical cyclone counts in the year following a warm ENSO event. For the eastern North Pacific, the formation point shifts farther west, more intense hurricanes are observed, and tropical cyclones track farther westward and maintain a longer lifetime in association with warm ENSO events. The central North Pacific sees more tropical cyclone counts in the El Niño year due to more tropical cyclone formation in this region and a tendency for tropical cyclones that originate in the eastern North Pacific to enter the central North Pacific. As in the North Pacific, tropical cyclones in the South Pacific originate farther east during El Niño years, resulting in more storms in the southeastern Pacific and fewer storms in the southwestern Pacific. The North Atlantic features fewer tropical cyclone counts, slightly weaker intense storms, and hurricane genesis farther north during El Niño years. Changes are approximately opposite in cold ENSO years.

Interest in the relationship between tropical cyclone activity in various ocean basins and the El Niño–Southern Oscillation (ENSO) phenomenon has grown over the last two decades. This interest is drawn from the fact that large-scale environmental conditions conducive to tropical cyclone activity (e.g., formation, track, frequency, life span, landfall, and/or intensity) during El Niño years differ profoundly from those of climatological or La Niña years. The El Niño phenomenon is manifested in the anomalous warming of the eastern and central tropical Pacific. La Niña refers to anomalous cooling of the tropical Pacific, or simply the opposite of El Niño. Because ENSO, as first recognized by Bjerknes (1969), is a powerful interplay between the tropical ocean and atmosphere in the Pacific Basin and because tropical cyclones form mainly in the tropics, modulation of tropical cyclone activity by ENSO is expected.

There is an extensive body of literature relating seasonal tropical cyclone activity in various ocean basins to ENSO. Gray (1984) and Gray and Sheaffer (1991) ascribed Atlantic seasonal hurricane frequency to El Niño and the Quasi-Biennial Oscillation of stratospheric wind. Through statistical analyses, Shapiro (1987) and Goldenberg and Shapiro (1996) further confirmed the dependence of tropical cyclone formation in the Atlantic on ENSO. For the western and eastern North Pacific, Chan (1985, 2000), Lander (1994), Chen et al. (1998), and Irwin and Davis (1999) noted a shift in tropical cyclone genesis location during El Niño and La Niña phases. Chu and Wang (1997), using actual tropical cyclone observations and statistical resampling techniques, found more tropical cyclone occurrences in the vicinity of Hawai'i in the central North Pacific when El Niño occurred as compared to non–El Niño years. Tropical cyclone activity in the South Pacific is likewise influenced by ENSO (e.g., Nicholls 1979, 1985; Sadler 1983; Revell and Goutler 1986; Hastings 1990; Basher and Zheng 1995; McBride 1995). More recently, Landsea (2000) reviewed the relationship between tropical cyclone activity and ENSO but focused on the Atlantic Ocean. In particular, he addressed the issue of seasonal predictability of tropical cyclones over the Atlantic Ocean.

This chapter first describes the ENSO phenomenon before discussing tropical cyclone activity in several ocean basins in relation to ENSO. The ocean basins include the western North Pacific and the South China Sea, the eastern and central North Pacific, the South Pacific, and the North Atlantic (figure 11.1). Because the ENSO signal in the Indian Ocean is weak, ENSO influences on tropical cyclone activity in the North and South Indian Oceans are not considered in this chapter but further investigations may discover significant correlations (but see Jury 1993).

General Descriptions of ENSO

Originally, the name El Niño (Spanish for the Christ child) was given to a weak coastal current that flows southward along the coast of Ecuador and Peru

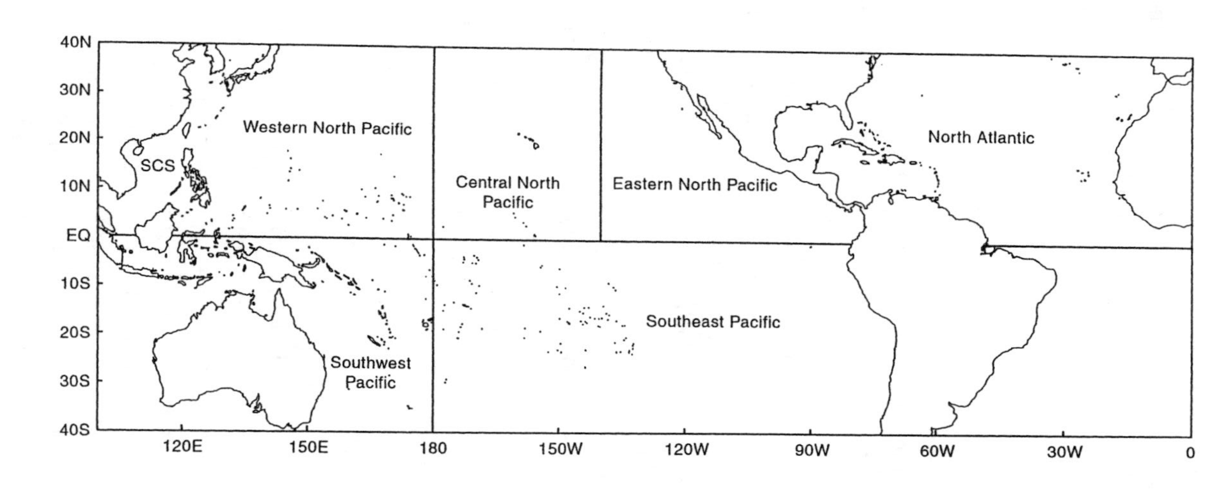

FIGURE 11.1 Orientation map of the western North Pacific and the South China Sea (denoted as the SCS), the eastern and central North Pacific, the southwestern Pacific, the southeastern Pacific, and the North Atlantic Basin.

around Christmastime. This occurs during the austral summer, when local winds are weak and the upwelling of cold waters that carry the primary food source for fish is reduced (Bjerknes 1969; Wyrtki 1975). As a result, the ocean surface along the west coast of tropical South America becomes anomalously warm. In some years the upwelling terminates abruptly, the ocean surface warms extensively, and fish starve from the lack of foods nourished by the nutrient-rich, upwelled cold water. These conditions are disastrous for the fishery industry and local economy in Ecuador and Peru.

In broad terms, the La Niña (Spanish for "the girl") phenomenon can be regarded as the opposite of the El Niño condition. During La Niña, easterly trade winds are strong and persistent, blowing from a region of high pressure over the southeastern Pacific toward a region of low pressure in the western Pacific where warm pools of water with light winds and convection prevail (e.g., Deser and Wallace 1990). Because surface winds in the tropics mainly follow the pressure gradient, easterly winds prevail. The easterly winds not only induce equatorial Ekman upwelling because of the Coriolis effect, creating a cold tongue in the equatorial eastern to central Pacific, but they also raise sea level in the west and lower it in the east. Thus, a west-east sea level slope occurs across the Pacific (figure 11.2A). The thermocline, an interface separating warm and relatively low-density water in the upper ocean from cold, high-density water in deep ocean, is relatively deep in the western Pacific (~ 220 m) but shallow off the west coast of South America (~30 m). The zonal difference in thermocline depth results in equatorial ocean dynamics that play a key role in ENSO perturbations (Cane and Zebiak 1985).

In the atmosphere, a zonal circulation along the equatorial Pacific occurs with rising air over the warm Indonesian region and sinking air over the cold eastern Pacific. The vertical air movement is connected by easterlies in the lower troposphere and westerlies in the upper troposphere. The zonal sea surface temperature gradient with cold water in the east and warm water in the west is considered the cause of this thermally driven direct circulation. Bjerknes (1969) named this circulation cell the Walker circulation. There is a positive feedback between the atmosphere and ocean in the tropics because surface winds drive ocean currents, and these currents redistribute surface thermal gradients that affect wind fields through hydrostatic effects in surface pressures.

The pioneering work by Bjerknes (1969) and Wyrtki (1975) laid the foundation for numerous theoretical and modeling studies pertinent to the ENSO phenomenon (e.g., McCreary 1983; Zebiak and Cane 1987). In a dynamical framework, the fundamental roles in the development of El Niño are played by oceanic Kelvin waves in the equatorial waveguide, off-equatorial Rossby waves, and reflections by the western boundary of the tropical Pacific (Schopf and

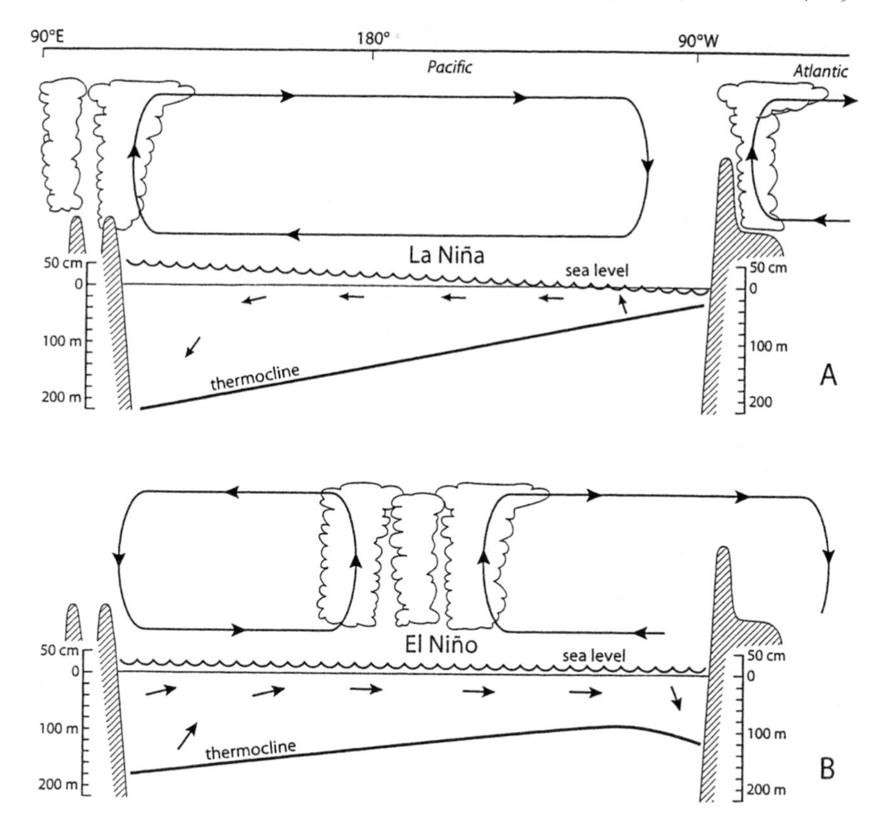

FIGURE 11.2 Schematics showing the near-equatorial atmosphere and ocean circulations in the Pacific and western Atlantic associated with the (*A*) La Niña and (*B*) El Niño conditions. During El Niño, major convection shifts eastward to the central Pacific, with subsidence over the western Pacific and the western Atlantic. During El Niño, oceanic currents flow eastward, the thermocline deepens along the South American coast, and sea level drops and the thermocline rises in the western Pacific (adapted from Wyrtki 1982).

Suarez 1988; Battisti and Hirst 1989). These phenomena interact and produce changes in sea-surface temperatures, the thermocline, and sea level. Note that Kelvin waves are equatorially trapped waves that propagate rapidly eastward. In contrast, Rossby waves propagate to the west relative to the zonal mean flow.

Wyrtki (1975) recognized from an analysis of observations that prior to an El Niño event there is a buildup of warm water between sea level and the thermocline in the western Pacific warm pool. As soon as easterly trade winds relax, the accumulated warm water flows eastward in the form of Kelvin waves to give rise to an El Niño event. The oceanic Kelvin waves, excited by episodic westerly wind bursts in the western and central Pacific, propagate rapidly eastward across the equatorial Pacific in a period of two to three months. These waves are responsible for the deepening of the thermocline in the eastern Pacific (figure 11.2B). Consequently, cold water from below cannot be upwelled efficiently and the normal steep slope of the transbasin thermocline levels off (e.g., Lukas, Hayes, and Wyrtki 1984). The eastward advection of warm water caused by changes in the zonal component of the surface winds in the western Pacific that trail the Kelvin waves, and a deepened thermocline induced by downwelling Kelvin waves cause sea-surface temperatures in the equatorial central to eastern Pacific to become anomalously warm. The wind anomalies also generate westward propagating Rossby waves in the off-equatorial Pacific that reflect at the western boundary and return as delayed Kelvin waves. In analyzing daily wind observations for a 30-year period, Chu, Frederick, and Nash (1991) noted an increased frequency of westerlies in the equatorial western Pacific during an El Niño period.

At the height of El Niño events, warm pools of seawater, low-level westerlies, and the attendant tropical convection shift from the western Pacific to the central or eastern Pacific (figure 11.2B). Accordingly, the rising branch of the Walker circulation is located in the equatorial central Pacific and the sinking branch is found over the western Pacific. Note that anomalies of large-scale flows are implied. As will be described later, these changes in atmospheric and oceanic circulation patterns during El Niño have profound impacts on regional tropical cyclone activity. The monsoon trough is regarded as the breeding ground of tropical cyclones and normally occurs near large land masses in the western Pacific where the monsoonal effects are pronounced. This trough is marked by a low-level wind shear line, with monsoon westerlies on its equatorward side and easterly trade winds on its poleward side. During an El Niño the monsoon trough is displaced to the east (Lander 1994; Clark and Chu 2002).

Barnston, Chelliah, and Goldenberg (1997) determined that the El Niño phenomenon is most reliably reflected in the equatorial sea-surface temperature from approximately 120°W westward to near the date line. As a result, one of the most popular indices used to monitor El Niño is the areally averaged sea-surface temperature in the Niño 3.4 region. The Niño 3.4 region covers an area between 5°N to 5°S and 170°W to 120°W and is close to the Pacific warm pool and the major center of convection during El Niño (figure 11.2B). Trenberth (1997) defined El Niño and La Niña events on the basis of the 5-month running mean of sea-surface temperature anomalies in the Niño 3.4 region exceeding positive and negative 0.4°C, respectively, for at least 6 consecutive months. Using this definition, El Niño events have been as short as 7 months (1951–1952) and as long as 19 months (1986–1988). Since 1950, on average, an event starts in May or June and ends in the following April. The average length of an event is about 11.8 months, or almost one year. For La Niña, the start and end months are similar to El Niño, but the average duration is 13.3 months, a little longer than El Niño. Furthermore, a neutral state of Pacific sea-surface temperatures occurs 45% of the time when El Niño or La Niña conditions are absent (Trenberth 1997).

Each El Niño has its own characteristic onset and demise time, duration, magnitude, exact place of maximum warming, phase propagation, and so on. Although each episode behaves differently, there is a tendency for the maximum amplitude of major events to occur near the end of the calendar year so it is phase-locked to the annual cycle. The El Niño and La Niña phases are also known to change preferentially around March-April when surface winds are weak, sea-surface temperatures in the equatorial cold tongue are warm, and the east-west sea-surface temperature gradient along the equatorial Pacific is slack (e.g., Lee et al. 1998). This is the time when atmosphere and ocean coupling is weakest. Prior to an El Niño, relatively large amounts of anomalously warm water accumulates between the sea level and the thermocline in the western Pacific warm pool (Wyrtki 1975, 1985). The excess heat content in this warm water is then discharged toward off-equatorial regions and into the atmosphere during El Niño. It has been suggested that the discharge and recharge of heat contained in the equatorial water is the key ingredient that controls the transition between El Niño and La Niña phases (Wyrtki 1975, 1985; Cane and Zebiak 1985; Jin 1997).

Changes in atmospheric pressure patterns between centers in the Pacific and Indian oceans are associated with changes in sea-surface temperatures and oceanic heat contents during El Niño. As pressures fall in the eastern South Pacific subtropical high, they tend to rise in the Indonesian low-pressure zone. The term "Southern Oscillation" (SO) was coined to describe the zonal, atmospheric mass exchanges across two southern oceans. To monitor the behavior of such a large-scale atmospheric circulation, the Southern Oscillation Index (SOI) is used. The standard SOI used by many operational weather and climate agencies and researchers throughout the world is derived from a difference in normalized sea level pressures between Tahiti and Darwin, Australia. Because the Southern Oscillation is closely linked with El Niño, both events are labeled collectively ENSO. Although not periodic, the ENSO phenomenon generally recurs every three to four years, but the time between past events has ranged between two and seven years (Trenberth 1976). Using a time-domain approach, Chu and Katz (1989) independently found a dominant spectral peak between three and four years in the SOI series.

Because ENSO has an immense impact on tropical cyclone activity, real-time ENSO forecasting has been performed by numerous researchers, institutes, and national meteorological centers around the world. In general, dynamical or statistical models are used in the forecasting enterprise, and these forecast results have been published routinely in the Experimental Long-Lead Forecast Bulletin (e.g., Kirtman 2001). For dynamical models, the degree of complexity varies from simple, linear shallow-water equations for both ocean and atmosphere, to intermediate coupled ocean-atmosphere models, to hybrid coupled models (e.g., statistical atmosphere and comprehensive ocean circulation), to fully coupled ocean-atmosphere models with multiple vertical layers. In comparison to dynamical models, statistical models are more simple and use less computer time and storage space.

By considering the persistence of initial conditions, trend, and climatology of past ENSO events, Knaff and Landsea (1997) developed a multiple regression model to forecast Niño 3 and Niño 3.4 region sea-surface temperature anomalies. They called it an ENSO-CLIPER model. The idea of the ENSO-CLIPER model is derived from the tropical cyclone community in which a simple CLIPER (i.e., Climatology plus Persistence) scheme has long been used as a benchmark against other more sophisticated models for storm-track prediction (e.g., Neumann 1977). Landsea and Knaff (2000) compared forecast skills of their ENSO-CLIPER baseline system with other dynamical and statistical models for the very strong 1997–1998 El Niño event. They noted that at short lead time (up to eight months ahead), the ENSO-CLIPER has the smallest root-mean square error among all models tested. This result is rather intriguing because forecasts made by the ENSO-CLIPER are considered as having no-skill. If forecasts for the 1997–1998 ENSO event are representative of other cases, then there is very little or no skill in ENSO prediction, despite the great efforts to develop sophisticated numerical models.

Throughout this chapter, the terms "El Niño," "warm ENSO phase," and "warm phase" are used interchangeably. Likewise, the terms "La Niña," "cold ENSO phase," and "cold phase" are used interchangeably. The term "neutral phase" describes conditions when sea-surface temperatures are near climatological averages. Note that a large and negative SOI lasting several months generally corresponds to a warm ENSO phase, and a large and positive SOI persisting several months is indicative of a cold phase. As a cautionary note, it should be mentioned that some abrupt changes occur occasionally in the monthly SOI (Chu and Katz 1985). For instance, the monthly SOI may fluctuate from a positive value in one month to a negative value in a second month, followed by a positive value in a third month, or vice versa. The Tropical Intraseasonal Oscillation (Madden and Julian 1971), which has a typical time scale of 30 to 50 days, may perturb the large-scale mass circulation in the SO regime on a short-term basis. As a result, the monthly SOI is occasionally contaminated by some transient circulation features (e.g., mid-latitude troughs and ridges during the austral winter) that are not inherent in the large-scale SO. In this regard, sea-surface temperature in the equatorial Pacific may serve as a more robust indicator of the state of ENSO phase than the SOI because of the well-known slow change in thermal content and large heat capacity of tropical oceans.

The Western North Pacific

Climatologically, tropical cyclone frequency (i.e., tropical storms and typhoons) in the western North Pacific is higher than in any other ocean basin, with an annual mean value of 26, based on 22-year statistics from 1968 to 1989 (Neumann 1993). The standard deviation of annual tropical cyclone counts is 4.1. The western North Pacific is also the only basin where tropical cyclone formation is observed throughout the 12 months of the year, although a majority of cyclones develop between June and November (Frank 1987).

Studies by Chan (1985, 2000, chapter 10 in this volume), Dong (1988), Lander (1994), Chen et al. (1998), and Kimberlain (1999) establish a relationship between tropical cyclone activity in the western North Pacific and El Niño and La Niña phenomena. Through spectral and cross-spectral analyses, Chan (1985) found that both the SOI and typhoon count series possess a dominant peak in the frequency band of 3 to 3.5 years and that the SOI leads typhoon series by about one year in this band. That is, a large and negative SOI (i.e., a warm ENSO phase) tends to be followed by an overall reduction in tropical cyclone frequency over the western North Pacific in the following year, and vice versa. In addition, Chan (1985) and Dong (1988) noted that tropical cyclone genesis location shifts eastward across 150°E in the western Pacific during warm ENSO years (figure 11.1). Therefore, more typhoons and tropical storms occurred in the eastern part (150°E to the date line) than the western part of the western North Pacific (120°E–150°E) during an El Niño event.

Lander (1994), however, only found a weak correlation between annual tropical cyclone counts in the entire western North Pacific Basin and ENSO, but he concurred with Chan and Dong in terms of the eastward displacement of the genesis location during an El Niño and the westward retreat in the genesis location during La Niña. This zonal displacement is intimately related to the low-level monsoon trough where its mean over-water position in August stretches from southeast to northwest over the Philippine Sea and southern Taiwan (figure 11.3). The monsoon trough is marked by moist, southwest monsoon flows to the south and easterly trades to the north of the trough. Tropical disturbances are often found in the trough where there is a weak cyclonic rotation. As the cyclonic spin in the trough increases, these systems tend to inten-

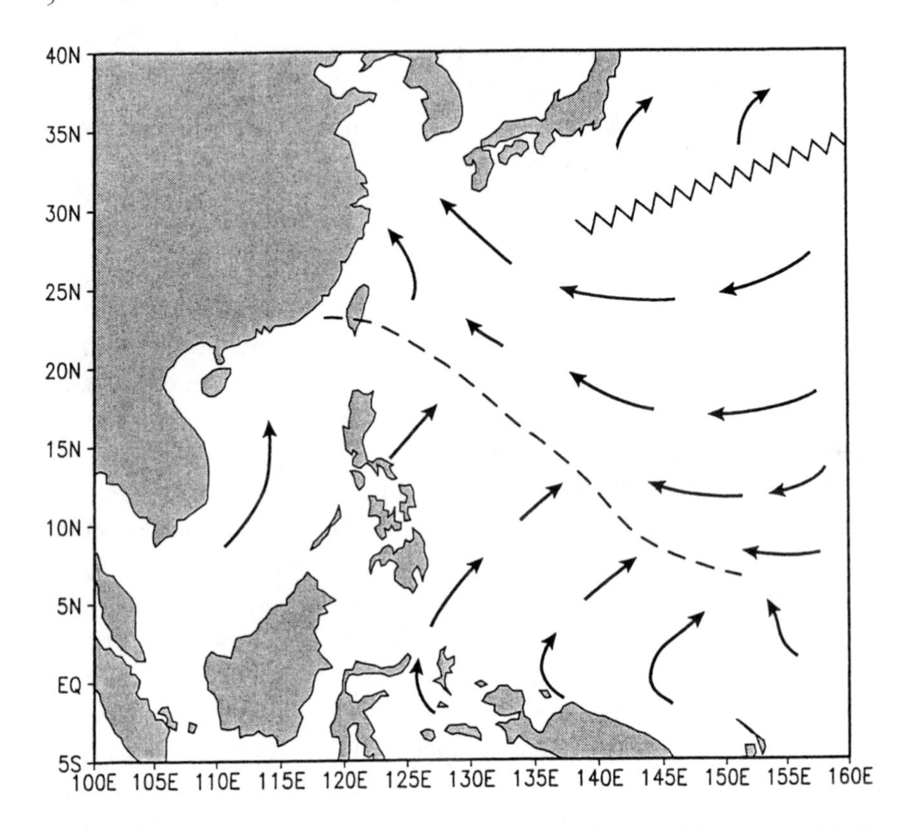

FIGURE 11.3 Schematic showing the long-term mean surface circulation in August in the western North Pacific. The monsoon trough axis is denoted by a broken line, and the ridge axis by a zigzag line. Wind directions are indicated by arrows.

sify into tropical storms or typhoons (Sadler 1967). It should be kept in mind that the position of the trough in figure 11.3 is only meant to represent the long-term mean condition. In any given summer month, this trough may deviate substantially from its mean position. For instance, at times the monsoon trough extends in an elongated east-west direction from the Philippine Sea to the date line, being reversed from its mean position (figure 11.3); or it is not identifiable at all (Lander 1996).

More recently, Chan (2000, chapter 10 in this volume) stratified tropical cyclone frequency month by month according to the ENSO cycle. During an El Niño year, tropical cyclone activity over the South China Sea is below normal in September and October but above normal in the eastern portion of western North Pacific. For the year immediately following an El Niño event, tropical cyclone formation is below normal over the entire western North Pacific Basin (Chan 2000). Conceivably, because of the eastward shift in the major convection and the rising branch of the Pacific Walker cell during El Niño, the western North Pacific sees a compensating subsidence that would be unfavorable for tropical cyclone formation. The cooler ocean surfaces and higher sea level pressures in the western Pacific that characteristically occur during El Niño years (Rasmusson and Carpenter 1982; Deser and Wallace 1990) may also contribute to the reduction of tropical cyclone frequency. For the La Niña composite, tropical cyclone frequency over the western North Pacific and South China Sea varies inversely to that during El Niño years.

The trough and monsoon westerlies extend eastward, in some years beyond the date line during a warm ENSO event; therefore it is no surprise that the genesis location of tropical cyclones shifts eastward. This is seen in figure 11.4, which portrays the easternmost location of the trough in boreal summer and autumn for each year. The monsoon trough is displaced farther eastward in autumn than in summer, which is consistent with the results described by Lander (1994) and Chen et al. (1998). In the years 1972, 1982, 1994, and 1997, the autumn mean location of the monsoon trough extends eastward past the date line into the central North Pacific (figures 11.4 and 11.1).

To illustrate the influence of monsoon trough on tropical cyclone development, figure 11.5 shows the genesis location of tropical cyclones during the six warmest and six coldest ENSO years using data from 1970 to 2000, a period when estimations of typhoon counts are thought to be more reliable (Kimberlain 2000). Here only tropical cyclones that reached at least tropical storm stage are considered. Typhoon season (June to November) means of the Nino 3.4 region sea-surface temperature anomalies were calculated and utilized to define extreme years. Out of the last 31 years, the six warmest ENSO years for which the mean sea-surface temperature during the typhoon season are highest include 1972, 1982, 1986, 1987, 1991, and 1997. The six coldest ENSO years, when the mean sea-surface temperatures during the typhoon season are lowest, are 1970, 1973, 1975, 1988, 1998, and 1999. The early season refers to April through June, peak season runs from July through October, and late season goes from November to December.

Relative to the La Niña samples, the eastward shift in genesis locations is more pronounced in early and late seasons during El Niño years (figure 11.5). For instance, in the early season, most formation points are in the Philippine Sea during the La Niña years, but during El Niño years one third of the formation points lies east of 150°E. In the late season, origin points during El Niño years can be found as far east as 170°E to 175°E in the Marshall Islands. For the La Niña composite, tropical cyclone origin points are confined to the west of 140°E in early season and to the west of 150°E in late season. There is also a tendency for tropical cyclones to form closer to the equator during El Niño years as compared to La Niña years, and this meridional shift is particularly clear in peak and late seasons. During La Niña years, tropical cyclone genesis

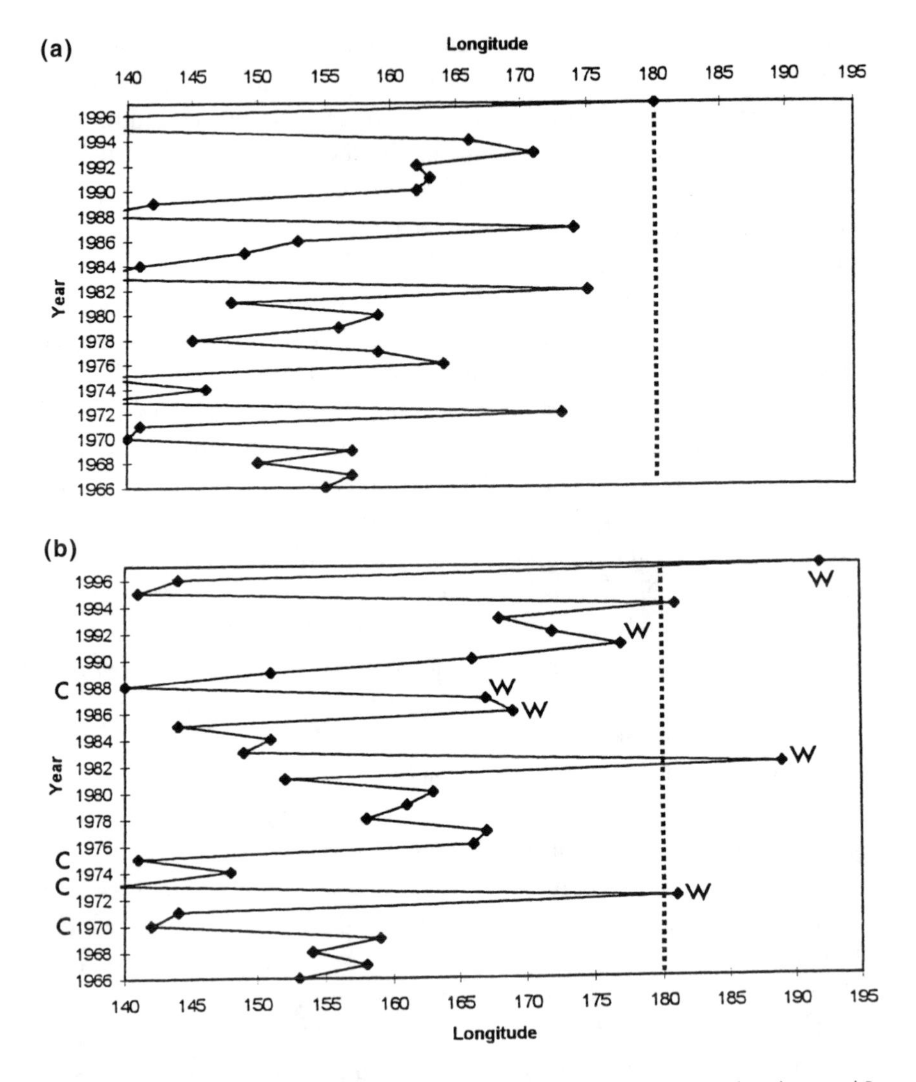

FIGURE 11.4 Time series of the farthest eastward extent of the monsoon trough at the 1,000 hPa level in the western and central North Pacific for the years 1966 to 1997 for (*a*) boreal summer (June–August) and (*b*) boreal autumn (September–November) means. The date line is indicated as a broken line. In (*b*), Ws denote six warmest ENSO years and Cs denote four out of six coldest ENSO years. Years refer to the period of June to November. The year 1994 was also a warm ENSO year, but the SST anomalies averaged during the typhoon season in that year do not qualify for the top six warmest years.

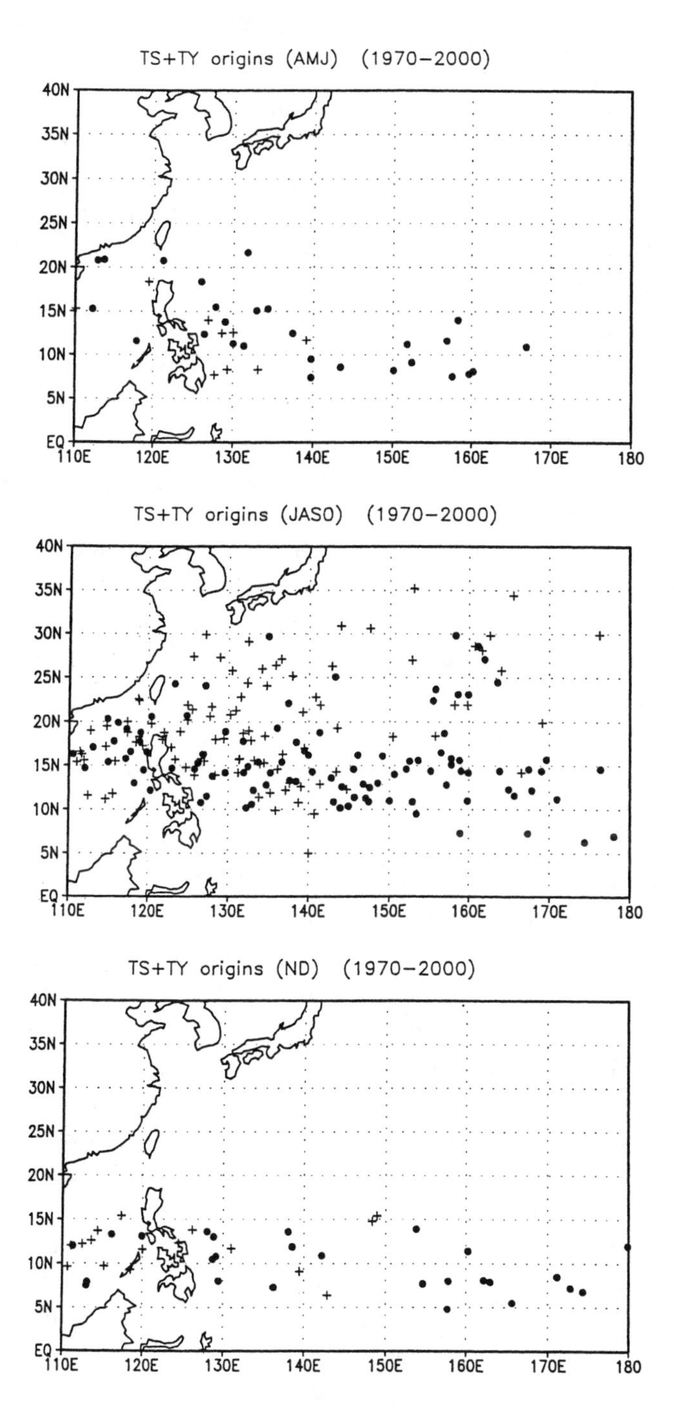

FIGURE 11.5 Origin points of tropical cyclones (tropical storms and typhoons) in the western North Pacific by season for six years during which the June through November mean sea-surface temperatures in the Niño 3.4 region are highest (represented by dots) and lowest (represented by crosses). Origin points refer to first tropical storm intensity location. The period used is 1970 to 2000. Early season refers to April to June (AMJ); peak season to July to October (JASO); and late season to November and December (ND).

locations in peak season stay at higher latitudes (20°N–30°N) over the western extreme of the North Pacific, being closer to the East Asian landmass. In the late season, most genesis points are found approximately equatorward of 15°N, regardless of warm or cold ENSO phases. In the South China Sea, more named storms formed in the early season of El Niño years but slightly more tropical cyclones formed in the late season during La Niña years.

An important environmental factor that modulates seasonal tropical cyclone activity is the vertical wind shear (Gray 1977). When tropical cyclones move into an area of strong vertical shear, the low-level center loses its upper level outflow channel and usually dissipates quickly (Gray 1968). Strong vertical shear also disrupts the organization of deep convection around the low-level center, which inhibits intensification of the incipient disturbance. Tropospheric vertical wind shear is hereafter defined as the magnitude of the difference between the zonal and meridional wind at 200 hPa and 850 hPa. Clark and Chu (2002) demonstrated a substantial reduction in vertical shear equatorward of 18°N over the eastern portion of the western North Pacific during an El Niño composite compared to the La Niña.

The eastward and equatorward shift in origin locations during El Niño years allow tropical cyclones to maintain a longer life span while tracking westward over open water. During La Niña years such as 1970, 1973, 1975, and 1988, when easterly winds prevailed in the western Pacific, the monsoon trough is short and confined in the western extreme of the North Pacific (figure 11.4). Accordingly, the genesis location is farther to the west and north. Being closer to the East Asia continent, cyclones that spawn along the monsoon trough are either on a collision course with land masses or being steered by a migrating upper-level trough away from the continent toward mid-latitude oceans. Once tropical cyclones move over a large land mass or over cold water, they lose intensity rapidly because the warm and moist air in a tropical cyclone is being cut off and the release of latent heat is greatly diminished. Furthermore, there is a substantial reduction in typhoon wind speeds over land because of increased surface roughness. Therefore, tropical cyclones during La Niña years will not be able to survive for as long as those mainly over lower latitude water during El Niño years. The number of named storm days and typhoon days in the western North Pacific Basin during El Niño years is nearly 1.5 times as large as that during La Niña years (Kimberlain 1999). Kimberlain (1999) applied a two-sample *t*-test and noted that the difference in tropical cyclone longevity between warm and cold ENSO years is statistically significant at the 5% level (a p-value of 0.01).

In addition to tropical cyclone frequency, genesis location, and longevity described previously, the tropical cyclone track in the western North Pacific varies considerably between El Niño and La Niña years. During peak season, tropical cyclones tend to recurve during El Niño years but they track farther northward after being formed at higher latitudes during La Niña years (Kimberlain 1999; Wang and Chan 2001) (figure 11.5). During the El Niño summer, the mid-tropospheric western Pacific subtropical high shifts eastward and upper-level troughs tend to deepen along the east Asian coast (Wang and Chan 2002). Accordingly, tropical cyclones from the western Pacific are likely recurved by upper-level troughs. Furthermore, because tropical cyclones during El Niño years have longer life spans, they have a better chance to interact with transient midlatitude synoptic systems, resulting in more recurved trajectories. In the late season, tropical cyclones continue to recurve during El Niño years, but during La Niña years they tend to move westward around the southern flank of the elongated subtropical high toward the Philippines and the South China Sea. In assessing the relative importance of typhoon landfalls (not origin points) associated with two contrasting climatic events, Saunders et al. (2000) found that typhoon impacts in Japan, South Korea, Taiwan, and China are more pronounced during El Niño than La Niña years. That is, a higher frequency of typhoon landfalls is observed in those countries during El Niño years. Given a longer lifespan and a tendency for recurved tracks of tropical cyclones during El Niño years, this result is not unexpected. Conversely, typhoon landfalls become more common in the northern Philippines and the South China Sea during La Niña years, a result consistent with Chan (2000) and Wu, Chang, and Leung (2004).

The Eastern and Central North Pacific

The average annual tropical cyclone number in the eastern and central North Pacific is 17, the second highest among the seven ocean basins (Neumann 1993). The standard deviation of the annual number of tropical cyclones is 4.1, the same as the western North Pacific. Given the smaller mean annual tropical cyclone counts in the eastern and central North Pacific, the same standard deviation in the two basins implies that there are larger interannual variations in tropical cyclone frequency in this basin. Unlike those in the western North Pacific, tropical cyclones in the eastern North Pacific do not occur in the cool season; the official hurricane season defined by the National Weather Service extends from May 15 to November 30 (OFCM 1999). A majority of the tropical cyclones form between the Mexican coast and Clipperton Island (~10°N, 110°W) and between 10°N and 15°N along the axis of the monsoon trough (figure 11.6).

In boreal summer, the strong southeast trades from the South Pacific cross the equator and turn into southwest currents in the eastern North Pacific. The

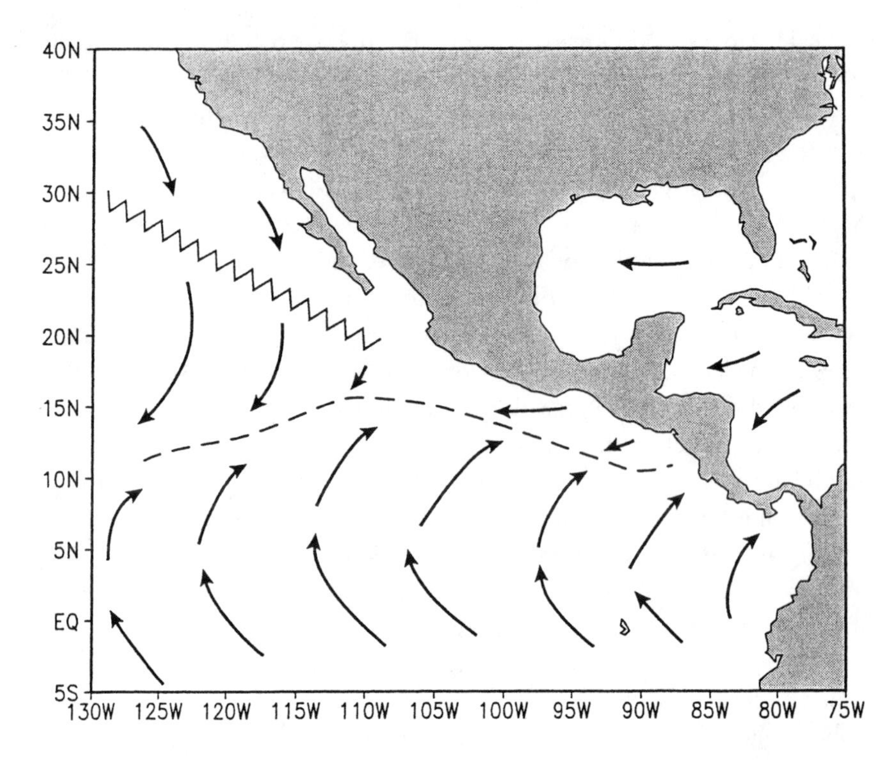

FIGURE 11.6 Schematic showing the long-term mean surface circulation in August in the eastern North Pacific. The monsoon trough axis is denoted by a broken line, and the ridge axis by a zigzag line. Wind directions are indicated by arrows.

low-latitude southwest monsoons meet the trade winds in the subtropics and form the monsoon trough where the sea-surface temperature is warm (≥ 28°C) and the vertical wind shear is weak. Tropical cyclones in the eastern North Pacific might also be triggered by tropical easterly waves from the North Atlantic (e.g., Rappaport et al. 1998). When tropical cyclones were active in the eastern North Pacific, they tended to be inactive over the North Atlantic and vice versa (Elsner and Kara 1999). Once formed, tropical cyclones generally track northwestward over the cooler water of the North Pacific and lose their strength gradually. Some tropical cyclones will occasionally curve northeastward and strike Mexico with lingering effects such as heavy rain and flooding in the southwest United States.

Whitney and Hobgood (1997) failed to find an ENSO impact on the overall tropical cyclone frequency in the eastern North Pacific. For instance, the average annual number of tropical cyclones (named storms) is 15.1 during El Niño but 15.0 for non–El Niño, based on records from 1963 to 1993. If only intense hurricanes (i.e., category three or above on the Saffir-Simpson hurricane disaster potential scale) for the last 30 years are considered, however, the ratio of intense hurricanes during warm to cold years is about 1.7. Note that the intense hurricane comparison uses six warm and cold ENSO years described in the previous section on the western North Pacific. This result is consistent with that of Gray and Sheaffer (1991), who found that the number of intense hurricanes (i.e., wind speeds of at least 50 m s^{-1}) during El Niño years increases by a factor of two compared to that during La Niña years. It is not yet clear what physical mechanisms are responsible for a higher number of intense hurricanes during El Niño years. Collins and Mason (2000), however, pointed out the need to study the eastern North Pacific by subregions because environmental parameters affecting tropical cyclone activity are different east and west of 116°W.

Although the overall cyclone frequency over the eastern North Pacific did not change appreciably during two opposite extreme climatic events, the cyclone track and its longevity have changed. In analyzing tropical cyclone tracks for warmest and coldest ENSO events, Schroeder and Yu (1995) and Kimberlain (1999) noted a westward expansion of tropical cyclone tracks during warm events and eastward retreat during cold events. Interestingly, the genesis location also appears to be changed from warm to cold events. According to Irwin and Davis (1999), the mean longitude of tropical cyclone origin points during the storm season shifted 5.7° west in the negative SOI phase (El Niño) relative to the positive phase (La Niña). During the positive SOI phase, tropical cyclones are more likely to form near the Mexican coast. Kimberlain (1999) also suggested that tropical cyclone lifetimes in the eastern and central North Pacific are longer during El Niño years relative to La Niña years, and this difference is statistically significant (a p-value of 0.02).

The central North Pacific covers an area between the date line and 140°W and north of the equator (figure 11.1). This domain coincides with the area of responsibility for the Central Pacific Hurricane Center, an entity of the U.S. National Weather Service Forecast Office in Honolulu, Hawaii. Tropical cyclone counts include storms that form within the domain of the central North Pacific as well as storms that form in the eastern North Pacific and subsequently propagate into the central North Pacific. In this regard, two types of tropical cyclones (i.e., propagated from the east and formed in situ) appear in the central North Pacific.

Time series of the annual number of tropical cyclone in the central North Pacific are displayed in figure 11.7. One notable feature is a tendency for a relative maximum of tropical cyclone occurrences during some of the El Niño years (e.g., 1972, 1982, and 1997). There is also an indication of decadal variations with fewer cyclones from 1966 to 1981 and more from 1982 to 1994 (Chu

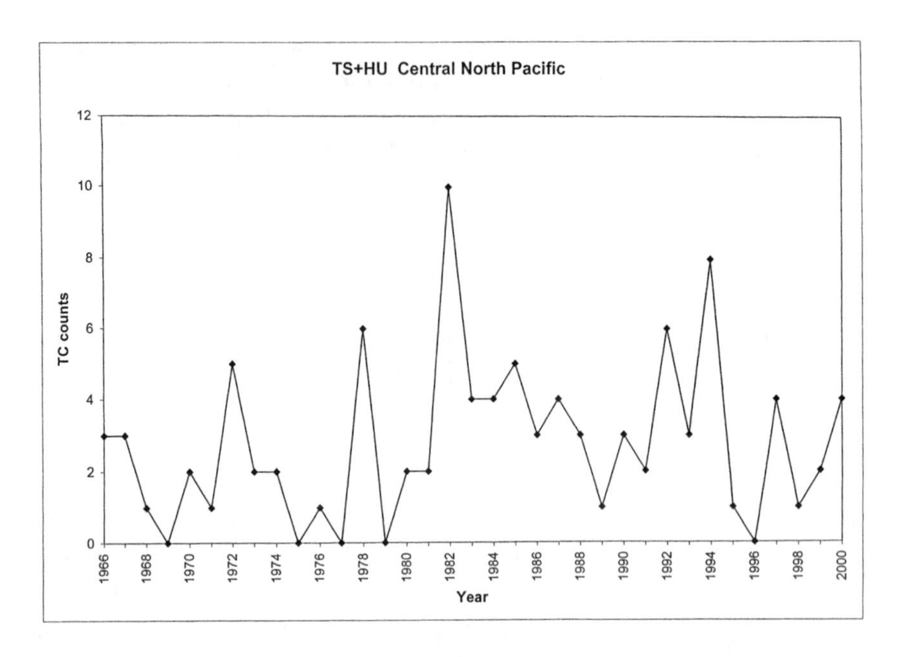

FIGURE 11.7 Time series of annual tropical cyclone numbers in the central North Pacific for the years 1966 through 2000. Only tropical storms and hurricanes are included.

and Clark 1999; Chu 2002). Moreover, the Quasi-Biennial Oscillation is also evident, particularly in the late 1970s and early 1990s. Although this is a simple time series, it reflects a multitude of various climate forcings on tropical cyclone activity. The ENSO influences on tropical cyclone frequency in the central North Pacific are further corroborated by the strong correlation coefficient between the SOI and tropical cyclone counts. The Pearson correlation between these two variables from 1966 to 1997 is –0.53, which is significant at the 1% level after climatological persistence is taken into account. Based on general circulation model simulations forced with observed monthly sea-surface temperatures in the tropical Pacific Ocean, Wu and Lau (1992) found that tropical storms in the central North Pacific form more often during El Niño events. Even if the focus is restricted to a smaller region near Hawai'i, the difference in the annual mean number of tropical cyclones between the El Niño and non–El Niño years is still statistically significant at the 5% level based on a two-sample permutation test (Chu and Wang 1997). It is surmised that tropical cyclone frequency during La Niña years is reduced relative to El Niño years.

To illustrate the difference in large-scale environmental conditions conducive to tropical cyclone development between extreme climatic events, figure 11.8 shows the low-level vorticity field in July-September for the El Niño and La Niña composites. This is the peak tropical cyclone season in the central

North Pacific. The vorticity data at the 1,000 hPa level are obtained from the NCEP/NCAR Reanalysis Project (Kalnay et al. 1996). The band of cyclonic (positive) relative vorticity in the El Niño composite is two to three times greater in the broad region from 150°E to 165°W to the south of Hawai'i when compared to the La Niña composite. This increase in cyclonic vorticity is mainly attributed to the eastward extension of the monsoon trough during El Niño years (figure 11.4). When coupled with other favorable environmental conditions such as a decrease in vertical wind shear (not shown) and a possible enhancement in moist layer depth due to boundary layer moisture convergence by the spin-up process, this increase in low-level cyclonic vorticity accounts for more tropical cyclone formation in the central North Pacific.

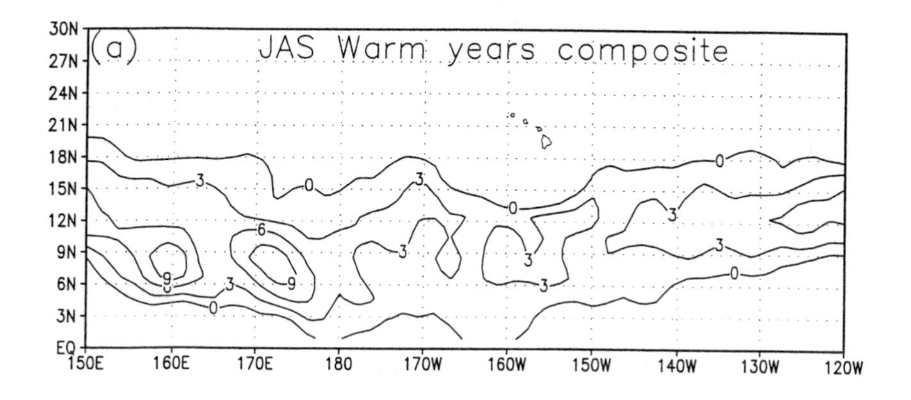

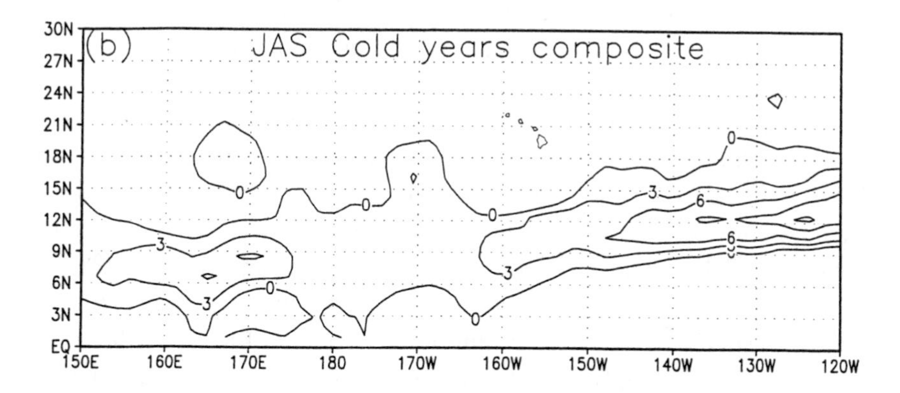

FIGURE 11.8 July through September (JAS) 1,000 hPa mean relative vorticity in the North Pacific for (*a*) El Niño (warm years) composite and (*b*) La Niña (cold years) composite. Units are 10^{-6} s^{-1}, and the contour interval is 3. Only positive values are contoured. Years considered for the El Niño batch include 1972, 1982, 1986, 1987, 1991, and 1997. For the La Niña batch, only five cases (1970, 1973, 1975, 1988, and 1998) are considered, as data for 1999 are unavailable. Due to topographical influences on surface winds near Hawai'i, the vorticity field in the vicinity of the Hawaiian Islands is masked.

The westward shift in the genesis location of tropical cyclones in the eastern North Pacific during El Niño years (Irwin and Davis 1999) would tend to propagate tropical cyclones farther west into the central Pacific. In addition, the decrease in vertical shear over the tropical central North Pacific in El Niño years would reduce the unfavorable conditions for tropical cyclones and make it more likely for tropical cyclones to propagate into the central North Pacific. Tropical cyclone tracks near the Hawaiian Islands also show marked differences that are associated with interannual climate variations. For non–El Niño years, most tropical cyclones follow a westward or northwestward track, but they become more erratic during El Niño years (Chu and Wang 1997).

The South Pacific

Nicholls (1979) first noted a strong correlation between the sea level pressure in Darwin, Australia, and tropical cyclone days around the Australian region (105°E–165°E). During the period from 1958–1959 to 1982–1983, the linear correlation between the preseason sea level pressure (July to September) and the number of tropical cyclone days in the cyclone season (October to April) is –0.68, which is significant at the 5% level. Because the Darwin pressure is a major component of the SOI (note that the standard SOI is inversely related to Darwin pressure), this strong and negative correlation implies that a reduction in tropical cyclone days during the cyclone season near Australia is preceded by an anomalously high pressure in Darwin, or a warm event. Higher sea level pressures, cooling of ocean surfaces, and the sinking branch of the Pacific Walker circulation during El Niño years combine to produce unfavorable conditions for tropical cyclone formation near Australia. Nicholls (1985) further suggested that seasonal tropical cyclone activity in the Australian region can be predicted provided Darwin pressures are known a few months prior to the cyclone season. Nicholls, Landsea, and Gill (1998) found an artificial bias in the Australian region storms before 1983. After accounting for this bias, a strong tropical cyclone-ENSO association still remains (Nicholls, Landsea, and Gill 1998).

Hastings (1990) and Evans and Allan (1992) also noted an increased tropical cyclone frequency near the date line as well as reduced activity to the northeast of Australia during El Niño years (figure 11.1). Tropical cyclone tracks in the tropical southwestern Pacific (west of the date line) became more zonal during El Niño years. In contrast, tropical cyclones tracked close to the coast of Queensland, Australia, and persisted southward with enhanced risk for coastal crossings during La Niña years. Basher and Zheng (1995) performed a similar study investigating the spatial patterns of tropical cyclones in the southwestern

Pacific in relation to the ENSO and regional sea-surface temperatures. They suggested that the incidence of tropical cyclones in the Coral Sea (west of 170°E) is influenced by local sea-surface temperature and east of 170°E the dominant control is not local sea-surface temperature but the eastward extent of ENSO-dependent atmospheric conditions (i.e., the monsoon trough).

As described earlier, tropical cyclones in the western Pacific generally spawn in the vicinity of the monsoon trough. For the South Pacific, the eastern terminus of the trough is usually located near 174°E (figure 11.9). Figure 11.10A shows a surface streamline analysis on November 15, 1982, at the height of the very strong 1982–1983 El Niño event. At that time, a pair of elongated monsoon troughs that extended as far east as 140°W were noted, one in each hemisphere. The South Pacific trough was almost 60° of longitude (~6,600 km) east of its November mean position. Embedded between this pair of troughs were equatorial westerlies that extended conspicuously all the way

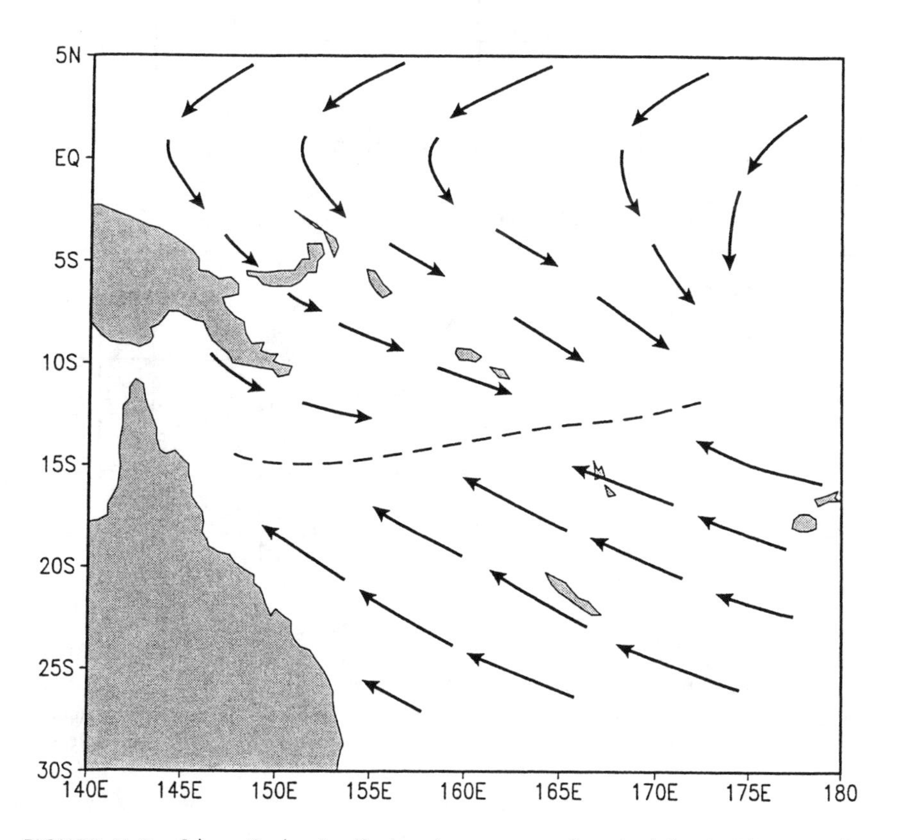

FIGURE 11.9 Schematic showing the long-term mean surface circulation in February in the southwestern Pacific. The monsoon trough axis is denoted by a broken line. Wind directions are indicated by arrows.

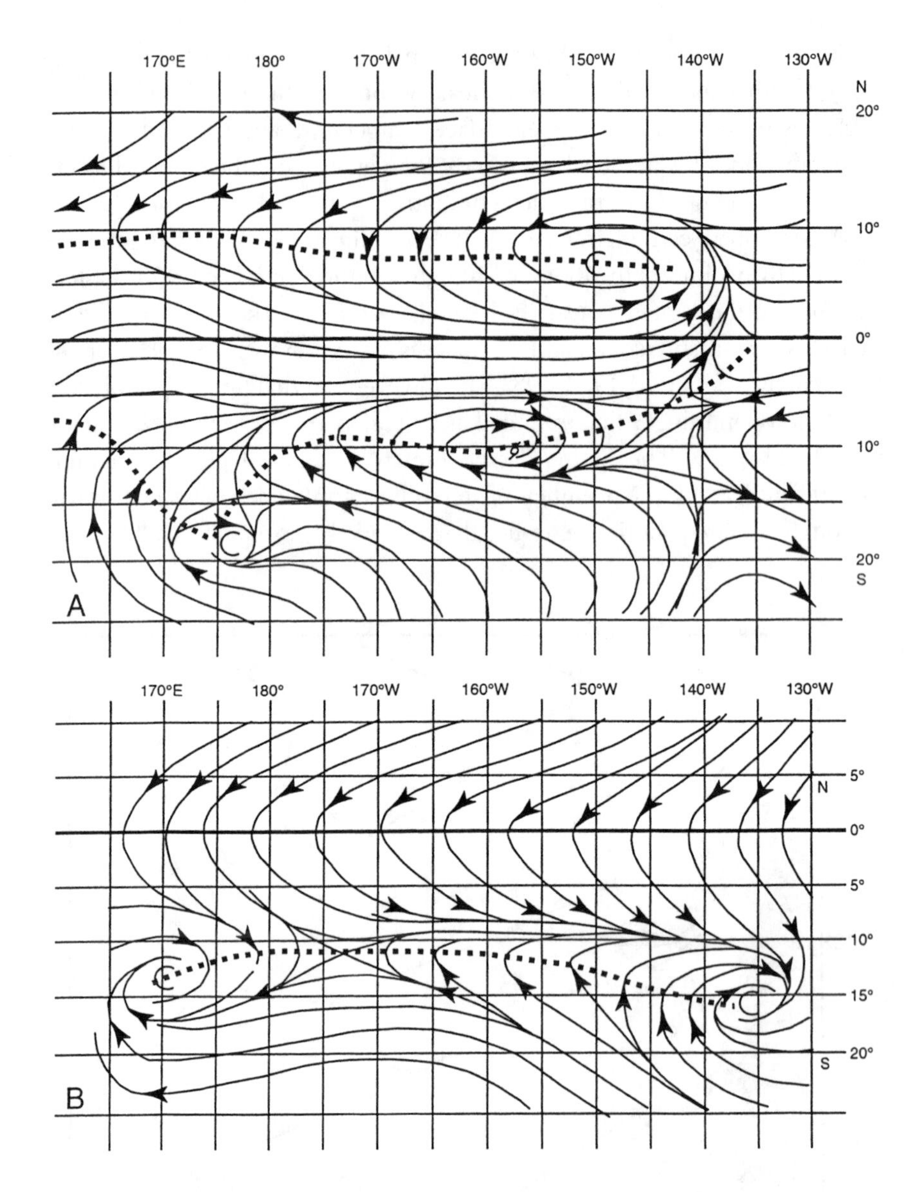

FIGURE 11.10 Surface streamline analyses for (*A*) November 15, 1982, and (*B*) March 21, 1983. The trough lines are indicated by dots. Note that in (*A*) the equatorial westerlies in the central Pacific are embedded between the double trough, one in each hemisphere, and that the tropical depression near Penryhn in the South Pacific (~158°W) is indicated. In (*B*), westerlies lie between the trough (dotted line) and the equator. Note the trough extends as far east as 135°W (adapted from Sadler, 1983, with permission).

from the western to the central and eastern Pacific; the westerlies were several thousand kilometers in extent. The double trough provided a favorable breeding ground for twin tropical cyclones on each side of the equator. The South Pacific cyclone formed near 8°S, 165°W and moved to the south of Penrhyn (figure 11.10A). The North Pacific counterpart, tropical storm Iwa, formed near 8°N, 167°W on November 18, 1982 (not shown). It then intensified to hurricane strength and inflicted major damages ($250 million) on Kauai, Hawai'i (Chu and Wang 1998). In March 1983, the South Pacific monsoon trough was still active and could be identified between 170°E and 135°W (figure 11.10B). Strong and persistent equatorial westerlies continued to move eastward, reaching beyond 110°W by May 1983. These El Niño related conditions were unusual because the southeastern Pacific is generally dominated by steady easterly trade winds.

Equatorial westerly winds and easterly trade winds in the subtropics generate low-level cyclonic shear and cyclonic relative vorticity. In doing so, they create an environment along the monsoon trough that is favorable for tropical cyclone formation. Moreover, the anomalously warm ocean surface over the central and eastern Pacific during El Niño years fuels the overlying atmosphere with additional heat and moisture, decreasing atmospheric stability and increasing the likelihood of atmospheric convection. Taken together, these dynamic and thermodynamic factors are instrumental in maintaining and generating tropical cyclones. Anomalous conditions during an El Niño may cause tropical cyclones to occur in a region that is not generally regarded as a cyclone-prone area, for example, the Hawaiian Islands in the North Pacific or French Polynesia in the South Pacific.

In accordance with the displacement of the monsoon trough and equatorial westerlies during the 1982–1983 El Niño, the genesis locations of tropical cyclones moved eastward with time (figure 11.11). Climatologically, this area is marked by strong vertical wind shears as the prevailing surface northeasterlies are overlain by southerlies in the upper troposphere. But from December 1982 through May 1983, 11 tropical cyclones were named in the southeastern Pacific. Three unnamed tropical depressions also developed in the eastern end of the South Pacific. Tropical cyclones east of 160°W are rare. During the 1982–1983 El Niño, however, six hurricanes struck French Polynesia (Sadler 1983). The recurvature of storms such as Veena and William is conspicuous.

In analyzing a 40-year sample (1939–1940 to 1978–1979) of tropical cyclone genesis locations in the South Pacific, Revell and Goulter (1986) noted eastward and equatorward displacements of the origin points during El Niño years as compared to non–El Niño years, a result subsequently confirmed by Basher and Zheng (1995). For instance, the climatological median location of genesis point

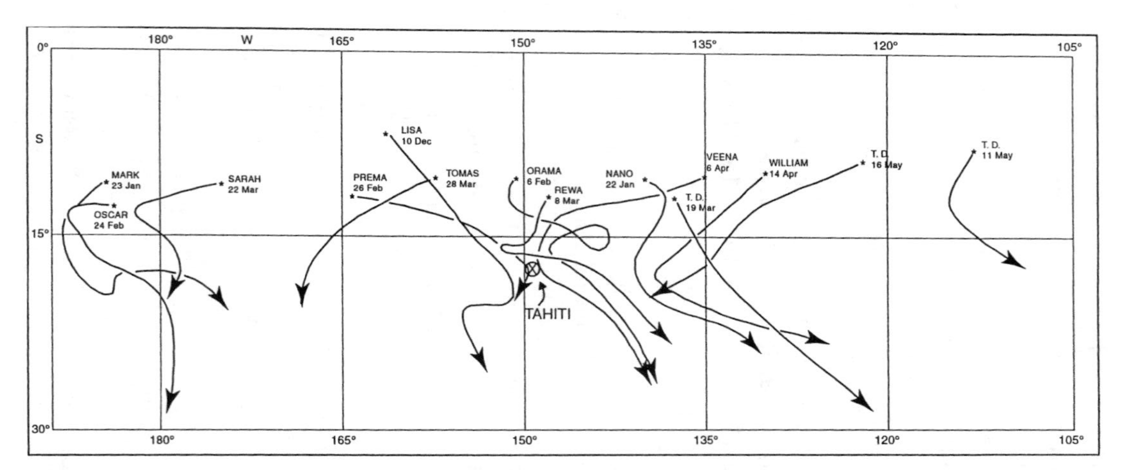

FIGURE 11.11 Tracks of southeastern Pacific tropical cyclones from December 1982 to May 1983. Asterisk indicates origin points of tropical depression (adapted from Sadler 1983, with permission).

is 14°S, 170°E near Vanuatu (Revell and Goulter 1986). During 1982–1983, the median origin point shifted northeastward to 11°S, 162°W, a remarkable eastward displacement by almost 28° longitude (~3,000 km) from its climatological median location. The meridional displacement of the median genesis location during the 1982–1983 El Niño event is 3° latitude equatorward from the climatological position. These dramatic displacements of origin points are intimately related to the migration of the South Pacific monsoon trough and the South Pacific convergence zone as suggested by Revell and Goulter (1986).

Figure 11.12 illustrates tropical cyclone origins and tracks for another very strong 1997–1998 warm ENSO event, as downloaded from the Australian Severe Weather Web site (http://australianservereweather.simplenet.com). During 1997 and 1998, the median location of genesis points for all named storms in the South Pacific was 12.2°S, 170.2°W, again a 20° longitudinal shift eastward from the climatological median position. During this warm event, tropical cyclones were more frequent in the South Pacific and they also formed in late season, as typified by Alan and Bart in late April 1998. This result is similar to what happened in 1983 (figure 11.11). The Cook Islands and French Polynesia were constantly under the threat of tropical cyclone strikes. There was also an unusual westward storm track between 10°S and 15°S to the west of 170°E (figure 11.12), a feature that is not uncommon during warm ENSO years (Evans and Allan 1992).

For the sake of comparison, it is also instructive to examine tropical cyclone activity in the South Pacific during the recent 1998–1999 La Niña years. From July 1998 to June 1999, only three cyclones with at least tropical storm strength were observed in the South Pacific to the east of the date line. In contrast, 11 cyclones occurred in the same region during the 1997–1998 El Niño episode (figure 11.12).

The North Atlantic

The mean annual number of tropical cyclones in the north Atlantic is 10, a smaller number in comparison to that found in the northwestern or northeastern Pacific. Most hurricanes in this basin occur between June and November, with 83% of the total annual numbers occurring in August, September, and October. Peak hurricane activity occurs in September with a 90% chance of at least one hurricane (Elsner and Kara 1999). In a series of papers, Gray (1984), Gray and Sheaffer (1991), and Gray et al. (1993), and Knaff (1997) found that interannual variations in the seasonal activity of Atlantic hurricanes can be correlated with several variables because of their effects on vertical wind shear in

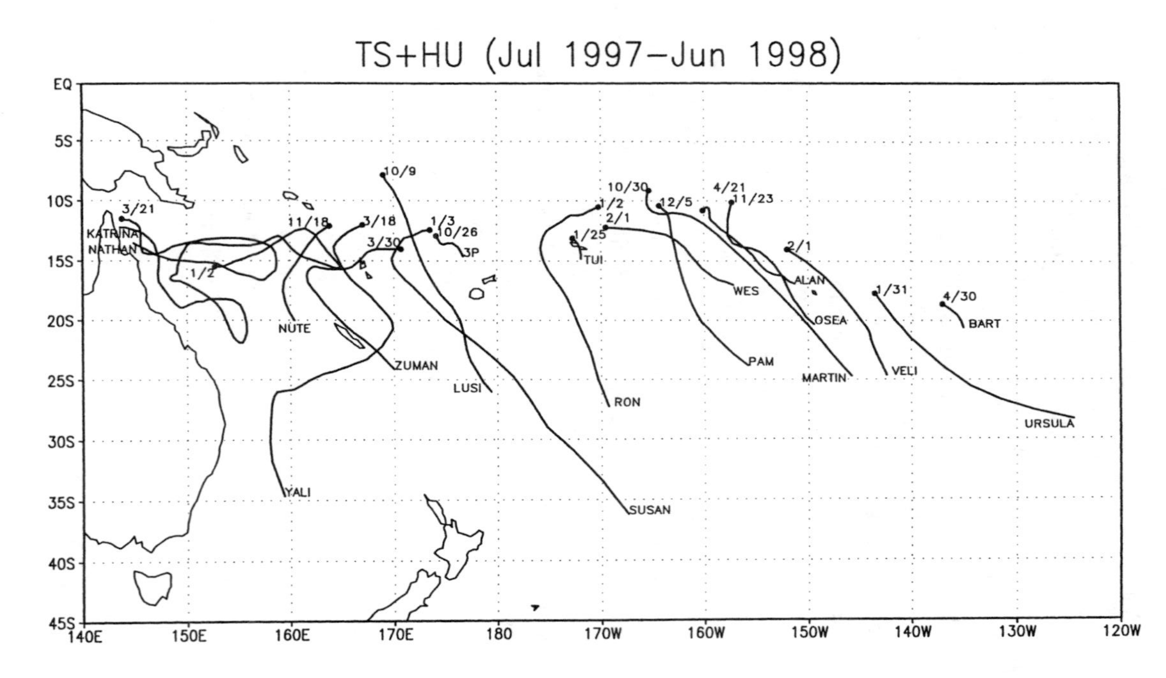

FIGURE 11.12 Tracks of South Pacific tropical cyclones from July 1997 to June 1998. Tropical depressions are omitted. Origin points are denoted by dots, and the month and date for each tropical cyclone are indicated by numbers. Termination points of track are marked by the name of cyclones.

the troposphere. These variables include, but are not limited to, ENSO, Atlantic Basin sea level pressure, west Sahel monsoon rainfall, and the Quasi-Biennial Oscillation of stratospheric wind. Shapiro (1987) was able to show a statistical relationship between vertical wind shear and an El Niño index. Subsequently, Goldenberg and Shapiro (1996) provided evidence that changes in vertical wind shear are the most important environmental factor modulating Atlantic hurricane activity on interannual time scales.

During El Niño, the warm pool of sea water and major tropical convection shift eastward to the eastern Pacific. ENSO alters Atlantic hurricane activity through shifts in the location of large-scale convection. The enhanced upper-level divergent outflows from deep cumulus convection cause upper tropospheric zonal winds over the Caribbean and tropical Atlantic to become more westerly. ENSO's influence on lower-tropospheric easterly winds over the tropical Atlantic is small so vertical wind shear over the tropical Atlantic and Caribbean region is enhanced during the El Niño hurricane season (Gray and Sheaffer 1991). Consequently, the number of hurricanes and hurricane days are reduced during El Niño years (Landsea et al. 1999). In addition to the wind-shear mechanism, the western North Atlantic is marked by subsidence during warm ENSO phases, another factor unfavorable for hurricane development (figure 11.2).

The ENSO phenomenon also affects the U.S. hurricane landfalls (Gray 1984). Assuming that the occurrence of U.S. hurricanes follows a Poisson distribution, Bove et al. (1998) showed that the probability of a landfalling hurricane is reduced during El Niño events but increased during La Niña events. For instance, the probability of observing two or more landfalling hurricanes during an El Niño is 28%, in contrast to 66% during La Niña (figure 11.13). Thus, the La Niña phase has a profound impact on U.S. hurricane landfalls. Landfall probability for the neutral phase lies between these two extremes. In terms of economic losses, the average damage per storm is $800 million during El Niño years but doubles to $1.6 billion during La Niña years (Pielke and Landsea 1999). More interestingly, there is a 20 to 1 ratio in median damage per year during two extreme climatic events: $3.3 billion in La Niña years versus $152 million in El Niño years.

Elsner and Kara (1999) further partitioned Atlantic hurricanes into tropical-only and baroclinically enhanced groups and noted that the ENSO influence occurs for tropical-only systems. In addition, Elsner and Kara (1999) found a difference in genesis locations throughout the six-month hurricane season for two contrasting ENSO phases (figure 11.14). During El Niño, most tropical cyclones formed over the Gulf of Mexico in early season (June–July), and

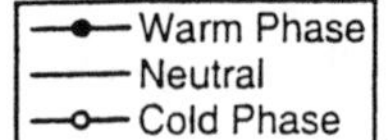

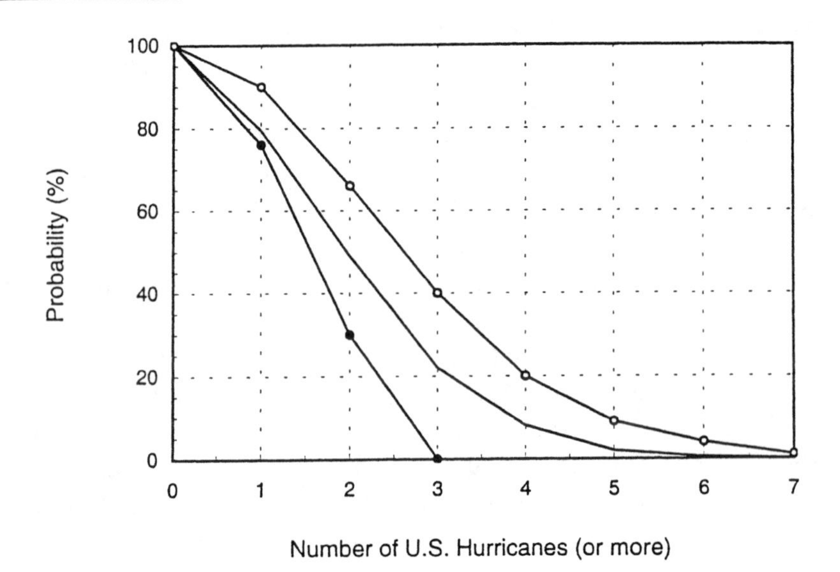

FIGURE 11.13 Inverse cumulative probability distributions for U.S. hurricane landfalls over the period from 1900 to 1997. Solid circle is for the warm ENSO phase, open circle is for the cold ENSO phase, and no circle is for the neutral condition (after Bove et al. 1998).

moved away from the continent to the western North Atlantic in mid-season (August–September) and late season (October–November). The mean latitudinal location of genesis point is 23.7°N in early season, 23.2°N in mid-season, and 24.4°N in late season during El Niño. In contrast, during La Niña, the origin points of tropical cyclones are found off the southeast U.S. coast in early season, moving toward the Gulf of Mexico in mid-season, and shifting equatorward over the Caribbean Sea in late season. On average, the position of tropical cyclone formation point is found in lower latitudes during La Niña years when compared to El Niño years.

During extreme climatic events, a shift in the mean longitudinal location of tropical cyclone origin points is even more pronounced than that for the mean latitudinal location. For instance, the mean longitude is 88.9°W in early season, 62.8°W in mid-season, and 72.3°W in late season during El Niño. There is also a considerable west-east movement of the mean origin points throughout the hurricane season during La Niña years, from the Gulf of Mexico in early season (82°W), to the western North Atlantic in mid-season (64.9°W), and to the southwestern North Atlantic in late season (72.3°W). In

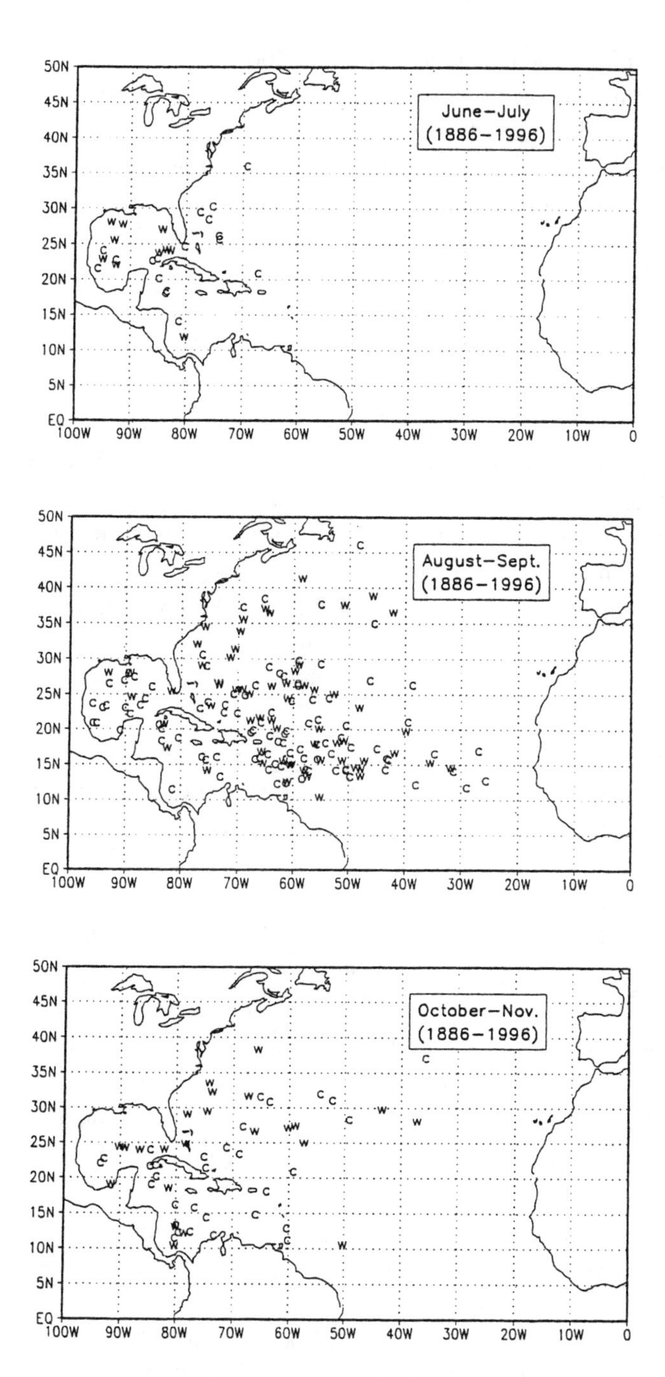

FIGURE 11.14 Origin points of North Atlantic hurricanes by season during warm (w) and cold (c) ENSO phases for the period 1886 to 1996 (from Elsner and Kara, 1999; reproduced with permission from Oxford University Press).

analyzing tropical cyclone tracks during the course of El Niño, Gray and Sheaffer (1991) noted fewer hurricanes crossing through the Caribbean Basin in a westward track during El Niño years compared to non–El Niño years. During the La Niña batch, the mean intensity, as measured by the average of the strongest winds for all named storms for a season, is only a bit stronger (6%) than the El Niño batch (Landsea et al. 1999).

Summary

This chapter has described the El Niño–Southern Oscillation (ENSO) phenomenon, which is the most dominant mode in year-to-year climate variations in the tropics, and its impact on tropical cyclone activity. It has been known for many years that on seasonal time scales tropical cyclone formation and development are intimately regulated by large-scale dynamic and thermodynamic environmental conditions (Gray 1977). An ENSO event alters large-scale environmental conditions and influences probability distributions that describe tropical cyclone attributes such as genesis location, frequency, track, lifespan, landfall, and intensity.

For the western North Pacific, the warm phase of ENSO shifts the monsoon trough to the east and reduces vertical wind shear. Low-level cyclonic vorticity associated with the monsoon trough nearly triples in the eastern portion during the warm phase. These changes are accompanied by a notable increase in tropical cyclone genesis and frequency in the eastern portion of the basin (150°E to the date line) (table 11.1). During the cold phase, tropical cyclone genesis locations at peak season stay at higher latitudes over the western extreme of the North Pacific. Tropical cyclone life spans are longer during the warm phase as most tropical cyclones form over lower latitude oceans and away from the large land masses. Tropical cyclones track mainly westward during La Niña years but they tend to recurve during El Niño years. As a result, typhoon impacts in Japan, South Korea, Taiwan, and China are more pronounced in El Niño than La Niña years.

The influence of two contrasting ENSO phases on the overall frequency of the eastern North Pacific hurricanes appears to be minimal (table 11.1), although more intense hurricanes are observed during the warm phase than in the cold one. Tropical cyclone genesis locations also shift westward and life spans are longer during El Niño years as opposed to La Niña years. Besides the monsoon trough (in-situ influence), tropical cyclone genesis in the eastern North Pacific is modulated by external forcings such as tropical easterly waves from the Caribbean.

TABLE 11.1 ENSO and Tropical Cyclone Activity in the Pacific and Atlantic Oceans

Basin	Frequency	Genesis location	Intensity	Track	Life span
Northwestern Pacific	No significant changes in the developing year, but is reduced in the following year	Farther east and south	Unknown	More likely to recurve	Longer
Northeastern Pacific	Overall frequency unchanged	Farther west	More intense hurricanes	Westward expansion	Longer
Central North Pacific (180°W–140°W)	More storms	More likely to form in the western part	Unknown	More erratic	Unknown
Southwestern Pacific (west of dateline)	Fewer storms	Farther north	Unknown	Likely to track westward in the tropics	Unknown
Southeastern Pacific	More storms	Farther east and north	Unknown	Unknown	Unknown
North Atlantic	Fewer storms	Farther north in mid- and late seasons	Slightly weaker	Fewer storms cross the lower Caribbean	Unknown

Note: Listed conditions are for El Niño; opposite changes are implied for La Niña.

The central North Pacific experiences an increased occurrence of tropical cyclones during the warm phase, and this increase is due to the eastward excursion of the monsoon trough from the western North Pacific and the combined effects of the weakening of the vertical wind shear as well as the westward shift of the genesis locations from the eastern North Pacific (table 11.1).

As in the western North Pacific, tropical cyclone frequency and genesis locations in the South Pacific undergo substantial changes during extreme ENSO events. Although most tropical cyclones form in the Coral Sea during non–El Niño years, they originate mainly to the east of the date line during the warm phase (table 11.1). Islands that are normally free from tropical cyclone risks, such as French Polynesia, are threatened more often by tropical cyclones during strong, warm ENSO events.

In the North Atlantic, El Niño years are associated with fewer hurricanes and major hurricanes, fewer hurricane days and a lower probability of U.S. hurricane landfalls (table 11.1). During the cold phase the U.S. coast is faced with a higher risk of hurricane landfalls. The tropical cyclone genesis locations shift equatorward as the hurricane season progresses; thus the Caribbean Sea becomes more vulnerable to tropical cyclone risks in late season. Hurricanes crossing the lower Caribbean Basin also become more prevalent during non–El Niño years.

Acknowledgments

I would like to express my thanks to Maria Rakotondrafa for performing data and graphic analyses and Di Henderson for editing. Constructive review comments by Chris Landsea, Rick Murnane, Tom Schroeder, and an anonymous reviewer led to significant improvement in the presentation of this chapter.

References

Barnston, A. G., M. Chelliah, and S. B. Goldenberg. 1997. Documentation of a highly ENSO-related SST region in the equatorial Pacific. *Atmosphere-Ocean* 35:367–83.

Basher, R. E., and X. Zheng. 1995. Tropical cyclones in the southwest Pacific: Spatial patterns and relationships to Southern Oscillation and sea surface temperature. *Journal of Climate* 8:1249–60.

Battisti, D. S., and A. C. Hirst. 1989. Interannual variability in a tropical atmosphere-ocean model: Influence of basic state, ocean geometry and non-linearity. *Journal of the Atmospheric Sciences* 46:1687–712.

Bjerknes, J. 1969. Atmospheric teleconnections from the tropical Pacific. *Monthly Weather Review* 97:163–72.

Bove, M. C., J. B. Elsner, C. W. Landsea, X. Niu, and J. J. O'Brien. 1998. Effect of El Niño on U.S. landfalling hurricanes, revisited. *Bulletin of the American Meteorological Society* 76:2477–82.

Cane, M. A., and S. E. Zebiak. 1985. A theory for El Niño and the Southern Oscillation. *Science* 228:1085–87.

Chan, J. C. L. 1985. Tropical cyclone activity in the northwest Pacific in relation to the El Niño/Southern Oscillation phenomenon. *Monthly Weather Review* 113: 599–606.

Chan, J. C. L. 2000. Tropical cyclone activity over the western North Pacific associated with El Niño and La Niña events. *Journal of Climate* 13:2960–72.

Chen, T.-C., S.-P. Weng, N. Yamazaki, and S. Kiehne. 1998. Interannual variation in the tropical cyclone formation over the western North Pacific. *Monthly Weather Review* 126:1080–90.

Chu, P.-S. 2002. Large-scale circulation features associated with decadal variations of tropical cyclone activity over the central North Pacific. *Journal of Climate* 15: 2678–89.

Chu, P.-S., and J. D. Clark. 1999. Decadal variations of tropical cyclone activity over the central North Pacific. *Bulletin of the American Meteorological Society* 80:1875–81.

Chu, P.-S., J. Frederick, and A. J. Nash. 1991. Exploratory analysis of surface winds in the equatorial western Pacific and El Niño. *Journal of Climate* 4:1087–102.

Chu, P.-S., and R. W. Katz. 1985. Modeling and forecasting the Southern Oscillation: A time-domain approach. *Monthly Weather Review* 113:1876–88.

Chu, P.-S., and R. W. Katz. 1989. Spectral estimation from time series models with relevance to the Southern Oscillation. *Journal of Climate* 2:86–90.

Chu, P.-S., and J. Wang. 1997. Tropical cyclone occurrences in the vicinity of Hawaii: Are the differences between El Niño and non-El Niño years significant? *Journal of Climate* 10:2683–89.

Chu, P.-S., and J. Wang. 1998. Modeling return periods of tropical cyclone intensities in the vicinity of Hawaii. *Journal of Applied Meteorology* 37:951–60.

Clark, J. D., and P.-S. Chu. 2002. Interannual variation of tropical cyclone activity over the central North Pacific. *Journal of the Meteorological Society of Japan* 80: 403–18.

Collins, J. M., and I. M. Mason. 2000. Local environmental conditions related to seasonal tropical cyclone activity in the Northeast Pacific basin. *Geophysical Research Letters* 27:3881–84.

Deser, C., and J. M. Wallace. 1990. Large-scale atmospheric circulation features of warm and cold episodes in the tropical Pacific. *Journal of Climate* 3:1254–81.

Dong, K. 1988. El Niño and tropical cyclone frequency in the Australian region and the northwest Pacific. *Australian Meteorological Magazine* 36:219–25.

Elsner, J. B., and A. B. Kara. 1999. *Hurricanes of the North Atlantic: Climate and society*. New York: Oxford University Press.

Evans, J. L., and R. J. Allan. 1992. El Niño/Southern Oscillation modification to the structure of the monsoon and tropical activity in the Australian region. *International Journal of Climatology* 12:611–23.

Frank, W. M. 1987. Tropical cyclone formation. In A *Global view of tropical cyclones*, edited by R. L. Elsberry, W. M. Frank, G. J. Holland, J. D. Jarrell, and R. L. Southern, 53–90. Chicago: University of Chicago Press.

Goldenberg, S. B., and L. J. Shapiro. 1996. Physical mechanisms for the association of El Niño and West Africa rainfall with Atlantic major hurricanes. *Journal of Climate* 9:1169–87.

Gray, W. M. 1968. Global view of the origin of tropical disturbances and storms. *Monthly Weather Review* 96:55–73.

Gray, W. M. 1977. Tropical cyclone genesis in the western North Pacific. *Journal of the Meteorological Society of Japan* 55:465–82.

Gray, W. M. 1984. Atlantic seasonal hurricane frequency. Part I: El Niño and 30 mb Quasi-Biennial Oscillation influences. *Monthly Weather Review* 112:1649–68.

Gray, W. M., C. W. Landsea, P. W. Mielke, Jr., and K. J. Berry. 1993. Predicting Atlantic basin seasonal tropical cyclone activity by 1 August. *Weather and Forecasting* 8:73–86.

Gray, W. M., and J. D. Sheaffer. 1991. El Niño and QBO influences on tropical cyclone activity. In *Teleconnections linking worldwide climate anomalies*, edited by M. H. Glantz, R. W. Katz, and N. Nicholls, 257–84. New York: Cambridge University Press.

Hastings, P. A. 1990. Southern Oscillation influences on tropical cyclone activity in the Australian/South-west Pacific region. *International Journal of Climatology* 10:291–98.

Irwin, R. P., and R. E. Davis. 1999. The relationship between the Southern Oscillation Index and tropical cyclone tracks in the eastern North Pacific. *Geophysical Research Letters* 20:2251–54.

Jin, F.-F. 1997. An equatorial ocean recharge paradigm for ENSO. Part I: Conceptual model. *Journal of the Atmospheric Sciences* 54:811–29.

Jury, M. 1993. A preliminary study of climatological associations and characteristics of tropical cyclones in the southwest Indian Ocean. *Meteorological and Atmospheric Physics* 51:101–15.

Kalnay, E., et al. 1996. The NCEP/NCAR 40-year reanalysis project. *Bulletin of the American Meteorological Society* 77:437–71.

Kimberlain, T. B. 1999. The effects of ENSO on North Pacific and North Atlantic tropical cyclone activity. In *Preprints of the 23rd Conference on Hurricanes and Tropical Meteorology*, 250–53. Boston: American Meteorological Society.

Kimberlain, T. B. 2000. Long-term trends in North Pacific tropical cyclone activity. In *Preprints of the 24th Conference on Hurricanes and Tropical Meteorology*, 472–73. Boston: American Meteorological Society.

Kirtman, B., ed. 2001. *Experimental Long-Lead Forecast Bulletin* 10, no. 1 (available at: http://www.iges.org/ellfb).

Knaff, J. A. 1997. Implications of summertime sea level pressure anomalies in the tropical Atlantic region. *Journal of Climate* 10:789–804.

Knaff, J. A., and C. W. Landsea. 1997. An El Niño–Southern Oscillation Climatology and Persistence (CLIPER) forecasting scheme. *Weather and Forecasting* 12:633–52.

Lander, M. A. 1994. An exploratory analysis of the relationship between tropical storm formation in the western North Pacific and ENSO. *Monthly Weather Review* 122:636–51.

Lander, M. A. 1996. Specific tropical cyclone track types and unusual tropical cyclone motions associated with a reverse-oriented monsoon trough in the western North Pacific. *Weather and Forecasting* 11:170–86.

Landsea, C. W. 2000. El Niño–Southern Oscillation and the seasonal predictability of tropical cyclones. In *El Niño: Multiscale variability and global and regional impacts*, edited by H. F. Díaz and V. Markgraf, 149–81. Cambridge: Cambridge University Press.

Landsea, C. W., and J. A. Knaff. 2000. How much skill was there in forecasting the very strong 1997–98 El Niño? *Bulletin of the American Meteorological Society* 81:2107–19.

Landsea, C. W., R. A. Pielke, Jr., A. M. Mestas-Nuñez, and J. A. Knaff. 1999. Atlantic Basin hurricanes: Indices of climatic changes. *Climatic Change* 42:89–129.

Lee, H.-K., P.-S. Chu, C.-H. Sui, and K.-M. Lau. 1998. On the annual cycle of latent heat fluxes over the equatorial Pacific using TAO buoy observations. *Journal of the Meteorological Society of Japan* 76:909–23.

Lukas, R., S. P. Hayes, and K. Wyrtki. 1984. Equatorial sea-level response during the 1982–83 El Niño. *Journal of Geophysical Reserach* C 6:10425–30.

Madden, R. A., and P. R. Julian. 1971. Detection of a 40–50 day oscillation in the zonal wind in the tropical Pacific. *Journal of the Atmospheric Sciences* 28:702–8.

McBride, J. L. 1995. Tropical cyclone formation. In *Global perspectives on tropical cyclones*, edited by R. L. Elsberry, 63–105. Report No. TCP-38. Geneva: World Meteorological Organization.

McCreary, J., Jr. 1983. A model of tropical ocean-atmosphere interaction. *Monthly Weather Review* 111:370–87.

Neumann, C. J. 1977. A critical look at statistical hurricane prediction models. In *Preprints of the 11th Conference on Hurricanes and Tropical Meteorology*, 375–80. Boston: American Meteorological Society.

Neumann, C. J. 1993. Global overview. In *Global guide to tropical cyclone forecasting*, edited by G. J. Holland, 1.1–1.56. Technical Document WMO/TC-No. 560, Report No. TCP-31. Geneva: World Meteorological Organization.

Nicholls, N. 1979. A possible method for predicting seasonal tropical cyclone activity in the Australian region. *Monthly Weather Review* 107:1221–24.

Nicholls, N. 1985. Predictability of interannual variations of Australian seasonal tropical cyclone activity. *Monthly Weather Review* 113:1144–49.

Nicholls, N. 1992. Recent performance of a method for forecasting Australian seasonal tropical cyclone activity. *Australian Meteorological Magazine* 21:105–10.

Nicholls, N., C. W. Landsea, and J. Gill. 1998. Recent trends in Australian region tropical cyclone activity. *Meteorological and Atmospheric Physics* 65:197–205.

Office of the Federal Coordinator for Meteorological Services and Supporting Research (OFCM). 1999. *National Hurricane Operations Plan (NHOP)*. FCM-P12–1999. Washington, D.C.: National Oceanic and Atmospheric Administration.

Pielke, R. A., Jr., and C. W. Landsea. 1999. La Niña, El Niño, and Atlantic hurricane damages in the United States. *Bulletin of the American Meteorological Society* 80:2027–33.

Rappaport, E. N., L. A. Avila, M. B. Lawrence, B. M. Mayfield, and R. J. Pasch. 1998. Eastern North Pacific hurricane season of 1995. *Monthly Weather Review* 126:1152–62.

Rasmusson, E. M., and T. H. Carpenter. 1982. Variations in tropical sea surface temperature and surface wind fields associated with the Southern Oscillation/El Niño. *Monthly Weather Review* 110:354–84.

Revell, C. G., and S. W. Goulter. 1986. South Pacific tropical cyclones and the Southern Oscillation. *Monthly Weather Review* 114:1138–45.

Sadler, J. 1967. *The tropical upper tropospheric trough as a secondary source of typhoons and a primary source of tradewind disturbances*. Report No. HIG-67-12.Honolulu: Department of Meteorology, University of Hawai'i.

Sadler, J. 1983. Tropical Pacific atmospheric anomalies during 1982–83. In *Proceedings of the 1982/83 El Niño/Southern Oscillation Workshop*, 1–10. Miami: National Oceanic and Atmospheric Administration and Atlantic Oceanographic and Meteorological Laboratory.

Saunders, M. A., R. E. Chandler, C. J. Merchant, and F. P. Roberts. 2000. Atlantic hurricanes and northwest Pacific typhoons: ENSO spatial impacts on occurrence and landfall. *Geophysical Research Letters* 27:1147–50.

Schopf, P. S., and M. J. Suarez. 1988. Vacillations in a coupled ocean-atmosphere model. *Journal of the Atmospheric Sciences* 45:549–66.

Schroeder, T. A., and Z.-P. Yu. 1995. Interannual variability of central Pacific tropical cyclones. In *Preprints of the 21st Conference on Hurricanes and Tropical Meteorology*, 437–39. Boston: American Meteorological Society.

Shapiro, L. J. 1987. Month-to-month variability of the Atlantic tropical circulation and its relationship to tropical storm formation. *Monthly Weather Review* 115:2598–14.

Trenberth, K. E. 1976. Spatial and temporal variations of the Southern Oscillation. *Quarterly Journal of the Royal Meteorological Society* 102:639–53.

Trenberth, K. E. 1997. The definition of El Niño. *Bulletin of the American Meteorological Society* 78:2771–77.

Wang, B., and J. C. L. Chan. 2002. How strong ENSO affect tropical storm activity over the western North Pacific. *Journal of Climate* 15:1643–58.

Whitney, L. D., and J. Hobgood. 1997. The relationship between sea surface temperatures and maximum intensities of tropical cyclones in the eastern North Pacific Ocean. *Journal of Climate* 10:2921–30.

Wu, G., and N.-C. Lau. 1992. A GCM simulation of the relationship between tropical storm formation and ENSO. *Monthly Weather Review* 120:958–77.

Wu, M. C., W. L. Chang, and W. M. Leung. 2004. Impacts of El Niño–Southern Oscillation events on tropical cyclone landfalling activity in the western North Pacific. *Journal of Climate* 17:1419–28.

Wyrtki, K. 1975. El Niño: The dynamic response of the equatorial Pacific Ocean to atmospheric forcing. *Journal of Physical Oceanography* 5:572–84.

Wyrtki, K. 1982. The Southern Oscillation, ocean-atmosphere interaction and El Niño. *Marine Technical Society Journal* 16:3–10.

Wyrtki, K. 1985. Water displacements in the Pacific and the genesis of El Niño cycles. *Journal of Geophysical Research* 90:7129–32.

Zebiak, S. E., and M. A. Cane. 1987. A model El Niño–Southern Oscillation. *Monthly Weather Review* 115:2262–78.

12

Hurricane Landfall Probability and Climate

James B. Elsner and Brian H. Bossak

This chapter discusses climatological aspects of U.S. hurricane activity during the twentieth century. It focuses on climate factors that are known to be related to the occurrence of hurricanes and major hurricanes along the coast. Statistical models are used to define and describe these relationships. In general it is found that coastal hurricane activity increases during episodes of La Niña, when the North Atlantic Oscillation is weak. Increases are most apparent along the central Gulf Coast, including New Orleans, as well as along portions of the Florida coast. Hurricane activity in the United States diminishes during El Niño episodes. Results of this study can be refined by incorporating additional hurricane information collected from historical proxies and geological records, which are discussed elsewhere in this volume.

Landfalling hurricanes are of important concern to society. In the United States, the potential for damage and loss of life from hurricanes rivals that from earthquakes (Díaz and Pulwarty 1997). In Florida, Hurricane Andrew (1992) caused more than $30 billion in direct economic losses, while Hurricane Floyd (1999) disrupted the lives of 2.5 million of its residents due to evacuation alone. Understanding the past record of hurricane activity provides clues about future frequency and intensity, which is important for land-use planning, emergency management, hazard mitigation, and insurance applications.

Empirical and statistical research has identified climate factors that contribute to conditions favorable for hurricanes over the North Atlantic Basin, which includes the Caribbean Sea and the Gulf of Mexico (Gray et al. 1992; Elsner, Kara, and Owens 1999; Elsner, Jagger, and Niu 2000). These factors influence the occurrence of hurricanes differently depending on the particular region. For instance, the effect of an El Niño on hurricane frequency over the entire North Atlantic Basin is significant, but El Niño's influence on the frequency of hurricanes forming over the subtropics is small. In fact, additional climate factors are usually needed to explain local variations in hurricane activity (Lehmiller, Kimberlain, and Elsner 1997; Jagger, Elsner, and Niu 2001). During some years there is a tendency for hurricanes to track westward through the Caribbean Sea and threaten Mexico and the United States. At other times, hurricanes tend to move parallel to the East Coast of the United States (Elsner, Bossak, and Niu 2001). To some extent, the degree to which the Gulf Coast is vulnerable to a hurricane in a given year is inversely related to the degree to which the East Coast is vulnerable.

In this chapter, we examine the occurrence of tropical cyclones that make landfall in the continental United States as hurricanes. These are termed "U.S. hurricanes." The focus on U.S. hurricanes allows the use of reliable data extending back to the beginning of the twentieth century, as well as emphasizing the socially relevant component of hurricane activity in this part of the world. Although tropical storms are capable of inflicting damage and loss of life, hurricanes cause significantly more wind destruction and flooding. The goal of this chapter is to describe quantitative changes in hurricane landfall probabilities in response to variations in climate. The assumption is that large-scale climate variations cause changes to hurricane activity.

The chapter is divided into two main sections. First we examine the annual frequency of U.S. hurricanes and show how the frequency is related to climate factors. In particular, we consider the annual occurrence of U.S. hurricanes and U.S. major hurricanes using the technique of generalized linear modeling. The modeling process is used to explore linkages between coastal hurricane activity and climate as well as to identify climate factors for the regional model. Climate factors (or covariates) are variables that have a statistical relationship to U.S. hurricane activity. Next, we make use of a technique for regional modeling of hurricane activity. The model uses covariates in assigning local probabilities of hurricane occurrence. The regions are coastal counties from Texas to North Carolina. Although our analysis is done for the coastal United States, the procedures described here can be applied in other tropical cyclone-prone regions of the world.

Annual Landfall Probabilities

Definitions and Hurricane Data

A hurricane is a tropical cyclone with maximum sustained (1-minute) 10-m winds of 65 kt (33 m s^{-1}) or greater. Major hurricanes of Category 3 or higher on the Saffir-Simpson damage potential scale have winds of 100 kt (51 m s^{-1}) or greater (Simpson 1974). Hurricane landfall occurs when all or part of the eye wall, the circular region of intense wind and rain surrounding the eye, passes directly over the coastline or over an adjacent barrier island. Because the eye wall can extend outward a distance of 50 km or more from the hurricane center, landfall may occur even when the precise center of lowest pressure remains offshore. A hurricane may make more than one landfall. In 1992, for example, Hurricane Andrew struck Florida and Louisiana. In this section we consider all tropical cyclones that make landfall in the continental United States at least once at hurricane intensity. In the next section we examine landfalls along the coast from Texas to North Carolina.

The HURDAT (best-track) dataset is the most complete and reliable source of North Atlantic hurricanes (Jarvinen, Neumann, and Davis 1984). The data set consists of the six-hourly position and intensity estimates of tropical cyclones back to 1886 (Neumann et al. 1999). These data are used to determine the annual frequency of hurricanes and major hurricanes reaching the United States over the period from 1900 to 1997. Additional U.S. hurricane information extending back to 1851 is now available through the hurricane data reanalysis project (Landsea et al., chapter 7 in this volume). This additional information is not considered in the present analysis.

The historical data indicate a total of 159 U.S. hurricanes and 63 U.S. major hurricanes over the 98-year period in question, for an average of 1.62 hurricanes and 0.64 major hurricanes per year (table 12.1). This amounts to approximately 13 hurricanes and 5 major hurricanes every eight years. But each year is different. In 1985, six hurricanes hit the United States, while in 1994 none did. More than half the years are without a major hurricane landfall. The 95% confidence intervals on the annual means are (1.38, 1.91) for hurricanes and (0.50, 0.80) for major hurricanes. The confidence intervals are based on bias-corrected bootstrapped samples. Whether a hurricane reaches the United States in a given year depends on formation and development

TABLE 12.1 U. S. Hurricane Statistics, 1900–1997

	Mean	Variance	Maximum number	Minimum number	BCMQ[1] 2.5%	BCMQ 97.5%
Hurricanes	1.62	1.660	6	0	1.38	1.91
Major hurricanes	0.64	0.582	3	0	0.50	0.80

[1]Bias-corrected mean bootstrapped (1,000 samples) quantiles.

mechanisms as well as steering currents, which are linked to large-scale climate factors including rainfall in western Africa, the El Niño cycle over the Pacific Ocean, and air pressure differences over the North Atlantic.

On average, the United States can expect one or two hurricanes each year. The distribution of first landfall dates in 10-day intervals is displayed in figure 12.1. The median strike date is September 5 for hurricanes and September 11 for major hurricanes. The interquartile range of U.S. hurricanes is 40 days, meaning that half of all strikes occur between August 17 and September 24. The most active period extends from approximately August 30 through September 28. During the period from 1900 to 1997, the earliest U.S. hurricane occurred on June 9 and the latest on November 30.

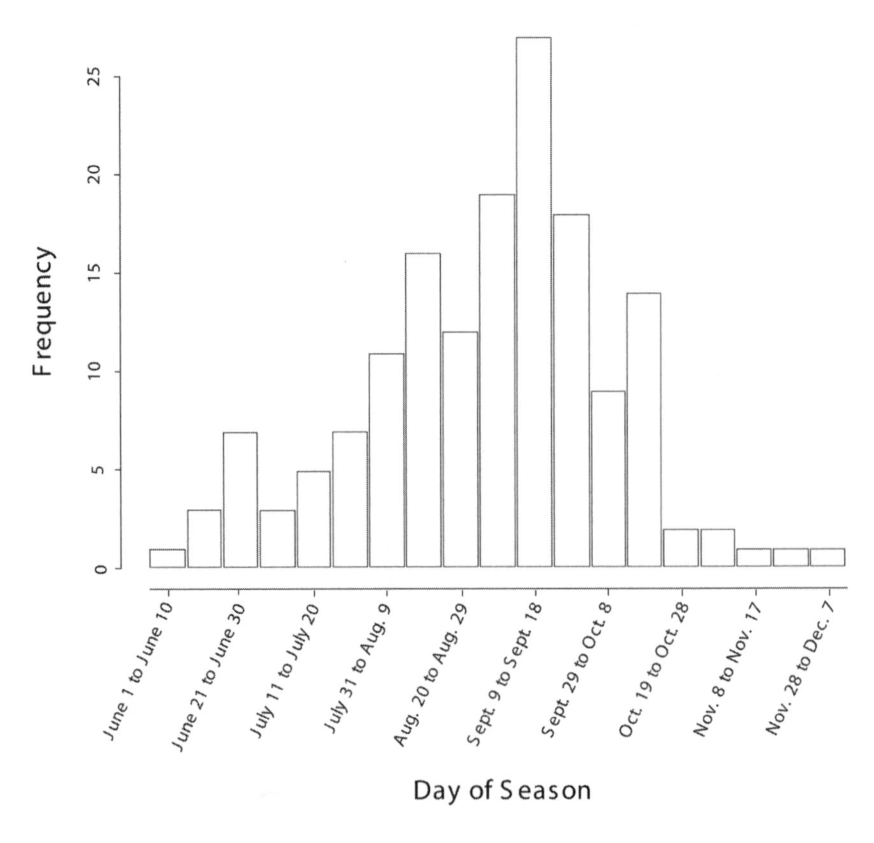

FIGURE 12.1 Histogram of the occurrence of U.S. hurricanes in 10-day intervals based on 159 hurricanes over the period from 1900 to 1997. For hurricanes that make more than one landfall in the United States, only the first landfall date is used. The median strike date is September 5.

Sources of Interannual Variability

It is well known that some of the year-to-year variability in U.S. hurricane activity is linked to the El Niño–Southern Oscillation (ENSO) cycle. In a year dominated by cooler than normal waters off the Peruvian coast (La Niña event), the United States is likely to see a greater number of hurricanes come onshore. In contrast, in a year with warm waters off South America (El Niño event) there tend to be fewer landfalls (Bove et al. 1998; Elsner and Kara 1999). It is speculated that El Niño creates stronger upper-atmospheric westerly winds over the Caribbean that lead to unfavorable wind shear and vorticity over the hurricane genesis and development region (Gray 1984). A greater amount of sinking air (Kimberlain and Elsner 1998) and lower surface air pressures (Knaff 1997) are additional inhibiting factors. During a mature El Niño event, the atmospheric sea level pressure pattern features negative anomalies (departures from average) across the central and eastern equatorial Pacific Ocean and positive anomalies over Australia and Indonesia. The pattern results in negative values of the Southern Oscillation Index (SOI). Formally, the SOI is defined as the normalized pressure difference between Tahiti and Darwin, Australia (Troup 1967). Monthly values of the SOI calculated on the basis of the method proposed by Ropelewski and Jones (1987) are used here to examine the relationship of El Niño to U.S. hurricanes.[1] In particular we focus on the period of August through October and use a cumulative value of the SOI in units of standard deviation (s.d.) over these three months. Cumulative SOI values during 98-year period range between −6.98 s.d. (1982) and +7.83 s.d. (1917).

A portion of the interannual variability of U.S. hurricanes is likely related to climate fluctuations over the Atlantic region. In particular, precipitation amounts over western areas of Africa are associated with the formation of hurricanes over the deep tropics (Gray 1990; Elsner and Schmertmann 1993). Heavy rainfall over the region falls from robust thunderstorm complexes, which are often precursors to easterly waves (African easterly waves) and tropical low pressure systems (lows) that move westward across the Sahel region of Africa toward the Cape Verde Islands. An easterly wave, or low pressure, that develops into a tropical cyclone (most do not) and remains over the warm waters at low latitudes can menace the Caribbean, Mexico, and the United States as a hurricane. A standardized index of Sahel regional rainfall (SRI) during the month of July (Janowiak 1988) is used here. July is the beginning of the rainy season in the Sahel region. The SRI is computed as an area average of standardized rainfall amounts from stations (approximately 14, depending on availability) within a box bounded by 8°N and 20°N latitudes and by 20°W and 10°E longitudes.[2] July values of the SRI range between −1.82 s.d. (1903) and +5.13 s.d. (1915).

An additional source of year-to-year variability in U.S. hurricane activity is the North Atlantic Oscillation (NAO). The NAO is a meridional difference in atmospheric sea level pressure between Iceland and the subtropics. It has been implicated in modulating tropical cyclone activity over the North Atlantic Basin and elsewhere (Elsner and Kocher 2000; Elsner, Bossak, and Niu 2001). The NAO is strongest when pressures are low over Iceland and high over the eastern and central subtropical North Atlantic Ocean. A normalized index of the NAO (NAOI) is calculated as the difference in monthly sea level pressures between Reykjavik, Iceland, and Gibraltar.[3] We use the May value of the NAOI as a compromise between signal strength and timing relative to the hurricane season. The signal-to-noise ratio of the NAO is largest during the boreal winter and spring, whereas the U.S. hurricane season begins in June. Values of the May NAOI range between −3.21 s.d. (1935) and +4.54 s.d. (1956). A recent study utilizing historical and geological data (Elsner, Liu, and Kocher 2000) finds that climatic conditions associated with strong hurricanes along the Gulf Coast occur with a negative (weak) phase of the NAO. Conversely, major hurricane activity along the northeast occur with a positive (strong) phase of the NAO.

Generalized Linear Modeling

The goal of this chapter is to understand the extent to which climate factors just mentioned are important in modeling U.S. hurricane activity from year to year. To help in this regard, we employed a data modeling approach. Tasks associated with data modeling include choosing the functional form of the model and determining the adjustable parameters. We chose the functional form that is most natural to the type of data being modeled.

Traditional linear regression models assume the response variable is continuous; categorical responses (annual hurricane counts) require a different approach. Here we consider generalized linear models (McCullagh and Nelder 1999). A generalized linear model is a probability model in which the mean of the response variable (μ) is related to the p covariates through a regression equation

$$g(\mu) = \alpha_0 + \alpha_1 x_1 + \ldots + \alpha_p x_p \qquad (1)$$

where $g(\mu)$ is called the *link function* that depends on the type of response variable. In the analysis presented here, the response variable μ is hurricane counts (or count) and the covariates $(x_1, \ldots, x_p)$ are the climate factors such as the SOI and the NAO. Values for the coefficients $(\alpha_0, \alpha_1, \ldots, \alpha_p)$ are estimated using the method of maximum likelihood. The logistic and Poisson regressions are special cases of the generalized linear model. Logistic regression is well suited for categorical responses such as whether or not a major hurricane hits the coast. Poisson regression is suited for describing the annual count of U.S. hurricanes.

Logistic Model for U.S. Major Hurricanes

The historical distribution of annual major U.S. hurricane counts shows that the maximum number in any one year is three. Therefore, we modeled the annual occurrence as a binary response variable. Either a year has at least one major hurricane make landfall or it does not. Logistic regression provides a natural general purpose modeling option in the case where there are only two possible responses. Let μ be the probability of at least one major hurricane, then the link function is called the logit function and is expressed as

$$g(\mu) = \log\left[\frac{\mu}{1-\mu}\right] = \text{logit}(\mu) \qquad (2)$$

The inverse of the logit function is the logistic function. We model $g(\mu)$ as a linear function of the covariates and use the SOI, NAOI, and SRI as the covariates.

Figure 12.2 shows the distribution of the three climate factors divided into years according to whether or not there was a major U.S. hurricane. In years of a major hurricane (Yes) the SOI tends to be positive (indicating a La Niña event) and the NAOI tends to be negative. In years without a major hurricane (No), the SOI tends to be negative (El Niño event) and the rainfall tends to be below normal. The SOI appears to have the strongest relationship to major U.S. hurricanes as the interquartile ranges (boxes) have the least overlap for "No" versus "Yes" years. Bivariate logistic regression models confirm SOI as the best single climate factor of the three, with NAO more important than Sahel rainfall.

The initial logistic model we entertain relates the probability of a U.S. major hurricane to these three covariates as

$$\begin{aligned}\text{logit}(\hat{\mu}) &= 0.0843 + 0.2681 \times \text{SOI} - 0.4572 \times \text{NAOI} \\ &\quad + 0.5233 \times \text{SRI} \qquad (3)\end{aligned}$$

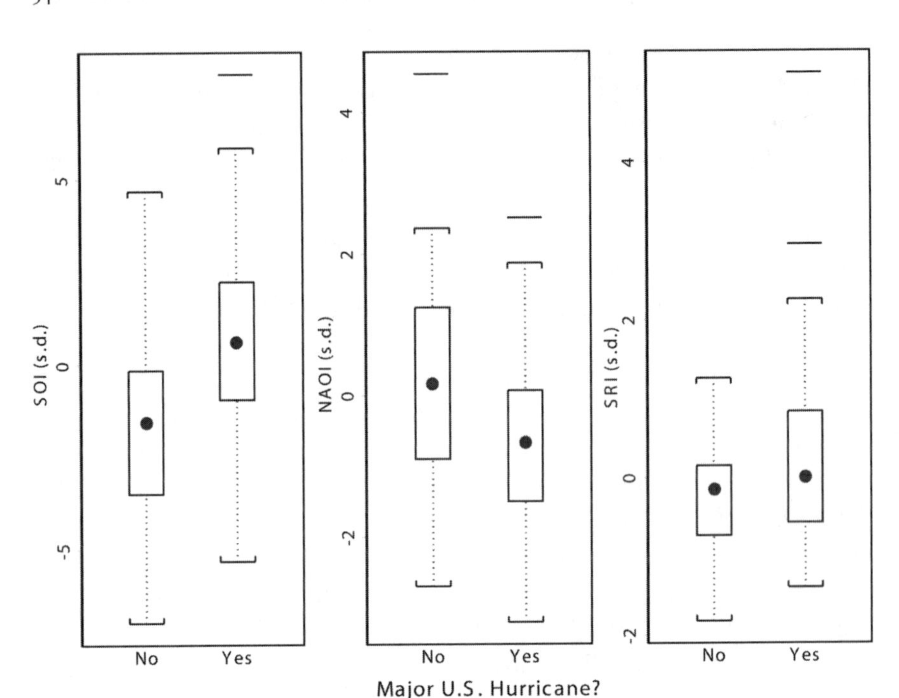

FIGURE 12.2 Box plots of SOI, NAOI, and SRI (covariates) grouped according to whether there was at least one major U.S. hurricane during the year. The circle is located at the median value of the covariate. The box height is equal to the interquartile distance (IQD). The whiskers (dotted lines) extend to the extreme values of the data or a distance of 1.5 IQD, whichever is less. Data points that fall outside the whiskers (outliers) are indicated by horizontal lines.

where a maximum likelihood procedure is used to estimate the coefficients. An analysis of model deviance (table 12.2) shows that the SOI is an important climate factor in the model (low p-value) and that the NAOI is important as a linear predictor after adjusting for the SOI. In contrast, SRI has a p-value that exceeds 0.05, and thus is not considered statistically important after accounting for SOI and NAOI. The deviance is a measure of the discrepancy between observations and fitted values. It serves as a generalization of the usual sum of squares. The magnitude of the deviance difference is proportional to model improvement when that term is added to the model. The deviance difference has a χ^2 distribution from which the p-value is estimated. A quantile plot (figure 12.3) validates the model assumptions by indicating near normal model residuals.

After removing the Sahel rainfall as a climate factor the final model is

$$\text{logit}(\hat{\mu}) = 0.0582 + 0.2903 \times \text{SOI} - 0.4655 \times \text{NAOI} \qquad (4)$$

TABLE 12.2 Analysis of Deviance for the U.S. Major Hurricane Logistic Model

Terms	Deviance difference	d.f.	Residual deviance difference	*p*-value
Null		97	135.86	
SOI	12.07	96	123.79	0.00051
NAOI	9.20	95	114.59	0.00243
SRI	3.26	94	111.34	0.07109

Note: SOI is the August through October cumulative value of the monthly normalized sea level pressure difference between Tahiti and Darwin, Australia; NAOI is the May value of the monthly normalized sea level pressure difference between Reykjavik, Iceland, and Gibraltar; and SRI is the July area-averaged and standardized index of rainfall over the Sahel region of western Africa. The magnitude of the deviance difference is proportional to model improvement when that term is added and has a χ^2 distribution from which the p-value is estimated.

The correlation between the SOI and NAOI is negligible [r(SOI, NAOI) = −0.06] so the NAO provides additional information about whether or not the United States will be hit by a major hurricane. The correlation between SOI and SRI is +0.25 so Sahel rainfall and U.S. major hurricanes are both, to some extent, related to the ENSO. The model applies to the 98-year data set and it may not generalize to other years.

Interpretation of the model is based on recognizing that the logit function is the logarithm of the odds of a U.S. major hurricane. For fixed values of NAOI the ratio of the odds when the SOI has a value A relative to the odds when the SOI has a value B is $e^{0.2903(A-B)}$. For A = +3 s.d. and B = −3 s.d. (for comparison, the 1997 value of the cumulative SOI during August through October was −5.6 s.d.), the odds ratio is 5.7, indicating that the odds of a U.S. major hurricane under a moderate La Niña event is more than five and a half times the odds of a major hurricane under a moderate El Niño event.

The modeling process verifies a relationship between the El Niño cycle and the fluctuating threat of a major hurricane along the U.S. coast. It also implicates the NAO as an additional independent factor in explaining the annual probability of these catastrophic events. Models of the annual frequency of U.S. hurricanes require a somewhat different statistical approach.

Poisson Model for U.S. Hurricanes

The Poisson distribution is a form of the binomial distribution for a large number of trials with small probabilities of an occurrence on any given trial (e.g., Elsner and Schmertmann 1993). The limiting form of the distribution sets no

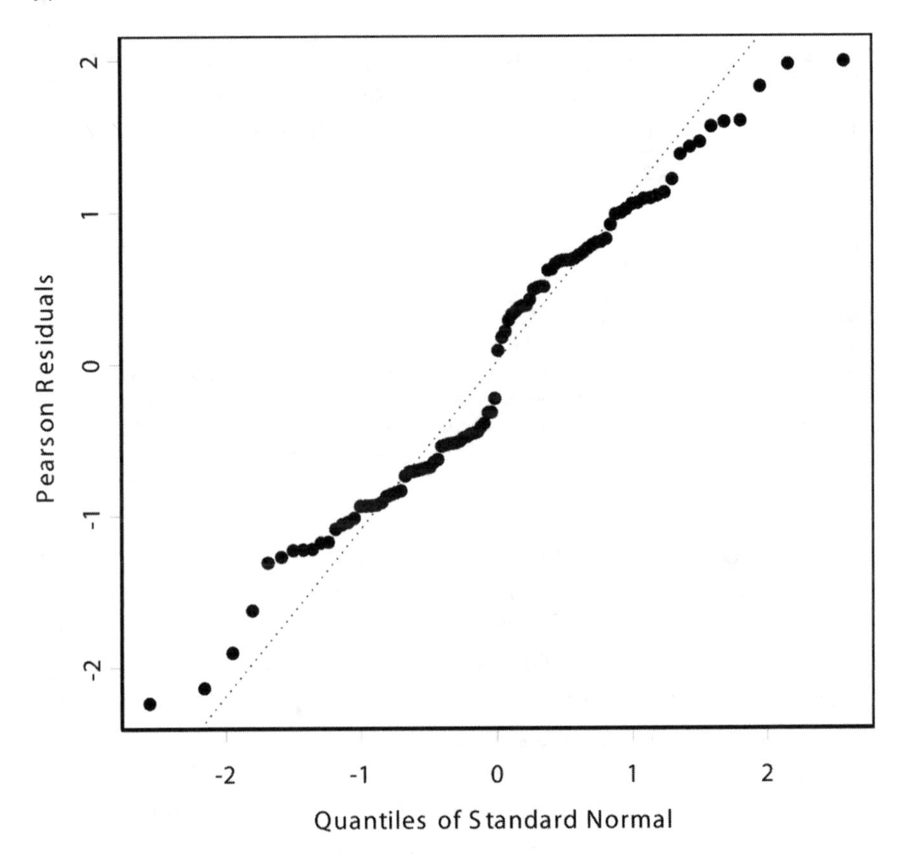

FIGURE 12.3 Normal quantile plot of standardized residuals from the major U.S. hurricane model. The model is a logistic regression with SOI, NAOI, and SRI as the covariates. The dashed diagonal represents a normal distribution. The points lie close to the diagonal indicating no evidence against the assumption of normally distributed residuals.

theoretical limit so it works well for modeling the annual count of U.S. hurricanes. Under this model, the probability of Y U.S. hurricanes is

$$\Pr\{Y\} = \exp(-\lambda)\lambda^{Y}/Y!, \text{ for } Y = 0, 1, 2, \ldots \quad (5)$$

where λ is the annual average. The Poisson distribution is skewed to the right with the skewness most pronounced for small λ. For large λ, the distribution is approximated by the normal distribution.

Figure 12.4 shows the annual distribution of U.S. hurricanes. With a count response (annual number of U.S. hurricanes) and covariates, the Poisson generalized linear model specifies that the distribution of U.S. hurricanes is Poisson (Elsner, Bossak, and Niu 2001) and that the natural logarithm of the mean (link

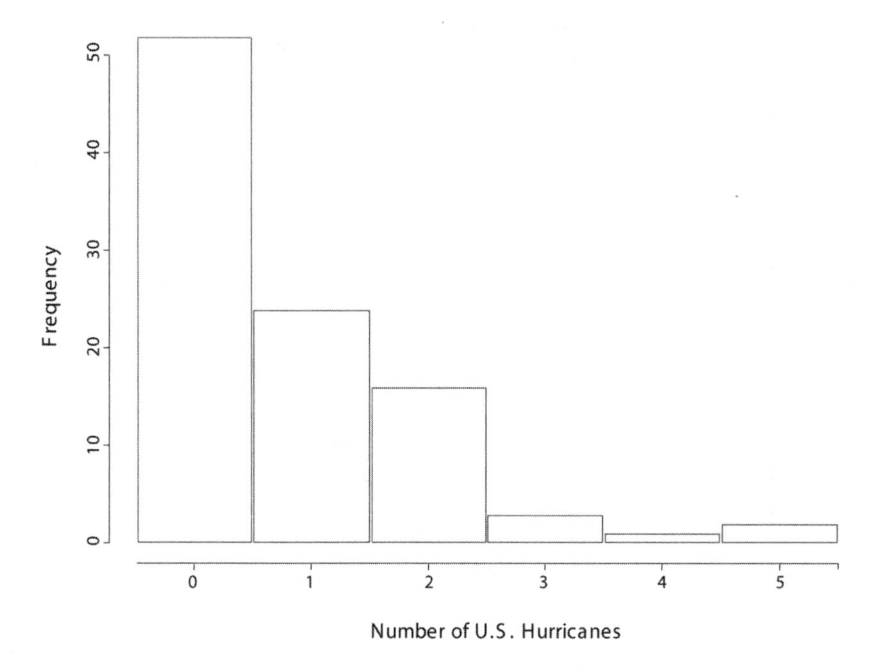

FIGURE 12.4 Histogram of the annual occurrence of U.S. hurricanes. The frequency is the number of years. Note the relative infrequency of years with three or more hurricanes.

function) is linear in the regression coefficients. Using the same covariates as in the major hurricane model, the initial Poisson generalized linear model is

$$\log(\hat{\lambda}) = 0.4597 + 0.0684 \times \text{SOI} - 0.0793 \times \text{NAOI} + 0.1031 \times \text{SRI} \quad (6)$$

where a maximum likelihood procedure is again used to estimate the coefficients. A quantile plot (figure 12.5) indicates a reasonable model. The analysis of deviance (table 12.3) shows that indices of NAO and Sahel rainfall are unimportant after adjusting for the influence of the SOI.

Thus we remove these two climate factors and fit a final model as

$$\log(\hat{\lambda}) = 0.4970 + 0.0828 \times \text{SOI} \quad (7)$$

This model indicates that the mean of the Poisson distribution increases (decreases) with a La Niña (El Niño) event. Accordingly, the probability of one or more U.S. hurricanes is 72% when the SOI is -3 s.d., but increases to 88%

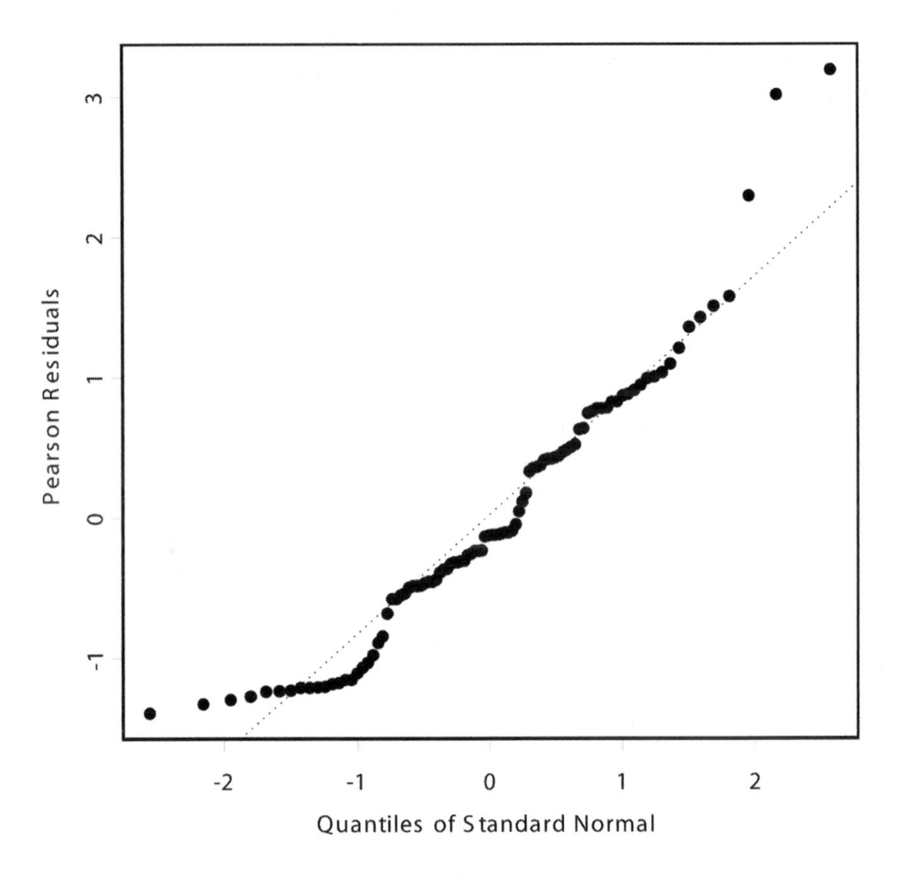

FIGURE 12.5 Normal quantile plot of standardized residuals from the U.S. hurricane model. The model is a Poisson regression with SOI, NAOI, and SRI as the covariates. The dashed diagonal represents a normal distribution. The points lie close to the diagonal indicating no evidence against the assumption of normally distributed residuals.

when the SOI is +3 s.d. These interpretations are strictly applicable only to the 98-year period.

In summary, the generalized linear models establish a statistical basis for the El Niño–Southern Oscillation and the North Atlantic Oscillation as independently important in determining U.S. hurricane activity. The models are based on hurricane landfalls over the entire U.S. coastline from Texas to Maine. The best single predictor of activity is the SOI. The situation is more complicated, however, as the occurrence of a La Niña event increases the probability of a hurricane, but additional factors are important in explaining the probability of a major hurricane. Thus, the annual probability distribution of hurricane winds along the U.S. coast is a function of both tropical cyclone intensity and climate factors. Moreover, the probability of hurricanes regionally

TABLE 12.3 Analysis of Deviance for the U.S. Hurricane Poisson Model

Terms	Deviance difference	d.f.	Residual deviance difference	*p*-value
Null		97	110.48	
SOI	9.00	96	101.48	0.00270
NAOI	2.90	95	98.58	0.08832
SRI	2.09	94	96.48	0.14795

Note: SOI is the August through October cumulative value of the monthly normalized sea level pressure difference between Tahiti and Darwin, Australia; NAOI is the May value of the monthly normalized sea level pressure difference between Reykjavik, Iceland, and Gibraltar; and SRI is the July area-averaged and standardized index of rainfall over the Sahel region of western Africa. The magnitude of the deviance difference is proportional to model improvement when that term is added and has a χ^2 distribution from which the p-value is estimated.

will likely depend upon different combinations of these two factors. In the next section we examine regional landfall probabilities using another data modeling approach.

Regional Landfall Probabilities

Various techniques for estimating annual probabilities of hurricanes locally are proposed in the literature (e.g., Neumann 1987). These approaches are useful in establishing a baseline climatology of extreme wind events, but are predicated on a static distribution of events over time. That is, the methods provide estimates of hurricane probabilities without regard to climate variations. Using these models, the annual probability of a hurricane strike along the Louisiana coast is the same regardless of El Niño. Results from the previous section suggest that regional landfall probabilities will likely change depending upon climate conditions associated with the ENSO and the NAO. In this section we examine results from a model that estimates landfall probabilities conditional upon the SOI and the NAOI (Jagger, Elsner, and Niu 2001).

Weibull Model for Regional Hurricane Winds

Based on a comparison with other distributions, Batts et al. (1980) suggest that the maximal wind speed over an area in a given year be modeled using a Weibull distribution. The survival function (one minus the cumulative distribution function) for the Weibull distribution is an exponential curve. Let V be the unknown yearly maximum wind speed, and v some known value, then the survival function for the Weibull distribution is

$$\Pr\{V > v\} = e^{-(v/b)^a} \quad (8)$$

where a is the shape parameter and b is the scale parameter.

Here an algorithm that extends the earlier work of Batts et al. (1980) is used. The extensions include (1) hurricane intensity estimates at the county level, and (2) conditioning of the Weibull parameters based on climate factors. The parameters are considered response variables with values changing from year to year, and are modeled with a linear regression. A detailed description of the algorithm including error estimates and model comparisons are given in Jagger, Elsner, and Niu (2001). The algorithm provides a parametric model of the annual probabilities at values of wind speeds corresponding to different hurricane intensity levels.

For each coastal county, a Weibull distribution is fit to the yearly maximum wind speed. The location and scale parameters of the Weibull distribution are estimated using linear regressions on the ENSO and NAO covariates using the maximum likelihood estimator. Using these parameters and the associated covariate information, the distribution of exceedence probabilities for any wind speed is estimated by calculating them directly from the Weibull distribution. Exceedence probabilities are related to the percentage chance of a hurricane strike in any one year. The model gives values that can be plotted on a wind speed versus probability graph.

The dynamic probability model from this algorithm can be used in two ways. First we show results using the model in the raw climatological mode. This means that the model provides annual exceedence probabilities of experiencing winds from a hurricane somewhere in the county at various hurricane intensities without regard to climate variations. The geographic distribution of probabilities for Category 1 and Category 3 hurricanes (categories are based on the Saffir-Simpson hurricane damage potential scale) are provided. We also examine results from the model run in the conditional climatogical mode. This means that the model provides exceedence probabilities conditioned on climate factors. Since the model can be run in a conditional climatological mode, it is referred to as a "dynamic" probability model. In general, the probabilities will be different from their raw climatological values depending on the strength and configuration of the climate anomalies. We show the geographic distribution of conditional probabilities for Category 1 and Category 3 hurricanes along with difference maps indicating the change in probabilities.

Raw Climatology

To examine the geographic distribution of annual exceedence probabilities we ran the model for hurricane winds and major hurricane winds (100 kt or greater) for coastal counties from Texas through North Carolina (plate 7). In counties with an insufficient number of hurricanes during the 98-year period the maximum likelihood estimator failed to converge for the scale and shape parameters and the county was assigned a probability in the lowest quintile.

As expected, the largest annual probabilities in the range of 12 to 25% for hurricane winds occurred over southern Florida and eastern North Carolina. Over eastern Texas, near Galveston Bay, counties also had an historically greater frequency of hurricane force winds. Moderately high probabilities were noted over the central Gulf Coast extending from eastern Louisiana through the Florida panhandle. Lowest probabilities, generally less than 10%, were noted over portions of South Carolina and Georgia and over the northern stretch of peninsular Florida. Florida counties along the northeastern Gulf of Mexico (the Big Bend region) also indicate fewer hurricanes. The geographic distribution of probabilities matches closely the variation of tropical cyclone frequencies given by Neumann et al. (1999), where it is noted that hurricane landfalls tend to be most frequent over eastern North Carolina, southern Florida, central Texas, and southeastern Louisiana.

The probabilities of major hurricane force winds are considerably lower than probabilities of Category 1 winds, although their geographic distribution is quite similar. Highest probabilities occur over Texas, parts of the central Gulf Coast, southern Florida, and eastern North Carolina. Dade and Monroe (which includes the Florida Keys) are the most likely counties to experience major hurricane force winds. In fact, the annual probability of hurricane winds in Pinellas and Hillsborough counties (Tampa–St. Petersburg area) of Florida is close to 10% but the probability of major hurricane force winds is less than 1%. Interestingly, the northern and central counties of peninsular Florida indicate some of the lowest probabilities for Category 3 winds.

Conditional Climatology

A useful addition to raw climatological frequencies of extreme winds are conditional frequencies (e.g., Murnane et al. 2000). Here we allow the model to generate exceedence probabilities conditioned on ENSO and the NAO. The effect of these two climate factors on annual hurricane probabilities in coastal counties are analyzed for parameter values that correspond to extremes of the covariates (±5.2 s.d. for SOI and ±3.1 s.d. for the NAOI). Plate 8 shows the differences in probabilities for each county (conditional probability minus raw probability) when the climate is favorable (positive values of the SOI and negative values of the NAOI) for U.S. hurricanes. The differences are smoothed using the values from two neighboring counties. Counties shaded in red indicate probabilities that exceed their climatological value.

As anticipated based on results from the previous section, most of the coastal counties show an increase in the probability of Category 1 winds when conditions of La Niña (positive SOI) and a weak NAO prevail. In particular, substantial increases in probability are noted for much of the Florida coast, especially the Big Bend region. Other regions of increased probability extend from Louisiana to the Florida panhandle as well as portions of the Carolinas. Using a different methodology, Saunders et al. (2000) noted that landfall probabilities along the central Gulf Coast and southern Texas are significantly enhanced during La Niña conditions.

Differences in exceedence probabilities for major hurricane force winds indicate a similar geographic pattern, although over Texas there are more counties with a decrease in probability compared with Category 1 probability differences. The largest increases are noted for portions of Florida and South Carolina. Over northeastern Florida, the probability of Category 1 winds is slightly decreased, but the probability of Category 3 winds is slightly increased under this climate scenario. Variations in probability suggest that the influence of large-scale climate anomalies on hurricane activity is regional. Caution is warranted, however, as much of the variation is likely due to a limited data set and random variability.

Plate 9 shows the differences in probabilities when the climate is unfavorable for U.S. hurricanes. Probabilities are generally lower than average. Many of the coastal counties indicate a drop in probability for both Category 1 and Category 3 hurricane winds, although there is considerable spatial variability. In particular, southeastern Florida and portions of the northern Gulf Coast show the largest decreases (between 10 and 15%) in probability for Category 1 winds. But increases in probabilities are noted over the Big Bend region of Florida and the eastern counties of Texas extending into Louisiana. When conditions are favorable for U.S. hurricanes, the largest probability increases are noted along the western counties of the Florida peninsula through the Big Bend region. When conditions are not favorable, decreases in probability occur in many of these same counties with the exception of the Big Bend region.

In summary, the dynamic probability model describes quantitative changes in hurricane landfall probabilities conditioned on climate variations. We find that positive SOI values combined with negative NAOI values are associated with an increase in the chance of a hurricane or major hurricane strike along the southeastern coastal regions of the United States. In contrast, negative SOI values combined with positive NAOI values are associated with a reduction in the probability of coastal hurricane activity, especially over southeastern Florida, although there is large spatial variability in these changes. This work is one of the first attempts to quantitatively understand the role of climate in modulating regional hurricane activity at the county level.

Future Improvements

The model can be improved in several ways. For instance it is possible to model the maximum intensity of each tropical cyclone as a Weibull distribution and use a Poisson distribution to model the occurrence of hurricanes. In this two-stage model, we can regress three parameters λ, a, b onto the predictors. We also could consider a four-parameter model by using a cutoff wind speed value, v_0, and replace v with $v - v_0$ in the Weibull distribution. In this case, λ is not the rate for the number of tropical cyclones of any velocity affecting the county, but only the rate for cyclones whose wind speeds are v_0 or higher. Another improvement is to incorporate information from adjacent counties into the model for a particular county. A Bayesian approach will work for adjusting the Weibull parameters in this case. Also, the choice of covariates in the model is based on a generalized linear model of activity for the entire coast. The dynamic probability model itself can be used to test the influence of additional climate factors on landfall probabilities if issues of statistical significance are addressed. Another potential improvement is to use the generalized Pareto distribution to model hurricane intensity (e.g., Holmes and Moriarty 1999). The advantage is that this distribution has an upper bound, which could be set using theoretical considerations of a hurricane's maximum potential intensity (e.g., Emanuel 1995; Holland 1997).

A major drawback to work of this kind is the length and quality of the data record. Limitations on sample size are connected to issues of statistical confidence and noise level in the data. Development of a model for prediction will require a cross-validation exercise (Elsner and Schmertmann 1994) with additional data. The hurricane data reanalysis project (Landsea et al., 2002, chapter 7 in this volume) and reconstructions of hurricane records from proxy sources (Liu and Fearn 2000; Donnelly et al. 2001; Liu, Shen, and Louie 2001; Donnelly and Webb, chapter 3, and Liu, chapter 2 in this volume) hold promise of additional data leading to model improvements. For instance, historical data on hurricanes along the coast during the nineteenth century (e.g., Mock, chapter 5 in this volume) suggest important low frequency changes in activity. Data with greater uncertainty can be combined with the modern record using Bayesian methods as illustrated in Elsner and Bossak (2001).

Differences in hurricane visits between the 1851 to 1900 period and the 1951 to 2000 period are mapped in plate 10. The analysis includes the paths of all tropical cyclones that hit the coast at hurricane intensity for the two 50-year periods. Here a county is considered to be hit if the track of the hurricane's eye wall intersects the county boundary. For inland counties, the winds are generally less than hurricane intensity. Counties with more (fewer) hits during the second half of the nineteenth century compared with the more recent period are shaded in blue (red). The possibility of a data bias exists because portions of southeastern Florida and southern Texas were undeveloped in 1851. In this case, the seemingly greater number of hurricanes during the more recent 50 years might be an artifact of undetected storms during the earlier period; however, many of the Gulf Coast counties show more hurricanes during the earlier, less reliable, period. Differences are most pronounced for the counties of northern Florida and southeastern Georgia. In contrast, eastern North Carolina apparently has become a bigger target for hurricanes in recent years.

The map underscores the need to consider the longest possible record of hurricane activity in assessing annual probabilities or return periods. Future work will incorporate these earlier data into the models to provide a more accurate picture of coastal hurricane vulnerability as it relates to the changing climate. The longer historical record will be used to study the robustness of the climatic relationships outlined in this chapter and how these relationships strengthen or weaken over time.

Acknowledgments

We thank T. Jagger for help with coding the dynamic probability model. Writing was improved with the help of R. Murnane, T. Jagger, and an anonymous referee. The National Science Foundation (ATM-9618913 and ATM-0086958) and the Risk Prediction Initiative of the Bermuda Biological Station for Research (RPI-99-001) provided funding for this research.

Notes

1. SOI values are available from the Climatic Research Unit of the University of East Anglia (www.cru.uea.ac.uk).

2. Monthly values of SRI were obtained from the Joint Institute for the study of the Atmosphere and Ocean of the University of Washington (www.jisao.washington.edu/science2.html).
3. Monthly values of the NAOI are available from the Climatic Research Unit of the University of East Anglia.

REFERENCES

Batts, M. E., M. R. Cordes, L. R. Russell, J. R. Shaver, and E. Simiu. 1980. *Hurricane wind speeds in the United States.* NBS Building Science Series No. 124. Washington, D.C.: National Bureau of Standards, Department of Commerce.

Bove, M. C., J. B. Elsner, C. W. Landsea, X. Niu, and J. J. O'Brien. 1998. Effect of El Niño on U.S. landfalling hurricanes, revisited. *Bulletin of the American Meteorological Society* 79:2477–82.

Díaz, H. E., and R. S. Pulwarty, eds. 1997. *Hurricanes: Climate and socioeconomic impacts.* Berlin: Springer-Verlag.

Donnelly, J. P., S. S. Bryant, J. Butler, J. Dowling, L. Fan, N. Hausmann, P. Newby, B. Shuman, J. Stern, K. Westover, and T. Webb III. 2001. A 700-year sedimentary record of intense hurricane landfalls in southern New England. *Bulletin of the Geological Society of America* 113:714–27.

Elsner, J. B., and B. H. Bossak. 2001. Bayesian analysis of U.S. hurricane climate. *Journal of Climate* 14:4341–50.

Elsner, J. B., B. H. Bossak, and X. Niu. 2001. Secular changes to the ENSO-U.S. hurricane relationship. *Geophysical Research Letters* 28:4123–26.

Elsner, J. B., T. Jagger, and X. Niu. 2000. Shifts in the rates of major hurricane activity over the North Atlantic during the 20th century. *Geophysical Research Letters* 27:1743–46.

Elsner, J. B., and A. B. Kara. 1999. *Hurricanes of the North Atlantic: Climate and society.* New York: Oxford University Press.

Elsner, J. B., A. B. Kara, and M. A. Owens. 1999. Fluctuations in North Atlantic hurricanes. *Journal of Climate* 12:427–37.

Elsner J. B., and B. Kocher. 2000. Global tropical cyclone activity: A link to the North Atlantic Oscillation. *Geophysical Research Letters* 27:129–32.

Elsner, J. B., K.-b. Liu, and B. Kocher. 2000. Spatial variations in major U.S. hurricane activity: Statistics and a physical mechanism. *Journal of Climate* 13:2293–305.

Elsner J. B., and C. P. Schmertmann. 1993. Improving extended-range seasonal predictions of intense Atlantic hurricane activity. *Weather and Forecasting* 8:345–51.

Elsner J. B., and C. P. Schmertmann. 1994. Assessing forecast skill through cross validation. *Weather and Forecasting* 9:619–24.

Emanuel, K. A. 1995. Sensitivity of tropical cyclones to surface exchange coefficients and a revised steady-state model incorporating eye dynamics. *Journal of the Atmospheric Sciences* 52:3969–76.

Gray, W. M. 1984. Atlantic seasonal hurricane frequency. Part I: El Niño and 30 mb Quasi-Biennial Oscillation influences. *Monthly Weather Review* 112:1649–68.

Gray, W. M. 1990. Strong association between West African rainfall and U.S. landfall of intense hurricanes. *Science* 249:1251–56.

Gray, W. M., C. W. Landsea, P. W. Mielke, Jr., and K. J. Berry. 1992. Predicting Atlantic seasonal hurricane activity 6–11 months in advance. *Weather and Forecasting* 7:440–55.

Holland, G. J. 1997. The maximum potential intensity of tropical cyclones. *Journal of the Atmospheric Sciences* 54:2519–41.

Holmes, J. D., and W. W. Moriarty. 1999. Application of generalized Pareto distribution to wind engineering. *Journal of Wind Engineering and Industrial Aerodynamics* 83:1–10.

Jagger, T., J. B. Elsner, and X. Niu. 2001. A dynamic probability model of hurricane winds in coastal counties of the United States. *Journal of Applied Meteorology* 40:853–63.

Janowiak, J. E. 1988. An investigation of interannual rainfall variability in Africa. *Journal of Climate* 1:240–55.

Jarvinen, B. R., C. J. Neumann, and M. A. S. Davis. 1984. *A tropical cyclone data tape for the North Atlantic Basin, 1886–1983: Contents, limitations, and uses.* NOAA Technical Memorandum, NWS NHC Report No. 22. Coral Gables, Fla.: National Oceanic and Atmospheric Administration.

Kimberlain, T. B., and J. B. Elsner. 1998. The 1995 and 1996 North Atlantic hurricane seasons: A return of the tropical-only hurricane. *Journal of Climate* 11:2062–69.

Knaff, J. A. 1997. Implications of summertime sea level pressure anomalies in the tropical Atlantic region. *Journal of Climate* 10:789–804.

Lehmiller, G. S., T. B. Kimberlain, and J. B. Elsner. 1997. Seasonal prediction models for North Atlantic Basin hurricane location. *Monthly Weather Review* 125:1780–91.

Liu, K.-b., and M. L. Fearn. 2000. Reconstruction of prehistoric landfall frequencies of catastrophic hurricanes in northwestern Florida from lake sediment records. *Quaternary Research* 54:238–45.

Liu, K.-b., Shen, C., and Louie, K.-s. 2001. A 1000-year history of typhoon landfalls in Guangdong, southern China, reconstructed from Chinese historical documentary records. *Annals of the American Association of Geographers* 91:453–64.

McCullagh, P., and J. A. Nelder. 1999. *Generalized linear models.* Second edition. New York: Chapman & Hall/CRC.

Murnane, R. J. and coauthors. 2000. Model estimates hurricane wind speed probabilities. *Eos, Transactions of the American Geophysical Union* 81:433, 438.

Neumann, C. J. 1987. *The National Hurricane Center risk analysis program (HURISK)* [reprinted with corrections, 1991]. NOAA Technical Memorandum, NWS NHC Report No. 38. Coral Gables, Fla.: National Weather Service, National Hurricane Center.

Neumann, C. J., B. R. Jarvinen, C. J. McAdie, and G. R. Hammer. 1999. *Tropical cyclones of the North Atlantic Ocean, 1871–1998*. NOAA Historical Climatology Series 6-2. Asheville, N.C.: National Climatic Data Center.

Ropelewski, C. F., and P. D. Jones. 1987. An extension of the Tahiti-Darwin Southern Oscillation index. *Monthly Weather Review* 115:2161–65.

Saunders, M. A., R. E. Chandler, C. J. Merchant, and F. P. Roberts. 2000. Atlantic hurricanes and northwest Pacific typhoons: ENSO spatial impacts on occurrence and landfall. *Geophysical Research Letters* 27:1147–50.

Simpson, R. H. 1974. The hurricane disaster potential scale. *Weatherwise* 27:169–86.

Troup, A. J. 1967. Opposition of anomalies of upper tropospheric winds at Singapore and Canton Island. *Australian Meteorological Magazine* 15:32–37.

13
Dynamical Seasonal Forecasts of Tropical Storm Statistics

Frédéric Vitart

This chapter explores the possibility of using dynamical methods to predict the statistics of tropical storms. Dynamical methods predict the evolution of the climate system by solving a set of numerical equations that simulate the basic physical processes in the atmosphere and ocean. These methods can be used to skillfully predict interannual changes in the tropical atmospheric and oceanic circulations a few months in advance. In particular, some dynamical models can be used to predict the occurrence and development of an El Niño event one season in advance. Because variability in tropical storm statistics is strongly related to changes in the tropical atmospheric and oceanic circulations, dynamical methods can be used for the seasonal forecast of tropical storm statistics through their forecasts of atmospheric and oceanic circulations.

Some dynamical atmospheric models produce tropical cyclonic systems that look similar to observed tropical storms; I have referred to these as "modeled tropical storms." Modeled tropical storms display a vertical structure, a genesis location, and a monthly variability generally consistent with observations, though their frequency is usually lower than observed, particularly over the north Atlantic and the eastern North Pacific. Most importantly, modeled tropical storms are sensitive to changes in the environmental atmospheric circulation and ocean surface temperatures as observed. For example, an El Niño event generally has the same significant impact on the statistics of modeled tropical storms as it has on the statistics of observed tropical storms. Therefore, modeled tropical storms can display an interannual variability consistent with the interannual variability of observed tropical storms.

The skill of a coupled atmosphere-ocean dynamical model in predicting the frequency of tropical cyclones a few months in advance has been evaluated. The dynamical model produces realistic seasonal forecasts of the tropical storm frequency in the Atlantic and the western North Pacific, but has been applied with less success for predicting storm frequency in other ocean basins. A comparison with the statistical forecasts issued by the Colorado State Hurricane Forecast Team indicates that dynamical seasonal forecasts can be competitive with current operational statistical forecasts. A study of tropical cyclone landfalls over Mozambique suggests that, in addition, to predicting the frequency of tropical cyclones, dynamical models can also be used to predict the risk of landfall a few months in advance.

Tropical storms give rise to some of the most devastating natural disasters. During the past 50 years, hundreds of thousands of lives have been lost because of tropical storm landfalls. Predicting their exact occurrence months in advance is out of reach, but predicting their probability may be possible and could help in preparing the population for such disasters.

The two main methods used to predict the seasonal variability of tropical storms are based either on empirical or dynamical methods. Empirical methods, which originated from studies by Riehl (1954) and Namias (1955), have been often used to issue seasonal forecasts of tropical storms (e.g., Gray et al. 1992, 1993, 1994; Nicholls 1992; Basher and Zheng 1995; Hess, Elsner, and LaSeur 1995). Empirical methods are based on the idea that there are predictors, up to one year in advance, of the main dynamical parameters that affect the genesis of tropical storms. Historical data are used to identify the predictors and estimate their weight in a statistical regression. The empirical methods suffer from two main limitations: (1) their performance is strongly restricted by the low number of past events and (2) the methods assume that future weather will behave in a manner similar to the past. This assumption may not be valid in the event of a significant change in climate due to natural or anthropogenic causes.

Dynamical methods use numerical models to predict from first-order principles, or established approximations to them, how the climate system will evolve. Dynamical methods have not been widely used for the seasonal forecasting of tropical storms, mostly because they are computationally expensive. Unlike most empirical methods, dynamical methods require substantial computer power that is at present accessible only in a few institutions.

This chapter focuses on dynamical methods for forecasting seasonal tropical cyclone activity. The first section will discuss the sensitivity of tropical storms to atmospheric circulation. The next section considers the different techniques that are used in dynamical seasonal forecasting. The third section focuses on the application of dynamical seasonal forecasting to the prediction of tropical storms, with a description of a method that tracks tropical disturbances created by an atmospheric model. The following section describes an example of a dynamical seasonal forecast of tropical storms and discusses its skill in predicting the frequency of tropical storms a few months in advance. This is followed by a discussion of seasonal prediction of tropical storm landfall. I close with a discussion of the prospects for dynamical methods.

Tropical Storm Variability

Gray (1968, 1975) documented the initial detection points of all cyclones observed over a 20-year period. During this period no tropical storm formed within 2.5° latitude of the equator, in the central North Pacific, in the South Pacific east of 160°E, or in the South Atlantic. In addition, tropical storms occurred only at specific times of the year. For instance, the Atlantic tropical storm season extends from June to November, with a peak in September. Approximately 80 tropical storms occur each year throughout the world. Remarkably, this number is very stable, although the number of tropical storms over one particular basin such as the Atlantic Ocean can display dramatic changes from one year to another. For instance, only four tropical storms were observed over the Atlantic in 1983, compared with 19 in 1995. Tropical storm intensity, location, and tracks can also display significant differences from one year to another. For example, tropical storms over the western North Pacific were more intense with a genesis location significantly more westward in 1999 than in 1998.

As tropical storms form in very specific locations and are highly seasonal, specific environmental conditions must be required to accomplish the transition from a loosely organized disturbance to an intense vortex. Gray (1968, 1975, 1979) related the frequency of tropical storms to six main environmental factors:

1. Above average vorticity in the low-level atmosphere. It is observed that tropical storms form only in regions of positive low-level vorticity. According to Gray (1975), such an environment favors the creation of a convergent wind toward the center of the storm because of friction. This convergent flow provides the supply of momentum, heat and moisture necessary to develop and maintain a tropical cyclone.
2. A location at least a few degrees poleward of the equator. Tropical storms do not form within a 4° to 5° latitude band around the equator. This is due to the earth's rotation, which, through the Coriolis effect, plays an important role in the generation of cyclones (Gray 1979).
3. Weak vertical shear of the horizontal wind, defined as the magnitude of the difference between the horizontal wind in the upper and lower tro-

posphere (the troposphere is the lowest layer of the atmosphere and extends up to 16 km in the tropical regions). The strong ascending motion near the center of the cyclone transports heat and moisture taken from the ocean to the upper part of the troposphere, where heat and moisture accumulate to form a "warm core," which plays a crucial role in the intensification of the cyclone. If the horizontal wind circulation in the upper troposphere is not coherent with the wind circulation in the lower troposphere, heat and moisture are carried away from the center of the storm, and therefore cannot accumulate in the upper troposphere. Under strong wind shear conditions, a tropical disturbance cannot intensify and become a tropical storm.

4. Sea-surface temperatures exceeding 26°C and a deep thermocline. Palmen (1956) observed that tropical cyclone genesis occurs only when sea-surface temperatures exceed 26°C. The top layer of the ocean provides the heat and moisture required for tropical storms. Therefore, warm sea-surface temperatures are necessary for the formation of tropical cyclones. This explains why tropical cyclones do not form in high latitudes or over the tropical southeastern Pacific, where sea-surface temperatures are generally below 26°C.
5. Conditional instability through a deep layer in the atmosphere. Tropical storms develop only if the lower and upper tropospheric flows are coherent. Riehl (1954) noticed that the flows in the tropical upper and lower troposphere are coherent only if there is strong vertical motion through the whole troposphere to produce a vertical coupling. This occurs only when the atmosphere is sufficiently unstable.
6. Above-average moisture in the middle levels of the atmosphere. Tropical storms form only in regions where the seasonally averaged values of humidity in the middle levels of the atmosphere are high (Gray 1979). A high level of humidity in the middle atmosphere is conducive to strong vertical motion through the whole troposphere, therefore increasing the coupling between the upper and lower troposphere. This provides more favorable conditions for the genesis of tropical cyclones as explained in item 5.

The first three parameters are dynamic parameters, the second three are thermodynamic parameters. As discussed by Gray (1975), the thermodynamic parameters vary slowly in time, and the dynamic parameters can change dramatically within the tropical cyclone season. Therefore, Gray hypothesized that cyclones form only during periods when dynamical parameters are perturbed beyond a threshold away from their climatological mean, because the thermodynamic parameters would be expected to remain beyond any threshold values necessary for tropical cyclone development throughout the entire tropical cyclone season.

A large part of the climatology and seasonality of tropical storms can be explained by these six parameters. For instance, the presence of very strong wind shear over the central Pacific may explain the scarcity of tropical storms over this region, whereas cold sea-surface temperatures over the southeastern Pacific likely prohibit formation of cyclones over this region.

Changes in the frequency of tropical cyclones from one year to another can also be largely explained by variability of these six parameters. For instance, the Atlantic tropical cyclone activity is significantly reduced during El Niño years (characterized by a significant increase of sea-surface temperatures over the eastern tropical Pacific) (Gray 1984; Shapiro 1987; Goldenberg and Shapiro 1996). The proposed mechanism is that the eastward shift of positive-sea-surface temperature anomalies associated with El Niño causes an increase of deep convection over the equatorial eastern Pacific. The increased convection enhances the upper-level westerly zonal winds and the vertical wind shear over the region where most Atlantic tropical storms develop. As discussed previously, the increased vertical wind shear reduces the Atlantic tropical storm activity. A more complete description of the variability of tropical cyclones is provided by McBride (1995).

The six large-scale parameters have a strong impact on tropical storm variability and can be derived from the large-scale circulation in the atmosphere. Therefore, predicting the statistics of tropical storms involves predicting the large-scale circulation conditions that affect the genesis of cyclones. A general circulation model can be used for this purpose.

Dynamical Seasonal Forecasts

Physical Justification

Seasonal forecasting (the production of forecasts ranging from a few months to one year) is a recent application of dynamical models and is based on a different physical justification than that for short-and medium-range-forecasts. Short- and medium-range forecasts, which do not usually include an analysis or forecast of the ocean, are atmospheric initial condition problems. This means that the initial state of the atmosphere is estimated from observations; then numerical equations that simulate physical laws of the atmosphere are applied in order to predict how these initial conditions of the atmosphere will evolve with time. Seasonal forecasts are atmospheric boundary-value problems. Energy and mass fluxes across the air-sea and air-land interfaces drive the long-term behavior of the atmosphere.

The 1950s and 1960s were years of strong optimism about the ability of humans to predict weather and even to control it. The emergence of computer technology and space satellites made scientists believe it was possible to predict the weather months in advance. The discovery of chaos, or the "butterfly effect," by Lorenz (1963) tempered this optimism. The butterfly effect recognizes that tiny errors in initial conditions can grow very quickly, degrading the quality of a weather forecast. It explains why it is impossible to predict individual weather systems months in the future in a deterministic way. Lorenz (1963) discovered, however, that although chaos implies a very strong limit to the deterministic forecasting of weather events, the unpredictable behavior of the weather is characterized by a limited range of patterns, called attractors. Seasonal forecasting can be understood as predicting changes in the attractor for weather, and hence shifts in the probability distribution function of weather (Palmer 1993).

Ocean temperatures are much more stable than atmospheric and land-surface temperatures, essentially because of the very large heat capacity of water and the efficiency of vertical mixing in the upper layers of the ocean. Anomalies of sea-surface temperatures (defined as the difference between observed sea-surface temperatures and their climatological mean) are of order of 1°C in magnitude, have a spatial scale of order of 1000 km, and can last several months. Due to their relatively large time and space scale, sea-surface temperature anomalies are the most important reason why the climate changes from one year to another. Anomalies of sea-surface temperatures in the tropics are particularly important because relatively small changes in tropical sea-surface temperatures can shift the deep convection in the tropical atmosphere over a large distance. Deep convection in the tropical atmosphere drives much of the global atmospheric circulation; therefore, anomalies in tropical sea-surface temperatures can have a global impact. For example, during an El Niño event, positive sea-surface temperature anomalies of a few degrees Celsius over the tropical eastern Pacific affect climate over most of the globe. Mid-latitude sea-surface temperature anomalies have a local influence, but their impact on the large-scale circulation is thought to be much weaker.

Land processes such as soil moisture and snow cover can also have an important impact on large-scale circulation on the seasonal time scale. Beljaars et al. (1996) showed that summer droughts and floods in the United States are influenced by previous soil-moisture anomalies. Yang and Lau (1998) pointed out the significant role of snow cover on the Summer Asian monsoon. Other factors like sea ice cover and vegetation may also impact the climate on the seasonal time scale, but their role is still not well understood. Stratospheric changes may also influence the large-scale circulation. The stratosphere displays a strong year to year variability with the Quasi-Biennial Oscillation (QBO), which is used as a predictor for hurricane activity over the Atlantic (Gray et al. 1992).

In summary, sea-surface temperature anomalies are not the only factors that affect large-scale atmospheric circulation, but the tropical ocean is the main driving force behind the atmospheric circulation on seasonal time scales. The anomalies of tropical sea-surface temperatures have a lifetime of order of several months, making it possible to issue forecasts of sea-surface temperatures for the next few months. From these forecasts, one may deduce, using empirical or dynamical methods, the impact of sea-surface temperature anomalies on the large-scale atmospheric circulation. The physical justification of seasonal forecasting is based on the slow evolution of ocean temperatures and the strong impact of tropical sea-surface temperature on the climate. The atmospheric boundary conditions (sea-surface temperatures, land surface, stratospheric circulation, etc.) are the source of predictability of the atmospheric circulation on a seasonal scale.

Different Strategies for Dynamical Seasonal Forecasting

A general survey of techniques for seasonal forecasting is given by Stockdale (2000), from which the present section is largely inspired. Dynamical seasonal forecasting consists of two steps. The first step involves the prediction of the sea-surface temperatures for the next few months. The second step requires the simulation of the impact of these temperature anomalies on the climate. Dynamical seasonal forecasting may include the use of a statistical method for one of these two steps.

Several methods may be applied to predict sea-surface temperature anomalies. The simplest of these methods is called "persistence." It assumes that the sea-surface temperature anomalies present at the initial time will persist for the next few months. Persistence produces a useful forecast at the seasonal time scale and because sea-surface temperatures vary slowly, it is difficult for a dynamical model to beat persistence during the first weeks of a forecast. The major problem of such a method is that sea-surface temperatures can display strong variability, as, for example, during an El Niño event.

Forecasts of sea-surface temperature anomalies can be produced using an ocean general circulation model coupled to an atmospheric model. The atmospheric component produces the forcing of the ocean model: wind stress, heat fluxes, precipitation and evaporation rates. It can be a statistical model (Barnett et al. 1993) or an atmospheric global circulation model. The main problem with a full global circulation model is that it produces an error in the mean state that can exceed the year-to-year changes that are being studied. Some methods for correcting these model errors include empirical corrections during the coupled model integrations (Ji, Leetmaa, and Kousky 1996; Kirtman et al. 1997), using only observed anomalies to initialize the model (Latif et al. 1993), or subtracting the model errors after the calculations (Stockdale 1997). The ability of coupled general circulation models to predict sea-surface temperature anomalies has significantly improved in the recent years, and some coupled models provided the best real-time numerical forecasts of the 1997–1998 El Niño (Trenberth 1998). However, most dynamical models are still unable to outperform a simple statistical forecast of the 1997 El Niño event (Landsea and Knaff 2000).

Once sea-surface temperature anomalies have been predicted, the next step in producing a seasonal forecast consists of translating the forecast of sea-surface temperature anomalies into prediction of the atmospheric large-scale circulation. Several experiments have demonstrated that an atmospheric general circulation model forced by observed sea-surface temperatures simulates some of the anomalies of the large-scale circulation (Palmer and Mansfield 1986; Lau and Nath 1994; Livezey et al. 1997). In particular, the impact of El Niño on the large-scale circulation seems to be well simulated by a general circulation model. Atmospheric general circulation models display teleconnection patterns that are broadly consistent with observations.

A strategy for seasonal forecasting consists of using a coupled ocean-atmosphere model to predict simultaneously both sea-surface temperatures and their impact on the large-scale circulation (Stockdale et al. 1998). The advantages with this approach are that the coupled interactions between atmosphere and ocean are better represented. A disadvantage is that the combination of ocean and atmospheric errors leads to a climate model that can be very different from the real world, and this may impact the atmospheric response. This coupled method approach, applied at the European Centre for Medium-range Weather Forecasts (ECMWF), has been successful in forecasting the El Niño event of 1997–1998 and the following La Niña event (Stockdale et al. 1998).

Two seasonal forecasts that differ only slightly in initial conditions can lead to very different predictions. This is due to the chaotic nature of the atmospheric system, as described by Lorenz (1963). By creating a large number of forecasts, it is possible to sample the probability distribution function of the atmospheric variable of interest. An analogy would be playing cards with a deck where some red cards have been removed. If someone draws randomly a card from the deck, it would be impossible to conclude that the deck is not a regular deck. But if the person puts the card back inside the deck and repeats the same operation a very large number of times, it will become clear that black cards are drawn more often than red cards. If the operation is repeated a sufficient number of times, then the person will be able to assess that statistically there is very little chance that the deck is a regular deck. The size of the ensemble for seasonal forecasting usually ranges between 5 and 50 operations and is limited by the cost of computer time. The obligation to create an ensemble of forecasts makes dynamical seasonal forecasting very expensive.

Dynamical seasonal forecasting is still in its early stages, and it is only recently that researchers used the method successfully to predict the occurrence of an El Niño event. This was an important step, because the El Niño–Southern Oscillation (ENSO) is the most important mode of climate variability from one year to another. Many other steps are necessary to improve dynamical seasonal forecasting in a significant way. For example, researchers using dynamical models have difficulty in realistically simulating organized convection in the tropics. This is of considerable concern, as tropical sea-surface temperature anomalies alter atmospheric circulation on a global scale through their impact on deep convection in the atmosphere. Improvements in model algorithms for organized convection would provide more realistic simulations of the impact of sea surface temperature on global atmospheric circulation. The drift of sea-surface temperatures from realistic values when an atmosphere and an ocean model are coupled is also an important limitation to seasonal forecasting. After a few months of integration, the drift in sea-surface temperatures can reach a few degrees Celsius. The drift is caused primarily by an imbalance in the heat flux exchange between the ocean and atmosphere. The model drift generates important systematic errors in the forecasts, and the elimination of those model errors is crucial to the success of dynamical seasonal forecasting.

Dynamical Seasonal Forecasts of Tropical Storms

Dynamical seasonal forecasting of tropical storm frequency has the same physical basis as seasonal forecasting. Tropical sea-surface temperatures have a predictability that can exceed a few months. Therefore, their impact on the statistics of tropical storms through large-scale atmospheric circulation and thermodynamic processes may be predictable months in advance. Dynamical seasonal forecasting of tropical storms involves three steps: (1) prediction of tropical sea-surface temperature anomalies, (2) simulation of the impact of the sea-surface temperature on the large-scale circulation in the tropics, and (3) simulation of the impact of the large-scale circulation on tropical storm statistics.

A combination of statistical and dynamical methods can be used to issue a forecast of tropical storm frequency. For instance, sea-surface temperature anomalies can be predicted using a coupled general circulation model and then a statistical model may be applied to translate the temperature anomalies into tropical storm forecasts.

Step 1: Prediction of Sea-Surface Temperature Anomalies

One measure of a model's seasonal forecast skill is the model's ability to correctly forecast ENSO events. A model's skill in forecasting ENSO is often based on a comparison between predicted and observed sea-surface temperature anomalies. Barnston et al. (1999) reviewed the performances of eight dynamical models in forecasting the 1997–1998 El Niño event. Dynamical models strongly underestimated the exceptional strength of this event, but one of the most successful forecasts was based on the ECMWF model (figure 13.1). The ECMWF model was used to successfully predict the occurrence of the 1997–1998 El Niño event, its duration, and its decay. It correctly predicted the propagation of equatorial waves and the shift in convective activity, but the model underestimated the intensification in June and July 1997 by more than 1°C (Stockdale et al. 1998). The underprediction of sea-surface temperature anomalies would probably degrade tropical cyclone forecasts.

Sea-surface temperatures in areas other than eastern tropical Pacific can affect the variability of tropical storm frequency. Goldenberg and Shapiro (1996) and Saunders and Harris (1997) documented the important role of Atlantic surface temperatures. Shapiro (1982) found positive correlation between Atlantic tropical storm activity and sea-surface temperatures just west of Africa. Zhang, Drosdowsky, and Nicholls (1990) found high correlations between the frequency of tropical cyclones over the western North Pacific and higher latitude sea-surface temperatures. Raper (1992) also confirmed an impact of local sea-surface temperatures on tropical storm frequency over the western North Atlantic, the Australian Basin, and possibly the eastern North Pacific. Raper (1992) argues that the link between these sea-surface temperatures and the tropical storm activity is most likely through the large-scale circulation rather than through a direct impact of sea-surface temperatures on tropical storms. These observational studies suggest that skill in forecasting ENSO

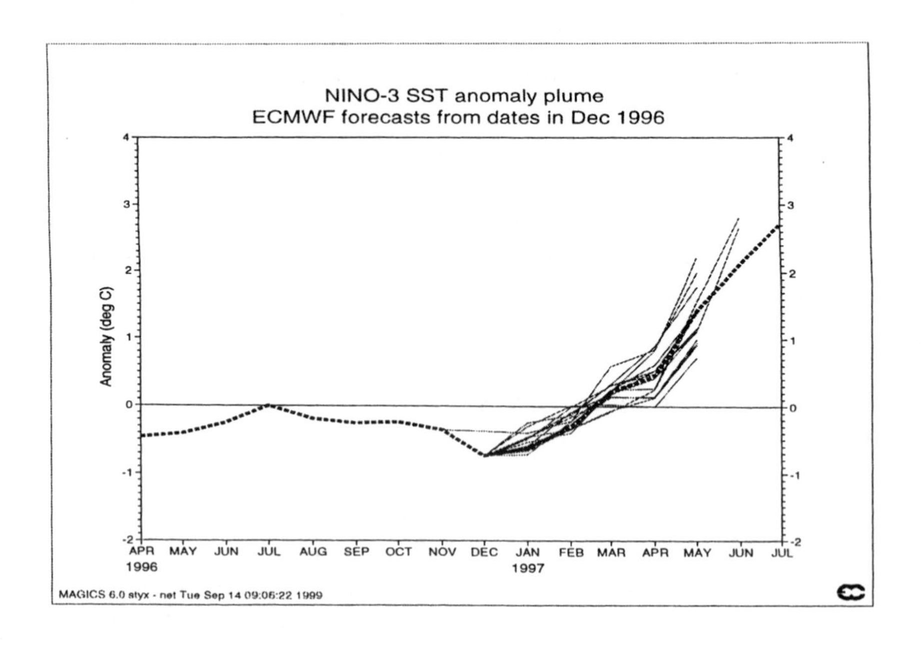

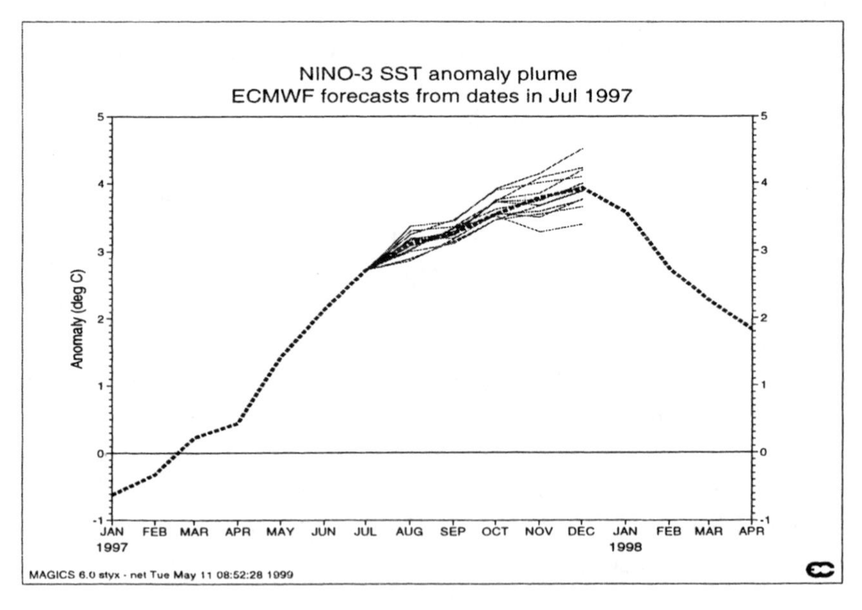

FIGURE 13.1 Plume of monthly mean sea-surface temperature anomalies predicted by the ECMWF seasonal forecasting system (Stockdale et al. 1998) for the Niño 3 region (5°S–5°N, 90°–150°W). Each light line represents one member of the ensemble. The heavy line shows the observed values. The initial date of the forecast is December 1, 1996 (*top left*), March 1, 1997 (*top right*), July 1, 1997 (*bottom left*), and October 1, 1997 (*bottom right*) (from Vitart and Stockdale 2001; copyright, American Meteorological Society, used with permission).

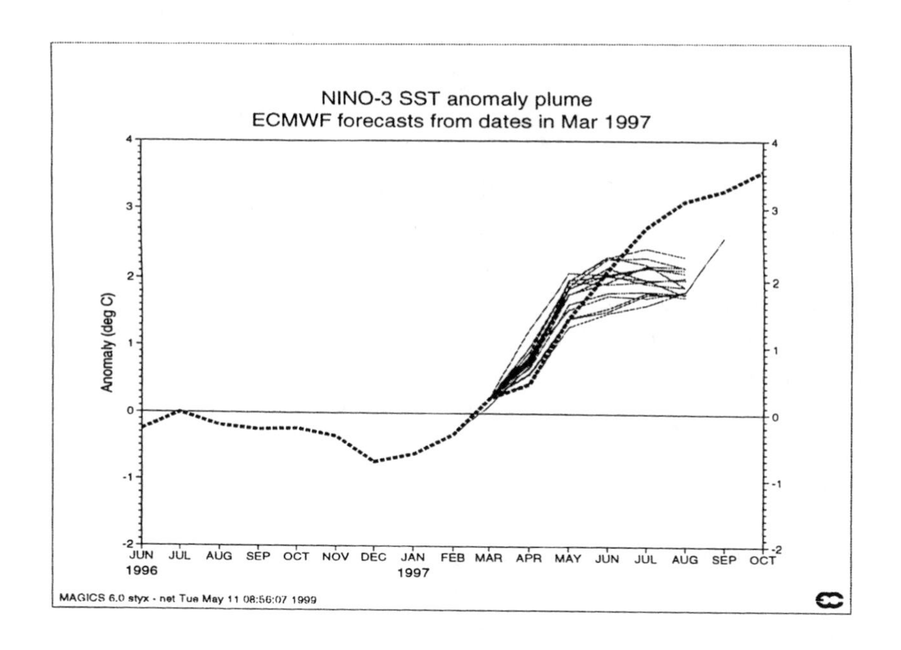
NINO-3 SST anomaly plume
ECMWF forecasts from dates in Mar 1997
Anomaly (deg C)
JUN JUL AUG SEP OCT NOV DEC JAN FEB MAR APR MAY JUN JUL AUG SEP OCT
1996
1997
MAGICS 6.0 styx - net Tue May 11 08:56:07 1999

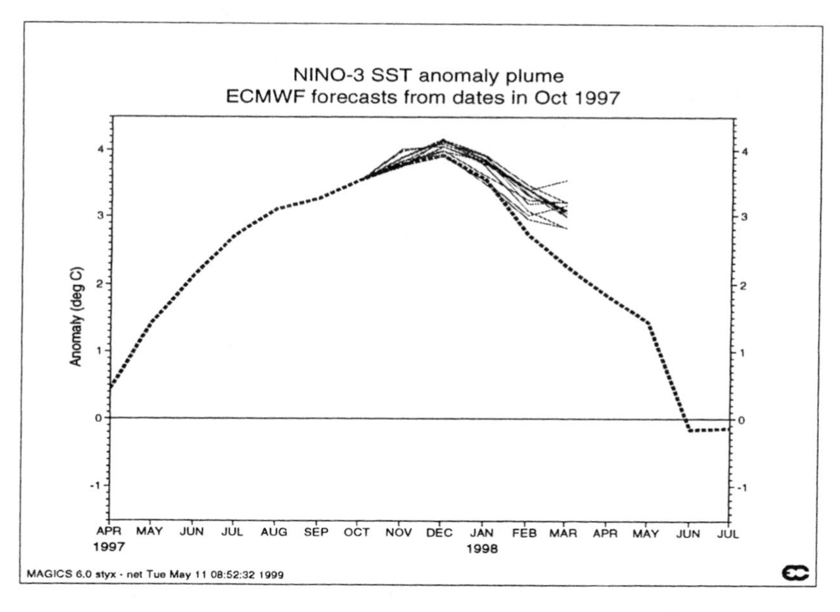
NINO-3 SST anomaly plume
ECMWF forecasts from dates in Oct 1997
Anomaly (deg C)
APR MAY JUN JUL AUG SEP OCT NOV DEC JAN FEB MAR APR MAY JUN JUL
1997
1998
MAGICS 6.0 styx - net Tue May 11 08:52:32 1999

is not sufficient in forecasting the interannual variability of tropical storms. The dynamical models should also show skill in predicting the variability of sea-surface temperatures over all basins where tropical storms develop and at extratropical latitudes.

An example of a coupled ocean-atmosphere general circulation model's skill in predicting sea-surface temperature anomalies is displayed in figure 13.2a. At this time scale, the ECMWF dynamical model (figure 13.2a) displays better skill than persistence (figure 13.2b) over most of the tropics. The skill of the coupled model is mostly concentrated in the tropical regions, where the linear correlation with the observed variability of sea-surface temperatures is on order of 0.5. Therefore, a coupled general circulation model can have useful skill in predicting the sea-surface temperatures for the next season. This is a significant improvement in forecast skill in comparison to coupled model skill 20 years ago. Errors in sea-surface temperature forecasts over the tropical regions are still significant, however, and are likely to degrade the prediction of tropical storms.

Errors in forecasts may originate from a combination of model errors, errors in the initial conditions, and lack of predictability. One can use ensemble forecasts to evaluate the potential predictability of sea-surface temperatures. Assume one member of a forecast ensemble as being the "truth." The other members of the forecast ensemble are used to predict the truth. The results shown in figure 13.2c imply that there is high potential predictability of sea-surface temperatures over most of the tropics. This is an encouraging sign that improvements in ocean and atmospheric modeling, and in generating initial conditions, should lead to significantly increased forecasting capability.

Step 2: Simulation of the Impact of Sea-Surface Temperature Anomalies on the Large-Scale Circulation in the Tropics

Two parameters, low-level vorticity and vertical wind shear, have been identified by Gray (1979) as being particularly important in the variability of tropical storms. For instance, the significant impact of El Niño on the tropical storm statistics over the Atlantic can be explained by its impact on vertical wind shear and low-level vorticity over the Atlantic Basin. Correctly simulating the impact of sea-surface temperature anomalies on the interannual variability of these two large-scale parameters is essential for dynamical seasonal forecasting of tropical storms.

ENSO is a major cause of tropical storm variability over several ocean basins, so it is important for a dynamical model to realistically simulate its

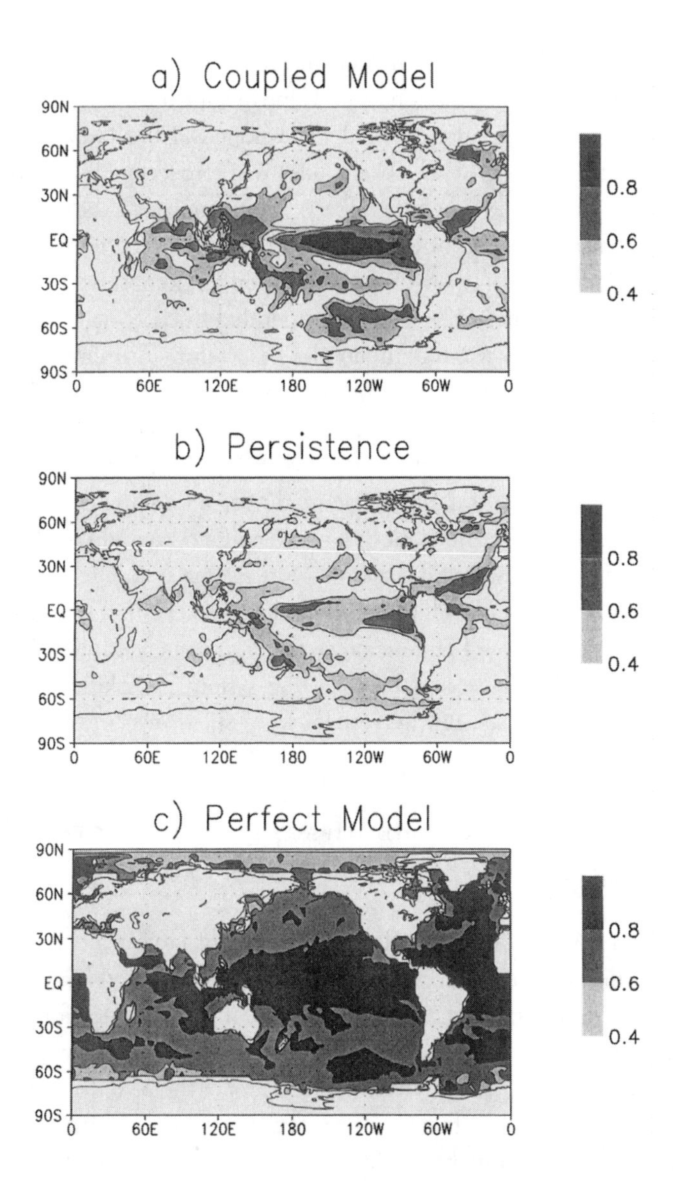

FIGURE 13.2 Point correlation map of seasonal mean sea-surface temperature anomalies for the period 1991 to 1999 among (*a*) ECMWF Seasonal Forecasting System and observations, (*b*) persistent sea-surface temperatures and observations, and (*c*) one member of ensemble forecast and the other members. The seasonal mean sea-surface temperatures have been averaged for months 3, 4 and 5 of each forecast. The contour interval is 0.2, with the first contour at 0.4.

impact on vertical wind shear and low-level vorticity. To test the impact of ENSO on large-scale circulation, Vitart and Anderson (2001) integrated an atmospheric general circulation model developed at the Geophysical Fluid Dynamics Laboratory (GFDL). The model was forced by climatological sea-surface temperatures everywhere except over tropical Pacific and Indian Ocean. In a first ensemble of experiments, the sea-surface temperatures over the tropical Pacific and Indian Ocean corresponded to observed El Niño conditions. In a second ensemble, they corresponded to La Niña conditions. The conclusion of this study was that the model successfully simulated a significant increase of vertical wind shear and a decrease of low-level vorticity over the western North Atlantic. Low-resolution atmospheric models can also successfully simulate the increase of anticyclonic vorticity and divergence in the lower troposphere over the central North Pacific during a La Niña event, along with opposite tendencies over the western North Pacific (Wu and Lau 1992).

These results suggest that if a seasonal forecasting system can skillfully predict the occurrence and development of an El Niño event, it may also be used to predict the variability of large-scale circulation associated with ENSO. For example, the ECMWF seasonal forecasting system can skillfully predict El Niño as well as its impact on the vertical wind shear over most of the tropics (figure 13.1).

Step 3: Simulation of the Impact of the Large-Scale Circulation on Tropical Storm Statistics

The simplest method for evaluating the impact of large-scale circulation on tropical storm statistics involves analyzing how the predicted large-scale circulation differs from climatology, and then deducing a tendency for the tropical storm statistics. For instance, if the model predicts more vertical wind shear and colder sea-surface temperatures over the Atlantic for the next tropical storm season, then it is likely that there will be fewer Atlantic tropical storms. This method does not give very precise forecasts, however, and usually works only if all the parameters indicate the same tendency. If the dynamical model predicts colder sea-surface temperatures over the Atlantic (conducive to less tropical storm activity) and less vertical wind shear (conducive to more tropical storm activity), then a more sophisticated method is needed to develop a forecast. This method can be empirical or dynamical.

A possible methodology for predicting tropical cyclone numbers is to use the environmental parameters predicted by the dynamical model as predictors in an empirical equation. Empirical methods have often been used to deduce the variability of small-scale quantities (in the present case, tropical storms) from large-scale tendencies. These methods are referred to as downscaling. Gray (1975, 1979) has pioneered such methods for evaluating the frequency of tropical storms with the creation of the genesis parameters described in the first section. Watterson, Evans, and Ryan (1995) used this approach with the large-scale parameters simulated by an atmospheric model forced by observed sea-surface temperatures. When applied to the climatology of the atmospheric model, the genesis parameters indicated regions of cyclonic activity that were broadly consistent with observations, but with an unrealistically high total number of tropical storms. Moderate correlations were found between the time series of the observed and simulated cyclogenesis numbers in the central Pacific, eastern North Pacific, and North Atlantic regions during the period from 1979 to 1988. In addition, both observations and genesis parameters indicated an increase in tropical cyclone activity in the central Pacific during El Niño years.

These results suggest that downscaling could lead to a powerful tool for the seasonal forecasting of tropical storms when applied to coupled ocean-atmosphere integrations. This methodology has the important advantage of not being directly affected by the low-resolution of the atmospheric model. Watterson, Evans, and Ryan (1995) provided evidence that a major limitation of this method is the deficiency of the genesis parameters defined by Gray (1979) in diagnosing climatological and interannual frequencies of tropical cyclones. The empirical equations are tuned to agree well with the statistics of tropical storms in the real-world climate. Because dynamical models simulate a climate that can be very different from observations, applying the empirical equations directly to model outputs is likely to result in an unrealistic frequency of tropical storms. In any case, this method would suffer from the same limitations as empirical methods in general (as already discussed in the introduction), and would be particularly vulnerable to climate changes.

Another strategy to deduce the statistics of tropical storms from the prediction of large-scale circulation considers the tropical disturbances created by the atmospheric component of the dynamical model as an analogy to observed tropical storms. The variability from one year to another of observed tropical storms is then deduced from the variability of model tropical disturbances. The rest of this chapter will be dedicated to this dynamical method.

Disturbances similar to observed tropical storms were detected in a low-resolution atmospheric circulation model for the first time by Manabe, Holloway, and Stone (1970). This was a surprising discovery as their model had only nine vertical levels and a horizontal resolution of about 400 km, which is about the scale of an observed tropical storm. The simulated tropical disturbances described by Manabe, Holloway, and Stone (1970) have a characteristic scale of 2000 to 4000 km, which is an order of magnitude larger than observed tropical storms. These tropical disturbances generated by a dynamical model are modeled tropical storms. Several studies (e.g., Bengtsson, Bottger, and Kanamitsu 1982; Wu and Lau 1992; Haarsma, Mitchell, and Senior 1993) have confirmed Manabe, Holloway, and Stone's findings. An interesting conclusion that resulted from these studies is that modeled tropical storms appear in all kinds of atmospheric circulation models, with different numerical schemes and different parameterizations of physical processes. Modeled tropical storms have been used to study the impacts of global warming (e.g., Broccoli and Manabe 1990; Bengtsson, Botzet, and Esh 1995, 1996; Tsutsui and Kasahara 1996) and El Niño (Wu and Lau 1992) on the frequency of tropical storms.

The ability of a low-resolution atmospheric model to simulate tropical storms remains controversial (McBride 1984; Evans 1992; Lighthill et al. 1994). Most criticisms question the ability of an atmospheric model with a horizontal resolution of a few hundreds of kilometers to capture physical processes with a size scale one order of magnitude lower. Modeled tropical storms display considerable differences with observations: they are much bigger than in reality and they do not have rainfall bands, an eye, or an eye wall. Therefore, modeled tropical storms lack physical processes at the core of the storm that are likely to play an important role in their intensification. A resolution on the order of less than 10 km would be necessary to capture the physical process in the eye wall. Unfortunately, such high resolution is presently not accessible for seasonal forecasting.

A way to simulate tropical storms more realistically is to use a very fine horizontal resolution over a targeted area of the globe and a coarser grid elsewhere (Knutson et al., chapter 15 in this volume). Walsh and Watterson (1997) adopted this approach with a regional model with a horizontal resolution of 125 km over the Australian Basin, nested within a coarse-resolution general circulation model. This approach increases the realism of the modeled tropical storms over the targeted area. Coarser resolution over the other regions (for instance, over the tropical eastern Pacific) is likely to degrade significantly the ability of the atmospheric model to predict a realistic variability of the large-scale circulation from one year to another. This would affect the seasonal forecast of tropical storms.

An alternative is to use a low-resolution global atmospheric model and make the assumption that the differences between modeled and observed tropical storms have no significant impact on their variability from one year to another. Observed tropical storms display a warm temperature anomaly in mid-troposphere above the center of the storm, which is known as the warm core. Strong upper-level wind prohibits the formation of a warm core and therefore the formation of a tropical storm by ventilating the warm air above the center of the storm. Modeled tropical storms also display a warm core, so it is likely that the simulated large-scale circulation will effect the formation of modeled tropical storms in a manner similar to what occurs in the observed atmosphere. If so, the statistics of modeled tropical storms may be analogues to those for observed tropical storms and useful for seasonal forecasting. In order to confirm if this is true, the modeled storms' vertical structure, genesis location, climatological frequency, seasonal cycle, and variability from one season to another have been compared with observations.

Mean Structure of Modeled Tropical Storms

Several studies (Wu and Lau 1992; Bengtsson, Botzet, and Esh 1995; Vitart, Anderson, and Stern 1997) have compared the mean structure of modeled tropical storms to observations. All show similarities between the simulated and observed tropical storm structures: convergence, high moisture content, strong upward motion, heavy precipitation at the lower levels of the tropical storm, and anticyclonic vorticity and divergence in the upper troposphere. An example of a modeled storm's characteristics in a two-dimensional cross-section of a tropical storm simulated by an atmospheric model from GFDL with a horizontal resolution of about 300 km is shown in figure 13.3 (Vitart, Anderson, and Stern 1997).

Modeled tropical storms exhibit a well-defined warm core. One striking difference between model and observed storms, however, is that modeled storms are too large. The modeled storms have their maximum winds at a distance of hundreds of kilometers from the center of the storm, instead of about 25 km, as in observations (Weatherford and Gray 1988). Bengtsson, Bottger, and Kanamitsu (1995) showed that increasing the resolution of the atmospheric model increases the realism of modeled tropical storms. With a horizontal resolution of 100 km, modeled tropical storms have a horizontal scale that is close to observed storms.

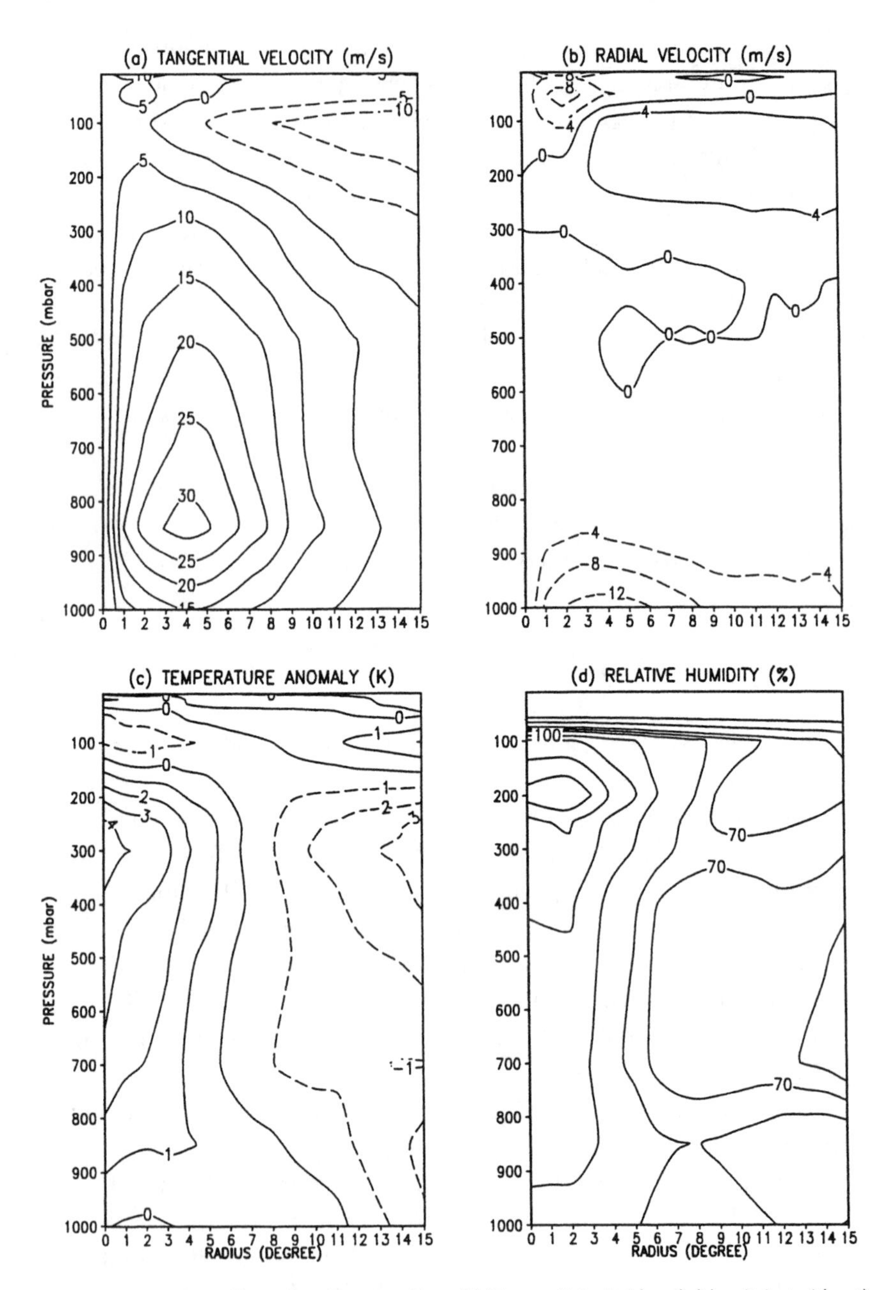

FIGURE 13.3 Two-dimensional cross-section of (*a*) tangential wind (ms^{-1}), (*b*) radial wind (ms^{-1}), (*c*) temperature anomaly (K), and (*d*) relative humidity for a modeled tropical storm simulated by a GFDL atmospheric model (from Vitart, Anderson, and Stern 1997; copyright, American Meteorological Society, used with permission).

Automatic Detection of Modeled Tropical Storms

Several articles have described algorithms for tracking modeled tropical storms (e.g., Haarsma, Mitchell, and Senior 1993; Bengtsson, Bottger, and Kanamitsu 1995; Vitart, Anderson, and Stern 1997). The algorithms are based on the basic principle that several meteorological criteria need to be satisfied in order to determine the presence of a tropical storm. The detection has two steps. First, the modeled tropical cyclones are detected for each output of the atmospheric model. Second, the modeled tropical storms from one model output are connected to another in order to determine the full trajectory of the model storm. Modeled tropical storms are defined by the presence of a local maximum of relative vorticity in the lower troposphere and a local minimum of sea level pressure, which will define the center of the storm. In addition, detection of modeled tropical storms includes testing for the presence of a warm core through, for example, the detection of a maximum of temperature anomaly in the upper troposphere above the center of the storm. This last criterion is very efficient for rejecting most extra-tropical systems in the process of detection.

The second step establishes the trajectories of the modeled tropical storms. For a given storm, the algorithm tests whether a modeled tropical storm will appear in the next model output at a distance smaller than the maximum distance a tropical storm can move during that interval of time. The major difficulty arises when there are several possible choices for a tropical storm trajectory. This may happen, for example, over the western North Pacific, where the density of observed and modeled tropical storms can be high during the peak period of the typhoon season. In those cases, the closest storm to the climatological track may be chosen. A more sophisticated technique calculates the steering wind at time $t = n$, and estimates the storm position for the time $t = n + 1$. The cyclone closest to this position can then be chosen as being part of the trajectory. Increasing the frequency of the atmospheric model outputs decreases the occurrence of several possible choices, and therefore facilitates tracking modeled tropical storms.

Using the algorithms to predict modeled tropical storms depends on the resolution of the atmospheric model. The finer the horizontal resolution, the more intense the modeled tropical storms are. The algorithm can be "tuned" by correlating it with observed tropical storms found in reanalysis data such as that produced by NCEP or ECMWF. The reanalyzed storm data are projected on the same grid as the atmospheric model. For instance, when applied to the 1998–1999 ECMWF global reanalyses projected on a 200-km resolution grid, Vitart, Anderson, and Stern (1997) used the detection algorithm to find up to

75% of all the tropical storms observed during that period. The detection algorithm also generated six non-observed tropical storms, with most of the false storms located over the North Indian Ocean.

Statistics of Modeled Tropical Storms

Manabe, Holloway, and Stone (1970) noticed that the tropical disturbances in their model appeared in the same regions and at the same period of the year as observed tropical storms. Several studies involving different atmospheric models confirmed the agreement between the genesis location of simulated and observed tropical storms (Haarsma, Mitchell, and Senior 1993; Bengtsson, Bottger, and Kanamitsu 1995; Tsutsui and Kasahara 1996; Vitart, Anderson, and Stern 1997). But some models simulated tropical storms at unrealistic locations such as the South Atlantic, over land, and at high latitudes.

These unrealistic tropical storms are scarce in the models. The model's simulation of tropical storms in high latitudes may be explained by the presence of polar lows with a warm core; such events are referred to as "arctic hurricanes" (Emanuel and Rotunno 1989). The simulated storms over the South Atlantic and over land suggest deficiencies in the physics of the atmospheric models.

Most studies of modeled tropical storms show good agreement between observations and the number of tropical storms simulated over each ocean (see, e.g., Bengtsson, Bottger, and Kanamitsu 1995). But the frequency of tropical storms over the eastern North Pacific is strongly underestimated in most models. It is not clear what causes this discrepancy. The mountains along Mexico's west coast probably have an impact on tropical storms over the eastern North Pacific. Simulating the impact of these mountain ranges on the large-scale atmospheric circulation is a challenge for numerical models, particularly when they have a low resolution. The tropical Atlantic is another basin where it is difficult to simulate a realistic frequency of tropical storms. The tropical Atlantic is characterized by a particularly strong climatological vertical wind shear that creates conditions that are often unfavorable for the formation of tropical storms. Therefore, small errors in the simulation of the Atlantic vertical wind shear can have strong consequences on modeling the climatological frequency of Atlantic tropical storms.

Bengtsson, Bottger, and Kanamitsu (1995) and Vitart, Anderson, and Stern (1997) reported a monthly variability of modeled tropical storms that is consistent with observed storms. Figure 13.4 displays the monthly variability of modeled tropical storms simulated at GFDL (Vitart 1999). The GFDL atmospheric model simulates tropical activity from May to December over the Atlantic, with a peak in September as in observations. Modeled tropical storms are present in the whole year over the western North Pacific, with a peak in August and September. The model is less successful over the North Indian Ocean, where the second peak of the bimodal distribution of tropical storm frequency is too weak.

Although a low-resolution atmospheric model can simulate tropical storms with a realistic frequency and genesis location, the simulated tracks can be unrealistically short and too poleward. With finer horizontal resolution, the tracks of the simulated tropical storms are significantly more realistic. Krishnamurti (1988) has shown that a reasonable representation of tropical storms is possible using high resolution spectral models (e.g., a T170 model, which has an equivalent resolution of about 80 km).

In modeled tropical storms, intensity is also very strongly dependent on the horizontal resolution of the atmospheric model. At low resolution, the storms

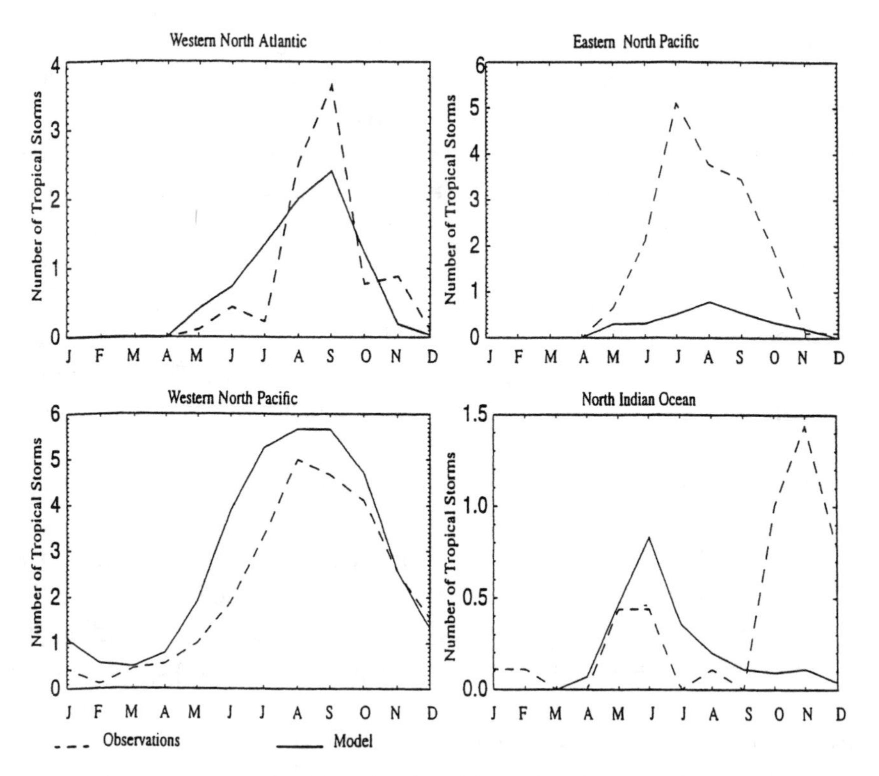

FIGURE 13.4 Monthly variability of observed (dashed line) and modeled tropical storm frequency (solid line) simulated by a GFDL model over the North Atlantic, eastern North Pacific, western North Pacific, and North Indian Ocean (from Vitart 1999).

are larger, but with a weaker intensity, than at high resolution. For instance, no modeled tropical storm in the GFDL atmospheric model has a minimum sea level pressure below 950 mb (Vitart, Anderson, and Stern 1997), which corresponds to the intensity of only a Category 2 hurricane. Bengtsson, Bottger, and Kanamitsu (1995) also reported that the lowest pressure in the Max Plank Institute model was 957 mb and the highest wind speed was 53.1 m s^{-1}.

Wu and Lau (1992) used an atmospheric model developed at GFDL to investigate if El Niño had the same impact on modeled tropical storms as it has on observed tropical storms. The model was integrated for 25 years and forced with observed sea-surface temperatures. They found a significant correlation between ENSO and the frequency of modeled tropical storms over the Atlantic, the eastern Pacific, the western North Pacific, and the South Pacific. They also noticed that in modeled storms, the genesis location over the western North Pacific is significantly shifted toward the east during El Niño years, as it is in observed tropical storms (Chan 1985).

A slight difference in the initial conditions in a model can lead to different frequencies of modeled tropical storms due to the chaotic nature of the atmosphere. Bengtsson, Bottger, and Kanamitsu (1995) integrated five years of the ECHAM3 atmospheric model from the Max Plank Institute and forced the model with climatological sea-surface temperatures. They then repeated the same experiment using observed sea-surface temperatures. The variability of modeled tropical storm frequency from one year to another was of the same order of magnitude in both cases. The variability of modeled tropical storm frequency due to sea-surface temperature anomalies is of the same order as the variability due to the chaotic nature of the atmosphere. The authors concluded that an ensemble of atmospheric integrations would be necessary to determine the significance of sea-surface temperature anomalies in modeling tropical storms.

An experiment using an ensemble of atmospheric model integrations was explored by Vitart, Anderson, and Stern (1997). In this experiment, the atmospheric model from GFDL was integrated for 10 years with nine different initial conditions. The first year of each integration was discarded to eliminate the direct effects of the initial conditions. A key result of this study was the significant variability of modeled tropical storm frequency from one year to another. For instance, all nine members of the ensemble simulated more Atlantic tropical storms in 1988 than were observed in 1982 and the smallest number of simulated storms for 1988 equaled the largest number of simulated storms for 1982 (figure 13.5). An ensemble simulation of a tropical storm forecast is said to have potential predictability when the ensemble distribution of the number of tropical storms for a given year can be distinguished from the modeled "climato-

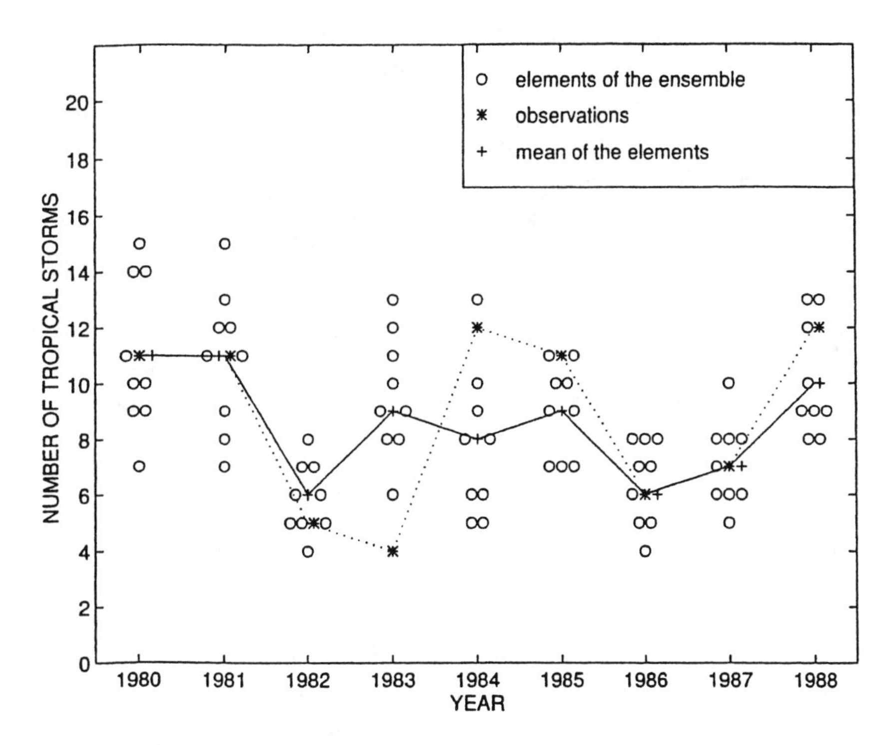

FIGURE 13.5 Interannual variability of tropical storm frequency over the North Atlantic for the period from 1980 to 1988. The dotted line represents observed tropical storm numbers. The solid line represents the mean tropical storm numbers for each element of the simulation ensemble. Each circle represents the number of tropical storms simulated by one member of the ensemble (from Vitart, Anderson, and Stern 1997; copyright, American Meteorological Society, used with permission).

logical" distribution in a statistically significant way. The potential predictability of the GFDL ensemble simulation was particularly strong over the western North Pacific, eastern North Pacific, and the Atlantic for most of the years. The only common point between the nine members of the ensemble was sea-surface temperatures, indicating that sea-surface temperatures have a significant impact on the variability of modeled tropical storms. In addition, the simulated variability of modeled tropical storms displayed reasonable agreement with the observed variability over the basins with strong potential predictability. Low potential predictability and low correlation with observed tropical storm variability were found for modeled storm frequencies in the North Indian Ocean, the South Indian Ocean, and the South Pacific.

It is possible that the consistency between the model and observed tropical storm variability was a coincidence. Therefore, it is important to understand what affects the interannual variability of modeled tropical storms and if the physical mechanisms are identical to those in observed storms. Vitart, Anderson, and Stern (1999) found that, for the 1980 to 1988 period, the interannual variability of tropical storms simulated by the GFDL model was very strongly correlated (correlation larger than 0.9) to the interannual variability of the atmospheric large-scale circulation, and weakly correlated to the variability of the local sea-surface temperatures. Therefore, sea-surface temperatures impact modeled tropical storms primarily through a nonlocal impact on the model's large-scale circulation, as in observations. Furthermore, the impact of the large-scale circulation on the modeled tropical storms seemed to be consistent with observations. The basins where the model failed to simulate a realistic variability of modeled tropical storms coincided with regions where the atmospheric model failed to simulate a realistic variability of vertical wind shear and low-level vorticity.

The horizontal resolution of the atmospheric model appears to be a key factor for simulating a realistic tropical storm mean state, intensity, and trajectory. Horizontal resolution may not, however, play such a crucial role in tropical storm frequency. An atmospheric model with a resolution of about 300 km can simulate a realistic climatological frequency, along with a realistic variability of tropical storm frequency (Vitart, Anderson, and Stern 1997). The frequency of modeled tropical storms can be very sensitive to the physical parameterization of the atmospheric model. To evaluate this sensitivity Vitart et al. (2001) analyzed modeled tropical storms simulated by four different models developed at GFDL. The models had identical dynamic cores and resolution (about 300 km) but had different parameterizations of deep convection. The results of these experiments indicate that the mean structure, intensity, and frequency of modeled tropical storms can vary dramatically with the model parameterization.

The intensity of modeled tropical storms seems to be particularly sensitive to the vertical mean-state of the simulated atmosphere. A warmer and dryer upper-level troposphere is conducive to more intense modeled tropical storms with a higher warm core, which is consistent with experiments realized with a high resolution regional model (Knutson and Tuleya 1999; Shen, Tuleya, and Ginis 2000). Therefore, the impact of model parameterizations on tropical storms seems to occur primarily through the parameterization's effect on mean vertical profiles of temperature and humidity. The mean frequency of modeled tropical storms is also very sensitive to the background thermodynamic mean state, especially the background Convective Available Potential Energy (CAPE). It is not clear at present if the impact of CAPE on modeled tropical storms is realistic, as there has been no observational evidence of such sensitivity.

Although the mean frequency and intensity of modeled tropical storms are strongly affected by the way the cumulus parameterization alters the atmosphere's thermodynamic mean state, the interannual variability of modeled tropical storms is mostly affected by the cumulus parameterizations's impact on the interannual variability of the atmosphere's large-scale circulation. Vitart et al. (2001) discuss a case where a change in the cumulus parameterization significantly degraded the realism of modeled tropical storms, reducing their intensity, and lowering their mean frequency, but improving significantly the realism of their interannual variability. Therefore, improving the realism of the physical structure of modeled tropical storms will not guarantee an improvement in the seasonal forecast of tropical storm frequency.

An Example of Dynamical Seasonal Forecast of Tropical Storms

To date there have been few attempts to use dynamical models for seasonal forecasting of tropical storms. One attempt undertaken at ECMWF used a coupled general circulation model (Vitart and Stockdale 2001). The present section will discuss the performance of this system in predicting the frequency of tropical storms.

The ECMWF seasonal forecasting system (Stockdale et al. 1998) is based on a coupled atmosphere-ocean model that has been integrated for about six months. The atmospheric component has a resolution of about 200 km and 30 forecasts are made each month from daily ocean and atmosphere analyses. An automatic procedure for tracking modeled tropical storms is then applied to the 30 forecasts. The modeled tropical storms are then counted over each ocean basin and for each member of the forecast ensemble. The probability distribution function, which is deduced from the ensemble of modeled tropical storm numbers, represents the seasonal forecast of tropical storm frequency.

This method is the highest level of modeling sophistication because all three steps described in the previous section are performed dynamically. The coupled dynamical system predicts sea-surface temperature anomalies. The atmospheric component of the coupled system simulates the impact of ocean temperature on the large-scale circulation. The simulated large-scale circulation and sea-surface temperatures control the model's tropical storms. A major problem with this method is that errors in the component models lead to a drift in the climate of the coupled system. This implies that modeled tropical storms develop in a different thermodynamical and dynamical environment than in observations and explains why the climatological frequency of modeled tropical storms differs greatly from the observations. The coupled model creates fewer and weaker tropical storms than observed over all the basins, except the North Indian Ocean. The deficit is particularly important over the North Atlantic and the eastern North Pacific.

A forecast is issued by comparing the probability distribution function of the 30-member ensemble forecast to the climatological probability distribution function. The predicted frequency is multiplied by a factor, so that the model climatology coincides with the observed climatology. To estimate the climatology of the coupled system, a set of forecasts has been made each month for an earlier period (1991 to 1996).

With this dynamical seasonal forecasting system, forecasts can be issued over each ocean basin a few months before the peak of the tropical storm season. For instance, forecasts are issued as early as April 1 and up to July 1 for the North Atlantic, where the peak season is August through September. Forecasts of tropical storm frequency over the North Atlantic issued on July 1 of each year for the period 1991 to 1999 are shown in figure 13.6. The predicted interannual variability is in good agreement with observations, with a linear correlation of 0.8, except for the year 1996, when the model strongly overestimates the number of tropical storms. For forecasts starting earlier (from April 1 to June 1), the agreement with observations is not as strong as when starting on July 1, but the linear correlation with observations consistently remains higher than 0.6.

Based on Trenberth's (1997) definition of El Niño, during the period from 1991 to 1999, one major (1997) and three moderate El Niño events (1991, 1992, and 1994) occurred. The ECMWF-coupled model had skill in predicting the occurrence of El Niño and its impact on the large-scale circulation as discussed in the previous sections (figure 13.1). The coupled model predicted a significant and realistic reduction of Atlantic tropical storm frequency during the El Niño events (figure 13.6) and an increase of Atlantic tropical storm activity as observed during the 1998–1999 La Niña event. The model also predicted a significant increase of tropical storm activity in 1995 as observed. Warmer sea-surface temperatures observed and correctly predicted by the coupled model over the Atlantic are the most likely cause for this increase of Atlantic tropical storm activity both in observation and in the coupled model.

The coupled model has the greatest skill in predicting the variability of tropical storms over the western North Pacific. The linear correlation between predicted and observed interannual variability of tropical storm frequency exceeds 0.7 for all the forecasts starting from April to July over this basin. In addition to frequency, the coupled model predicts a realistic variability of tropical storm genesis location. The linear correlation between the variability of predicted and observed mean longitude and latitude of tropical storm genesis is larger than 0.8 over the western North Pacific. For instance, the model predicts a signifi-

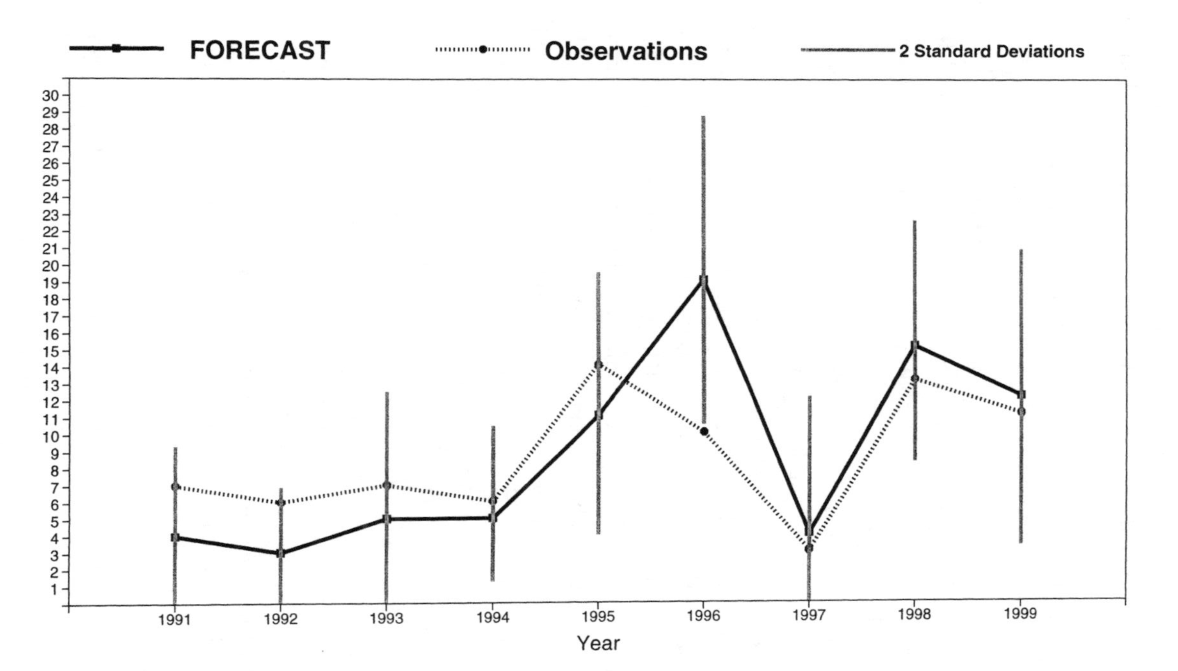

FIGURE 13.6 Interannual variability of tropical storm frequency over the North Atlantic for the period from 1991 to 1999. The dashed line represents observations. The solid line represents the mean of the ensemble of forecasts (after scaling) starting on July 1 and covering the period from August to November. The vertical lines represent 2 standard deviations in the ensemble of forecasts (from Vitart and Stockdale 2001; copyright, American Meteorological Society, used with permission).

cant westward shift of tropical storm location in 1998 and an eastward shift in 1997 (figure 13.7).

The coupled model is less skillful in predicting storms over the eastern North Pacific, North and South Indian Ocean, Australian Basin, and the South Pacific. The linear correlation between observed and predicted tropical storm frequency is always positive, but it rarely exceeds 0.5 in these regions. Vitart, Anderson, and Stern (1997) noticed the same discrepancy from one basin to another with the GFDL atmospheric model forced by observed sea-surface temperatures, and the realism of the simulated large-scale circulation was identified as an explanation (Vitart, Anderson and Stern 1999). To check if this is also the case for the present coupled system, the atmospheric component of the ECMWF-coupled model has been integrated while forced by observed sea-surface temperatures. The linear correlation between observed and simulated variability of tropical storm frequency exceeds 0.9 over the Atlantic, which means that the atmospheric model is able to simulate more than 80% of the variance of Atlantic tropical storms. The correlation is much weaker over the other ocean basins, as it is in the coupled model. Furthermore, the basins where the coupled model poorly predicts the variability of tropical storms coincide with the basins where researchers have difficulty using the atmospheric model to predict the variability of low-level vorticity and vertical wind shear. The limited skill of the atmospheric model component in simulating realistic variability of the large-scale circulation is likely to be the strongest limitation to the skill of the coupled model's ability to predict the frequency of tropical storms. It is not clear if this limited skill is mostly due to model errors or to the poor predictability of the large-scale circulation over certain areas.

Forecasts of Atlantic tropical storm frequency are issued each year at Colorado State University (CSU) (Gray et al. 1992, 1993, 1994). These forecasts, available since 1991, are based on statistical methods and are produced as early as December of the year prior to the Atlantic tropical storm season. A comparison between this forecast and the dynamical forecasts previously described indicates that dynamical seasonal forecasts can be competitive with a well-established empirical method (figure 13.8).

Present seasonal forecasting systems have a horizontal resolution of order of 200 km, which corresponds to the resolution used for operational short and medium-range forecasts about 15 years ago. Thanks to the constant improvement in computer capabilities, short- and medium-range forecasts currently have a horizontal resolution on the order of 10 km. If computers improve as rapidly as they did during the last decades, such high resolution should be accessible for operational seasonal forecasting in the future. At such high resolution,

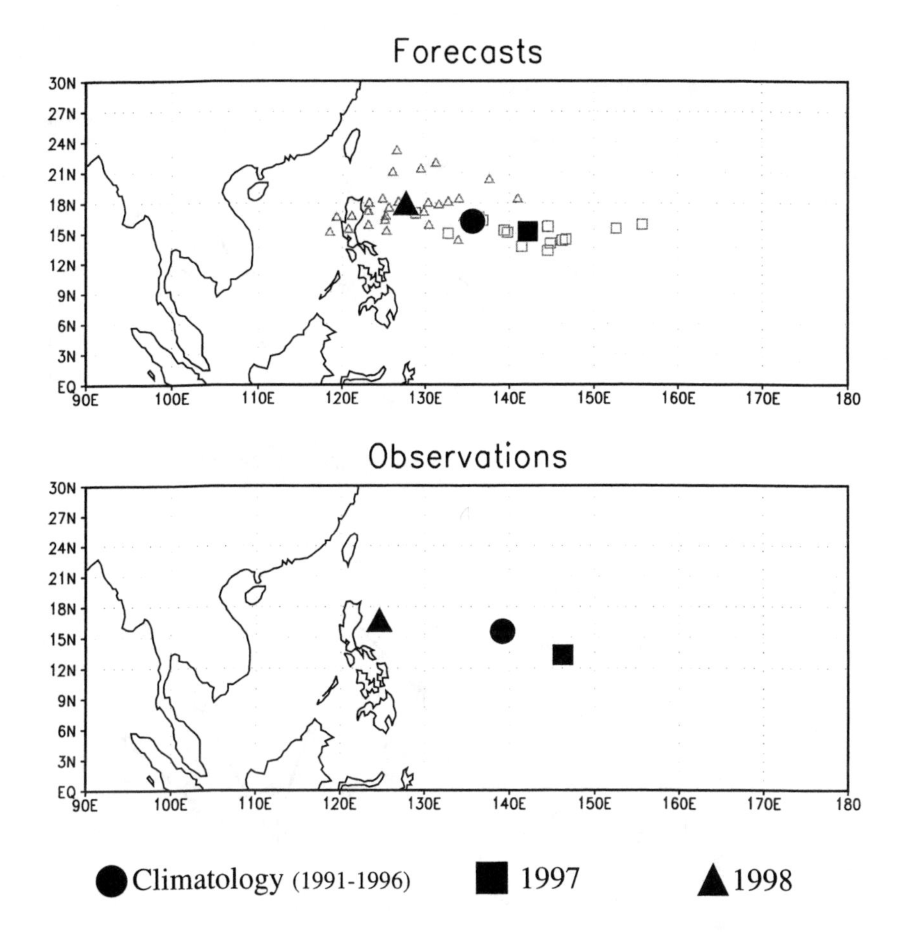

FIGURE 13.7 Mean genesis position of all the tropical storms that occurred during one year over the western North Pacific. In the top panel, the large black square and triangle represent the mean of all the members of the ensemble in 1997 and 1998, respectively. The small open symbols represent one member of the ensemble. In the bottom panel, the large black square and triangle represent the mean genesis position for tropical storms observed during 1997 and 1998, respectively (from Vitart and Stockdale 2001; copyright, American Meteorological Society, used with permission).

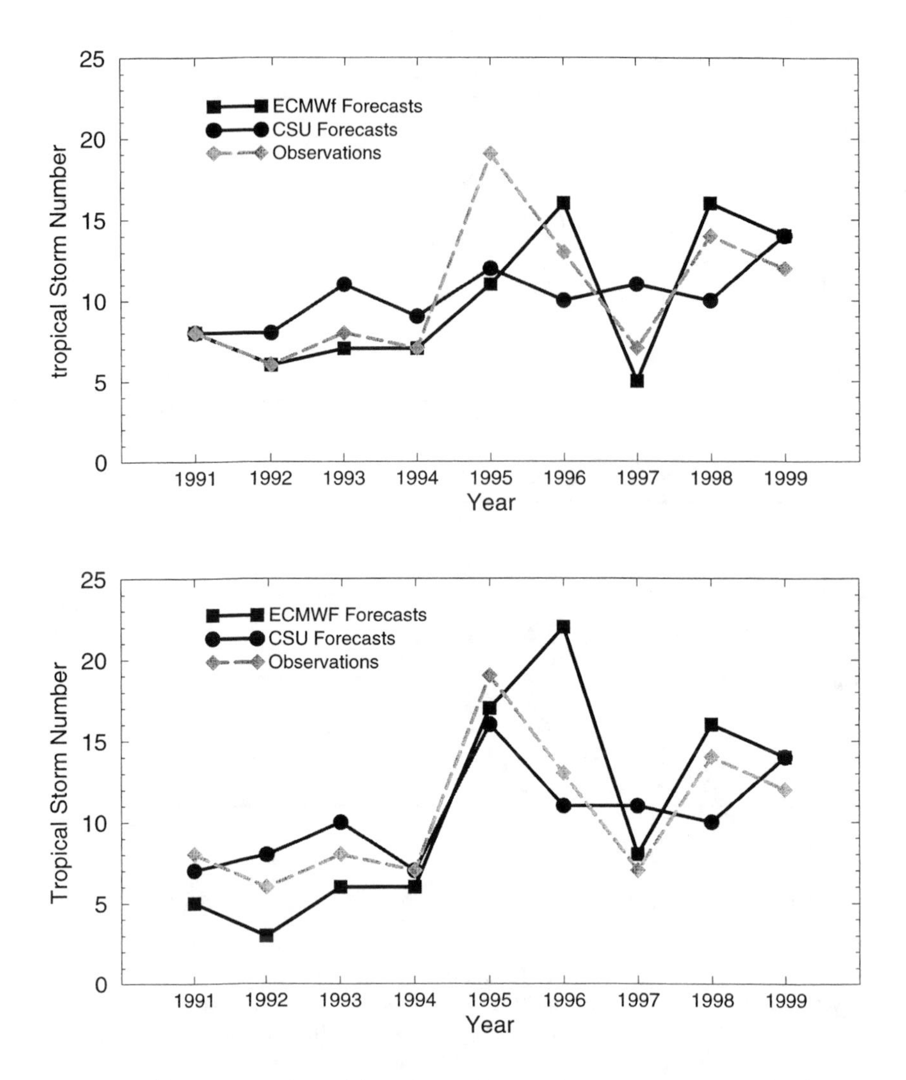

FIGURE 13.8 Interannual variability of tropical storm frequency over the North Atlantic in observations (dashed line), predicted with the ECMWF seasonal forecasting system, and then scaled (full line with squares) and predicted by the Colorado State University Forecast Team (full line with circles). Forecasts are issued in June (*top*) and August (*bottom*) (from Vitart and Stockdale 2001; copyright American Meteorological Society, used with permission)

modeled tropical storms look significantly more realistic, and most of the current criticisms about modeled tropical storms will no longer be valid. Therefore, the method presented here has good potential, and is likely to be more widely accepted in the coming years.

Tropical Storm Landfall

Seasonal forecasting of tropical storm landfall means forecasting a change in the probability of landfall over a large area. It is not the deterministic forecast of a particular event. A tropical storm landfall is a rare event (the majority of tropical storms do not have landfall) and sometimes can be associated with synoptic conditions that cannot be foreseen months in advance. There are, however, some cases where tropical storm landfall can be favored by large-scale conditions with a predictability exceeding a few months. In this instance, it could be possible to use dynamical seasonal forecasts to predict a risk of tropical storm landfall. The rest of this section will discuss such an example.

The landfall of tropical cyclone Eline over Mozambique on February 2000 created a catastrophic situation by bringing more than 20 cm of precipitation per day over regions that had already been flooded. In the beginning of April, a second cyclone, Hudah, made landfall over the northern part of Mozambique. Such strong tropical cyclone activity over Mozambique is very rare according to the historical record (Neuman et al. 1993). In most years, tropical cyclones over the South Indian Ocean tend to recurve to the south well before reaching the African coast. But wind circulation over the South Indian Ocean during La Niña years favors more zonal tropical storm tracks. Therefore, La Nina conditions may explain why tropical cyclone Eline, with a genesis location in the eastern part of the basin, could strike Mozambique after two weeks of an almost perfectly westward track. In addition, sea-surface temperatures over the South Indian Ocean were not as cold as typical during a La Niña season. As a result, tropical cyclones were not as weak and scarce as during the previous La Niña season of 1998–1999. Therefore, a combination of La Niña conditions and warmer local sea-surface temperatures may have contributed to the exceptional tropical cyclone history of 1999–2000.

The ECMWF dynamical seasonal forecasting system described in the previous section predicted correctly the persistence of La Niña conditions and sea-surface temperature anomalies over the South Indian Ocean. It predicted that the mean tracks of tropical storms would be significantly more westward during

TABLE 13.1 Tropical Cyclones with Hurricane Intensity over the South Indian Ocean

	1997–1998	1998–1999	1999–2000	Climatology[1]
Observations	5	3	8	6 (2)
Model[2]	5.5	4	7	6 (0.5)

[1] The standard deviations in the climatology are in parentheses.
[2] The numbers for model predictions are issued on November 1 from the ECMWF seasonal forecasting system. The numbers have been scaled by a factor that adjusts model to observed climatology.

the two La Niña seasons of 1998–1999 and 1999–2000. The frequency of intense modeled tropical storms was also significantly higher in 1999–2000 and significantly lower in 1998–1999 than in climatology, as observed (table 13.1).

Prospects for Dynamical Methods

Although dynamical forecasting systems are still in their infancy, they already display useful skill in predicting the frequency of tropical storms over at least some ocean basins, such as the Atlantic and the western North Pacific. The example of the ECMWF seasonal forecasting system indicates that a dynamical model can be competitive with current statistical methods for seasonal predictions.

Dynamical seasonal forecasting systems are presently mature enough to predict the occurrence of El Niño or La Niña events. More generally, dynamical models have some skill in predicting the year to year variability of sea-surface temperatures over most of the tropical regions. Dynamical models still suffer strong limitations that are likely to affect the seasonal forecasting of tropical storms. For instance, most dynamical models have poor skill in predicting the intensity of an El Niño event (Barnston, Glantz, and He 1999). The Madden-Julian Oscillation (MJO) is also a source of concern, because it is believed to have an impact on Pacific and Atlantic tropical storms (Maloney and Hartmann 2000; Mo 2000). Most atmospheric models fail to simulate a realistic MJO, even when forced by observed sea-surface temperatures (Sling et al. 1996). Recent modeling efforts have also not been successful in predicting the interannual variability of the Australian-Asian monsoon and the observed ENSO-monsoon relationships have been difficult to replicate with an atmospheric model (Webster et al. 1998). Finally, atmospheric models still have difficulties in simulating the Quasi-Biennial Oscillation (QBO), which according to Gray (1984), has a significant impact on the frequency of intense Atlantic tropical storms. Improved dynamical models would likely increase the skill of dynamical seasonal forecasts of tropical storms.

The quality of the seasonal prediction of tropical storms may be improved by using combined ensemble forecasts produced by different models (multi-model ensemble forecasts). This method is efficient in filtering model errors present in the individual ensemble forecasts. In the United States and in Europe, the best way of combining different forecasts is being actively explored. Results from the PROVOST project in Europe demonstrate that the reliability of a seasonal forecast can be strongly enhanced by the use of multi-model ensembles (Palmer and Shukla 2000). Krishnamurti et al. (2000) demonstrated that a multi-model ensemble outperforms all the individual models for hurricane track and intensity forecasts.

Finally, some important questions remain unanswered. Can dynamical seasonal forecasting produce useful forecasts more than a few months in advance? This question is particularly important for the insurance industry, since reinsurance premiums in the United States are fixed more than six months prior to the start of the Atlantic tropical storm season. Some statistical models, for instance Gray et al. (1992) and University College, London (http://forecast.mssl.ac.uk/), produce seasonal forecasts of Atlantic tropical storms as early as December or November. The skill of the ECMWF seasonal forecasting system has been discussed only for predictions a few months before the start of the tropical storm seasons. The system's skill for longer ranges remains unexplored. A second important question concerns the limit of predictability of tropical storm variability. For instance, the ECMWF dynamical model displays very poor skill in predicting storms in the North Indian Ocean. It is not clear whether this is due to model errors, errors in the initial conditions, or to a lack of predictability. In the coming years, model errors are likely to be reduced, and limits in the predictability are likely to become the major limitation of the dynamical seasonal forecast of tropical storms.

REFERENCES

Barnett, T. P., M. Latif, N. Graham, M. Flugel, S. Pazan, and W. White. 1993. ENSO and ENSO-related predictability: part I-prediction of equatorial Pacific sea surface temperatures with a hybrid coupled ocean-atmosphere model. *Journal of Climate* 6:1545–66.

Barnston, A. G., M. H. Glantz, and Y. X. He. 1999. Predictive skill of statistical and dynamical climate models in SST forecasts during the 1997–1998 El Niño episode and the 1998 La Niña onset. *Bulletin of the American Meteorological Society* 80:217–43.

Basher, R. E., and X. Zheng, 1995. Tropical cyclones in the southwest Pacific: Spatial patterns and relationship to Southern Oscillation and sea surface temperatures. *Journal of Climate* 8:1249–60.

Beljaars, A. C. M., P. Viterbo, M. J. Miller, and A. K. Betts. 1996. The anomalous rainfall over the United States during July 1993: sensitivity to land surface parameterization and soil moisture anomalies. *Monthly Weather Review* 124:362–83.

Bengtsson, L., H. Bottger, and M. Kanamitsu. 1982. Simulation of hurricane-type vortices in a general circulation model. *Tellus* 34:440–57.

Bengtsson, L., M. Botzet, and M. Esh. 1995. Hurricane-type vortices in a general circulation model. *Tellus* 47A:175–96.

Bengtsson, L., M. Botzet, and M. Esh. 1996. Will greenhouse gas–induced warming over the next 50 years lead to higher frequency and greater intensity of hurricanes? *Tellus* 48A:57–73.

Broccoli, A. J., and S. Manabe. 1990. Can existing climate models be used to study anthropogenic changes in tropical cyclone climate? *Geophysical Research Letters* 17:1917–20.

Chan, J. C. L. 1985. Tropical cyclone activity in the northwest Pacific in relation to the El Niño/Southern Oscillation phenomenon. *Monthly Weather Review* 113:599–606.

Emanuel, K. A., and R. Rotunno. 1989. Polar lows as arctic hurricanes. *Tellus* 41A: 1–17.

Evans, J. L. 1992. Comments on "Can existing climate models be used to study anthropogenic changes in tropical cyclone climate?" *Geophysical Research Letters* 19: 1523–24.

Goldenberg, S. B., and L. J. Shapiro. 1996. Physical mechanism for the association of El Niño and West African rainfall with Atlantic major hurricane activity. *Journal of Climate* 9:1169–87.

Gray, W. M. 1968. Global view on the origin of tropical disturbances and storms. *Monthly Weather Review* 96:669–700.

Gray, W. M. 1975. *Tropical cyclone genesis.* Paper No. 323. Fort Collins: Department of Atmospheric Sciences, Colorado State University.

Gray, W. M. 1979. Hurricanes: Their formation, structure and likely role in the tropical circulation. In *Meteorology over the tropical oceans,* edited by D. B. Shaw, 155–218. London: Royal Meteorological Society.

Gray, W. M. 1984. Atlantic seasonal hurricane frequency. Part I: El Niño and 30 mb Quasi-Biennial Oscillation influences. *Monthly Weather Review* 112:1649–68.

Gray, W. M, C. W. Landsea, P. W. Mielke, Jr., and K. J. Berry. 1992. Predicting Atlantic Basin seasonal hurricane activity 6–11 months in advance. *Weather and Forecasting* 7:440–55.

Gray, W. M, C. W. Landsea, P. W. Mielke, Jr., and K. J. Berry. 1993. Predicting

Atlantic Basin seasonal tropical cyclone activity by 1 August. *Weather and Forecasting* 8:73–86.

Gray, W. M, C. W. Landsea, P. W. Mielke, Jr., and K. J. Berry. 1994. Predicting Atlantic Basin seasonal tropical cyclone activity by 1 June. *Weather and Forecasting* 9:103–15.

Haarsma, R. J., J. F. B. Mitchell, and C. A. Senior. 1993. Tropical disturbances in a GCM. *Climate Dynamics* 8:247–57.

Hess, J. C., J. B. Elsner, and N. E. LaSeur. 1995. Improving seasonal hurricane predictions for the Atlantic Basin. *Weather and Forecasting* 10:425–432.

Ji, M., A. Leetmaa, and V. E. Kousky. 1996. Coupled model forecasts of ENSO during the 1980s and 1990s at the National Meteorological Center. *Journal of Climate* 9:3105–20.

Kirtman, B. P., J. Shukla, B. Huang, Z. Zhu, and E. K. Schneider. 1997. Multiseasonal predictions with a coupled tropical ocean global atmosphere system. *Monthly Weather Review* 123:3103–13.

Knutson, T. R., and R. E. Tuleya. 1999. Increased hurricane intensities with CO2-induced global warming as simulated using the GFDL hurricane prediction system. *Climate Dynamics* 15:503–19.

Krishnamurti, T. N. 1988. Some recent results on numerical weather prediction over the tropics, *Australian Meteorological Magazine* 36:141–70.

Krishnamurti, T. N., C. M. Kishtawal, Z. Zhang, T. LaRow, D. Bachiochi, and E. Williford. 2000. Multimodel ensemble forecasts for weather and seasonal climate. *Journal of Climate* 13:4196–216.

Landsea, C. W., and J.A Knaff. 2000. How much skill was there in forecasting the very strong 1997–98 El Niño? *Bulletin of the American Meteorological Society* 81:2107–20.

Latif, M., A. Sterl, E. Maier-Reimer, and M. M. Junge. 1993. Structure and predictability of the El Niño/ Southern Oscillation phenomenon in a coupled ocean atmosphere general circulation model. *Journal of Climate* 6:700–708.

Lau, N.-C., and M. J. Nath. 1994. A modeling study of the relative roles of tropical and extratropical SST anomalies in the variability of the global atmosphere-ocean system. *Journal of Climate* 7:1184–1207.

Lighthill, J., G. Holland, W. Gray, C. Landsea, G. Graig, J. Evans, Y. Kurihara, and C. Guard. 1994. Global climate change and tropical cyclones. *Bulletin of the American Meteorological Society* 75:2147–57.

Livezey, R. E., M. Masutani, A. Leetmaa, H. Rui, M. Ji, and A. Kumar. 1997. Teleconnective response of the Pacific-North American region atmosphere to large central equatorial Pacific SST anomalies. *Journal of Climate* 10:1787–820.

Lorenz, E. N. 1963. Deterministic nonperiodic flow. *Journal of the Atmospheric Sciences* 20:130–41.

Maloney, E. D., and D. L. Hartmann. 2000. Modulation of hurricane activity in the Gulf of Mexico by the Madden-Julian oscillations. *Science* 287:2002–4.

Manabe, S., J. L. Holloway, and H. M. Stone. 1970. Tropical circulation in a time-integration of a global model of the atmosphere. *Journal of the Atmospheric Sciences* 27:580–613.

McBride, J. L. 1984. Comments on "Simulation of hurricane-type vortices in a general circulation mode." *Tellus* 36A:92–93.

McBride, J. L. 1995. Tropical cyclone formation. In *Global perspectives on tropical cyclones*, edited by R. L. Elsberry, 63–105. Report No. TCP-38. Geneva: World Meteorological Organization.

Mo, K. C. 2000. The association between intraseasonal oscillations and tropical storms in the Atlantic Basin. *Monthly Weather Review* 128:4097–107.

Namias, J. 1955. Secular fluctuations in vulnerability to tropical cyclone activity in and off New England. *Monthly Weather Review* 83:155–62.

Neumann, C. J., B. R. Jarvinen, C. J. McAdie, and J. D. Elms. 1993. *Tropical cyclones of the North Atlantic, 1871–1992*. NOAA Historical Climatology Series 6-2. Asheville, N.C.: National Climatic Data Center.

Nicholls, N. 1992. Recent performance of a method for forecasting Australian seasonal tropical cyclone activity. *Australian Meteorological Magazine* 21:105–10.

Palmen, E. 1956. A review of knowledge on the formation and development of tropical cyclones. In *Proceedings of the Tropical Cyclone Symposium*, 213–32. Melbourne: Australian Bureau of Meteorology.

Palmer, T. N. 1993. Extended range atmospheric prediction of the Lorenz model. *Bulletin of the American Meteorological Society* 74:49–65.

Palmer, T. N., and D. A. Mansfield. 1986. A study of wintertime circulation anomalies during past El Niño events using a high resolution general circulation model. II, variability of the seasonal mean response. *Quarterly Journal of the Royal Meteorological Society* 112:639–60.

Palmer, T. N., and J. Shukla. 2000. Editorial of the special issue DSP/PROVOST. *Quarterly Journal of the Royal Meteorological Society* 126:1989–90.

Raper, S. 1992. Observational data on the relationships between climatic change and the frequency and magnitude of severe tropical storms. In *Climate and sea level change: Observations, projections, and implications*, edited by R.A. Warrick, E. M. Barrow, and T. M. L. Wigley, 192–212. Cambridge: Cambridge University Press.

Riehl, H. 1954. *Tropical meteorology*. New York: McGraw-Hill.

Saunders, M. A., and A. R. Harris. 1997. Sea warming as a dominant factor behind near-record number of Atlantic hurricanes. *Geophysical Research Letters* 24:1255–58.

Shapiro, L. J. 1982. Hurricane climatic fluctuations. Part II: Relation to large-scale circulation. *Monthly Weather Review* 110:1014–23.

Shapiro, L. J. 1987. Month-to-month variability of the Atlantic tropical circulation and its relationship to the tropical cyclone formation. *Monthly Weather Review* 115: 2598–614.

Shen, W., R. Tuleya, and I. Ginis. 2000. A sensitivity study of the thermodynamic environment on GFDL model hurricane intensity: Implication for global warming. *Journal of Climate* 13:109–21.

Slingo, J. M., K. R. Sperber, J. S. Boyle, J.-P. Ceron, M. Dix, B. Dugas, W. Ebisuzaki, J. Fyfe, D. Gregory, J.-F. Gueremy, J. Hack, A. Harzallah, P. Inness, A. Kitoh, W. K.-M. Lau, B. McAvaney, R. Madden, A. Matthews, T. N. Palmer, C.-K. Park, D. Randall, and N. Renno. 1996. Intraseasonal oscillations in 15 atmospheric general circulation models: Results from an AMIP diagnostic subproject. *Climate Dynamics* 12:325–57.

Stockdale, T. N. 1997. Coupled ocean-atmosphere forecasts in the presence of climate drift. *Monthly Weather Review* 125:809–18.

Stockdale, T. N. 2000. An overview of techniques for seasonal forecasting. *Stochastic Environmental Research and Risk Assessment* 14:305–18.

Stockdale, T. N., D. L. T. Anderson, J. O. S. Alves, and M. A. Balmaseda. 1998. Global seasonal rainfall forecasts using a coupled ocean-atmosphere model. *Nature* 392:370–73.

Trenberth, K. E. 1997. The definition of El Niño. *Bulletin of the American Meteorological Society* 78:2771–77.

Trenberth, K. E. 1998. Development and forecasts of the 1997/1998 EL Niño: CLIVAR scientific issues. *Exchanges* [CLIVAR newsletter]. Hamburg: Max-Planck-Institute for Meteorology.

Tsutsui, J.-I., and A. Kasahara. 1996. Simulated tropical cyclones using the National Center for Atmospheric Research community climate model. *Journal of Geophysical Research* 101:15013–32.

Vitart, F. 1999. Tropical storm interannual variability in an ensemble of GCM integrations. Ph.D. diss., Princeton University.

Vitart, F., and J. L. Anderson. 2001. Sensitivity of tropical storm frequency to ENSO and interdecadal variability of SST's in an ensemble of GCM integrations. *Journal of Climate* 14:533–45.

Vitart, F, J. L. Anderson, J. Sirutis, and R. E Tuleya. 2001. Sensitivity of tropical storms simulated by a GCM to changes in cumulus parameterization. *Quarterly Journal of the Royal Meteorological Society* 127:25–51.

Vitart, F., J. L. Anderson, and W. F. Stern. 1997. Simulation of interannual variability of tropical storm frequency in an ensemble of GCM integrations. *Journal of Climate* 10:745–60.

Vitart, F, J. L. Anderson, and W. F. Stern. 1999. Impact of large-scale circulation on tropical storm frequency, intensity and location simulated by an ensemble of GCM integrations. *Journal of Climate* 12:3237–54.

Vitart. F., and T. N. Stockdale. 2001. Seasonal forecasting of tropical storms using coupled GCM integrations. *Monthly Weather Review* 129:2521–37.

Walsh, K., and I. G. Watterson. 1997. Tropical cyclone-like vortices in a limited area model: Comparison with observed climatology. *Journal of Climate* 10:2240–59.

Watterson, I. G., J. L. Evans, and B. F. Ryan. 1995. Seasonal and interannual variability of tropical cyclogenesis: diagnostics from large-scale fields. *Journal of Climate* 8:3053–66.

Weatherford, C. L., and W. M. Gray. 1988. Typhoon structure as revealed by aircraft reconnaissance. Part II: Structural variability. *Monthly Weather Review* 116: 1044–56.

Webster, P. J., V. O. Magaña, T. N. Palmer, J. Shukla, R. A. Tomas, M. Yanai, and T. Yasunari. 1998. Monsoons: Processes, predictability, and the prospects for prediction. *Journal of Geophysical Research* 103:14395–451.

Wu, G., and N. C. Lau. 1992. A GCM simulation of the relationship between tropical-storm formation and ENSO. *Monthly Weather Review* 120:958–77.

Yang, S., and K.-M. Lau. 1998. Influence of sea surface temperatures and ground wetness on Asian Summer Monsoon. *Journal of Climate* 11:3230–46.

Zhang, G. Z., W. Drosdowsky, and N. Nicholls. 1990. Environment influences on northwest Pacific tropical cyclone numbers. *Acta Meteorologica Sinica* 4:180–88.

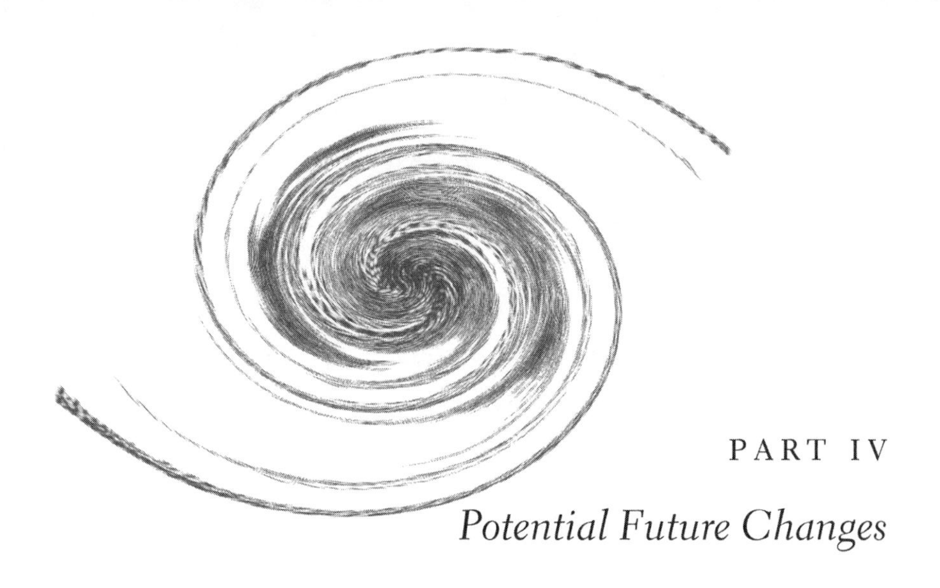

PART IV

Potential Future Changes

14

Response of Tropical Cyclone Activity to Climate Change: Theoretical Basis

Kerry Emanuel

The normal state of the tropical atmosphere is unstable to cumulus clouds, which serve to transport heat to the high atmosphere, where it is then radiated to space. This convection drives the atmosphere to a state that is very nearly neutrally stable to cumulus clouds. In this equilibrium, the absorption of solar radiation by the surface is approximately balanced by the turbulent transport of heat away from the surface; this in turn requires a degree of thermodynamic disequilibrium between the tropical oceans and atmosphere.

When the concentration of greenhouse gases is increased, the downwelling infrared radiation to the surface increases, requiring an increased heat flux from the surface to the atmosphere to maintain the oceans' heat balance. In general, increased surface heat flux requires either stronger surface winds or greater thermodynamic disequilibrium between the surface and the atmosphere. Increased thermodynamic disequilibrium is consistent with an increase in the thermodynamic upper bound on hurricane intensity, known as the "potential intensity." Theory and models both point to noticeable increases in this upper bound with anthropogenic global warming, at a rate of about 3.4 m s^{-1} for each 1°C rise in tropical ocean surface temperature. At the same time, a statistical analysis of storms in the present climate strongly suggests that hurricane intensity obeys a universal distribution function that is nearly linear. This in turn suggests that increasing the potential intensity increases the actual intensity of all storms by the same percentage. This idea is supported by numerical simulations of tropical cyclones using coupled models.

Although basic physics and models point to increased hurricane intensity in warmer climates, the factors governing the frequency of events are poorly understood, and models have produced inconsistent results. Neither models nor theory support the contention that global warming would increase the area of the earth's surface subject to tropical cyclones.

Finally, the response of tropical cyclone activity to climate change has always been regarded as passive. Recent work by the author suggests, however, that global tropical cyclone activity strongly affects the magnitude of the poleward heat transport by the oceans, thus providing a new feedback mechanism that stabilizes tropical climate but destabilizes climate at higher latitudes.

Each year, roughly 80 tropical cyclones develop around the globe. These storms reach varying intensities (but have 1-minute sustained winds greater than 17.5 m s^{-1}, according to the accepted definition of "tropical cyclone") and then decay. The number, duration, and intensity of tropical cyclones constitute what one may refer to broadly as *global tropical cyclone activity*. It is of basic and practical interest to understand how tropical cyclone activity depends on climate. It may also prove necessary to account for changes in tropical cyclone activity when predicting climate change.

In trying to understand the effect of climate change on hurricane activity, it is tempting to begin by observing how climate change affects the activity of storms simulated in global climate models. But there are a number of potentially serious pitfalls to this approach. In the case of tropical cyclones, very fine horizontal resolution is required to resolve the all-important eyewall, and experience with several regional models has demonstrated that grid-spacing on the order of 1 to 2 km is required for numerical resolution. For example, simulations of actual tropical cyclones using the NCAR MM5 model show substantial differences in going from grid-spacing of 15 km to 1.6 km (S. Chen, personal communication). But global models typically have grid spacings on the order 100 km, and although they do simulate storms with characteristics of tropical cyclones, these storms are pale shadows of the real phenomenon. One partial solution is to run regional models with boundary conditions supplied by global climate models. This is the approach taken by Tom Knutson and collaborators, as described in the next chapter of this volume.

For all their potential utility, mere simulations of a phenomenon by no means constitute a satisfying understanding of the problem; that is what I seek here. Quite aside from providing a more fulfilling aesthetic under which to perform science, such an understanding may point to important physical processes that modelers may overlook.

In the following section, I review the fundamental relationships between climate and atmospheric thermodynamics, with an emphasis on the potential intensity of hurricanes. In the next sections I present some calculations of the effect of global warming on potential intensity and then show how the statistical distribution of hurricane intensity in the present climate is related to the potential intensity. A brief discussion of the problem of storm frequency and geographic extent is addressed in the subsequent section, and in the penultimate section I consider a possible feedback of changing tropical cyclone activity on climate itself. A summary is provided in the final section.

The Greenhouse Effect and Hurricane Intensity

Hurricanes are driven by evaporation of ocean water, which transfers heat from the ocean to the atmosphere. (That is why hurricanes dissipate so quickly upon moving over land.) By treating hurricanes as heat engines, which absorb heat from the ocean at relatively high temperature and export heat to the environment at the relatively low temperatures of the upper atmosphere, I derived an expression governing the maximum surface wind speed that can be achieved in mature hurricanes (Emanuel 1986; Bister and Emanuel 1998):

$$V^2 = \frac{C_k}{C_D} \frac{T_s - T_o}{T_o} (k_s - k_a) \qquad (1)$$

where V is the maximum wind speed, C_k and C_D are dimensionless exchange coefficients for enthalpy and momentum, T_s and T_o are the absolute temperature of the sea surface and storm top, and k_s and k_a are the specific enthalpies of the air at saturation at the ocean surface and ambient boundary layer air, respectively.[1] That the outflow rather than inflow temperature appears in the denominator of (1) is owing to the fact that the dissipative heating in the storm's boundary layer returns some of what would otherwise be waste heat back to the front end of the heat engine (Bister and Emanuel 1998). To calculate the potential intensity using (1) one needs to know the sea-surface temperature, which enters directly through the thermodynamic efficiency factor and indirectly through its determination of k_s, the temperature profile of the troposphere and lower stratosphere, which determine T_o, and the humidity of the atmospheric boundary layer which, together with its temperature, determine k_a. All these quantities are easily estimated from the broad-scale climate state of the tropical atmosphere. Experiments with a variety of numerical models demonstrate the validity of (1) (Emanuel 1995). In a subsequent section, I present observational evidence for the importance of this upper bound in determining the intensity of actual hurricanes.

Assuming that the dimensionless exchange coefficients are not strong functions of climate, the main determining factors in the potential intensity given by (1) are the ocean surface temperature, the mean outflow temperature, and the thermodynamic disequilibrium ($k_s - k_a$) that exists under normal conditions between the tropical oceans and atmosphere. In the long-term average, and ignoring lateral heat transport by the oceans and atmosphere, the heat balance at the ocean surface may be written

$$\frac{S}{4}(1 - \alpha_p) = \sigma T_s^4 - F\downarrow + \rho C_k |V_a| (k_s - k_a) \qquad (2)$$

where S is the solar constant, α_p is the planetary albedo (the fraction of sunlight reflected back to space by the earth, including the ocean, atmosphere and clouds), σ is the Stefan-Boltzmann constant, $F\downarrow$ is the downward flux of infrared radiation at the surface, ρ is the density of air at the surface, and $|V_a|$ is the magnitude of the wind speed near the surface. Equation (2) simply states that, in equilibrium, the sunlight absorbed by the surface is balanced by the emission of infrared radiation by the surface (σT_s^4) and the turbulent flux of heat carried by the atmosphere away from the surface—the last term in (2)—minus the infrared radiation emitted by the atmosphere back down toward the surface. This last is a measure of the greenhouse effect. The fact that the atmosphere contains greenhouse gases means that it emits infrared radiation both upward to space and downward toward the surface.

One can use (2) to calculate the thermodynamic disequilibrium between the tropical atmosphere and ocean, ($k_s - k_a$), and use this in (1) to get

$$V^2 = \frac{T_s - T_o}{T_o} \frac{\frac{S}{4}(1 - \alpha_p) + F\downarrow - \sigma T_s^4}{\rho C_D |V_a|} \qquad (3)$$

Increasing the concentration of greenhouse gases increases $F\downarrow$, but this is partially offset by increases in the surface temperature. On the other hand, increasing the surface temperature in the tropics increases the temperature of the free atmosphere, and this also increases $F\downarrow$. The result is that the net increase in the surface temperature only partially offsets the net increase in $F\downarrow$, so that the numerator of (3) increases. Detailed calculations of actual global warming also uniformly exhibit a decrease in tropopause temperatures, T_o. So increasing the concentration of greenhouse gases leads, through two different mechanisms, to an increase in the potential intensity given by (3). These could be offset, though, by an increase in the average surface wind in the tropics, V_a. Global models differ somewhat in their predictions of V_a, but none show any appreciable increase. Thus it is likely that increasing the concentration of greenhouse gases in the atmosphere increases the potential maximum wind speed of hurricanes. The question is, by how much?

Quantitative Estimates of the Effect of Global Warming on Hurricane Potential Intensity

A first estimate of the effect of global warming on tropical cyclone intensity may be obtained by running a one-dimensional column model into a state of radiative-convective equilibrium. Such a model attempts to account in a realistic way for radiative and convective heat transfer in the atmosphere, while omitting horizontal heat and moisture transport. I quantified the effect of increasing trace gases on potential intensity through the use of a single-column model run to radiative-convective equilibrium and through examination of the output of a general circulation model (GCM) run to statistical equilibrium with double CO_2 (Emanuel 1987). One may characterize the dependence of potential intensity on climate through its relationship with tropical sea-surface temperature. This relationship is shown in figure 14.1.

On average, the potential intensity increases at a rate of 3.4 m s^{-1} for each 1°C increase in sea-surface temperature brought about by greenhouse warming. This rate of increase is also supported by numerical experiments with a tropical cyclone model, as described by Knutson, Tuleya, and Kurihara (1998). It is important to note that the increase in potential intensity shown in figure 14.1 fully accounts for the atmospheric warming that accompanies increasing greenhouse gases. It is substantially smaller than the rate of increase one would obtain by fixing the atmospheric temperature while increasing the sea-surface temperature.

Implications of Increasing Potential Intensity for Actual Hurricane intensity

What do increases in hurricane potential intensity imply about changes in the intensity of actual storms? Numerically simulated hurricanes typically intensify right up to their potential intensity, but real storms rarely achieve their potential. To examine the relationship between actual and potential intensities of hurricanes, I looked at the best-track record of tropical cyclones in the North Atlantic and western North Pacific regions for a period of time for which the

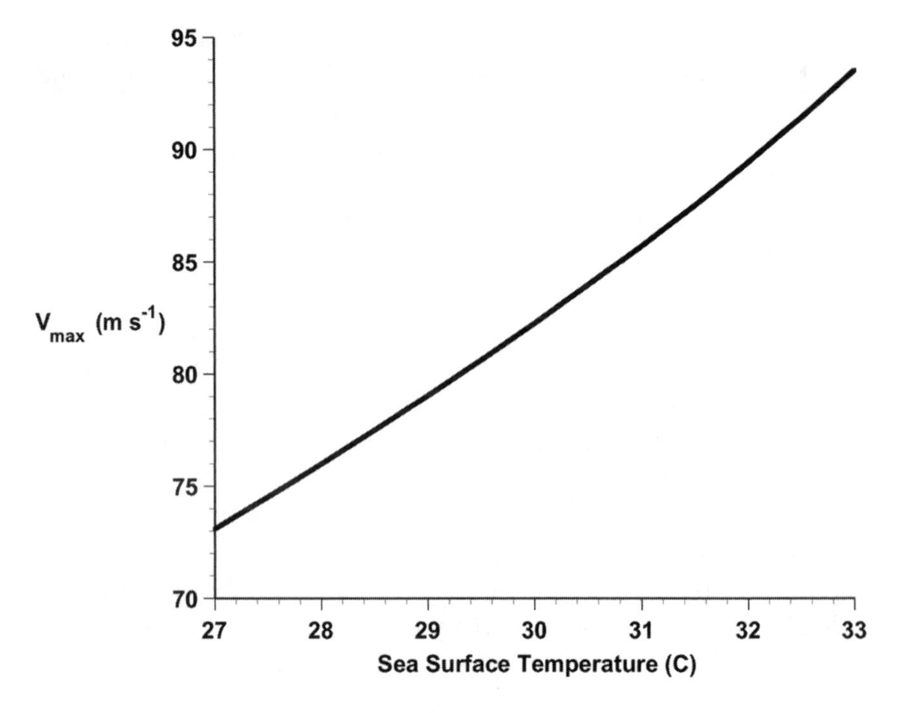

FIGURE 14.1 Average change in potential intensity with sea-surface temperature in global-warming experiments.

intensity estimates are regarded as being reasonably reliable (Emanuel 2000). In some cases, these were adjusted to account for changes in measuring or reporting practices. The maximum wind speed of each storm was normalized by the monthly climatological value of the potential intensity, linearly interpolated to the day in question and to the actual position of the storm center. These climatological values were calculated assuming that C_k/C_D is equal to unity.

One can see in figure 14.2 the cumulative distributions of the maximum normalized intensity achieved by each tropical cyclone that achieved a maximum 1-minute sustained wind speed of at least 33 m s^{-1} in the Atlantic and western North Pacific during a period of several decades. Here I have intentionally excluded those storms whose maximum intensity was limited by landfall or passage over cold ocean waters by insisting that the potential intensity remain higher than the actual intensity for 72 hours after maximum intensity was achieved.

The distributions are distinctly linear in both basins, intercepting the *x*-axis at a normalized intensity of around 0.9, suggesting that the actual value of

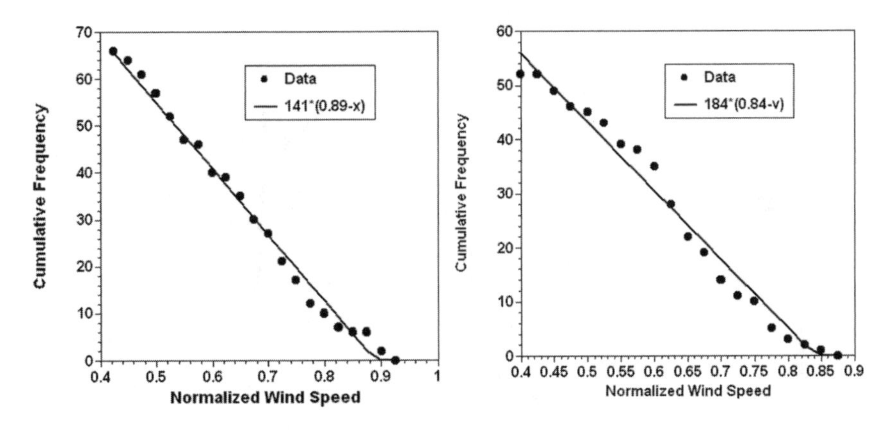

FIGURE 14.2 Cumulative frequency of occurrence of lifetime maximum normalized intensity of tropical cyclones in the North Atlantic (*left*) and western North Pacific (*right*) (from Emanuel 2000).

C_k/C_D is close to 0.8 (i.e., the square of 0.9). I have calculated distributions like these for a number of subsets of the data, for different regions and times of year. All of them are nearly linear, suggesting that linearity is a universal property of tropical cyclone intensity distributions. Thus, the statistical distribution of tropical cyclones by normalized intensity can be characterized by a single number: the overall frequency of events. To get the distribution of actual storm intensities, one needs to multiply by the potential intensity.

Note, in particular, that *increasing the potential intensity increases the intensity of all storms in proportion*. Thus a 10% increase in potential intensity can be expected to increase the intensity of *all* storms by 10%. The universality of the cumulative distribution function of tropical cyclone intensity allows us to characterize the effect of global climate change on tropical cyclone activity as consisting of changes in frequency and potential intensity in any given region.

The effect of increasing potential intensity can also be tested using numerical models of tropical cyclones. One such model is that of Emanuel (1999), which couples an axisymmetric atmospheric model with a very simple ocean model. The atmospheric model is phrased in "potential radius" coordinates, giving very high spatial resolution where it is needed, in the eyewall region of the storm. This model has been shown to give surprisingly accurate "hindcasts" of the intensity of real events. Figure 14.3 shows the evolution of the hindcast intensity of Hurricane Andrew, which struck south Florida and then the northern Gulf Coast in 1992. One curve shows the intensity evolution of the model

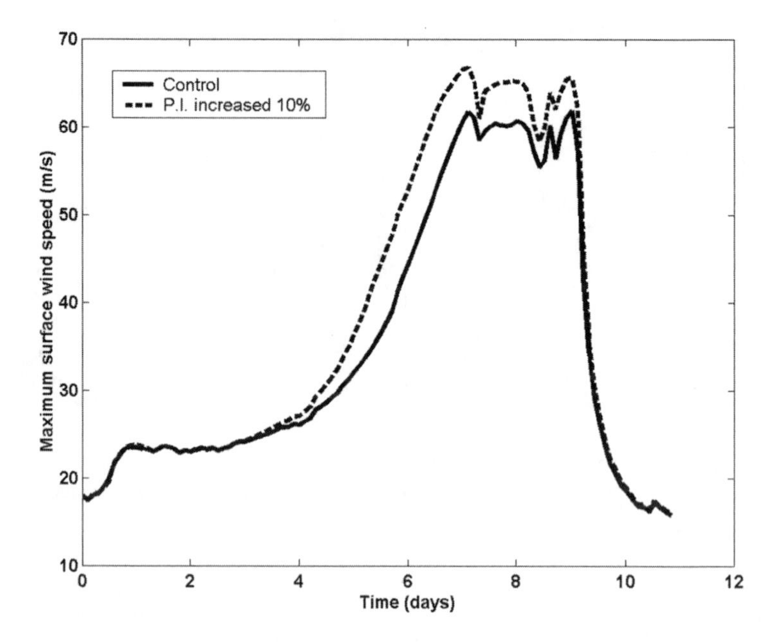

FIGURE 14.3 Evolution of the maximum wind speed in Hurricane Andrew, using the coupled model of Emanuel (1999). The solid curve is control run; the dashed curve shows the effect of increasing the climatological potential intensity by 10%.

run with normal climatological conditions; the second shows the effect of uniformly increasing the climatological potential intensity by 10%. The storm produces about 10% greater wind speeds in the second case. This provides further support of the notion that increasing potential intensity increases the actual intensity of tropical cyclones by the same proportion.

Can one detect an actual increase in global tropical cyclone intensity? Actual tropical sea-surface temperatures have increased on average by about 0.3°C since 1950. According to figure 14.1, this should result in an average increase in potential intensity of about 1 m s^{-1} or about 1.3%. Given that the average hurricane-strength tropical cyclone reaches an intensity of around 40 m s^{-1}, one would expect to have observed an average increase in intensity of around 0.5 m s^{-1} or 1 knot. Because tropical cyclone maximum wind speeds are only reported at 5-knot intervals and are not believed to be accurate to better than 5 to 10 knots, and given the large interannual variability of tropical cyclone activity, *such an increase would not be detectable*. Thus any increase in hurricane intensity that may have already occurred as a result of global warming is inconsequential compared to natural variability.

Effects of Climate Change on Frequency and Geographic Extent of Tropical Cyclones

In contrast to the issue of intensity, climate control of the frequency of tropical cyclones is poorly understood. We do not know why about 80 tropical cyclones develop worldwide every year, or why that number is not much greater or much smaller. In part, this lack of understanding is traceable to our collective ignorance of the physics of tropical cyclogenesis. Although the necessary conditions for the maintenance of tropical cyclones are present in much of the tropics during much of the year, the storms themselves are rare occurrences and we have had little success in predicting individual instances of genesis. Attempts to use global models to simulate the effect of global warming on tropical cyclone frequency have produced conflicting results. For example, the study of Haarsma, Mitchell, and Senior (1993), using the GCM run by the United Kingdom Meteorological Office, shows an increase in both the intensity and frequency of tropical cyclones with CO_2-induced warming. In contrast, the analysis by Broccoli and Manabe (1990), using the Princeton/GFDL (Geophysical Fluid Dynamics Laboratory) model, shows ambiguous results, with an increase in tropical cyclone activity if cloud-radiation feedback is not included, and a decrease in activity otherwise. Nowhere is the need for better understanding more pressing than in the case of tropical cyclone frequency. Probably the best hope for improved understanding of the relationship between climate and tropical cyclone frequency is the systematic application of the paleotempestology techniques discussed earlier in this volume.

In the current climate, hurricanes develop over tropical ocean waters whose sea-surface temperature exceeds about 26°C, but, once developed, they may move considerably poleward of these zones. An oft-stated misconception about tropical cyclones is that were the area enclosed by the 26°C sea-surface temperature isotherm to increase, so, too, would the area experiencing tropical cyclogenesis. Regions prone to tropical cyclogenesis are better characterized as places where the atmosphere is slowly ascending on the largest scales. Because about as much atmosphere is descending as ascending, it is hard to change the total area experiencing ascent. Thus there is little basis for believing that there would be any substantial expansion or contraction of the area of the world prone to tropical cyclogenesis. This is borne out by the GCM simulations performed by Haarsma, Mitchell, and Senior (1993), who show that although there is a substantial increase in the area enclosed by the 26°C SST isotherm in a double CO_2 environment, there is no perceptible increase in the area experiencing tropical cyclones.

Feedback of Tropical Cyclone Activity on Climate Change

The response of global tropical cyclone activity to climate change may have an important feedback on climate, through its modification of the ocean's overturning circulation. As was recognized by Jeffreys (1925), the magnitude of the overturning circulation is limited by downward diffusion of heat in the tropics. Cold water sinking near the poles flows equatorward at depth and must be reheated as it returns to the surface. Experiments with ocean models show that the magnitude of the poleward heat flux by the overturning circulation varies as the 2/3 power of the coefficient of vertical heat diffusion in the model (see, e.g., Marotzke and Scott 1999). There is considerable debate among oceanographers about the source of vertical mixing required to drive the overturning circulation. Munk and Wunsch (1998) proposed that much of the mixing is owing to turbulence caused by breaking of tidally generated internal waves. I calculated the amount of surface heat input to the ocean required to re-warm the cold wakes left behind by tropical cyclones in one year (1996) (Emanuel 2001). The average rate of heat input in that year, $1.4 \pm 0.7 \times 10^{15}$ W, is comparable to the observed poleward heat flux by the ocean, suggesting that tropical cyclones may provide much of the mixing required to drive the ocean's overturning circulation. The fact that most tropical cyclones occur in the western North Pacific and most deep-water formation is in the North Atlantic is consistent with the observed northward heat flux in the South Atlantic Ocean.

If tropical cyclones do indeed drive the ocean's overturning circulation, then changing the intensity and or frequency of storms will influence the strength of the poleward heat flux by the oceans. For example, if global warming increases the intensity of storms, as suggested previously, the poleward heat flux by the oceans would increase, cooling the tropics and warming higher latitudes. This constitutes a negative feedback on tropical climate change, but a positive feedback on temperature change outside the tropics. This might help explain why tropical temperatures have remained quite stable over the last few million years, while extratropical temperatures went through large swings associated with the advance and retreat of polar ice caps.

Summary

Tropical cyclones are powered by the thermodynamic disequilibrium that exists between the tropical oceans and atmosphere. The degree of this disequilibrium is directly related to the concentration of greenhouse gases in the atmosphere. As a result, the thermodynamic upper bound on hurricane intensity is related to the concentration of greenhouse gases in the atmosphere, as shown by equation (3). Experiments with a wide range of models show that the potential intensity of tropical cyclones increases with the concentration of greenhouse gases in the atmosphere, at a rate of about 3.4 m s^{-1} for each 1°C rise in tropical ocean surface temperature. Thus, if global warming were to result in an increase of 2°C in tropical sea-surface temperature, the potential intensity of tropical cyclones would increase by about 7 m s^{-1}, or nearly 10%. In turn, the statistical distribution of hurricane intensity in the present climate, as shown in figure 14.2, together with numerical simulations of hurricanes, suggest that the intensity of actual storms would increase in proportion. While the relationship between tropical cyclone intensity and tropical temperature is fairly well established, there is, of course, great uncertainty in the response of global climate in general, and tropical sea-surface temperature in particular, to increasing concentrations of greenhouse gases in the atmosphere. There appears to have been an average increase in tropical sea-surface temperature of about 0.3°C over the last century; the corresponding increase in tropical cyclone intensity would not be detectable given the large natural variability in tropical cyclone activity, and the relative inaccuracy of measurements of hurricane strength.

It is sometimes stated that because natural variability of tropical cyclone activity is large compared to expected changes owing to global warming, we may safely disregard the effects of the latter. This is true only if one takes a short-term view of the tropical cyclone hazard. The longer one's time horizon, the more consequential the effects of global warming. Thus if global warming were to increase the intensity of tropical cyclones by 10% while leaving their frequency unchanged, damages from these storms would increase by 20 to 30%, assuming no change in either sea level or the average value of coastal property. (The force of the wind increases as the square of the wind speed, while engineering studies suggest that damage tends to rise more nearly as the cube of the wind speed.) Although damages owing to tropical cyclones may change by 100% from year to year, the global increase would be readily apparent in decadal averages of storm damages.

Unfortunately, there is little understanding of the effect of climate change on the frequency of tropical cyclone occurrence. Paleotempestology probably offers the best hope for discovering a relationship between storm frequency and climate. At the same time, there is strong theoretical and modeling evidence that the total surface area of the globe affected by tropical storms is approximately invariant with climate change, though there may of course be shifts of activity within tropical cyclone-prone regions.

Finally, there is some evidence that changes in global tropical cyclone activity may have a noticeable feedback on the climate system itself, through their regulation of the strength of the oceans' overturning circulation. Through their collective effect in mixing the upper ocean and driving the poleward heat flux by the ocean, tropical cyclones stabilize tropical climate, but may destabilize the climate at middle and high latitudes.

NOTE

1. The specific enthalpy of air is a measure of its heat content and is given by $k = c_p T + L_v q$, where c_p is the heat capacity at constant pressure, L_v is the latent heat of vaporization, and q is the concentration of water vapor. In the definition of the saturation enthalpy, the saturation water vapor concentration replaces the actual concentration.

REFERENCES

Bister, M., and K. A. Emanuel. 1998. Dissipative heating and hurricane intensity. *Meteorological and Atmospheric Physics* 50:233–40.

Broccoli, A. J., and S. Manabe. 1990. Can existing climate models be used to study anthropogenic changes in tropical cyclone climate? *Geophysical Research Letters* 17:1917–20.

Emanuel, K. A. 1986. An air-sea interaction theory for tropical cyclones. Part I. *Journal of the Atmospheric Sciences* 42:1062–71.

Emanuel, K. A. 1987. The dependence of hurricane intensity on climate. *Nature* 326:483–85.

Emanuel, K. A. 1995. Sensitivity of tropical cyclones to surface exchange coefficients and a revised steady-state model incorporating eye dynamics. *Journal of the Atmospheric Sciences* 52:3969–76.

Emanuel, K. A. 1999. Thermodynamic control of hurricane intensity. *Nature* 401: 665–69.

Emanuel, K. A. 2000. A statistical analysis of tropical cyclone intensity. *Monthly Weather Review* 128:1139–52.

Emanuel, K. A. 2001. The contribution of tropical cyclones to the oceans' meridional heat transport. *Journal of Geophysical Research* 106:14771–82.

Haarsma, R. J., J. F. B. Mitchell, and C. A. Senior. 1993. Tropical disturbances in a GCM. *Climate Dynamics* 8:247–57.

Jeffreys, H. 1925. On fluid motions produced by differences of temperature and humidity. *Quarterly Journal of the Royal Meteorological Society* 51:347–56.

Knutson, T. R., R. E. Tuleya, and Y. Kurihara. 1998. Simulated increase of hurricane intensities in a CO_2-warmed climate. *Science* 279:1018–20.

Marotzke, J. and J. R. Scott. 1999. Convective mixing and the thermohaline circulation. *Journal of Physical Oceanography* 29:2962–70.

Munk, W., and C. Wunsch. 1998. Abyssal recipes II: Energetics of tidal and wind mixing. *Deep-Sea Research I* 45:1977–2010.

15

Impact of Climate Change on Hurricane Intensities as Simulated Using Regional Nested High-Resolution Models

Thomas R. Knutson, Robert E. Tuleya, Weixing Shen, and Isaac Ginis

The use of a nested high-resolution regional modeling strategy to investigate the impact of CO_2-induced warming on tropical cyclone intensities is reviewed here based on a series of investigations with the Geophysical Fluid Dynamics Laboratory (GFDL) hurricane model. In one approach, a series of tropical storm case studies from a global climate model (for both present-day and high-CO_2 conditions) are re-run, nesting the storms within the high-resolution hurricane model. In another approach, hurricane simulations are done using highly idealized flow fields along with large-scale time-averaged thermodynamic conditions (sea-surface temperatures, temperature, and moisture) from a global climate model. Using either approach, more intense hurricanes are simulated for high-CO_2 conditions than for present-day conditions. The magnitude of this effect is roughly a 3 to 10% increase in maximum wind speeds for a warming of the tropical environment induced by a century-long build-up of atmospheric CO_2 at 1% per year compounded. A series of sensitivity experiments demonstrate that the increased intensity results from a tropical sea-surface temperature increase of 2.2 to 2.7°C, counteracted to some extent by enhanced warming of the upper troposphere (relative to the surface). The storm intensification is found to be robust to such factors as inclusion of ocean coupling beneath the storms or changes in storm initialization methods. In addition to greater intensity, the high-CO_2 storms had substantial increases (~18 to 32%) in near-storm precipitation. The question of a possible impact of climate change on the frequency or location of occurrence of tropical cyclones is not addressed by the nested model methodologies reviewed in this chapter and remains an unresolved issue.

Climate models project a substantial increase in global mean surface temperatures over the next century (on the order of 2 to 4°C), according to a recent science assessment by the Intergovernmental Panel on Climate Change (IPCC) (Houghton et al. 2001) due primarily to the buildup of greenhouse gases in the earth's atmosphere. Although there are large uncertainties in such climate change projections due to uncertainties in future radiative forcing and in the climate system's response to such forcing, it is important to begin to assess how various aspects of the climate system would respond to such a warming if it occurred. The impact of such a warming on tropical cyclones is of particular interest, as these storms rank among the most destructive of natural disasters, and they are a characteristic feature of warm tropical ocean regions in the present climate. A greenhouse-gas–induced warming could potentially affect tropical cyclones in a number of ways including intensity, frequency or location of occurrence, storm size, tracks, and rainfall amounts.

Studies of possible changes in tropical cyclone frequency have been attempted using global climate models (Broccoli and Manabe 1990; Haarsma, Mitchell, and Senior 1993; Bengtsson, Botzet, and Esch 1996; Krishnamurti et al. 1998; Tsutsui 2002; Sugi, Noda, and Sato 2002) and regional nested models (Nguyen and Walsh 2001). Typically, these studies are based on counting the occurrence of tropical storm-like vortices in extended simulations under present-day and high-CO_2 climate conditions. These various climate model studies have yielded conflicting results, such that there is no consensus among models as to even the sign of future changes in tropical cyclone frequency. There is little indication from such studies, however, that the regions in which tropical storms occur would change markedly in a doubled CO_2 world. For example the so-called "threshold temperature" (e.g., ~26.5°C) for tropical storm formation in the present day climate appears to be climate dependent, as this apparent threshold shifts toward higher values in greenhouse-warming scenarios (e.g., Haarsma, Mitchell, and Senior 1993; Royer et al. 1998; Henderson-Sellers et al. 1998). As an alternative to the direct simulation of tropical storm-like vortices, several investigators have examined empirical tropical storm genesis parameters (following Gray 1975) in control and high-CO_2 climate model simulations (Ryan, Watterson, and Evans 1992; Druyan, Lonergan, and Eichler 1999; Royer et al. 1998). But the empirical genesis parameters appear to require significant modification for use in a climate change context (Royer et al. 1998). In assessing the current state of the science, Henderson-Sellers et al. (1998) and the IPCC (Houghton et al. 2001) conclude that future changes in tropical cyclone frequency are uncertain.

With regard to tropical cyclone intensities, there is some indication that the upper limit or potential intensity of tropical cyclones would increase in a greenhouse-warmed climate. This assessment is based on theoretical analyses (Emanuel 1987; Holland 1997; Tonkin et al. 1997; Henderson-Sellers et al.

1998) and on experiments with a nested high-resolution hurricane model (Knutson, Tuleya, and Kurihara 1998; Knutson and Tuleya 1999; Shen, Tuleya, and Ginis 2000; Knutson et al. 2001). The theoretical basis of this assessment is reviewed in the previous chapter of this volume. In the present chapter, modeling studies aimed at this question are reviewed.

The modeling studies described in the present chapter use a high-resolution regional hurricane model to simulate the impact on tropical cyclone intensities of CO_2-induced changes in large-scale environmental thermodynamic conditions (i.e., sea-surface temperatures, atmospheric temperatures and moisture).[1] The methodology involves "nesting" the high-resolution model either within a global climate model simulation of much lower resolution, or within an idealized atmospheric environment. For example, a global climate model can provide the large-scale thermodynamic conditions for either present-day or high-CO_2 climates. These environments are then used as boundary conditions/initial conditions for relatively short duration (a few days) hurricane simulations using the nested high-resolution model.

The use of such a modeling framework allows for tests of the impact of additional physical processes on hurricane intensities under altered climate conditions. For example, it is now well-established that the coupling between tropical cyclones and the underlying ocean can have an important impact on the intensity of the cyclones, primarily through the mixing of cooler sub-surface ocean waters to the surface in the vicinity of the storm (Bender, Ginis, and Kurihara 1993; Schade and Emanuel 1999; Emanuel 1999; Bender and Ginis 2000). This "cool wake" induced by the storm can then reduce the intensity of the storm. The impact of this effect on CO_2-induced storm intensification can be accessed by coupling a high-resolution hurricane model to a high-resolution ocean model in order to simulate the effect of the storm-generated "cool sea-surface temperature wake."

The modeling studies reviewed here focus primarily on the impact of various thermodynamic factors, such as sea-surface temperatures and atmospheric temperature and moisture, on future tropical cyclone intensities. Although we recognize that a number of factors contribute to determining the intensity of real world tropical cyclones (e.g., Elsberry et al. 1992), we view the thermodynamic state as setting the potential or upper-limit intensity that a tropical cyclone can attain. Included in this chapter is some additional discussion of this issue, including a more focused discussion of the potential impact of vertical wind shear on the intensity results for CO_2-induced climate change.

Background on Upper Limits of Tropical Cyclone Intensities

Observations

A number of investigators (Merrill 1988; Evans 1993; DeMaria and Kaplan 1994a; Whitney and Hobgood 1997; Baik and Paek 1998; Kuroda, Harada, and Tomine 1998) have examined the statistical relationship between observed intensities of tropical cyclones and environmental factors including sea surface temperature. These studies suggest that the upper limit of observed tropical cyclone intensities increases with increasing sea-surface temperature. Such an empirical sea-surface temperature–intensity relationship cannot, however, be reliably extrapolated to the question of hurricane intensity changes under CO_2-induced warming because several other environmental factors besides sea-surface temperature can affect intensities (e.g., wind shear, lapse rates, and large-scale regions of ascent and descent) (Holland 1997; Shen, Tuleya, and Ginis 2000). These additional factors could also change in various ways regionally as a result of CO_2-induced warming.

In terms of historical trends, Henderson-Sellers et al. (1998) report no clear evidence for long-term trends in storm intensities for either the North Atlantic or western North Pacific. Goldenberg et al. (2001) documented a pronounced multi-decadal variation in the occurrence of intense Atlantic hurricanes in recent decades, apparently linked to fluctuations in North Atlantic sea-surface temperatures and local vertical windshear in the primary development region. Landsea et al. (1996) found essentially no trend in the Atlantic Basin time series of maximum hurricane intensity for each year. Concerning historical sea-surface temperatures in the North Atlantic tropical storm basin, Knutson et al. (1999:plates 5 and 7) computed observed regional sea-surface temperature trends over the period 1949 to 1997 and found little evidence for a significant trend of either sign during the period. Their comparison of observed and model-simulated regional sea-surface temperature trends indicated that the North Atlantic tropical storm basin is an area with significant discrepancies between observed and simulated sea-surface temperature trends, based on a global climate model forced by estimated past concentrations of greenhouse gases and the direct effect of anthropogenic sulfate aerosols (Knutson et al. 1999:plates 3, 6, and 7). The cause of these model-observation discrepancies in regional trends remains a topic of investigation. Significant warming trends were found in their study in other tropical storm basins, including much of the tropical Indian Ocean and Northeast Pacific basins.

Theoretical Frameworks

Theoretical models of hurricane intensity predict that the potential intensity of hurricanes will increase in a warmer climate (Emanuel 1987; Tonkin et al. 1997; Henderson-Sellers et al. 1998) based on thermodynamical considerations. These theories are discussed the previous chapter of this volume. Since potential-intensity theories contain assumptions and caveats (Emanuel 1986, 1988, 1995; Holland 1997; Henderson-Sellers et al. 1998) it is important to test their conclusions using alternative methods. Rotunno and Emanuel (1987) have shown that potential-intensity theory agrees well with simulation results from a convection-resolving non-hydrostatic axisymmetric model. Further support for the potential-intensity theories, as applied to the greenhouse-warming problem, is provided by the results discussed in the present chapter.

An important recent finding regarding Emanuel's potential-intensity theory is that observed tropical cyclones, once attaining hurricane strength, are apparently equally likely to attain any intensity from minimal hurricane strength up to—but not exceeding—their potential intensity (Emanuel 2000). This finding, which applied to storms whose intensity was not limited by declining potential intensity, suggests that potential-intensity theories can provide empirical information on the entire distribution of tropical cyclone intensities under climate change—not just the upper-limit intensity as has been assumed previously. We return to this point in our concluding discussion on interpretation of the model simulation results.

Global Climate Model Simulations

Global climate models in principle could be used directly to examine thermodynamical and dynamical (e.g., wind shear, storm interaction) influences on storm intensities including regionally dependent influences. Unfortunately the resolution of the global models used to date for CO_2-induced climate change studies has been too coarse to allow a simulation of realistic hurricane structure or of the most intense storms. For example, Bengtsson, Botzet, and Esch (1995) report that even using a T106 global atmospheric model (grid resolution of 1.1°) forced by sea-surface temperatures derived from a lower resolution coupled climate model, the lowest central surface pressure simulated was 957 mb. This is much weaker than the observed record low central pressure of 870 mb, or the 906 mb storm simulated in Hamilton and Hemler's (1997) exploratory 1/3° resolution global model experiment.

As a further illustration of this problem, figure 15.1 shows the radial profile of "typical" surface winds (composites over a large sample of events) for tropical cyclones as simulated in a current-generation global coupled climate model (approximately 4° longitude resolution) versus that simulated using the regional triply nested GFDL hurricane model discussed in this chapter, which has a maximum resolution of 1/6°, or about 18 km. The radius of maximum winds occurs one gridpoint out from the storm center in the coarse-resolution model. In contrast, in the hurricane model the radius of maximum winds occurs less than 1° from the storm center, and the maximum surface winds are about twice as strong as in the coarse-resolution model. Owing to their inability to resolve important features of hurricanes, the reliability of global climate models used to date to

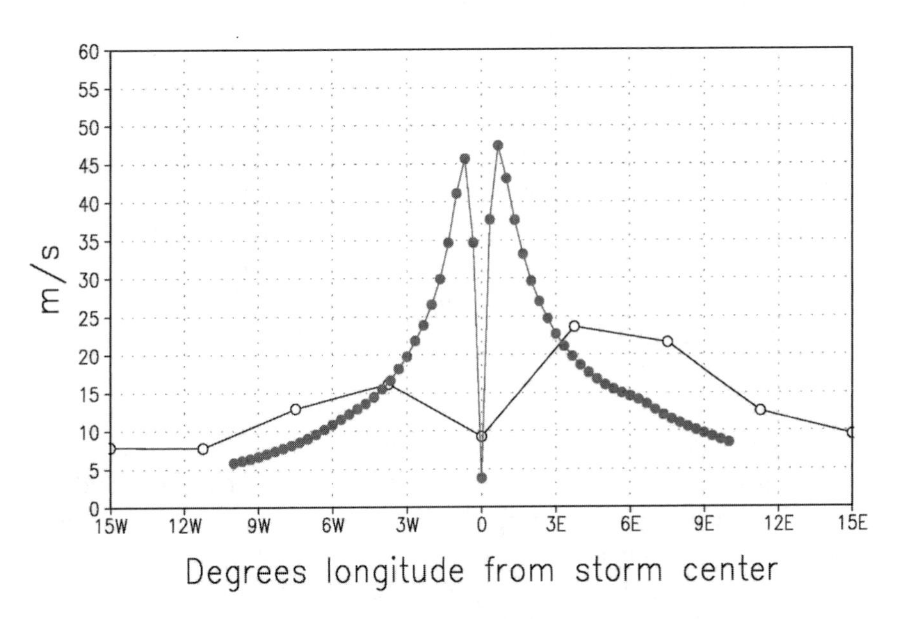

FIGURE 15.1 Surface wind profiles for tropical cyclones in a coarse-resolution global climate model compared with a high-resolution hurricane prediction model. The surface wind speeds are plotted as a function of distance from storm center along an east-west transect through a composite tropical storm as simulated by the low-resolution climate model (open circles) and the high-resolution hurricane model (solid circles). Note the more intense, compact structure of the storms in the high-resolution model. The climate model composite is based upon a large sample of tropical Northwest Pacific storms having maximum surface wind speed of 26 m s^{-1} or higher. The hurricane model composite is based on 51 case studies described in more detail in the text. The hurricane model winds are plotted uniformly at 1/3° resolution, although the model's resolution within 2.5° of the storm center is actually 1/6°. The surface winds are for the lowest model level ($p/p_{surface}$ = 0.997, where p is pressure, for the climate model and $p/p_{surface}$ = 0.995 for the hurricane model).

address possible climate-related hurricane intensity changes has been questioned (Henderson-Sellers et al. 1998 and references therein).

Despite the limitations of global climate models in simulating realistic hurricane intensities and structure, they appear to have some ability to simulate the statistical occurrence of tropical storms for given large-scale climate conditions. For example, the climatological distribution of simulated tropical storms in the Bengtsson, Botzet, and Esch (1995) global model appears fairly realistic. As a second example, the global model used by Vitart, Anderson, and Stern (1997, 1999) appears to have some skill in hindcasting interannual and even interdecadal fluctuations of tropical storm occurrence in the Northwest Atlantic Basin when forced by observed sea-surface temperatures (Vitart and Anderson 2001; see also Vitart, chapter 13 in this volume).

Regional Model Simulations

Regional models have a long history in explorations of the influence of sea-surface temperature changes on storm intensities (e.g., Ooyama 1969; Tuleya and Kurihara 1982; Evans, Ryan, and McGregor 1994). These early studies mostly involved altering only sea-surface temperature, without changes to the atmospheric temperature profile. This contrasts with the theoretical methods discussed previously, which either explicitly (Holland) or implicitly (Emanuel) incorporate enhanced warming of the upper troposphere, relative to lower levels, with increased sea-surface temperature. Drury and Evans's (1993) numerical experiments explored the impact on hurricane intensity of increasing sea-surface temperature under different atmospheric temperature change conditions. Although their study was limited to a few cases, they found that the degree of sea-surface temperature–induced storm intensification was considerably reduced if atmospheric temperatures were adjusted in such a way that convective available potential energy was unchanged in the warm sea-surface temperature scenario.

Several recent studies (Knutson, Tuleya, and Kurihara 1998; Knutson and Tuleya 1999; Walsh and Ryan 2000; Shen, Tuleya, and Ginis 2000; Knutson et al. 2001) have used variations on a nested high-resolution modeling approach to investigate tropical cyclone intensity change under different climate conditions. The influence of various large-scale thermodynamic conditions (i.e., sea-surface temperatures, atmospheric temperatures and moisture) have been incorporated in these studies. One approach (Knutson, Tuleya, and Kurihara 1998; Knutson and Tuleya 1999) involves taking a sample of tropical storms from a coarse-resolution global climate model and re-running these storm cases individually using 5-day simulations of a high-resolution GFDL hurricane model. The intent of this approach is to sample across a wide range of synoptic situations from each climate, including both thermodynamical and dynamical influences, such as wind shear. Walsh and Ryan (2000) employed a similar methodology for the Australia region—although based on a different nested regional model. A second general approach is to simulate storm cases using a more idealized environment, such as a uniform easterly flow field. The thermodynamic environment for these idealized experiments can be derived either from observations (Shen, Tuleya, and Ginis 2000) or from temporally and spatially averaged fields from a global climate model (Knutson and Tuleya 1999). The basic profile can then be altered to investigate the sensitivity of the simulated storm intensities to the thermodynamic environment. This can be done either by systematically examining a wide parameter space (as in Shen, Tuleya, and Ginis 2000) or by obtaining high-CO_2 environmental conditions from a global climate model CO_2 perturbation experiment (as in Knutson and Tuleya 1999). The idealized approach has also been extended by incorporating an interactive three-dimensional ocean underneath the storm (Knutson et al. 2001). This ocean coupling process is included to simulate the effect of the storm-generated "cool sea-surface temperature wake" on the simulated storm intensities and their sensitivity to climate change. Results from the series of studies using the GFDL hurricane model are reviewed in detail in the following sections.

Hurricane Model Case Studies

Several tropical storm cases from a coarse-resolution global climate model have been identified and re-run as a series of five-day "time-slice" experiments using a high-resolution regional hurricane model. For these nested high-resolution experiments, the large-scale environmental conditions are taken from the global model. This procedure is repeated for 51 storm cases, each from the present-day and high-CO_2 simulations of the global climate model. The statistics of the 51-member samples are then compared. The high-CO_2 conditions correspond to about 100 years from present day with a 1% per year compounded increase in atmospheric CO_2. The tropical sea-surface temperatures in this scenario increase by about 2.2 to 2.7°C. The cases are selected from the northwestern tropical Pacific region, where the strongest tropical cyclones are observed in the present climate. The procedure used to initialize the storm simulations in the regional hurricane model is analogous to that presently used operationally for hurricane prediction at the U.S. National Centers for Environmental Prediction (NCEP), except that in the operational case the initial disturbance strength is based on actual storm observations and the global fields are derived from operational atmospheric analyses, rather than from a global climate model. Further details on the "host" global climate model, the nested regional hurricane model, the method of storm case selection, and the initialization procedure are given in Knutson, Tuleya, and Kurihara (1998), Knutson and Tuleya (1999), and references cited therein.

Storm Intensities

The spatial distribution and magnitude of the wind speeds in the present-day case studies (plate 11b) appear fairly realistic to first order in comparison to the observed maximum wind speeds (plate 11a). In particular, note the decrease of maximum intensities over higher latitudes (with cooler sea-surface temperatures), near the equator, and over land regions. Some differences between the simulated and observed distributions are also apparent: simulated intensities are too high west of the Philippines (where a lack of ocean coupling may be important) and in the northwestern part of the domain in the vicinity of Japan. Another shortcoming of the simulations is that wind speeds in the very strongest storms appear to be slightly underpredicted in the present-day simulations (plate 11b) compared with actual observations (plate 11a). This underprediction of high wind speeds for intense storms is a known bias of the hurricane model, although the model nonetheless simulates surface pressure minima at least as low as the observed record (870 mb). The high-CO_2 distribution model (plate 11c) has more areas that experienced very intense (>70 m s^{-1}) wind speeds than the control distribution model (plate 11b), suggesting a modest increase in maximum surface winds in response to CO_2-induced warming.

In figure 15.2b, the frequency distribution of maximum wind speeds (one value per storm) is compared for the control and high-CO_2 case studies. The

FIGURE 15.2 Observed (*a*) and simulated (*b–e*) maximum surface wind speeds (one value per storm) for Northwest Pacific Basin tropical cyclones. Observations (*a*) are from the Joint Typhoon Warning Center (Guam) as compiled by C. J. Neumann as of 1993 and were obtained from NCAR (http://dss.ucar.edu/datasets/ds824.1). The data period analyzed was 1971 to 1992 for the months of July to November and within the region 8° to 26°N, 124° to 161°E. Each of the storms shown had a reported intensity of at least 17.5 m s^{-1} for at least four (6-hourly) observations while located within the study region. Simulated distributions in (*b–d*) are from the hurricane model, and in (*e*) from the global climate model. Dashed and solid lines in (*b–e*) correspond to control and high-CO_2 experiments, respectively. Note the higher intensity storms simulated for high-CO_2 conditions in comparison to control conditions. The initial storm disturbances in the hurricane model simulations were specified as follows: (*b*) synthetic initial vortex with a maximum wind speed of 17.5 m s^{-1} and a radius of maximum wind of 175 km; (*c*) synthetic initial vortex with a maximum wind speed of 10 m s^{-1}, a radius of maximum wind of 25 km, and a relatively shallow vertical depth; and (*d*) no synthetic vortex substitution, since the initial disturbance was taken from the global climate model.

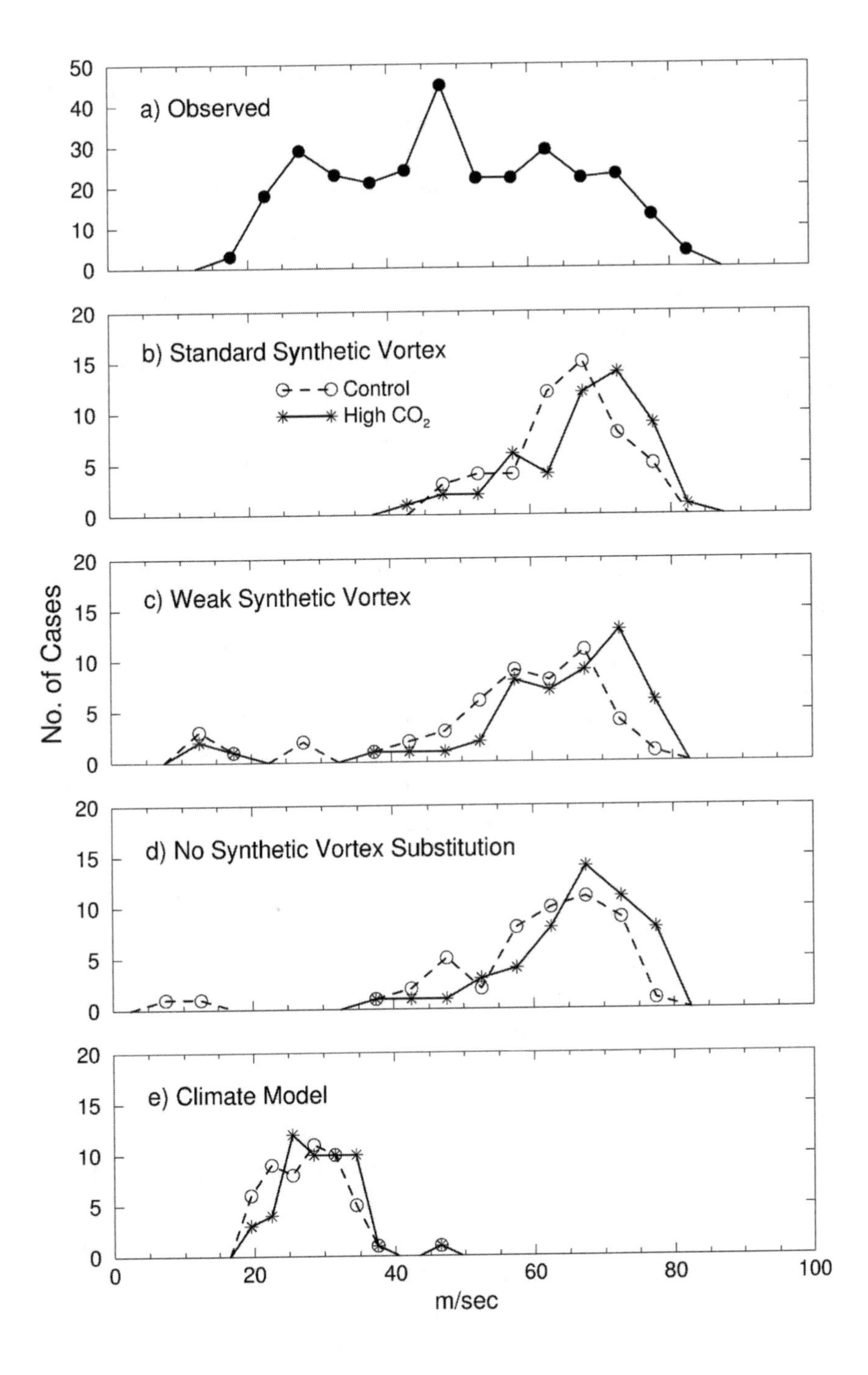

a) Observed
b) Standard Synthetic Vortex
Control
High CO2
c) Weak Synthetic Vortex
d) No Synthetic Vortex Substitution
e) Climate Model
No. of Cases
m/sec

high-CO_2 distribution is shifted toward higher intensities. The median of the wind speed distribution is 3.2 m s^{-1} higher for the high-CO_2 case than for the control, and the median central surface pressure (not shown) is 6.6 mb lower (i.e., more intense). Statistical tests indicate that the difference between the high-CO_2 and control-climate intensities are statistically significant, though only marginally so (Knutson and Tuleya 1999).

Figure 15.2a shows a distribution of observed maximum storm intensities (one value per storm) for the Northwest Pacific Basin. This distribution is much broader than the intensity distribution of the storms simulated by the regional hurricane model (figure 15.2b). Although the high-intensity end of the simulated distributions is in reasonable agreement with the observed high-end, there are relatively few occurrences of lower intensity storms (e.g., less than 40 m s^{-1}) in the simulated distribution.

The reason for the lack of weak simulated storms in the hurricane model case studies is not clear. One possibility is that the storm case selection procedure, by emphasizing relatively strong and long-lived climate model storm cases from the climate model (Knutson and Tuleya 1999), preferentially selects unusually favorable environments (e.g., low shear) for storm intensification. A second possibility is the lack of ocean coupling in the regional model—a physical process that is discussed in a subsequent section. Other possibilities include the regional model's tendency to over-develop weak systems (Kurihara, Tuleya, and Bender 1998), a lack of sensitivity to vertical wind shear, or the possibility that the large-scale environments from the climate model are overly conducive to development in the hurricane model.

The potential biases introduced by the possibilities just mentioned are present for both control and CO_2 case studies, whereas the CO_2-sensitivity study emphasizes the systematic *differences between* the control and CO_2 case study ensembles. Nonetheless, the hurricane model samples appear to be most representative of the intensity of the strongest storms in the region, rather than of the overall distribution or mean value of storm intensities.

To examine the robustness of the results in figure 15.2b, different hurricane vortex initialization procedures have been tested. The base case comparison discussed earlier (figure 15.2b) uses a synthetic initial vortex generated by the hurricane model, following the procedure currently used for operational hurricane forecasting. The cases shown in figure 15.2d use the original coarsely resolved initial vortex from the global climate model as the hurricane model initial condition. The results without using the synthetic initial vortex (figure 15.2d) are quite similar to the base case (figure 15.2b), again showing a slightly higher intensity for the high-CO_2 storms relative to the control. The intensity changes shown in figure 15.2d are more statistically significant than the base case in figure 15.2b (Knutson and Tuleya 1999); the median of the maximum wind speed values was 4.1 m s^{-1} (7%) higher, and the median surface pressure was 11.2 mb lower (more intense). A third initialization method (results shown in figure 15.2c) uses a much weaker initial synthetic vortex for each case. This method shows a few more occurrences of weaker storms than with the stronger initial vortex cases (figure 15.2b). The CO_2-induced intensification "signal" is more pronounced and statistically significant (Knutson and Tuleya 1999) for the weak initial vortex comparison (figure 15.2c) than for the strong initial vortex comparison (figure 15.2b). For the weak initial vortex cases, the median intensity is 6.6 m s^{-1} (11%) higher for the high-CO_2 sample; the median central surface pressure is 24 mb lower.

The results in figure 15.2b–d indicate a more statistically significant intensity change than was simulated by Walsh and Ryan (2000) for the Australia region. Although Walsh and Ryan's results are qualitatively similar to those in figure 15.2b–d and Knutson, Tuleya, and Kurihara (1998), their simulated intensity increases are mostly not statistically significant. It should be noted that Walsh and Ryan (2000) used a somewhat coarser resolution regional model (30 km grid spacing) as compared with the GFDL hurricane model (~18 km).

The maximum wind speeds for the original storms from the much lower resolution global climate model (two 51-case samples) are shown for comparison in figure 15.2e. As expected, the global climate model storms are generally much weaker than their higher-resolution counterparts simulated using the nested hurricane model. In terms of CO_2 sensitivity, the high-CO_2 global climate model cases were slightly more intense than the control cases, but the difference is not statistically significant (Knutson and Tuleya 1999).

Other Storm Characteristics

In addition to intensity, the simulated storms in these experiments can be examined for changes in other important measures. For example, in the hurricane model case studies, near-storm precipitation is 28% greater for the high-CO_2 storms than the control storms (7.79 cm vs. 6.08 cm), based on the 51-case average, 6-hour accumulated rainfall for a $3.5° \times 3.5°$ region in the vicinity of the storm center. The two frequency distributions of this measure are also quite statistically distinct (Knutson and Tuleya 1999). Although not discussed here for brevity, the hurricane model case studies have also been analyzed with respect to track changes, composite structure, climatological background environmental fields, and the relationship between simulated intensities and various measures of the near-storm environment, such as potential intensity or vertical wind shear (Knutson and Tuleya 1999).

Idealized Experiments: Parameter Space Exploration

Experimental Design

As an alternative to the case study approach, in this section the sensitivity of hurricane intensity to changes in environmental conditions is explored over a wide parameter space of sea-surface temperature and atmospheric lapse rate conditions by simulating storms embedded in idealized environmental fields. This idealized approach is helpful for distinguishing changes in storm intensity caused by changes in sea-surface temperatures and environmental lapse rates from changes due to the noise of synoptic weather variability that occurs in the case studies, particularly for relatively limited samples (order 50) of storms. The approach does have limitations, however. For example, the idealized flow fields, being devoid of synoptic scale weather disturbances, are less realistic than those of the case studies; and the combinations of sea-surface temperature and lapse rate changes used are not as physically based as the fields from the case studies, which are derived directly from climate model simulations.

The same triply nested version of the hurricane model is used as for the case studies in the previous section. For each idealized experiment (integrated for 72 hours), an initial disturbance is embedded into an environment with no mean flow, no other initial disturbances, no land, and approximately uniform sea-surface temperature. The sensitivity of the results to different background flow fields was addressed in Shen, Tuleya, and Ginis (2000). The vertical distributions of temperature and relative humidity for the base case, derived from observations taken during the Global Atmospheric Research Program Atlantic Tropical Experiment (GATE) III, are shown in figure 15.3a.

Using this idealized framework, a range of combinations of sea-surface temperature and lapse rate changes has been explored, recognizing that there is uncertainty in climate model projections of changes in these quantities due to uncertainties in future radiative forcing, climate sensitivity, and the spatial structure (vertical and horizontal) of the climate response to different radiative forcings. The atmospheric temperature anomalies, relative to the base case, for the different experiments are illustrated in figure 15.3b. The profiles with tropospheric warming are patterned after global climate model CO_2-increase scenarios showing enhanced warming in the upper tropical troposphere relative to

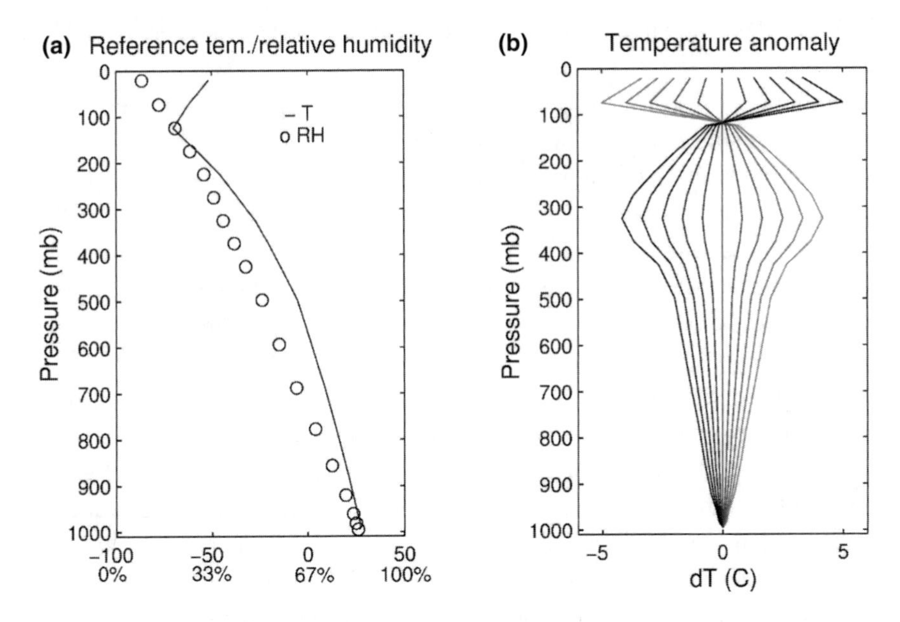

FIGURE 15.3 Some variables used in idealized hurricane simulations. (*a*) Vertical distributions of temperature (solid line) and relative humidity (circles) used in the base/control case (with SST of 28.5°C). The surface air-sea temperature difference is 1.2°C. Surface relative humidity is 84%. In the cases with different SSTs and no lapse rate anomaly, the same surface air-sea temperature difference and lapse rate, –dT/dz, profile of this temperature distribution are used. (*b*) Temperature anomalies, relative to the surface temperature change, used in the experiments.

the surface, along with a CO_2-induced cooling of the stratosphere. The tropospheric cooling–stratospheric warming profiles (symmetrical about zero with the "CO_2-increase scenario" profiles) are idealized perturbations constructed to explore a wider range of parameter values. An identical relative humidity profile is used for all cases, because climate models typically simulate relatively small changes in tropical relative humidity in global warming scenarios. Further details on these experiments are contained in Shen, Tuleya, and Ginis (2000).

Storm Intensity Results

The dark contours in plate 12 show the experimental results in terms of model-attained maximum intensity as a function of both specified sea-surface temperature and the upper tropospheric temperature anomaly initial condition relative to the surface. For the intensity comparisons, the central surface pressure is averaged over the final 24 hours of each experiment (i.e., hours 49–72). The base case (sea-surface temperature of 28.5°C, no upper tropospheric temperature anomaly) is located in the center of the diagram. Positive upper tropospheric anomalies denote a warming of the upper troposphere relative to the surface. An offsetting relationship is apparent between the effects of sea-surface temperature increase and enhanced upper tropospheric warming relative to the surface; it is monotonic and quite systematic in the parameter space investigated. For example, the effect on intensity of a sea-surface temperature increase of 1.5°C can be compensated by the effect of a positive upper-tropospheric temperature anomaly (relative to the surface) of about 3 to 4°C.

The blue dots in plate 12 denote sea-surface temperature/upper tropospheric temperature anomaly combinations as simulated in a control experiment of a GFDL global climate model. The red dots denote those factors for high-CO_2 conditions. These are superimposed on the parameter space to illustrate that most of the combinations of sea-surface temperatures and lapse rate anomalies in the parameter space are attainable (at least temporarily) in climate model integrations, although some parts of the parameter space are more likely, according to the model. The upper tropospheric temperature anomalies denoted by the dots are the deviations of $T_{320mb}–T_{1000mb}$ from the mean profiles from the Northern Hemisphere Summer control climatology. Although the individual anomaly values show considerable spread, the high-CO_2 (red) values are generally characterized by higher sea-surface temperatures and larger positive temperature anomalies than control climate (blue) values. Many of the values from the high-CO_2 climate lie off of the parameter space shown.

Using time mean conditions for the Northwest Pacific Basin from the global climate model (i.e., a 2.2°C increase in sea-surface temperatures and an upper tropospheric warming that is about 2.8°C larger than the surface air warming), an increase in storm intensity of roughly 8 mb for high-CO_2 conditions can be inferred from plate 12. Thus, the intensity enhancement due to increased sea-surface temperatures is larger than the intensity reduction effect of the enhanced upper tropospheric warming, according to the GFDL hurricane model using GFDL climate model initial conditions. Without the enhanced upper tropospheric warming, an intensity increase of more than 15 mb would be inferred, indicating the important role that enhanced upper tropospheric warming plays in the simulations in moderating the response of tropical cyclone intensities to a CO_2-induced increase in sea-surface temperatures. It should be noted that Emanuel's potential-intensity theory (discussed in the previous chapter) implicitly incorporates upper tropospheric warming, because it assumes the background atmospheric state is moist adiabatic.

Inclusion of Hurricane-Ocean Coupling

One limitation of the experiments described in the previous sections is the neglect of the effect of the tropical cyclone on the underlying ocean. Hurricane-ocean coupling can have an important impact on the intensity of the storm through the mixing of cooler, subsurface ocean waters with surface waters in the vicinity of the storm. This "cool wake" induced by the storm can reduce the intensity of the storm, with the degree of reduction of intensity dependent on many factors, such as the hurricane size, intensity, translation speed, and the thermodynamic structure of the upper ocean (Ginis 1995). An example of a cool wake induced by a hurricane as simulated in a version of the GFDL hurricane model coupled to an ocean model is illustrated in plate 13. The cooling induced in the sea-surface temperature field by the hurricane is clearly evident, with a local magnitude of up to 4°C or more in the wake of the moving storm. Comparisons between observed cool wakes and those simulated by the coupled model (e.g., Bender and Ginis 2000:figures 4 and 14) suggest that the model simulates hurricane-generated cool wakes fairly realistically in terms of their magnitude and spatial extent.

Experimental Design

The coupled tropical cyclone modeling approach follows that used by Shen, Tuleya, and Ginis (2000) and Knutson et al. (2001). For these idealized experiments, the large-scale thermodynamic environments are derived from the seasonal climatology of the GFDL global climate model for each tropical storm basin. The regional hurricane prediction model is coupled to a regional ocean model for a series of 72-hour coupled experiments. For the background winds, an idealized easterly flow field of 5 m s^{-1} is used unless otherwise noted. By using such idealized environmental flow fields, this approach considers neither the influence of storm interaction with transient flow features, nor the wide range of sea-surface temperature–lapse rate parameter space, both of which were explored in earlier sections. The initial ocean state is derived from observed upper ocean temperature profiles from the U.S. Navy Generalized

Digital Environmental Model (GDEM), with the ocean temperature stratification for the high-CO_2 experiments adjusted based on perturbations from the global climate model high-CO_2 scenario. Further details are provided in Knutson et al. (2001).

Simulated Storm Intensities

The simulated storm intensities are compared for various basins, ocean thermal stratifications, and climate boundary conditions (control vs. high CO_2) (figure 15.4). For these intensity comparisons, the central surface pressure averaged over the final 24 hours of each experiment (i.e., hours 49–72) is used. Similar results (not shown) are obtained using the maximum surface wind speed as the intensity measure.

Each diagram in figure 15.4 shows the results for one of the six tropical storm basins studied. For each diagram (basin) there are three sets of results, identified by the horizontal axis labeling: (1) without ocean coupling, (2) with ocean coupling using the basin-mean ocean temperature vertical stratification, and (3) with ocean coupling using a highly stratified ocean temperature profile. "Highly stratified" here refers to both a shallow mixed layer and a relatively strong vertical temperature gradient in the upper 200 m of the ocean (Knutson et al. 2001:figure 2). The circles linked by a vertical line segment in figure 15.4 show the ensemble means of the central surface pressures for a given ocean stratification for either high-CO_2 (solid circles) or control (open circles) condi-

FIGURE 15.4 Assessment of the impact of greenhouse warming on simulated tropical cyclone intensities in a model that includes ocean coupling. More intense storms are simulated for the warmer climate conditions, with ocean coupling having only a minimal impact on this enhancement. Minimum central surface pressure (mb) are shown for idealized experiments using average thermodynamic conditions (sea-surface temperature, atmospheric temperature and moisture, vertical profile of ocean temperature) for the various tropical storm basins (*a–f*), as described in Knutson et al. (2001). A 5 m s^{-1} easterly environmental atmospheric flow and the same initial storm vortex—aside from a small random perturbation—is used for each experiment shown. The horizontal axis qualitatively identifies the ocean thermal stratification for each group of experiments (i.e., no coupling, average temperature stratification, or strong temperature stratification). The ensemble mean central surface pressures for the high-CO_2 and control experiments are depicted by the solid and open circles, respectively. Each + represents a single integration or element of the ensemble. The + symbols to the left and right of the ensemble mean that results correspond to the control and high-CO_2 experiments, respectively. Although the absolute central pressure scale (vertical axis labeling) varies among the different diagrams, the relative scaling (pressure difference between tic marks) is the same for all diagrams. The central surface pressures are averages over the final 24 hours of each experiment (i.e., hours 49–72).

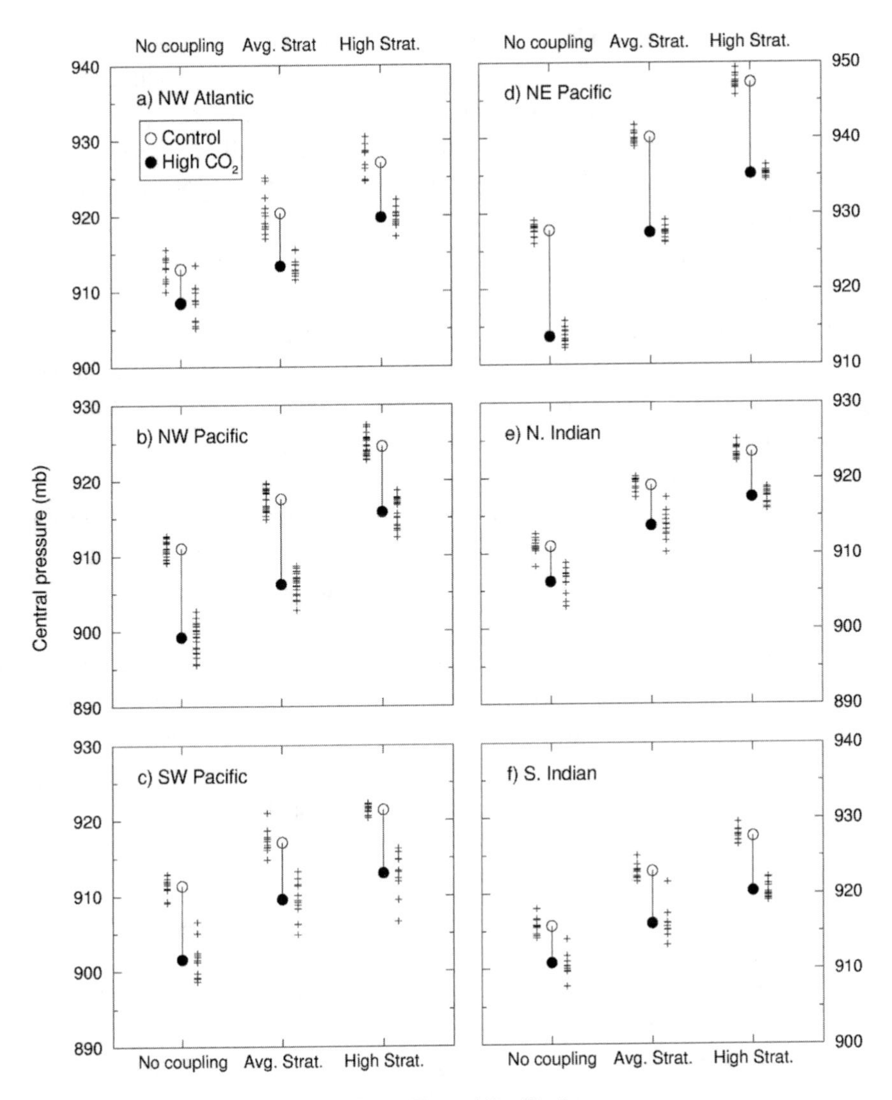
No coupling
Avg. Strat
High Strat.
a) NW Atlantic
Control
High CO_2
b) NW Pacific
c) SW Pacific
d) NE Pacific
e) N. Indian
f) S. Indian
Avg. Strat.
Central pressure (mb)
Ocean Thermal Stratification
940
930
920
910
900
890
950

tions. Each + symbol depicts the central surface pressure, averaged over hours 49 to 72, for a single model integration. The + symbols to the left and right of each connected set of circles represent the individual elements of the ensembles for the control and high-CO_2 samples, respectively. The individual elements of the ensembles differ only by small changes in the specified initial vortex intensity (Knutson et al. 2001). The ensemble technique was used to assess the robustness of the results to small changes in initial conditions.

Several notable results of the experiments are clearly shown in figure 15.4, of which four are mentioned here. First, the storms are generally weaker for the ocean coupling cases than for the uncoupled cases, as expected. For example, the control run storms (open circles) for the coupled cases (either average or strong stratification) have higher central surface pressures than the control run storms without ocean coupling. Second, stronger vertical gradients of ocean temperature associated with shallower-mixed layers lead to greater coupling-induced reductions in hurricane intensity, as seen by comparing the average and strong stratification cases. Third, for each individual type of ocean coupling condition (no coupling, average-ocean stratification or highly stratified ocean profile) the high-CO_2 storms (solid circles) tend to be more intense than the control storms (open circles). Thus, a CO_2 warming–induced intensification of hurricanes still occurs even when the hurricane-ocean coupling effects are included. Fourth, the CO_2-induced increase of intensity varies from basin to basin and is largest in the northeastern Pacific. The largest sea-surface temperature increase (2.7°C) occurs for the Northeast Pacific Basin, where, as noted in Knutson and Tuleya (1999:figure 3), a slightly larger enhancement of the lower tropospheric equivalent potential temperature occurs than in the other basins.

The main purpose of this set of experiments is to determine whether ocean coupling alters the CO_2-induced intensification of hurricanes as simulated in previous experiments, which did not include ocean coupling. If such an effect were present, the separation between the control and high-CO_2 ensemble mean intensities (solid and open circles connected by a line segment in figure 15.4) would be different in the coupled and uncoupled cases. Inspection of figure 15.4 suggests that a slight reduction of the CO_2-induced intensification may occur with ocean coupling for three of the six basins (Northwest Pacific, Northeast Pacific, and Southwest Pacific) but the effect is relatively small and nonsystematic. Such an effect is not even present for the Northwest Atlantic, North Indian, and South Indian cases.

The intensity results for the basins are combined to form a six-basin average set of results. The six-basin mean intensity change (high CO_2 minus control) is –8.3 mb (uncoupled) versus –8.4 mb (average stratification) and –8.3 mb (strong stratification). In terms of maximum surface wind speed (again averaged over hours 49 to 72 for each case) the six-basin mean intensity change (high CO_2 minus control) is 2.6 m s^{-1} or 5.3% (uncoupled) versus 2.8 m s^{-1} or 5.9% (average stratification) and 2.8 m s^{-1} or 6.1% (strong stratification). The six-basin mean results suggest that the effect of ocean coupling on the CO_2-induced intensification is essentially negligible, at least for the case of 5 m s^{-1} easterly environmental flow. Other environmental flow regimes examined yield broadly similar conclusions (Knutson et al 2001).

In summary, from these experiments, we conclude that, although ocean coupling reduces the intensity of simulated storms, it has only a minor impact on the simulated CO_2-induced intensification of storms.

Precipitation Changes

As mentioned in the discussion of the Northwest Pacific case studies, a substantial increase in near-storm precipitation occurs for the high-CO_2 storms. In this section, we show some additional storm precipitation results for the more idealized simulations, both with and without ocean coupling.

Figure 15.5 shows instantaneous precipitation rates, averaged within 100 km of the storm center (surface pressure minimum) at hour 72 of the same idealized experiments shown in figure 15.4 (no coupling and highly stratified ocean profile cases only). Note that the vertical axis extends to zero on these figures—an indication that the fractional changes in precipitation are substantial. The average percentage increase for the six basins is 17% and 19% for the no coupling and highly stratified ocean experiments, respectively, although it varies from 9% up to 48% in individual basins. There is a fair amount of noise or scatter among the ensemble members for a given basin, reflecting in part the inherent variability of instantaneous precipitation fields (as opposed to time-averaged ones). Nonetheless, a consistent enhancement of precipitation is readily apparent in the high-CO_2 versus control ensemble mean results.

Other measures of precipitation yield different fractional increases than those shown in figure 15.5 (which are based on an area average within 100 km of storm center). For example, a comparison of the maximum precipitation rates at any gridpoint in the storms, averaged over hours 49 to 72, indicates that the maximum precipitation in the high-CO_2 storms is enhanced by about 34% (with a range of 28% to 45% across the six basins) in the runs without ocean coupling. For the cases with ocean coupling (highly stratified temperature profiles) the maximum precipitation rate in the high-CO_2 storms is enhanced by 31% (range of 28% to 34% across the six basins).

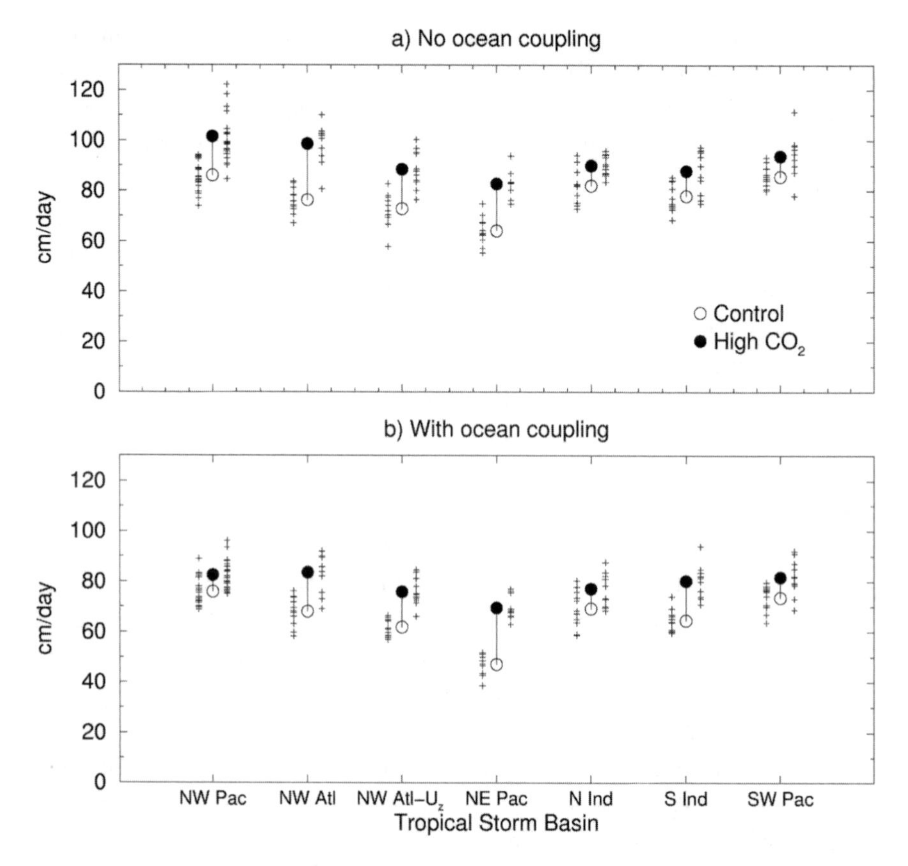

FIGURE 15.5 Assessment of the impact of greenhouse warming on simulated tropical cyclone precipitation for idealized experiments (*a*) without ocean coupling or (*b*) with ocean coupling using a "strong" temperature vertical stratification in the ocean. The individual tropical storm basins are identified along the horizontal axis. The model statistic analyzed is the instantaneous precipitation rate, area-averaged within 100 km of the storm center (surface pressure minimum), at hour 72 of each experiment. The ensemble means for the high-CO_2 and control experiments are depicted by the solid and open circles, respectively. Each + symbol represents a single integration or element of the ensemble. The + symbols to the left and right of the ensemble mean that results correspond to the control and high-CO_2 experiments, respectively.

The basin labeled "NW Atl-Uz" in figure 15.5 corresponds to an additional set of experiments in which the area-averaged time-mean zonal wind profiles from the climate model for the Atlantic Basin are used, rather than a uniform 5 m s^{-1} easterly profile as for the standard experiments. Inclusion of the mean environmental wind profiles has relatively little impact on the CO_2-sensitivity results: substantially more precipitation still occurs for the high-CO_2 cases.

The simulated increase of hurricane-related precipitation in figure 15.5 is reminiscent of recent observational evidence (Karl and Knight 1998) of an upward trend in the intensity of extremely heavy precipitation events in the United States since 1910. Karl and Knight's observational study did not, however, distinguish between hurricane-related and other types of precipitation events. It remains to be seen whether Karl and Knight's observations are related to the hurricane model simulations of precipitation presented here. For example, a similar physical mechanism, such as the increased moisture-holding capacity of a warmer atmosphere leading to enhanced moisture convergence and precipitation, could be operating in both cases.

A caveat to the results presented here is that real-time precipitation forecasts with the GFDL hurricane model have not been as thoroughly evaluated in operational use as track or intensity forecasts. Therefore, further assessment of the impact of CO_2-induced warming on tropical cyclone-related precipitation is called for, using a variety of approaches including theory and alternative models.

Discussion of Uncertainties in the Simulations

A number of important physical processes have been incorporated into the simulations discussed in this chapter. As illustrated in figure 15.1, nested high-resolution hurricane models offer the potential advantage of being able to simulate more realistic wind speeds and tropical cyclone structure than is possible with current-generation global climate models. A number of caveats remain regarding the nested approaches described in this chapter. In general, the simulation of hurricane intensities remains a challenging problem, even using relatively high-resolution models and assuming that future large-scale environmental conditions are perfectly known, which is far from the case. As model uncertainties are narrowed, future studies should revisit the results presented in this chapter using refined climate model scenarios or nested high-resolution models.

Model Dependence of Large-Scale Boundary Conditions

The large-scale boundary conditions (e.g., sea-surface temperatures, lapse rates) used for the experiments described previously are from a single global climate model forced by increasing greenhouse gases alone. There is, however, significant uncertainty in both the future course of radiative forcing and in the climate sensitivity to that forcing (e.g., Houghton et al. 2001). Other climate models and other combinations of radiative forcings could produce different large-scale climate responses from the climate-change scenario used for the nested model studies reviewed here. For example, in several past assessments the global climate sensitivity to a doubling of CO_2 has been estimated to be in the range of 1.5 to 4.5°C, with cloud feedback usually considered the most important source of uncertainty. Despite these uncertainties, current global climate models consistently produce both significantly warmer tropical sea-surface temperatures and enhanced warming of the tropical upper troposphere as greenhouse-gas concentrations increase. The large-scale boundary conditions from the GFDL global climate model used in the present study reflect this general aspect of anticipated climate change, although the climate sensitivity of this model (3.4°C) lies in the upper half of the often-cited 1.5 to 4.5°C range. It should be recalled that the storm intensity sensitivity studies discussed in the section on idealized experiments are not dependent on a single global climate model. The basic state for these experiments is derived from tropical Atlantic (GATE) observations, and the perturbations are obtained by systematically exploring a wide parameter space of sea-surface temperature and lapse rate changes.

Regional aspects of future climate change present additional challenges related to tropical cyclones, an example of which is the question of future changes in El Niño as a result of increased greenhouse gases. In the present climate, it appears that tropical Pacific sea-surface temperature variability (i.e., El Niño and La Niña) exerts a significant remote influence on Atlantic Basin hurricane activity through changes in vertical wind shear in the Atlantic (Gray 1984; Goldenberg and Shapiro 1987; Vitart, Anderson, and Stern 1999). Therefore, future changes in El Niño or the tropical Pacific mean "background state" would likely influence Atlantic hurricane activity. Idealized coupled ocean-atmosphere models indicate that El Niño is potentially sensitive to greenhouse-gas–induced climate change if the background state is altered (e.g., Fedorov and Philander 2000). Adequate simulation of future radiatively forced changes in even the background state of the tropical Pacific poses large research challenges. Current climate models continue to have limitations in their ability to realistically simulate both present-day El Niño variability and the tropical Pacific mean climate (Latif et al. 2001; Davey et al. 2002). Reviewing climate change simulations to date, a number of current models suggest that as greenhouse warming occurs, the tropical Pacific mean state will become more El Niño-like as the climate warms (Houghton et al. 2001), which would favor reduced Atlantic hurricane activity. On the other hand, Cane et al. (1997) argue that the tropical Pacific may in fact become more La Niña-like. With regard to El Niño variability, there is presently no clear consensus among climate models as to whether or how the amplitude of El Niño/La Niña "swings" would change in a warmer climate (Houghton et al. 2001). Reducing the uncertainty in assessments of future changes in regional tropical cyclone activity must await improved modeling and understanding of the tropical Pacific sea-surface temperature response (both time-mean and variability) to anthropogenic forcing.

Aside from sea-surface temperatures, CO_2-induced changes in tropospheric temperatures (i.e., lapse rates) differ among climate models (e.g., Henderson-Sellers et al. 1998), partly owing to differing treatment of convective parameterization. As shown in plate 12, these differences in climate response could be quite important for the tropical cyclone intensity issue. Similarly for the ocean, the vertical temperature structure (e.g., mixed-layer depths) are only crudely resolved in the GFDL global climate model used to provide CO_2-induced ocean temperature profile changes for the present study. Although no dramatic changes in near-surface tropical ocean temperature stratification are indicated in these low-resolution climate model simulations, if more pronounced changes occurred in the real climate, they could have a significant impact on storms intensities.

Nested Simulation Approach

In the nested model experiments, the high-resolution nested model does not evolve its own high-CO_2 climate. The climate sensitivity of a hypothetical global version of a given regional model could be different from that of the global climate model in which the regional simulation is nested.[2] In fact, the various differences between the regional and global models, such as spatial resolution, physics, and so on, can be expected to lead to differences between the models' "present-day" climates. Therefore, the environmental fields in the regional model will tend to adjust toward the regional model's climate, even during relatively short (three- to five-day) integrations. Because these adjustments occur in both the control and high-CO_2 cases in the experiments, it is assumed that their net effect on the *sensitivity* results (high-CO_2 minus control intensity) is small compared with the CO_2-induced changes in intensity. Nonetheless, it will be important to test the nested regional model results using global models of comparable resolution at some point in the future when such high-resolution models become viable for long-term climate simulation.

Role of Vertical Wind Shear

Vertical wind shear has been identified as an important negative factor in tropical storm intensification (Gray 1968; DeMaria and Kaplan 1994b). Therefore, changes in vertical wind shear as a result of climate change could reduce or enhance the increases in intensity due to thermodynamical changes (i.e., SSTs, atmospheric temperatures and moisture), which have been the primary focus of this chapter. Some of the studies with the GFDL hurricane model reviewed in this chapter have attempted to address the potential impact of vertical wind shear on the climate change results. For example, the Northwest Pacific case studies cited earlier sample a variety of synoptic settings from control and high-CO_2 climate simulations of a global model, and thus include environmental vertical wind shear in the initial conditions. For these cases, Knutson and Tuleya (1999:table 1) found little statistical correlation between the environmental wind shear and nested model-simulated storm intensities. In terms of real-time hurricane forecasts, Rhome, Raman, and Pasch (2002) and Mark DeMaria (personal communication) have found that the GFDL model has difficulty forecasting hurricane intensities in situations where environmental shear is apparently playing an important role. These results suggest that the GFDL hurricane model may underestimate the impact of vertical wind shear on storm intensity.

Using a more idealized experimental design, Knutson and Tuleya (1999:figure 11) have found that vertical wind shear has at least some negative effect on intensity in the GFDL model. They performed idealized hurricane simulations for the Northwest Atlantic Basin that included the time-mean vertical shear of the zonal wind in the initial environmental condition. These were compared with experiments using a uniform 5 m s^{-1} zonal environmental flow with no vertical shear. The storms with the sheared environment were about 12 mb weaker than those in the no-shear environment. Five-day time series of central pressures (not shown) from these experiments indicate that the negative impact of vertical shear on intensity is even more pronounced during the first two days of storm development (when the storms are weaker) than is indicated by comparing the minimum pressures achieved during the entire five-day integrations. Noting that the vertical wind shear in the Northwest Atlantic Basin was slightly enhanced in the high-CO_2 global climate model experiment compared to the control (Knutson and Tuleya 1999:figure 2b), they also compared the amount of CO_2-induced storm intensification in the cases with no shear to the cases with shear (i.e., incorporating the slight increase of vertical shear in the high-CO_2 environments). A similar CO_2-induced increase in storm intensity was found for the shear cases (9 mb) as for the no-shear cases (10 mb).

To summarize, vertical wind shear appears to have an important impact on tropical cyclone intensity. There is some evidence that this physical effect may be underestimated by the GFDL hurricane model used for the studies reviewed in this chapter. At present, however, there is no strong indication from the GFDL global climate model runs examined that substantial changes in climatological vertical wind shear will occur in response to increasing greenhouse gases (Knutson and Tuleya 1999). The role of vertical wind shear should be revisited in future studies, with both improved hurricane simulation models and revised global model simulations of climate changes in the various tropical storm basins.

Summary

In this chapter, we summarized several attempts to simulate the impact of CO_2-induced climate change on tropical cyclone intensities using the GFDL hurricane prediction system. In one approach, a high-resolution regional hurricane model is embedded in a global climate model without feedback of the high-resolution model onto the global model. A series of 51 Northwest Pacific storm case studies were then simulated for both the control (present-day) and high-CO_2 climates. Surface wind speed increases of about 3 to 7 m s^{-1} (5–11%) and central surface pressure decreases of 7 to 24 mb were simulated using a variety of vortex initialization choices.

A more idealized approach was used to explore a wide parameter space of sea-surface temperature and upper tropospheric temperature anomaly changes. These results indicate that in the GFDL hurricane model, increased sea-surface temperature leads to stronger tropical cyclones, whereas enhanced upper tropospheric warming—relative to the surface warming—acts to suppress the intensity increase. For the greenhouse-warming conditions as simulated by the GFDL global climate model, the sea-surface temperature effect is larger than the enhanced upper tropospheric warming effect, so that a net increase in tropical cyclone intensities is simulated. Using these sensitivity tests, a decrease in central surface pressure (intensity increase) of about 8 mb for high-CO_2 conditions is estimated for the Northwest Pacific Basin.

Ocean coupling is a physical process that can act to limit the intensity of tropical cyclones. An attempt has been made to quantify how the inclusion of ocean coupling affects the CO_2 warming–induced enhancement of hurricane intensities. When ocean coupling is included, the simulated hurricanes produce a cold wake in the sea-surface temperature field as they propagate over the open ocean. Stronger storms produce stronger wakes, as do storms that propagate more slowly or storms that occur over ocean regions with stronger thermal stratification. The question examined is whether the intensification of storms due to CO_2-induced warming would be mitigated by ocean coupling, because the stronger high-CO_2 storms should produce stronger cold wakes that should act as a negative feedback, limiting the intensity of the storms. In the experiments performed, the high-CO_2 storms do have slightly stronger cold wakes (Knutson et al. 2001), but the effect is not strong enough to have more than a minor impact on the CO_2-induced intensification. Therefore, the inclusion of ocean coupling in the GFDL coupled hurricane modeling system has only a minor impact on the simulated CO_2-induced intensification of hurricanes, estimated in these experiments to be about 3 to 10% for maximum surface wind speeds in the various basins.

Although statistically significant impacts of CO_2-induced warming on hurricane intensity have been found in these simulations, the statistical significance of the results is fairly marginal (particularly for the Northwest Pacific case studies), even for climate conditions based on a century of greenhouse gas increases at 1% per year compounded. This suggests that increased tropical cyclone intensities such as those simulated here may be quite difficult to detect in the real world data for many decades to come (Knutson et al. 2001).

The tropical cyclones simulated using the case study approach were generally quite strong storms with only a few weak ones. Furthermore, the idealized approaches generally do not incorporate factors such as vertical wind shear, which potentially can limit storm intensification. Owing to these limitations, the intensity results reviewed in this chapter are probably most relevant to the question of the potential intensity of tropical cyclones as determined by their thermodynamic environment (in some cases including effect of ocean coupling). In that regard, the recent statistical analysis of tropical cyclone intensities by Emanuel (2000) suggests that an increase of the potential intensity of tropical cyclones implies an increase in average of intensity of the storms as well. To the extent that Emanuel's statistical finding holds in a climate change context, the simulation results reviewed here probably imply an increase in the mean intensity of future tropical cyclones, although the effect of future changes in vertical wind shear requires further evaluation, and future changes in tropical cyclone frequencies (which have not been evaluated here) remain uncertain.

In conclusion, there is a strong indication in current nested high-resolution simulations that a CO_2-induced warming of the type simulated in the GFDL global climate model would lead to potentially more intense tropical cyclones. The magnitude of this effect in the GFDL hurricane model is estimated as a 3 to 10% increase in maximum wind speeds for a CO_2-induced increase in tropical sea-surface temperatures of 2.2 to 2.7°C. A number of caveats to the simulation results were discussed in the previous section. These should be borne in mind when assessing the uncertainty levels of these simulation results. Aside from storm intensity, the high-CO_2 storms simulated in this chapter tend to show rather large (~18 to 32%) increases in hurricane-related precipitation, relative to storms simulated for present-day conditions. Because storm-related precipitation can cause severe damage during tropical cyclones, as in the recent example of Hurricane Mitch, we recommend further examination of this issue.

Acknowledgments

We thank J. D. Mahlman for originally encouraging our work on this project and three reviewers for helpful suggestions for improving the manuscript. We also wish to acknowledge the model development contributions of Y. Kurihara, M. Bender, S. Manabe, R. Stouffer, and numerous other colleagues.

Notes

1. The word "hurricane" refers to a tropical cyclone with 1-minute average surface (10 m) winds of at least 64 knots in the Western Hemisphere. Other terms (e.g., "typhoon," "severe tropical cyclone") are used to refer to these storms in various regions of the world.
2. A related limitation applies to high-resolution global atmospheric model "time-slice" experiments forced by sea-surface temperature changes from lower resolution coupled climate models. In this case, the sea-surface temperature changes are derived from a model that may have different climate sensitivity characteristics from the high-resolution global model.

References

Baik, J.-J., and J.-S. Paek. 1998. A climatology of sea surface temperature and the maximum intensity of western North Pacific tropical cyclones. *Journal of the Meteorological Society of Japan* 76:129–37.

Bender, M. A., and I. Ginis. 2000. Real-case simulations of hurricane-ocean interaction using a high-resolution coupled model: Effects on hurricane intensity. *Monthly Weather Review* 128:917–46.

Bender, M. A., I. Ginis, and Y. Kurihara. 1993. Numerical simulations of tropical cyclone-ocean interaction with a high-resolution coupled model. *Journal of Geophysical Research* 98:23245–63.

Bengtsson, L., M. Botzet, and M. Esch. 1995. Hurricane type vortices in a general circulation model. *Tellus* 47A:175–96.

Bengtsson, L., M. Botzet, and M. Esch. 1996. Will greenhouse gas-induced warming over the next 50 years lead to higher frequency and greater intensity of hurricanes? *Tellus* 48A:57–73.

Broccoli, A. J., and S. Manabe. 1990. Can existing climate models be used to study anthropogenic changes in tropical cyclone climate? *Geophysical Research Letters* 17:1917-20.

Cane, M. A., A. C. Clement, A. Kaplan, Y. Kushnir, R. Murtugudde, D. Pozdnyakov, R. Seager, and S. E. Zebiak. 1997. 20th century sea surface temperature trends. *Science* 275:957–60.

Davey, M. K., M. Huddleston, K. R. Sperber, P. Braconnot, F. Bryan, D. Chen, R. A. Colman, C. Cooper, U. Cubasch, P. Delecluse, D. DeWitt, L. Fairhead, G. Flato, C. Gordon, T. Hogan, M. Ji, M. Kimoto, A. Kitoh, T. R. Knutson, M. Latif, H. Le Treut, T. Li, S. Manabe, C. R. Mechoso, G. A. Meehl, S. B. Power, E. Roeckner, L. Terray, A. Vintzileos, R. Voss, B. Wang, V. M. Washington, I. Yoshikawa, J.-Y. Yu, S. Yukimoto, and S. E. Zebiak. 2002. STOIC: A study of coupled model climatology and variability in tropical ocean regions. *Climate Dynamics* 18:403–20.

DeMaria, M., and J. Kaplan. 1994a. Sea surface temperature and the maximum intensity of Atlantic tropical cyclones. *Journal of Climate* 7:1324–34.

DeMaria, M., and J. Kaplan. 1994b. A statistical hurricane intensity prediction scheme (SHIPS) for the Atlantic basin. *Weather and Forecasting* 9:209–20.

Drury, S., and J. L. Evans. 1993. Sea surface temperature and CAPE: Importance for tropical cyclone intensity. In *Preprints of the 20th Conference on Hurricanes and Tropical Meteorology*, 89–92. Boston: American Meteorological Society.

Druyan, L. M., P. Lonergan, and T. Eichler. 1999. A GCM investigation of global warming impacts relevant to tropical cyclone genesis. *International Journal of Climatology* 19:607–17.

Elsberry R. L., G. J. Holland, H. Gerrish, M. DeMaria, C. Guard, and K. Emanuel. 1992. Is there any hope for tropical cyclone intensity prediction?—A panel discussion. *Bulletin of the American Meteorological Society* 73:264–75.

Emanuel, K. A. 1986. An air-sea interaction theory for tropical cyclones. Part I: Steady-state maintenance. *Journal of the Atmospheric Sciences* 43:585–604.

Emanuel, K. A. 1987. The dependence of hurricane intensity on climate. *Nature* 326:483–85.

Emanuel, K. A. 1988. The maximum intensity of hurricanes. *Journal of the Atmospheric Sciences* 45:1143-55.

Emanuel, K. A. 1995. Sensitivity of tropical cyclones to surface exchange coefficients and a revised steady-state model incorporating eye dynamics. *Journal of the Atmospheric Sciences* 52:3969–76.

Emanuel, K. A. 1999. Thermodynamic control of hurricane intensity. *Nature* 401: 665–69.

Emanuel, K. 2000. A statistical analysis of tropical cyclone intensity. *Monthly Weather Review* 128:1139–52.

Evans, J. L. 1993. Sensitivity of tropical cyclone intensity to sea surface temperature. *Journal of Climate* 6:1133–40.

Evans, J. L., B. F. Ryan, and J. L. McGregor. 1994. A numerical exploration of the sensitivity of tropical cyclone rainfall intensity to sea surface temperature. *Journal of Climate* 7:616–23.

Fedorov, A. V., and S. G. Philander. 2000. Is El Niño changing? *Science* 288: 1997–2002.

Ginis, I. 1995. Ocean response to tropical cyclones. In *Global perspectives on tropical cyclones,* edited by R. L. Elsberry, 198–260. Report No. TCP-38. Geneva: World Meteorological Organization.

Goldenberg, S. B., C. W. Landsea, A. M. Mestas-Nuñez, and W. M. Gray. 2001. The recent increase in Atlantic hurricane activity: Causes and implications. *Science* 293:474–79.

Goldenberg, S. B., and L. J. Shapiro. 1996. Physical mechanisms for the association of El Niño and West African rainfall with Atlantic major hurricane activity. *Journal of Climate* 9:1169–87.

Gray, W. M. 1968. Global view of the origin of tropical disturbances and storms. *Monthly Weather Review* 96:669–700.

Gray, W. M. 1975. *Tropical cyclone genesis.* Paper No. 323. Fort Collins: Department of Atmospheric Sciences, Colorado State University.

Gray, W. M. 1984. Atlantic seasonal hurricane frequency. Part I: El Niño and 30 mb Quasi-Biennial Oscillation influences. *Monthly Weather Review* 112:1649–68.

Haarsma, R. J., J. F. B. Mitchell, and C. A. Senior. 1993. Tropical disturbances in a GCM. *Climate Dynamics* 8:247–57.

Hamilton, K., and R. S. Hemler. 1997. Appearance of a supertyphoon in a global climate model simulation. *Bulletin of the American Meteorological Society* 78:2874–76.

Henderson-Sellers, A., H. Zhang, G. Berz, K. Emanuel, W. Gray, C. Landsea, G. J. Holland, J. Lighthill, S.-L. Shieh, P. Webster, and K. McGuffie. 1998. Tropical cyclones and global climate change: A post-IPCC assessment. *Bulletin of the American Meteorological Society* 79:19–38.

Holland, G. J. 1997. The maximum potential intensity of tropical cyclones. *Journal of the Atmospheric Sciences* 54:2519–41.

Houghton, J. T., Y. Ding, D. J. Griggs, M. Noguer, P. J. van der Linden, X. Dai, K. Maskell, and C. A. Johnson, eds. 2001. *Climate change 2001: The scientific basis.* Contribution of Working Group I to the Third Assessment Report of the Intergovernmental Panel on Climate Change. Cambridge: Cambridge University Press.

Karl, T. R., and R. W. Knight. 1998. Secular trends of precipitation amount, frequency, and intensity in the United States. *Bulletin of the American Meteorological Society* 79:231–41.

Knutson, T. R., T. L. Delworth, K. W. Dixon, and R. J. Stouffer. 1999. Model assessment of regional surface temperature trends (1949–1997). *Journal of Geophysical Research* 104:30981–96.

Knutson, T. R., and R. E. Tuleya. 1999. Increased hurricane intensities with CO_2-induced warming as simulated using the GFDL hurricane prediction system. *Climate Dynamics* 15:503–19.

Knutson, T. R., R. E. Tuleya, W. Shen, and I. Ginis. 2001. Impact of CO_2-induced warming on hurricane intensities as simulated in a hurricane model with ocean coupling. *Journal of Climate* 14:2458–68.

Knutson, T. R., R. E. Tuleya, and Y. Kurihara. 1998. Simulated increase of hurricane intensities in a CO_2-warmed climate. *Science* 279:1018–20.

Krishnamurti, T. N., R. Correa-Torres, M. Latif, and G. Daughenbaugh. 1998. The impact of current and possibly future sea surface temperature anomalies on the frequency of Atlantic hurricanes. *Tellus* 50A:186–210.

Kurihara, Y., R. E. Tuleya, and M. A. Bender. 1998. The GFDL hurricane prediction system and its performance in the 1995 hurricane season. *Monthly Weather Review* 126:1306–22.

Kuroda, M., A. Harada, and K. Tomine. 1998. Some aspects on sensitivity of typhoon intensity to sea-surface temperature. *Journal of the Meteorological Society of Japan* 76:145–51.

Landsea, C. W., N. Nicholls, W. M. Gray, and L. A. Avila. 1996. Downward trends in the frequency of intense Atlantic hurricanes during the past five decades. *Geophysical Research Letters* 23:1697–1700.

Latif, M., K. Sperber, J. Arblaster, et al. 2001. ENSIP: The El Niño simulation intercomparison project. *Climate Dynamics* 18:255–76.

Merrill, R. E. 1988. Environmental influences on hurricane intensification. *Journal of the Atmospheric Sciences* 45:1678–87.

Nguyen, K. C., and K. J. E. Walsh. 2001. Interannual, decadal, and transient greenhouse simulation of tropical cyclone-like vortices in a regional climate model of the South Pacific. *Journal of Climate* 14:3043–54.

Ooyama, K. 1969. Numerical simulation of the life cycle of tropical cyclones. *Journal of the Atmospheric Sciences* 26:3–40.

Rhome, J. R., S. Raman, and R. J. Pasch. 2002. An analysis of the forecast performance of the GFDL model during Debby (2000). In *Preprints of the 25th Conference on Hurricanes and Tropical Meteorology*, 126–27. Boston: American Meteorological Society.

Rotunno, R., and K. A. Emanuel. 1987. An air-sea interaction theory for tropical cyclones. Part II: Evolutionary study using a nonhydrostatic axisymmetric numerical model. *Journal of the Atmospheric Sciences* 44:542–61.

Royer, J. R., F. Chauvin, B. Timbal, P. Araspin, and D. Grimal. 1998. A GCM study of the impact of greenhouse gas increase on the frequency of occurrence of tropical cyclones. *Climatic Change* 38:307–43.

Ryan, B. F., I. G. Watterson, and J .L. Evans. 1992. Tropical cyclone frequencies inferred from Gray's yearly genesis parameter: validation of GCM tropical climates. *Geophysical Research Letters* 19:1831–34.

Schade, L. R., and K. A. Emanuel. 1999. The ocean's effect on the intensity of tropical cyclones: Results from a simple coupled atmosphere-ocean model. *Journal of the Atmospheric Sciences* 56:642–51.

Shen, W., R. E. Tuleya, and I. Ginis. 2000. A sensitivity study of the thermodynamic environment on GFDL model hurricane intensity: Implications for global warming. *Journal of Climate* 13:109–21.

Sugi, M., A. Noda, and N. Sato. 2002. Influence of global warming on tropical cyclone climatology: An experiment with the JMA global model. *Journal of the Meteorological Society of Japan* 80:249–72.

Tonkin, H., C. Landsea, G. J. Holland, and S. Li. 1997. Tropical cyclones and climate change: A preliminary assessment. In *Assessing climate change: Results from the Model Evaluation Consortium for Climate Assessment*, edited by W. Howe and A. Henderson-Sellers, 327–60. Amsterdam: Gordon and Breach.

Tuleya, R. E., and Y. Kurihara. 1982. A note on the sea surface temperature sensitivity of a numerical model of tropical storm genesis. *Monthly Weather Review* 110:2063–69.

Tsutsui, J. 2002. Implications of anthropogenic climate change for tropical cyclone activity: a case study with the NCAR CCM2. *Journal of the Meteorological Society of Japan* 80:45–65.

Vitart, F., and J. L. Anderson. 2001. Sensitivity of Atlantic tropical storm frequency to ENSO and interdecadal variability of SSTs in an ensemble of AGCM integrations. *Journal of Climate* 14:533–45.

Vitart, F., J. L. Anderson, and W. F. Stern. 1997. Simulation of interannual variability of tropical storm frequency in an ensemble of GCM integrations. *Journal of Climate* 10:745-60.

Vitart, F., J. L. Anderson, and W. F. Stern. 1999. Impact of large-scale circulation on tropical storm frequency, intensity, and location, simulated by an ensemble of GCM integrations. *Journal of Climate* 12:3237–54.

Walsh, K. J. E., and B. F. Ryan. 2000. Tropical cyclone intensity increase near Australia as a result of climate change. *Journal of Climate* 13:3029–36.

Whitney, L. D., and J. S. Hobgood. 1997. The relationship between sea surface temperatures and maximum intensities of tropical cyclones in the eastern North Pacific Ocean. *Journal of Climate* 10:2922–30.

16

Conclusion

Richard J. Murnane and Kam-biu Liu

Hurricanes and typhoons have always been part of our climate. The high winds and intense precipitation produced by hurricanes and typhoons represent the release of huge amounts of energy. Anything unfortunate enough to experience the energy produced by the most intense storms will be harmed; however, most natural systems have evolved ways to cope with the winds, rain, and waves produced by most hurricanes and typhoons. The best way for humankind to deal with hurricanes and typhoons is through prudent planning, well-executed emergency management, and sophisticated financial systems. Even when all these resources are present, the potential for human death and destruction from hurricanes and typhoons will continue to increase as populations grow along vulnerable coastlines. We can help minimize future death and destruction by applying improvements in our understanding of the extent and sources of hurricane and typhoon variability to societal efforts such as emergency management and zoning regulations.

The energy released by hurricanes and typhoons originates from the thermodynamic disequilibrium of the ocean and atmosphere. Atmospheric circulation, weather in general, and hurricanes and typhoons in particular represent an attempt by the coupled ocean-atmosphere system to reach thermodynamic equilibrium. Therefore, much of the past, present, and future variability in hurricane and typhoon activity reflects changes in how the ocean and atmosphere attempt to achieve thermodynamic equilibrium.

The analyses of hurricane and typhoon variability discussed in the chapters in this book provide at least two important insights to climate and the coupled ocean-atmosphere system. The first is that analyses of past temporal variability in tropical cyclone activity can provide knowledge of how climate has changed. The second is that information on how tropical cyclones respond to climate variability, when coupled with projections of how climate will respond to increasing concentrations of atmospheric greenhouse gases, permits projections of future tropical cyclone activity. There is still much to learn, although our knowledge of hurricane and typhoon variability has grown significantly since 1954 when Tor Bergeron published his paper "The Problem of Tropical Hurricanes," the source of the quote that starts this book.

The first two chapters demonstrate that tropical cyclone activity varies on centennial to millennial time scales. They present innovative techniques that allow the reconstruction of prehistoric hurricane landfalls. When coupled with the historical record, the results permit analyses of hurricane variability beyond that feasible with the historical record alone. Liu's results (chapter 2) suggest that the preferred areas for intense tropical cyclone landfall could be related to long term shifts in the location of the subtropical high in the North Atlantic Ocean. Donnelly and Webb's work (chapter 3) suggests that the characteristic path of hurricanes as they recurve in the North Atlantic induces an east-west gradient in hurricane landfall probability with the probability increasing to the east.

Extreme events such as hurricane landfall can produce significant effects that are preserved in biological and geological records. Expanding studies to other geographic areas using these coring techniques or other methods and developing new proxies could provide additional insights on changes in large-scale climate features. Examples of other proxies of hurricane landfall that offer promise and are under development include analyses of variations in tree and coral rings and the oxygen isotopic content of tree rings and carbonate deposits. Developing proxy-based records of hurricane landfall tends to focus on the more extreme, less common events associated with landfalls of intense tropical cyclones. This approach complements studies of historical archives.

Part II of the book presents five efforts aimed at extending the historical record of hurricane and typhoon landfall. Louie and Liu (chapter 8) explore the longest historical record of tropical cyclone landfall. Boose (chapter 4) combines a model and historical records extending back to the colonial era as an aid in understanding the ecological impacts of landfalling hurricanes and reconstructing an approximation of the hurricane's structure using a simple model. Mock (chapter 5) and García Herrera et al. (chapter 6) demonstrate the potential of archival research for reconstructing hurricane records in the North Atlantic. Landsea et al. (chapter 7) discuss their work on reanalyzing and extending the best-track data set for the North Atlantic. Reanalysis efforts such as those described chapter 7 benefit from the knowledge recovered through the study of archival records. Reanalyzed and extended best-track data sets will allow for more detailed and reliable analyses of tropical cyclone variability on a variety of time scales.

The chapters in part III provide examples of current studies of tropical cyclone variability and their use for forecasting tropical cyclone activity and hurricane and typhoon landfall. Chan (chapter 10) examines tropical cyclone variability in the western Pacific Ocean on time scales that range from intraseasonal to interannual. An understanding of this variability can be exploited through the development of improved forecasts of seasonal tropical cyclone activity. Chu (chapter 11) summarizes the response of tropical cyclone activity in different ocean basins to changes in the El Niño–Southern Oscillation (ENSO).

A combination of better knowledge on how tropical cyclones respond to changes in ENSO and other modes of climate variability and improvements in climate forecasts should result in better estimates of landfall probability. Elsner et al. (chapter 12) show how knowledge of the effects of ENSO and the North Atlantic Oscillation can be used to improve landfall probability estimates. Vitart (chapter 13) discusses the potential of dynamical models for developing seasonal forecasts of tropical cyclone activity. As dynamical models improve and observational systems become more extensive, the skill of dynamical forecasts should improve. Improved dynamical models will also lead to a better understanding of how hurricane and typhoon activity will change in the future.

The two chapters in part IV explore different aspects of future tropical cyclone variability. Emanuel (chapter 14) summarizes the theoretical basis for changes in tropical cyclone intensity. Both sea-surface temperatures and upper tropospheric temperatures are expected to increase, but the parallel increases will tend to counteract their impact on maximum potential intensity. Nevertheless, the effect of warming the ocean surface is likely to dominate the effect of warming the upper troposphere and result in a modest increase in tropical cyclone intensity. Model simulations by Knutson et al. (chapter 15) suggest that tropical cyclone intensity and rainfall will increase as a result of increased greenhouse-gas concentrations. Knutson and his colleagues point out that these results do not necessarily account for the effects of atmospheric dynamics on tropical cyclone intensity.

The only way to identify upcoming changes in hurricane and typhoon activity will be through a comparison of future observations to records of past storms. In chapter 9, Murnane points out some of the difficulties and limitations involved with developing best-track data sets. The quality of most information in the existing best-track data sets is such that it will be many years before any changes in tropical cyclone intensity can be discerned with certainty. Better and more extensive best-track data will allow for more reliable identification of future changes in tropical cyclone intensity.

A concerted effort should be made to enhance and maintain the best-track data on tropical cyclones. Best-track data form the basis of many studies of tropical cyclone variability and will be the standard by which any future changes in activity will be judged. The forecasters who monitor and forecast tropical cyclone activity generally are responsible for developing best-track data. Although this arrangement is logical because forecasters have intimate knowledge of a storm's behavior, it is sometimes impractical because of time demands on the forecaster in an operational environment and the difficulty of collecting and analyzing all observations in a timely manner.

Some of the best opportunities for improving our understanding of tropical cyclone variability involve the development of new proxy techniques and the refinement of existing techniques. Estimates of intense hurricane landfall probability based on proxy records are extremely valuable because they provide information independent of model estimates derived from the historical record of landfalling storms. In addition, proxy records are the only way to extend significantly our record of hurricane and typhoon landfall in areas with little or no potential for archival work. Finally, as mentioned by Emanuel (chapter 14), proxy studies offer what is perhaps the best hope for understanding the relationship between climate and tropical cyclone frequency.

Liu (chapter 2) and Donnelly and Webb (chapter 3) provide an excellent overview of the state of the art in developing hurricane landfall probability estimates from overwash deposits. This approach can be applied to a variety of coastal locations in different parts of the world. So far, only a handful of coastal lakes and marshes on the U.S. Gulf and Atlantic coasts have been studied, but results to date have demonstrated that reliable records of past hurricane strikes extending back hundreds to thousands of years can be obtained from overwash deposits. Obviously more work should be done in the United States as well as other parts of the world to produce more long-term records of past tropical cyclone activities. A different approach for reconstructing tropical cyclone landfall involves the analysis and dating of coastal ridges of coral rubble and sand deposited by wave action during a storm. This approach has been used to reconstruct tropical cyclone landfall in several locations in northern Australia. Consistent with the results of Liu (chapter 2), this work suggests that there are centennial to millennial scale fluctuations in the frequency of tropical cyclone landfall.

Chapters in part II of the book highlight the potential of archival research for extending the best-track record and provided a sample of the wide range of material that contains information on tropical cyclone landfall. The work by

Louie and Liu (chapter 8) probably represents the limit for a relatively complete record of landfalling storms. Prior to about 1500, the number of storm accounts in the Guangdong records drops precipitously. Historical reconstruction efforts should be feasible for other coastal locations in China as well as for Japan, Korea, and the Philippines, where writing and culture have supported the collection and maintenance of documentary records.

The work by García Herrera et al. (chapter 6) presents an example of the wealth of material contained in the colonial records of Spain and Great Britain. The records of other colonial powers could be mined for similar information. Archival work is very time intensive, and success requires a different set of skills than those typically associated with a meteorologist, but extending the historical record is one of the best ways to increase our understanding of tropical cyclone variability on decadal to centennial time scales.

Analyses of a relatively complete record of hurricane and typhoon tracks, central pressures, and wind speeds provide the most direct method of deciphering how climate variations alter tropical cyclone activity. Given the limitations of the historical record, dynamical models offer an attractive approach for improving the skill of seasonal and interannual forecasts of tropical cyclone activity and for assessing how tropical cyclone activity varies with climate. Improvements in dynamical models will come from increased computational resources, better observational networks, and advances in our understanding of the theory and dynamics of tropical cyclones and the climate system.

The chapters in part IV examine the potential future variability in hurricanes and typhoons. Emanuel (chapter 14) hypothesizes that there is a feedback loop between tropical cyclone activity and climate. This possibility should be explored, but, for a number of reasons, the exploration will be difficult. Perhaps one of the most important reasons is that current climate models do not adequately resolve tropical cyclones. A better representation of future changes in tropical cyclone activity will require improvements in observations, experiments, and theory. For example, there needs to be larger computational resources and better parameterizations of small-scale processes not resolved by current climate models so that atmospheric and oceanic dynamics will be represented in a more realistic manner.

Hurricane and typhoon activity varies over a continuous range of time scales. This book documents a number of efforts to extend the record of the tropical cyclone activity and understand the past, present, and potential future variability of the storms. Variations in tropical cyclone activity are of interest in their own right because of the inherent attraction of one of the most powerful phenomena in nature. But beyond the esoteric attraction of tropical cyclones, society has social and economic reasons for understanding and predicting their variability. Landfalling tropical cyclones produce some of nature's deadliest and costliest natural disasters. Human and monetary losses could be minimized through proper planning, accurate forecasting, and efficient emergency management. All these efforts will be easier with a better understanding of the range in the natural variability of tropical cyclones and how the range and variability will adjust to future climate change.

CONTRIBUTORS

Craig Anderson
NOAA/Climate Diagnostics Center
Boulder, Colorado 80305

Emery R. Boose
Harvard Forest
P.O. Box 68
Petersham, Massachusetts 01366
Phone: (978) 724-3302
Fax: (978) 724-3595
E-mail: boose@fas.harvard.edu

Brian H. Bossak
Mendenhall Fellow
United States Geological Survey
Center for Coastal and Watershed Studies
600 Fourth Street South
St. Petersburg, Florida 33701
Phone: (727) 803-8747, ext. 3046
Fax: (727) 803-2031
E-mail: bbossak@usgs.gov

Johnny C. L. Chan
Laboratory for Atmospheric Research
Department of Physics and Material Sciences
City University of Hong Kong
83 Tat Chee Avenue
Kowloon, Hong Kong
China
Phone: (852) 2788-7820
Fax: (852) 2788-7830
E-mail: Johnny.Chan@cityu.edu.hk

Noel Charles
767 NW 104th Avenue
Pembroke Pines, Florida 33026
Phone: (786) 457-7834

Pao-Shin Chu
Department of Meteorology
2525 Correa Road
H1G350
Honolulu, Hawaii 96822
Phone: (808) 956-2567
Fax: (808) 956-2877
E-mail: chu@hawaii.edu

Gilbert Clark
304 Margo Drive
Pearsall, Texas 78061
Phone: (830) 334-4414

Jeffrey P. Donnelly
Geology and Geophysics Department
Woods Hole Oceanographic Institute
MS 22
Woods Hole, Massachusetts 02543
Phone: (508) 289-2994
Fax: (508) 457-2187
E-mail: jdonnelly@whoi.edu

Jason Dunion
NOAA AOML/Hurricane Research Division
4301 Rickenbacker Causeway
Miami, Florida 33149
Phone: (305) 361-4530
Fax: (305) 361-4402
E-mail: Jason.Dunion@noaa.gov

Francisco Rubio Durán
Departamento Física de la Tierra II
Facultad de Ciencias Físicas
Universidad Complutense de Madrid
Ciudad Universitaria
28040 Madrid
Spain
Phone: (34) 913-944-456
Fax: (34) 913-945-195

James B. Elsner
Department of Geography
Florida State University
Tallahassee, Florida 32306
Phone: (850) 644-8374
Fax: (850) 644-5913
E-mail: jelsner@garnet.fsu.edu

Kerry Emanuel
Professor of Meteorology
Massachusetts Institute of Technology
Room 54-1620
77 Massachusetts Avenue
Cambridge, Massachusetts 02139
Phone: (617) 253-2462
Fax: (425) 740-9133
E-mail: emanuel@texmex.mit.edu

José Fernández-Partagás (deceased)
National Hurricane Center (volunteer)
University of Miami
Miami, Florida

Ricardo García Herrera
Departamento Física de la Tierra II
Facultad de Ciencias Físicas
Universidad Complutense de Madrid
Ciudad Universitaria
28040 Madrid
Spain
Phone: (34) 913-944-490
Fax: (34) 913-945-195
E-mail: rgarciah@fis.ucm.es

Luis Gimeno
Universidad de Vigo (Campus de Ourense)
Facultad de Ciencias
As Lagoas-32004-Ourense
Spain
Phone: (34) 988-387-208
Fax: (34) 988-387-214
E-mail: l.gimeno@uvigo.es

Isaac Ginis
Graduate School of Oceanography
University of Rhode Island
Narragansett, Rhode Island 02882

Phone: (401) 874-6484
Fax: (401) 874-6728
E-mail: iginis@gso.uri.edu

Paul Hungerford
3321 East Red Bridge Road
Apt. 5
Kansas City, Missouri 64137
Phone: (816) 550-3239

Thomas R. Knutson
Geophysical Fluid Dynamics Lab/NOAA
P.O. Box 308
Forrestal Campus
U.S. Route 1N
Princeton, New Jersey 08542
Phone: (609) 452-6509
Fax: (609) 987-5063
E-mail: Tom.Knutson@noaa.gov

Christopher W. Landsea
NOAA AOML/Hurricane Research Division
4301 Rickenbacker Causeway
Miami, Florida 33149
Phone: (305) 361-4357
Fax: (305) 361-4402
E-mail: Chris.Landsea@noaa.gov

Kam-biu Liu
James J. Parsons Professor of Geography
Department of Geography and Anthropology
Louisiana State University
227 Howe-Russell Geoscience Complex
Baton Rouge, Louisiana 70803
Phone: (225) 578-6136
Fax: (225) 578-4420
E-mail: kliu1@lsu.edu

Kin-sheun Louie
Honorary Research Fellow
Hong Kong Institute of Asia-Pacific Studies
Chinese University of Hong Kong
Shatin, Hong Kong
China
Phone: (852) 2304-6282
Fax: (852) 2660-9129
E-mail: kslouie@cuhk.edu.hk

Emiliano Hernández Martín
Departamento Física de la Tierra II
Facultad de Ciencias Físicas
Universidad Complutense de Madrid
Ciudad Universitaria
28040 Madrid
Spain
Phone: (34) 913-944-457
Fax: (34) 913-945-195
E-mail:
emiliano@6000aire.fis.ucm.es

Cary J. Mock
Department of Geography
University of South Carolina
Columbia, South Carolina 29208
Phone: (803) 777-1211
Fax: (803) 777-4972
E-mail: mockcj@sc.edu

Richard J. Murnane
Risk Prediction Initiative
Bermuda Biological Station for Research

P.O. Box 405
Garrett Park, Maryland 20896
Phone: (443) 622-6484
Fax: (301) 942-1886
E-mail: rmurnane@bbsr.edu

Charlie Neumann
6855 Southwest 104th St.
Pinecrest, Florida 33156
Phone: (305) 661-0316
E-mail: CBNeum@aol.com

María Rosario Prieto
Instituto Argentino de Nivologia, Glaciologia y Ciencias Ambientales
Avenida Adrián Ruiz Leal, s/n
Parque General San Martín (5500)
Mendoza, Argentina
Phone: (54) 261-428-7029
Fax: (54) 261-428-5940
E-mail: mrprieto@lab.cricyt.edu.ar

Weixing Shen
Environmental Modeling Center
National Centers for Environmental Prediction
5200 Auth Road
Camp Springs, Maryland 20746
E-mail: Weixeng.Shen@noaa.gov

Robert E. Tuleya
Environmental Modeling Center
National Centers for Environmental Prediction
5200 Auth Road
Camp Springs, Maryland 20746
E-mail: Robert.Tuleya@noaa.gov

Frédéric Vitart
ECMWF
Shinfield Park
Reading RG2 9AX
England
Phone: (44) 118-949-9650
Fax: (44) 118-949-9450
E-mail: nec@ecmwf.int

Thompson Webb III
Department of Geological Sciences
Brown University
Providence, Rhode Island 02912
Phone: (401) 863-3128
Fax: (401) 863-2058
E-mail:
Thompson_Webb_III@brown.edu

Dennis Wheeler
School of Health, Natural and Social Sciences
University of Sunderland
Sunderland SR1 3PZ
United Kingdom
Phone: (44) 191-515-2233
Fax: (44) 191-515-2229
E-mail:
dennis.wheeler@sunderland.ac.uk

Mark Zimmer
9225 SW Sixty-eighth Street
Miami, Florida 33173
Phone: (305) 271-0273

INDEX

Italic page numbers indicate figures, and boldface page numbers indicate tables.